S0-AVN-727

THE METALLIC AND NONMETALLIC STATES OF MATTER

Dedicated to
Professor Sir Nevill Mott, F.R.S.
Nobel Laureate in Physics and Emeritus Cavendish Professor of Physics,
University of Cambridge
Doyen of condensed matter science, on the occasion of his 80th birthday.

THE METALLIC AND NONMETALLIC STATES OF MATTER

Edited by

P.P. EDWARDS

University of Cambridge

and **C.N.R. RAO**

Indian Institute of Science, Bangalore

Taylor & Francis
London and Philadelphia
1985

UK Taylor & Francis Ltd, 4 John St, London WC1N 2ET

USA Taylor & Francis Inc., 242 Cherry St, Philadelphia, PA 19106–1906

British Library Cataloguing in Publication Data

The metallic and non-metallic states of matter
1. Matter—Properties
I. Edwards, P.P. II. Rao, C.N.R. III. Mott, N.F.
530.4 QC173.3
ISBN 0-85066-321-0

Library of Congress Cataloging in Publication Data is available

Filmset by Mid-County Press, 2a Merivale Road, London
Printed in Great Britain by Redwood Burn Ltd, Trowbridge, Wilts.

CONTENTS

PREFACE

A wide variety of systems undergo a transition between the two limiting electronic regimes in condensed matter; namely, the metallic and the non-metallic states. Indeed, the metal–nonmetal transition in both crystalline and disordered systems is currently a subject of intense experimental and theoretical investigation. Typical systems of interest range from elemental metals (e.g. Cs, Hg) near their liquid–vapour critical point, doped semiconductors (e.g. Si:P, Ge:As, etc.), transition metal oxides and sulphides, organic metals and semiconductors, and even liquids (e.g. the alkali metals in liquid ammonia and alkali metal–metal halide melts). In addition, the venerable periodic system of elements represents what, perhaps, is the most fundamental example of the metal–nonmetal transition in the condensed phase.

We therefore felt that it was most timely to attempt a comprehensive presentation of this vast array of experimental systems traversing the transition between the metallic and non-metallic states. This book is the result of such an effort. We have attempted to limit the volume to a reasonable length, and have at the same time included aspects of both materials and phenomena. We believe that the articles in this volume provide an up-to-date survey of this important area of condensed matter science.

Editing of a book which one wishes to be comprehensive—but not encyclopaedic—may lead to omission of certain areas. For example, we have made no attempt to provide a detailed discussion of the electronic properties of certain low-dimensional inorganic compounds. However, we feel that the material covered in this volume will provide a sufficient and useful framework for any discussion of the metallic and the non-metallic states of condensed matter.

We anticipate that this collection of articles will be useful for students with a reasonable background in chemistry, physics or materials science. The book could be used as a supplementary text for a graduate course in solid-state science or as a primary text for a specially designed course on the metallic and the non-metallic states of matter.

One of us (C.N.R.R.) wishes to express his thanks to the University of Cambridge for the Jawaharlal Nehru Visiting Professorship during which tenure this book was planned.

June 1985

P.P. Edwards
C.N.R.Rao

CHAPTER 1

metals, nonmetals and metal–nonmetal transitions†

Sir Nevill Mott

Cavendish Laboratory, University of Cambridge, Madingley Road, Cambridge CB3 0HE, UK

I remember attending an undergraduate lecture course on 'Electron theory of metals' given by Ebenezer Cunningham in, I think, 1926. This was based on O.W. Richardson's book with the same title. We realized then that in metals electrons were free and the Hall effect showed that this was so and that, in the monovalent metals at any rate, each atom had lost approximately one electron. We also realized that in insulators there were no free electrons. I remember asking the lecturer why electrons were free in some materials and not in others. Of course he did not know. We just had to think that in insulators the electrons were 'stuck'.

I entered Cambridge to read for the mathematical tripos in 1924, always meaning to do research in quantum theory if I got the chance. When quantum mechanics burst upon us in 1926, nobody in Cambridge, apart from Dirac, understood it very well and there were no lectures on it. I learned how to use the new theories from the papers of Schrödinger and Born, with a German dictionary in one hand and a treatise on differential equations in the other. After my first degree, a term in Copenhagen and a year in Manchester with W.L. Bragg, I became in 1930 a teaching fellow at Gonville and Caius College at Cambridge with a university lectureship, and with close but unofficial connections with Rutherford's Cavendish. It was not a place to encourage an interest in what we now call solid-state physics, though R.H. Fowler's knowledge of statistical mechanics led him in this direction, and under his influence A.H. Wilson (now Sir Alan) wrote his Adams Prize essay which led to his well-known book *The Theory of Metals*. Wilson first put forward the band-theory explanation of the difference between metals and insulators which, for me, gave the answer to my question of 1926; I remember Fowler explaining it in my presence to C.D. Ellis—Rutherford's collaborator in the book *Rutherford, Chadwick and Ellis*—and Ellis replying 'very interesting' in a tone of voice that implied that he was hardly interested at all. Nor indeed was I; at that time, Ellis and I were looking at the

† In part reproduced, with permission, from *Rep. Prog. Phys.* **47**, 909 (1984)

facts of beta-decay which should, had we been bolder, have forced us to propose a neutrino.

It is perhaps worth reminding the reader what a remarkable theory Wilson's was. The electrons in a crystalline insulator are not 'stuck'. No current can flow, but this is because of the Pauli principle. Only two electrons can occupy one quantized state, and in a full band exactly as many electrons must be moving from left to right as from right to left. If one is removed, by absorption of a quantum of radiation for instance, the remaining electron *can* move; we call the state from which the electron is removed a mobile 'positive hole'. Electrons are only 'stuck' if there are impurity levels. It is perhaps typical of the almost instantaneous acceptance of the new quantum mechanics that no one doubted this theory, even before the present overwhelming evidence was obtained for the existence of mobile positive holes.

In 1933 I moved to the chair of theoretical physics at the University of Bristol, where a physics laboratory rather out of proportion with the rest of the university had been built and endowed through munificent gifts from the Wills tobacco family. There I felt that my job was to apply quantum mechanics to the experimental work in progress. I couldn't do anything on these lines with Cecil Powell's cosmic rays, but with the help of Harry Jones, I certainly could for Herbert Skinner's work on soft X-ray emission in metals. It came about in the following way.

In 1933 Hans Bethe had already written his astonishingly complete article in the *Handbuch der Physik* about electrons in solids, particularly metals, and with Bethe in Bristol for a while after leaving Nazi Germany it was easy enough to get the hang of the subject. Bethe's article, it seemed to me, answered most of the questions that can be answered without considering electron–electron interactions. It surprises me, looking back, how easily we were satisfied with a model of a metal in which electrons occupy states up to a limiting energy, or in k space (wavevector space) up to a limiting surface, without taking into account the very large interaction between them. Bethe's article contained drawings of this surface, as calculated; we now call it the Fermi surface. In Bristol we wondered, was this a mathematical fiction, and would it disappear and become fuzzy when interaction was taken into account? Skinner, a brilliant experimenter, always covered with cigarette ash, seemed to have three pairs of hands as he pulled out from his spectrometer the radiation from the L_{III} emission in sodium, magnesium and aluminium. They showed bands, just as electron theory predicted, and what is more these bands showed a sharp upper limit. There was no fuzziness due to electron–electron interaction. With Harry Jones and Skinner in 1934 (Jones *et al.* 1934) we gave an explanation of this, which in fact depended on this interaction; it was, in principle, the same as the more complete treatment of Landau later on, which demonstrated the physical reality of the Fermi surface. The explanation was simple. An electron excited into a state separated from the limiting Fermi energy E_F by a small energy ΔE would have a lifetime determined by the Auger processes, in which it lost energy to one of the conduction electrons. The electron, however, could not lose a value of the energy greater than ΔE, because to do so would involve a transition to a state already occupied by other electrons. Therefore the only electrons to which it could lose energy were those in states in a range of energies ΔE below E_F. It followed that the probability $(1/\tau)$ per unit time of a collision would tend to zero with ΔE; the lifetime τ of an electron in a state just above E_F would be very long. This meant, using the uncertainty principle, that the limiting energy E_F would be sharply defined.

This work gave our group at Bristol confidence that the sharp Fermi energy was a physical concept, that a metal was to be described by a partly filled band, and that the introduction of the interaction e^2/r_{12} could be manageable and useful. Baber in 1937 showed that inclusion of this term led to a T^2 term in the resistivity, a result obtained independently in the previous year in the Soviet Union by Landau & Pomeranchuk (1936). Stoner's collective electron treatment of magnetism fitted into these concepts and we were able to apply it to some properties of transition metals, showing that it could be used even for narrow d bands. (The present author (Mott 1974) believes it can also be used for narrow 4f bands in the 'intermediate valency' metallic compounds.) After the war, Jacques Friedel, working for a PhD in Bristol, described the screening of an impurity by electrons near the Fermi limit and produced the 'Friedel sum rule'.

At a conference in Bristol in 1937, sponsored and published by the Physical Society, two visitors from Holland, de Boer & Verwey (1937) pointed out that the Wilson theory of metals and insulators, when applied to such crystalline materials as nickel oxide, gave incorrect results. The nickel ion, Ni^{2+}, contained eight electrons in 3d states; a 3d band would have room for ten electrons and the cubic field could split it into the e_g band with four states and the t_{2g} band with six. It was impossible that eight electrons could lead to full and empty bands; one band would be partly filled, so metallic behaviour was predicted. But, in fact, the material is a transparent insulator. Peierls at the meeting pointed out that the electron–electron interaction (e^2/r_{12}) would have to be taken into account to explain this (Peierls 1937). For a current to pass it would be necessary to form states with the configurations Ni^{3+} and Ni^{+} which could move, and this interaction could prevent their formation.

At that time we were unaware of the existence of antiferromagnetism and that the moments on the nickel ions in NiO were arranged in an antiferromagnetic way. It was not till 1957 that John Slater pointed out that the antiferromagnetic superlattice would split the e_g band in nickel oxide into two equal halves, with a gap between them; one-half would be occupied and the other half empty, so the insulating and transparent nature was explained. But this was not the complete story, because if the superlattice were indeed the origin of the insulating property, this should disappear at the Néel temperature, and above this temperature nickel oxide should be a metal. This is far from being the case. A full understanding, then, of the insulating property goes some way beyond band theory and does depend on electron–electron interaction, as Peierls surmised. A better understanding really dates from Hubbard's (1964) introduction of a new

Hamiltonian and of the concept of 'Hubbard bands' in the early 1960s. Stripped of the mathematics, the physical concept is this. In nickel oxide, as already stated, a current can be carried if Ni^{+} or Ni^{3+} ions are present. These electronic configurations can move through the lattice of Ni^{2+} ions, with a definite wavenumber for the wavefunction, just as an electron, hole or exciton can. The motion of the Ni^{3+} configuration is described as 'in the lower Hubbard band', that of the Ni^{+} as 'in the upper Hubbard band'. For both, bandwidths B_1 and B_2 can be defined; we suppose these to be of the tight-binding type, and so with a density of states symmetrical about the mid-point. The energy required to create these two configurations, if the ions are a long way from each other, is called the Hubbard U and is

$$U = I - E$$

I in this case is the energy required to remove an electron from Ni^{2+} to form Ni^{3+} and E is the energy gained when a free electron at rest is added to Ni^{2+} to form Ni^{+}. If, however, the ions are in the crystal, the energy required to produce the pair of carriers is

$$U - \tfrac{1}{2}(B_1 + B_2)$$

If this is positive, an energy gap exists and the material is an insulator; if not, it is a metal.

This distinction between a metal and an insulator, perhaps more 'chemical' than the formulation of Wilson, is applicable whether or not the atoms or ions involved have a magnetic moment. If they do, while the moments at low temperatures will be ordered, for instance antiferromagnetically, the insulating property does not depend on the ordering. If not, the carriers are just the familiar holes and electrons.

My interest in the nature of a metal, dating from my memory of Cunningham's lectures and the discussion in 1937 with Peierls, came alive again during the post-war period, stimulated by the intense interest at that time in semiconductors, and particularly in silicon and germanium. Before the war it was thought that these materials were 'bad metals' because the specimens available showed a comparatively high residual conductivity at low temperatures. It was only the wartime work, notably that of Lark-Horowitz and his group in Purdue, that established that the pure materials were nonmetallic and that their conducting properties were a consequence of 'doping' by impurities such as phosphorus, which introduced levels just below the conduction band as in Wilson's model of 1931. It was also clear that, if there were too many of these, the material began to behave like a metal. I realized in the late 1940s that a donor was like a swollen hydrogen atom; Ronald Gurney and I had described donors in that way in our book *Electronic Processes in Ionic Crystals* (1940). So here in doped germanium the facts showed that a random array of one-electron centres, if spaced far apart, produced levels below the conduction band, but if in sufficient concentration they showed the properties of an amorphous metal. This reminded me of the nickel oxide problem; the nickel ions were far enough apart for 'correlation'—i.e. electron–electron interaction—to stop metallic behaviour, but the semiconductors showed that, when the concentration increased, metallic behaviour set in. A 'metal–nonmetal transition' occurred. This is commonly called the Mott transition, though of course it was not I who observed it.

The way I treated the problem (Mott 1949) was one which now has historic interest only as applied to doped semiconductors. I considered the electron gas as screening the positive charge on the donors, so that the potential energy seen by an electron in a heavily doped semiconductor is of the form

$$V(r)=\frac{e^2}{\varepsilon r}\exp(-qr) \tag{1}$$

where ε is the dielectric constant of the semiconductor and q is the Thomas–Fermi screening constant. I asked whether a bound state was possible for an electron for which the expression (1) was the potential energy; if q (which depends on the electron density) is large enough, it is not. Then I supposed the electron gas would have metallic properties. But as the density (n) of electrons drops, at some point bound states become possible. I calculated the value of n_c at which this occurs to be given by the equation

$$n_c^{1/3}a_H=0\cdot 25 \tag{2}$$

where a_H is the Bohr radius for a donor. I predicted, then, that a *discontinuous* transition would occur from a state when all electrons were bound to donors to one in which they were all free. This kind of transition has been called a 'Mott transition' and the right-hand side of equation (2) the 'Mott ratio'. As Edwards & Sienko (1978) pointed out, it works for a very large number of materials.

I remember speaking on these ideas at the Massachusetts Institute of Technology in the early 1950s. Slater did not like them at all—they did not fit into the scheme of band theories. I now sympathise with Slater; a derivation within the concepts of band theory was highly desirable and came later (Slater 1957).

Of course, my early work did not take into account the fact that the donors were at random sites in a crystal. Any quantitative treatment had to await Anderson's paper in 1958 on the 'Absence of diffusion in certain random lattices'. In 1961 I wrote (with W.D. Twose) a review of the subject, in which we considered the transition again, also taking into account the hypothesis that it took place in an impurity band which remained separate from the conduction band. Whether this is so for silicon has been queried very recently as we shall see (Mott & Kaveh 1983).

A conference on Metal–Insulator Transitions was held in San Francisco in 1968, organized by J.C. Thompson of the University of Texas, whose interests lay in the application of these ideas to solutions of alkali metals in liquid ammonia. It was, however, after this that a description of the transition was given which lay nearer to band theory.

In my paper of 1961 with W.D. Twose I pointed out that a discontinuous jump from zero to a finite value was to be expected in a nonmetal if, under pressure, or in some other way, an indirect gap should disappear. The most elegant proof of this was given by Brinkman and Rice of the Bell Laboratories in 1973. This formulation arose from their work, in co-operation with Gordon Thomas and others in the same laboratory, on electron–hole droplets in silicon and germanium. These are indirect-gap semiconductors. If electrons and holes are produced by irradiation, then at low temperatures they will combine to form excitons, which for an indirect gap have a comparatively long lifetime. Excitons attract each other and 'droplets' of an electron–hole gas may form. In principle this can have two forms, that of an 'excitonic insulator' (a Bose–Einstein gas of excitons) or an electron–hole *metallic* gas. Which has the lowest free energy is a matter for detailed calculations but it seems that the electron–hole gas usually has. As regards the excitonic insulator, 15 years ago there was much discussion of its expected properties (see, for instance, Keldysh & Kupaev (1965) and Halperin & Rice (1968)) but only recently have materials been identified which seem to have the predicted behaviour. One, according to M.H. Cohen and co-workers, is expanded fluid mercury (Turkevich & Cohen 1983). As regards electron–hole droplets, the argument was as follows. Suppose a droplet contains n electrons and n holes per unit volume. At zero temperature the kinetic energy per particle for a single-valley or nondegenerate band is of the form

$$3\hbar^2 n^{2/3}/5m_{\mathrm{eff}}$$

where m_{eff} is an effective mass; for the actual material we may write

$$A\hbar^2 n^{2/3}/m$$

where m depends on the masses of electrons and holes and A is a constant. Attraction between electrons and holes will give rise to a term

$$-Cn^{1/3}e^2/\varepsilon$$

where ε is a dielectric constant. The sum of these has a minimum:

$$E_0 = -\text{constant} \times me^4/\hbar^2\varepsilon^2 \tag{3}$$

at a value of n_c given by

$$n_c^{1/3}a_H = \text{constant} \qquad a_H = \hbar^2\varepsilon/me^2 \tag{4}$$

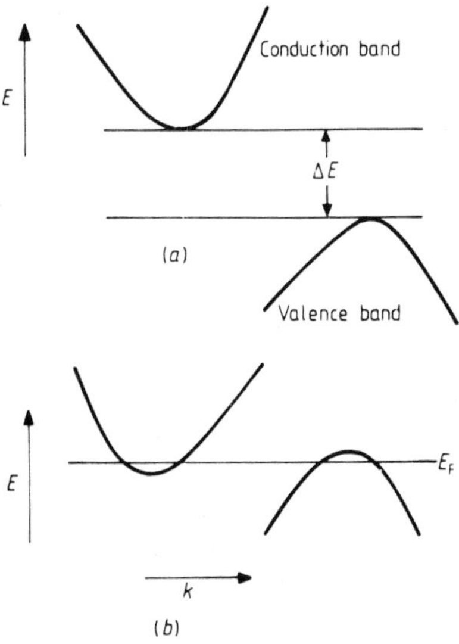

Figure 1. Showing the band structure of an indirect gap material (*a*) before and (*b*) after a metal–insulator transition.

so electron–hole droplets will form at this density. To obtain the constant, evaluation of the effects of correlation is essential.

The relevance of this to the metal–insulator transition as pointed out by Brinkman & Rice in 1973 is as follows. Suppose the gap ΔE in an indirect-gap semiconductor (Figure 1) decreases, for instance, as the result of pressure or change of composition. As already pointed out, a metal–insulator transition can occur if and when the two bands overlap (Figure 1(*b*)). But we now see that a *sudden* jump in the number of carriers, from zero to a value given by (4), will occur when the gap ΔE reaches the value E_0 given by (3). A band crossing metal–insulator transitions must always be discontinuous.

The argument can be applied to a 'Mott' transition, the bands now being Hubbard bands. For one-electron centres, the two bands are as follows. The upper band represents the motion of an extra electron placed on one of the centres, the lower band that of a hole when an electron is removed. As before, overlap, *in the absence* of the electron–hole interaction, is expected when

$$\tfrac{1}{2}(B_1 + B_2) \simeq U$$

One would expect, then, that the Mott criterion (2) would follow from this equation. Cyrot (1972) was the first to show that this was so. The quantities B (particularly for the lower band) must contain the factor $\exp(-R/a_{\mathrm{H}})$ where R is the distance between atoms (or centres) and $a_{\mathrm{H}} = \hbar^2 \varepsilon / me^2$. Thus, again without interaction, a transition should occur when

$$R = a_H \times \ln(\ldots)$$

and the logarithm turns out to be large, about 4, so that

$$n_c^{1/3} a_H \sim 0{\cdot}25$$

is the result. Mott & Davies (1980) showed that improvements to the theory, affecting terms within the logarithm, made little difference to the result. Neither will the expected discontinuity, though Mott (1974) gave reasons for believing that the discontinuity in n would be greater for 'Mott' transitions (systems with one-electron centres) than for those of the band-crossing type.

Here then is a theory of metal–insulator transitions. Needless to say, there have been alternative theories. Castner (1980), among others, has argued that the Clausius–Mossotti relationship:

$$\varepsilon = 4\pi\alpha(1 - 4\pi\alpha/3)^{-1}$$

between ε and the polarisability α plays an essential role, a transition occurring when ε diverges. For doped silicon, arguments against this view were givte large, Mott & Davies (1980). Also certain materials, such as NiS and Ti_2O_3, need in my view a different description (White & Mott 1971, Mott 1981).

The theory outlined here, however, is, we believe, applicable to many systems, which include most transition-metal oxides, certain fluids which show a transition for low densities achieved at high temperatures and some doped semiconductors—though not apparently doped silicon for which it was originally developed, as we show at the end of this article. We shall describe these in turn. But first we mention an important consequence of the discontinuous nature of the transition.

It was pointed out by the present author (Mott 1967) (see also Cyrot & Lacour-Gayet 1972) that, as a consequence of the discontinuous change in the number of current carriers predicted here, the plot of free energy at zero temperature against any variable x which affects the band gap ΔE must show a kink at the transition, as at the point X in Figure 2. Thus values of x between A and B are unstable. If x is the volume, then under increasing pressure x will jump from B to A; such a discontinuous change of volume is observed at a transition from metal to nonmetal, for instance in the intermediate valence compound SmS. If x is the composition in an alloy P_xQ_{1-x}, then between A and B there should be a two-phase region. As the temperature rises a critical point is expected as in Figure 3. So the only possibility of observing a discontinuity in the number of carriers is through the use of *quenched* alloys. We expect a jump at low T from semiconducting behaviour with σ proportional to $\exp(-\Delta E/kT)$, with a finite

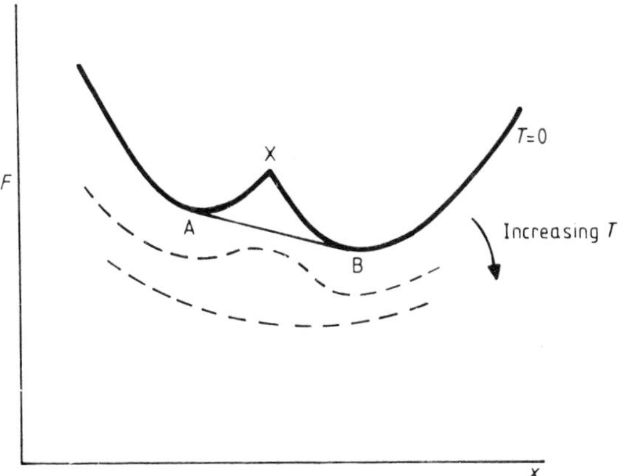

Figure 2. Showing schematically the free energy F against some parameter x which may be volume or composition. The full curve refers to zero temperature and the broken curves to finite T.

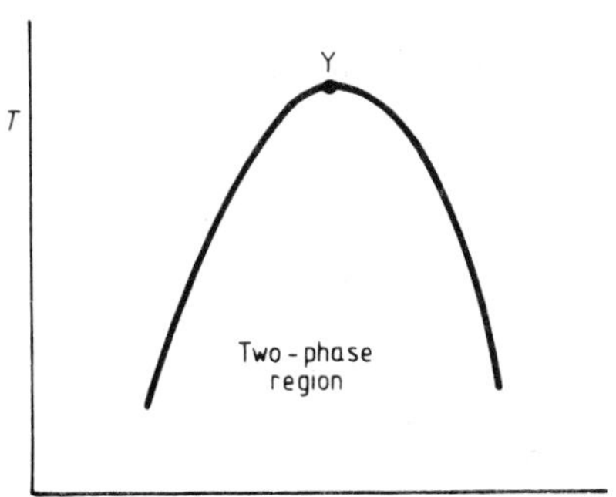

Figure 3. Showing the consulate (critical) point for an alloy.

value of ΔE at the transition, to metallic behaviour. As far as we know, observation of this effect has not been achieved with certainty.

Of the crystalline materials showing a metal–insulator transition, certain vanadium oxides have been of interest since the war and were investigated in some detail in the Bell Laboratories in the early 1970s. The case of V_2O_3 is typical; since 1946 the transition from semiconducting behaviour to metallic at about 200 K had been known and the Bell group showed that the metallic form was stable at all temperatures and pressures above 24 kbar, and also on alloying with a few per cent of titanium. A narrow two-phase region did indeed exist at the transition, showing a critical point; the activation energy E for conduction was quite large, suggesting that if the investigations could have been made in a quenched region there would have been a discontinuity in ΔE. But this was not done.

In 1970 also, this group showed that the semiconducting phase was antiferromagnetic, with a moment of 1·2 μ_B on each vanadium atom. Was it then to be considered a typical Mott transition? The doubt lay in the fact that the nonmetallic phase had a different structure. John Goodenough, the chemist whose work on magnetism and on the transition-metal oxides has been very influential, believed that the magnetic behaviour should be treated as a consequence of the change of structure, while the Bell Laboratory theorists looked at it the other way round. I am not aware that this controversy is settled yet. On the side of the Bell physicists is perhaps their observation that metallic V_2O_3 shows certain properties that they predict for a metal when in a state near a Mott transition (Brinkman & Rice 1973), such as an abnormally large electronic specific heat.

I now turn to fluids. Work at high temperatures and pressures, notably by Hensel and co-workers first at Karlsruhe and then at Marburg, has shown that fluids such as mercury and caesium undergo a metal–semiconductor transition with increasing volume†. But the system most intensively investigated has been the solution of alkali metals in ammonia. These have been known since the work of Weyl in 1863, and since the war there has been a triennial international conference on the subject, known as the 'Colloque Weyl'. Concentrated solutions in the range 7–20 mol litre^{-1} have a metallic bronze colour and the high electrical conductivity characteristic of a liquid metal; dilute solutions have a bright blue colour, and for concentrations below 10^{-3} mol litre^{-1} behave as ideal electrolytes. Somewhere between a metal–insulator transition must occur. The San Francisco conference of 1969 was arranged by J.C. Thompson whose research work has been mainly on these materials (see Thompson 1976) and I think this conference established an interest in the question: is this a 'Mott transition'? Certainly all these materials (except Cs–NH_3) show a solubility gap as in Figure 3 and a critical point (the consulate point) somewhat below room temperature. The present author suggested that this is related to the transition—though the idea of two kinds of critical point, one related to such a transition and one depending simply on interatomic forces as in the rare gases, goes back to a paper in 1943 by Landau & Zeldovich.

Transitions in liquids, and perhaps even more in doped semiconductors such as silicon–phosphorus, cannot be treated properly without taking into account the fact that these materials are not crystalline, and so to go further I will give an outline of the beginnings of our understanding of electrons in noncrystalline materials, particularly as it relates to the subject matter of this article. I must begin with the remarkable article already mentioned, published by P.W. Anderson in 1958 entitled 'Absence of diffusion in certain random lattices', described with truth by the author as 'often quoted but rarely read'. In this paper Anderson took a tight-binding model of an electron moving in the potential of Figure 4, each potential well having a random depth as illustrated. That such a fluctuating potential could introduce traps was hardly surprising, but what was new was that,

† References are given by Mott & Davis (1979) and Mott (1982).

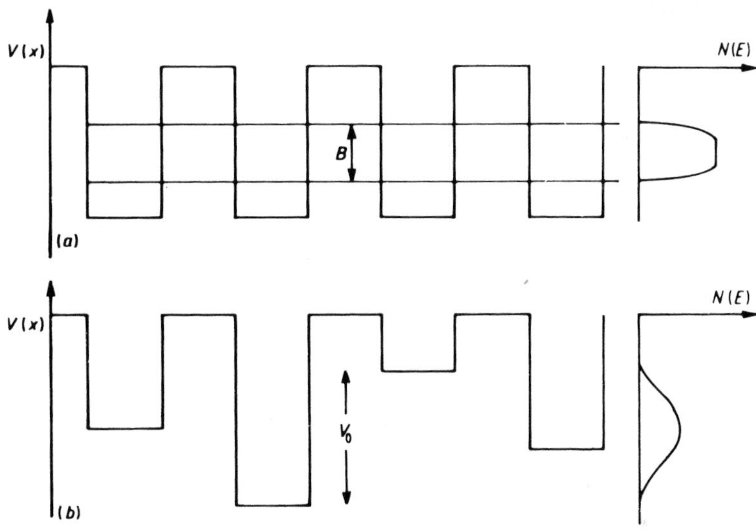

Figure 4. Potential energy used by Anderson (1958) (*a*) without a random potential and (*b*) with such a potential. The density of states is also shown.

if the fluctuations were large enough, *all* sites become traps. For almost a decade little was done with Anderson's work, either by its author or anyone else, and it was widely disbelieved. The paper was written with spin diffusion in mind rather than electrons; what was surprising was that it predicted localization with a continuous distribution of traps, in spite of the fact that an electron could tunnel from one to another, and if it went far enough it could find a state with practically the same energy. I remember a distinguished American theorist on sabbatical leave in the Cavendish telling me that he neither understood nor believed Anderson's paper. At the international conference on Amorphous and Liquid Semiconductors held in Cambridge in 1969 there was still considerable scepticism.

Personally, though I must confess that I am among those who have quoted rather than read Anderson's paper, I believed its result because of my work on the Mott transition and my knowledge of the phenomenon of 'impurity conduction' in doped semiconductors, discovered by Hung & Gliessman in 1950 and investigated in derail by Fritzsche and co-workers. I had to ask how electrons could pass from one donor to another with a small activation energy in, say, doped germanium. I knew that this could not occur for concentrations below that of the Mott transition if all donor states were occupied, and so that the phenomenon must depend on 'compensation', that is, the presence of some acceptors, so that some of the donors were unoccupied. In a compensated semiconductor, then, the effect of the Hubbard U would not be of major importance and impurity conduction could occur. That compensation was necessary for impurity conduction was at first controversial; I remember two American colleagues who visited Bristol before 1953 strongly denying it; but both

Esther Conwell and I, independently in 1956, published the hypothesis and it came to be accepted. In 1956 I also gave a model for the activation energy in impurity conduction, but after Anderson's paper appeared I realised that if his conclusions were *not* true a compensated impurity band would necessarily be metallic. As this was not so, I could and did accept Anderson (1958) and used it in my review paper (with W.D. Twose) in 1961, as did Miller and Abrahams implicitly in their theory of hopping conduction between occupied and empty donor states in 1960. Also a good deal later (1967) I realised that, if in Anderson's theory the disorder was not great enough to create localization throughout the band, this would nonetheless lead to localized states in the band tail, and the energies at which this occurred would be separated from the energies of the so-called extended states by a sharply defined energy E_c. This was later called a 'mobility edge' (Cohen *et al.* 1969). All this was very much in evidence, queried and supported, at the Cambridge Conference of 1969. It was used by Le Comber & Spear (1970) to explain their results on the mobility of electrons in glow-discharge-deposited amorphous silicon. I remember a plaintive remark from Phil Anderson at the meeting: 'My name is being used to describe too many phenomena of which I have never heard'.

At the 1969 conference the existence of amorphous semiconductors was already well known, thanks mainly to the work of Kolomiets in Leningrad on the chalcogenide glasses dating from the 1950s, the use of these materials for a threshold switch by Ovshinsky (1968) and the work of Spear and Le Comber on the mobility of electrons in amorphous silicon. Nonetheless, some surprise was expressed at the conference at their existence, because the Wilson theory to which I have referred earlier depended on the reflection of electron waves within a crystal by the atoms of the lattice. In a glass, sharp reflection could not occur and it was therefore surmised that the bands of allowed energies should have 'tails'. In an influential paper Cohen *et al.* (1969) suggested that, in the chalcogenide glasses for instance, both conduction and valence bands would have tails of 'localized' states, separated from the conduction and valence bands by values of the energy which they called 'mobility edges'—identical with my E_c. They also suggested that these bands would overlap, pinning the Fermi energy near mid-gap; the term 'pseudo-gap' arose from this concept. I remember, however, some correspondence with Professor Cohen pointing out that this could not be a universal feature of non-crystalline materials; glasses such as vitreous SiO_2 are, after all, transparent and so must have a very wide gap, not a pseudo-gap. It is now believed that band tails do indeed exist as illustrated in Figure 5(*b*), but that states deep in the gap, if any, are a consequence of defects, such as dangling bonds.

Kolomiets' work established an important property of the chalcogenide glasses, namely that they cannot be doped; impurities have little effect on their conductivity. In Leningrad Gubanov, in his book published in 1963, proposed that this could be due to gap states; any donor would lose its electron to a state

† See *Proc. J. Non-crystalline Solids*, **6** (1969).

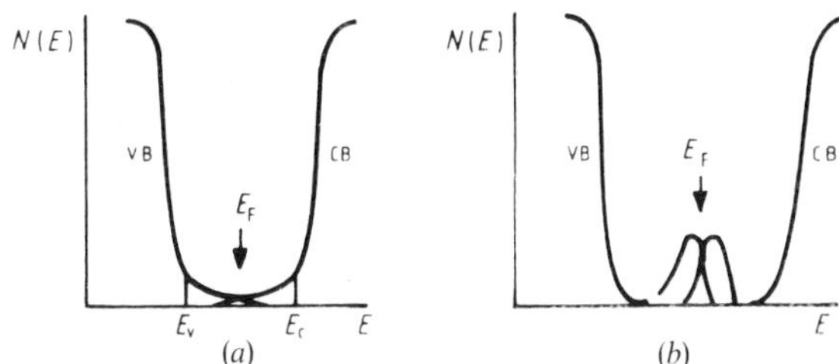

Figure 5. (*a*) Density of states in an amorphous semiconductor, as suggested by Cohen *et al.* (1969). (*b*) As proposed here. CB and VB denote the conduction and valence bands, E_F is the Fermi energy and E_c and E_v are mobility edges.

near the Fermi energy. I suggested in 1969 the '8-*N* rule'. According to this, an atom is incorporated in a glass in such a way that all electrons are taken up in bonds and so are not free to move. Both concepts now play a role in understanding crystalline materials.

The 8-*N* rule shows how far we have come from the Wilson concept of full and empty bands; in glasses, electrons are unable to move because they are stuck in bonds—a concept that we might have proposed prior to quantum mechanics. Needless to say, with the help of the concept of Anderson localization, these and the older ideas can be unified within quantum mechanics, but it is useful in glasses to retain these more chemical insights.

For a few years it was thought that the 8-*N* rule was valid not only for glasses but for all amorphous materials, until Spear and Le Comber in 1975 showed that amorphous silicon prepared from silane (SiH_4) in a glow discharge *could* be doped, for instance by including PH_3 in the discharge; some of the phosphorus turned out to be coordinated with three silicon atoms and so was inactive, but some went in with fourfold coordination as in the crystal and could act as a donor. This enabled p–n junctions to be made and was the origin of a new industry—the production of solar cells from amorphous silicon; at the conference on amorphous and liquid semiconductors in Tokyo (August 1983), two thirds of the papers were on hydrogenated amorphous silicon and its applications. Actually, in an ingenious paper Street (1982) has shown that a form of the 8-*N* rule, taking into account dangling bonds, can account for the ratio of active to inactive phosphorus.

Next, I shall describe the kind of metal–insulator transition which can arise in non-crystalline systems through the application of the concept of localization described by Anderson in 1958 (the Anderson transition). If, at zero temperature, a degenerate electron gas is in a field such as that illustrated in Figure 4, or in any non-periodic field, and if the states at the Fermi energy are localized in the Anderson sense, the material is not a metallic conductor and the conductivity tends to zero with temperarure, although the density of states is finite at the Fermi energy. This concept was inherent in the paper on impurity conduction published by Mott & Twose in 1961 and was developed in greater detail in papers before 1970. Such a material has been called (I think by P.W. Anderson) a 'Fermi glass'; a

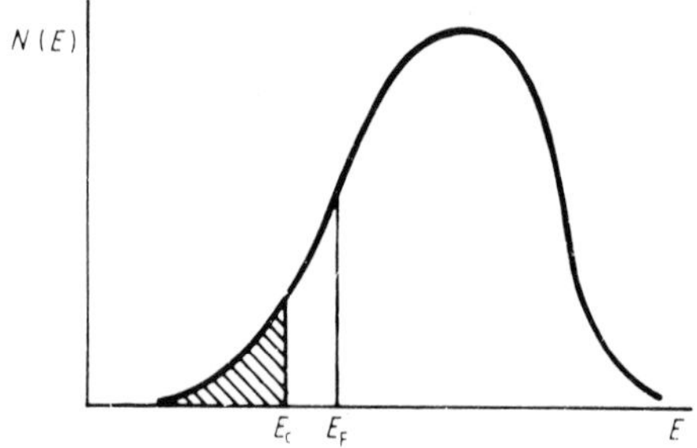

Figure 6: Density of states in a material with non-periodic potential showing mobility edge E_c and Fermi energy E_F. Occupied states are shaded.

good example is a doped and compensated semiconductor at low temperatures, when transport is in an impurity band. Two mechanisms are available for charge transport; one is thermally activated hopping, introduced by Miller & Abrahams in 1960 and leading, at low temperatures, to the law of variable-range hopping (Mott 1969):

$$\sigma = A \exp(-B/T^{1/4}) \tag{5}$$

a behaviour which has frequently been observed, though as shown by Efros & Shklovskii (1975) the electron–electron interaction should change the expression to $\exp(-B'/T^{1/2})$ at low enough temperatures (see also Davies *et al.* 1982, 1984). The second is excitation to a mobility edge, if one exists in the band under consideration. The density of states in such a band is illustrated in Figure 6. The second mechanism must give for the conductivity

$$\sigma = \sigma_0 \exp[-(E_c - E_F)/kT] \tag{6}$$

and will be predominant at high temperatures if $\sigma_0 \gg A$. The impurity band of a doped and compensated semiconductor is an example, if a mobility edge exists in the band.

We now suppose that E_c or E_F or both can be moved. This can be done by a change in composition, or in other ways. If E_F lies above E_c, the material is a metal; if it lies below, it is a nonmetal with a conductivity given by (6) or, if at lower temperatures, by (5). A metal–insulator transition is predicted. This, the Anderson transition, unlike the Mott transition discussed earlier, does not depend on the electron–electron interaction. It is a consequence of a theory of non-interacting electrons, as was the model of Wilson which, from 1931, has been used to describe the difference between metals and nonmetals.

The present author, having realised from about 1970 that Anderson's paper predicted that such a transition was possible, found examples of it in the extensive work of Fritzsche in the previous decade on the electrical properties of doped and

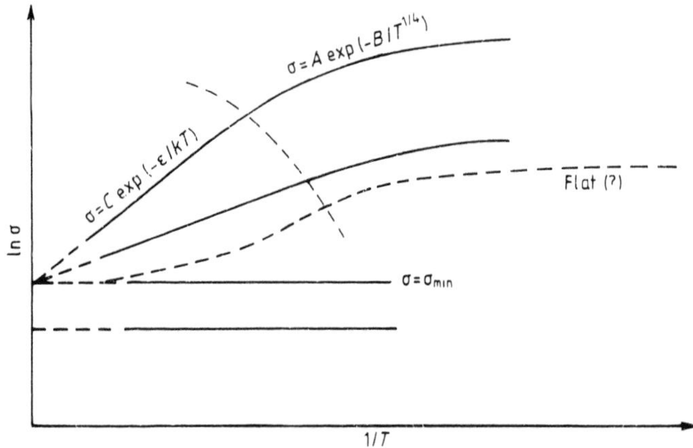

Figure 7. Typical behaviour of the resistivity of a material undergoing an Anderson transition plotted as a function of $1/T$. If there is no minimum metallic conductivity, below 1 K behaviour as shown in the broken curve must occur.

compensated germanium, and also in that of Davis & Compton (1965) carried out in the University of Illinois. E.A. Davis, now Professor of Physics at Leicester, worked in the Cavendish during the 1970s and discussions of this work led to our collaboration on our book on *Electronic Properties of Non-Crystalline Materials* (1971, 2nd edn 1979). What was observed in these materials, in experiments carried out at temperatures down to about 1 K, is shown in Figure 7. The different curves were obtained with different concentrations of donors, or varying compensation, or in other work (for other semiconductors such as InP and InSb) with increasing magnetic fields, which by decreasing the radius of donors and thus the width of the impurity bands, moved any mobility edge upwards towards the middle of the band. The lowest unactivated mobility I called the 'minimum metallic conductivity'; it was supposed to be the conductivity when the Fermi energy lay at the mobility edge. That it should also be the pre-exponential factor σ_0 in (6) is obvious, since both are equal to $N(E_c)\, e\mu kT$, μ denoting the electron mobility and $N(E)$ and μ being the values at E_c.

The concept of a minimum metallic mobility has had a chequered history. I introduced it in 1970, with a detailed discussion in 1972; in 1975 in co-operation with Michael Pepper and others (Mott *et al.* 1975), I summarised the extensive evidence for its existence, behaviour as illustrated in Figure 7 (without the broken curve) being shown by a wide variety of systems, both solid and liquid. The observed values seemed to agree well with a value that I calculated, namely

$$\sigma_{\min} = 0{\cdot}03\, e^2/\hbar a \tag{7}$$

the value of the constant depending on the Anderson localization criterion. The phenomenon was observed in two-dimensional systems too, particularly the inversion layer at the interface between silicon and its oxide; here σ_{min} was $0{\cdot}1\ e^2/\hbar$, not depending on any length scale. Numerical work by Licciardello & Thouless (1975) confirmed that the quantity should exist.

However, in 1979 a paper by Abrahams, Anderson and others used a scaling theory to show that the conductivity at zero temperature should always go continuously to zero, probably linearly as the Fermi energy approaches E_c; no discontinuity was to be expected. The first experiments to show that this was so were those of Gordon Thomas and co-workers (Rosenbaum *et al.* 1980) who used silicon doped with phosphorus and temperatures down to a few degrees millikelvin to obtain by extrapolation the value at $T=0$. They found that, with decreasing concentration, σ dropped below σ_{min} very sharply, but nonetheless decreased *continuously* to zero in a range δn of n of about 1%. Here n is the concentration of phosphorus. Since the material was uncompensated, I thought that this might be a discontinuous 'Mott transition' broadened by disorder (Mott & Kaveh 1983), but there is much evidence now that this is not so, and moreover the continuous drop, either linear or as $(n-n_c)^s$ with s between 0·5 and 1, has been observed in many other systems. The only case where I now believe that a discontinuous transition is to be observed is when a transition is induced by a strong magnetic field, when the behaviour illustrated in Figure 7 (without the broken curve) has been observed in InP, both compensated and uncompensated and in experiments down to 30 mK, in work performed independently by Long & Pepper (1984) and Biskupski (1984).

Some of the earlier results of Biskupski and co-workers are shown in Figure 8; more recent work done to 30 mK confirms that, as the field increases, a discontinuous change of the conductivity from a value of order given by (7) to zero is observed. Since both concentration of dopants and magnetic field are variables, it is possible to plot the observed value of σ_{min} (essentially the pre-exponential in equation (6)) against a, the mean distance between donors. As will be seen from Figure 9, equation (7) is confirmed with the best value of the constant 0·0235.

I will now attempt a summary of our present understanding of metal–insulator transitions in disordered materials, following the work of Abrahams *et al.* (1979), to which I have already referred. From the theoretical point of view, I am talking about a degenerate electron gas with a potential energy similar to that in Figure 4(*b*). The experimental situation closest to this is that in a doped semiconductor. Random fields give a distribution of well-depths and in addition the centres have a random distribution in space, a situation which increases the tendency towards localization, and has been treated mathematically (Debney 1977). As in the theory of metals, there are different levels of approximation in the theory. As in that of crystalline metals developed in the twenties, one can neglect electron–electron interactions altogether; the results were surprisingly successful. The introduction of electron–electron interaction made little difference for normal metals, and showed (Jones *et al.* 1934) that the sharp cut-off at the Fermi energy remained but that electron–electron collisions give a small term in the

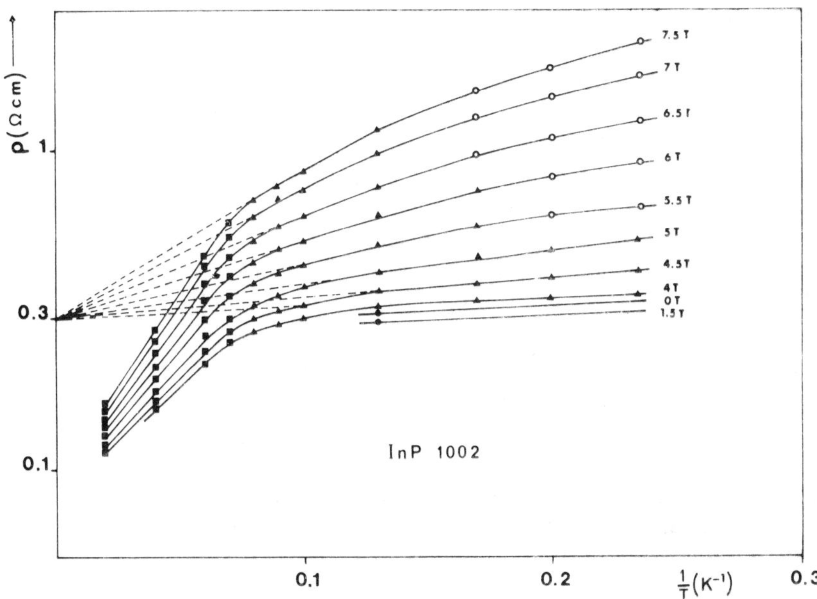

Figure 8. Showing the behaviour of ρ (Ω cm) in InP. The high-temperature behaviour is ε_2 conduction in the magnetic fields marked on the diagram in tesla (Biskupski 1982).

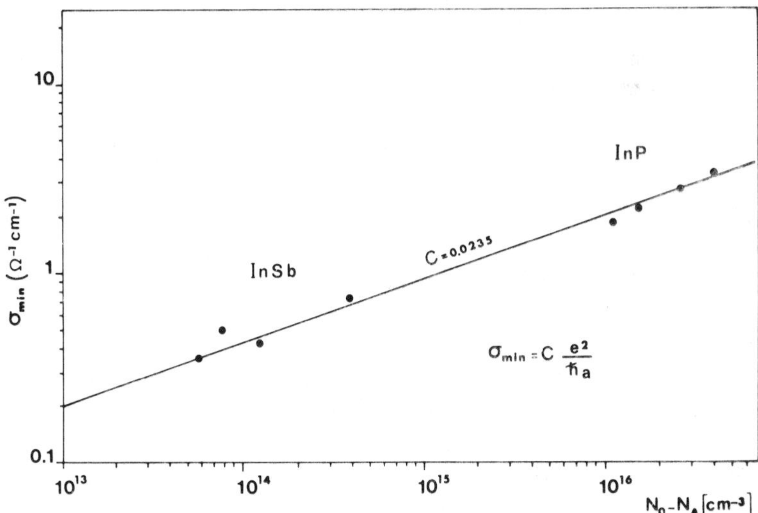

Figure 9. Plot of experimental values of $\sigma_{\min}(\Omega\ \mathrm{cm})^{-1}$ in InP against $N_0 - N_A$, using varying magnetic fields (Biskupski 1982). The straight line is Ce^2/ha with $C = 0{\cdot}0235$.

resistivity varying as T^2. If the tight binding model of Figure 4(*a*) is used, the interaction can be divided into (a) intra-atomic; this is described by the Hubbard U (Hubbard 1964) already introduced, and given by

$$\iint |\psi(r_1)|^2 |\psi(r_2)|^2 (e^2/r_{12})\, \mathrm{d}^3x_1\, \mathrm{d}^3x_2$$

over a single well, (b) long-range inter-centre coulomb interaction. As we have already seen, if the Hubbard U is comparable with the bandwidth, a first order transition to an insulating state will occur, and great enhancement of the specific heat and magnetic susceptibility near the transition (Brinkman & Rice 1973).

Turning now to electrons in the field of Figure 4(*b*), in the impurity band of an amorphous or crystalline semiconductor, one should consider a *compensated* material; otherwise for a narrow band it is likely that the Hubbard U will be of major importance and lead to an insulating state. In what follows, then, I shall discuss either an impurity band with less than one electron per centre, or electrons in a conduction band under the influence of some kind of disorder, such as the scattering by the donors.

The scaling theory of Abrahams *et al.* shows that, in a theory of non-interacting electrons, there is no minimum metallic conductivity but that, as the Fermi energy E_F tends to the mobility edge E_c, the conductivity σ tends continuously to zero. This is at zero temperature and in the absence of a magnetic field. According to the present author (1984), when E_F is near the mobility edge the conductivity will be given by

$$\sigma = 0{\cdot}03\, e^2/\hbar\xi \qquad (8)$$

where ξ is the localization length at an energy $|E_F - E_c|$ *below* the mobility edge. How ξ depends on $|E_F - E_c|$ is not precisely known, but it appears probable that the behaviour is linear. We thus expect that $\sigma \propto [E_F - E_c]$.

This behaviour depends on the interference of scattered waves with infinite space available (Bergmann 1983), and if the conductivity of a volume L^3 is envisaged, (8) should be replaced by

$$\sigma = 0{\cdot}03\, e^2/\hbar L \qquad (9)$$

This concept of the conductivity of a block of size L, introduced by Thouless (1977), is essential to scaling theory. Its practical result is that, at a finite temperature, (8) must be replaced by

$$0{\cdot}03\, e^2/\hbar L_i$$

where L_i is the inelastic diffusion length, that is, the distance that an electron will diffuse before an inelastic collision with another electron or a phonon. This is valid if $L_i < \xi$. For electron–electron collisions L_i varies as $1/T$, so the conductivity near the transition, rather surprisingly, increases with temperature. This is indeed observed in doped silicon at low temperatures.

The present author (1984) has proposed that in a magnetic field the conductivity at the mobility edge is

$$0{\cdot}03\, e^2/\hbar L'$$

where L' is the cyclotron radius, and

$$L' = L_H = (\hbar c/eH)^{1/2} \quad \text{for } L_H > a$$

or $L' = a$ for $L_H = a$. With this assumption the results of Long & Pepper (1984) and Biskupski *et al.* (1984) are explained. But from a theoretical point of view the matter is controversial, and at any rate for weak fields many authorities believe that σ tends to zero at E_c (see review by Mott and Kaveh (1985)).

We turn now to the effects of interaction. Early in these new developments the Moscow group (Altshuler & Aronov 1979) showed that long-range interaction could change the density of states near the Fermi energy, producing a cusp as Figure 10. The effect is only important when the mean free path is short. It leads to a temperature-dependence of the conductivity of the form

$$\sigma = \sigma_0 + AT^{1/2}$$

where A can have either sign. They showed that in bismuth, which normally shows the Landau–Baber T^2 correction, after the introduction of defects by cold work this $T^{1/2}$ term, which at low T is much the bigger, is observed. The $T^{1/2}$ term is observed in doped silicon and other semiconductors below ~ 1 K, and recently in amorphous metallic films at much higher temperatures.

In InP, investigated by Long & Pepper (1984), this term does not appear near the transition. We do not know why, but, since the constant A may have

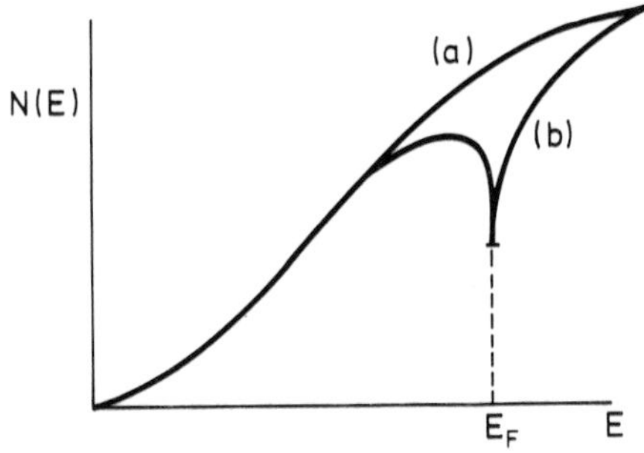

Figure 10. Density of states of a degenerate electron gas (*a*) without, and (*b*) with long-range electron–electron interaction.

either sign, a very small value is of course possible. In InSb (Mansfield *et al.* 1985), on the other hand, the term is very large, and here the observations in a magnetic field show that $\sigma_{\min}$, if it exists, is very small. This suggests the possibility that $\sigma_{\min}$ in a field exists only in a theory of non-interacting electrons.

The extensive evidence for the existence of a minimum metallic conductivity reviewed by the author and others in various books and papers before 1979 (Mott *et al.* 1975, Mott & Davis 1979, Fritzsche 1978) may in part be understood in terms of these temperature-dependent terms. It is highly probable that, as E_F approaches E_c, a negative temperature coefficient of resistance will develop, either on account of the term L_i or that varying as $T^{1/2}$. Unless experiments are made at sufficiently low temperatures, the negative temperature coefficient could be (and was) interpreted as an entry into the non-metallic state (see Mott & Kaveh 1983).

References

Abrahams, E., Anderson, P.W., Licciardello, D.C. and Ramakrishnan, T.V., 1979, *Phys. Rev. Lett.* **42**, 693.
Altshuler, B.L. and Aronov, A.G., 1979, *Solid St. Commun.* **36**, 115.
Anderson, P.W., 1958, *Phys. Rev.* **109**, 1492.
Baber, W.G., 1937, *Proc. R. Soc.* A **158**, 383.
Bergmann, G., 1983, *Phys. Rev.* B **28**, 2914.
Biskupski, G., 1982, *Thesis*, Lille.
Biskupski, G., Dubois, H., Wojkievicz, J.L., Briggs, H. and Remenyi, G., 1984, *J. Phys. C: Solid St. Phys.* **17**, L411.
Brinkman, W.F. and Rice, T.M., 1973, *Phys. Rev.* B **7**, 1508.
Castner, T.G., 1980, *Phil. Mag.* B **42**, 873.
Cohen, M.H., Fritzsche, H. and Ovshinsky, S.R., 1969, *Phys. Rev. Lett.* **22**, 1665.
Conwell, E.M., 1956, *Phys. Rev.* **103**, 57.
Cyrot, M., 1972, *Phil. Mag.* **25**, 1031.
Cyrot, M. and Lacourt-Gayet, P., 1972, *Solid St. Commun.* **11**, 1767.
Davies, J.H., Lee, P. and Rice, T.M., 1982, *Phys. Rev. Lett.* **49**, 1950.
Davies, J.H., Lee, P. and Rice, T.M., 1984, *Phys. Rev.* B **29**, 426.
Davis, E.A. and Compton, W.D., 1965, *Phys. Rev.* **140**, A2183.
Debney, B.T., 1977, *J. Phys. C: Solid St. Phys.* **10**, 4719.
de Boer, J.H. and Verwey, E.J.W., 1937, *Proc. Phys. Soc.* A **49**, 59.
Edwards, P.P. and Sienko, M.J., 1978, *Phys. Rev.* B **17**, 2573.
Efros, A.L. and Shklovskii, B.I., 1975, *J. Phys. C: Solid St. Phys.* **8**, L45.
Fritzsche, H., 1978, The metal–insulator transition, in *Proc. 19th University Summer School in Physics*, ed. L.R. Friedman and D.P. Tunstall, p. 193 (Edinburgh: SUSSP).
Gubanov, A.L., 1963, *Quantum Electron Theory of Amorphous Conductors* (English translation) (New York: Consultants Bureau).
Halperin, B.I. and Rice, T.M., 1968, *Rev. Mod. Phys.* **40**, 758.
Howson, M.A., 1984, *J. Phys. F.* **14**, L25.
Hubbard, J., 1964, *Proc. R. Soc.* A **277**, 237.
Hung, C.S. and Gliessman, J.R., 1950, *Phys. Rev.* **79**, 726.
Jones, H., Mott, N.F. and Skinner, H.W.B., 1934, *Phys. Rev.* **45**, 379.
Keldysh, L.V. and Kupaev, Yu.V., 1965, *Sov. Phys.-Solid St.* **6**, 2219.

Landau, L. and Pomeranchuk, I., 1936, *Phys. Z. Sov. Union*, **10**, 649.
Landau, L. and Zeldovich, G., 1943, *Acta Phys. Chim. USSR* **18**, 194.
Le Comber, P.G. and Spear, W.E., 1970, *Phys. Rev. Lett.* **25**, 509.
Licciardello, D.C. and Thouless, D.J., 1975, *J. Phys. C: Solid St. Phys.* **8**, 415.
Long, A.P. and Pepper, M., 1984, *J. Phys. C: Solid St. Phys.* **17**, 3391.
Mansfield, R., Abdul-Gader, M. and Foozoni, P., 1985, *J. Appl. Electronics* **28**, 109.
Miller, A. and Abrahams, E., 1960, *Phys. Rev.* **120**, 745.
Mott, N.F., 1949, *Proc. Phys. Soc.* A **62**, 416.
Mott, N.F., 1956, *Can. J. Phys.* **34**, 1356.
Mott, N.F., 1967, *Adv. Phys.* **16**, 49.
Mott, N.F., 1969, *Phil. Mag.* **19**, 835.
Mott, N.F., 1972, *Phil. Mag.* **26**, 1015.
Mott, N.F., 1974, *Metal–Insulator Transitions* (London: Taylor & Francis).
Mott, N.F., 1981, *J. de Phys. (Paris)* **42**, 277.
Mott, N.F., 1982, *Proc. R. Soc.* A **382**, 1.
Mott, N.F., 1984, *Phil. Mag.* B **49**, L75.
Mott, N.F., 1985, *Phil. Mag.* **51**, 14.
Mott, N.F. and Davis, E.A., 1979, *Electronic Processes in Non-crystalline Materials*, 2nd edn. (Oxford: Clarendon Press).
Mott, N.F. and Gurney, R.W., 1940, *Electronic Processes in Ionic Crystals* (Oxford: Clarendon Press).
Mott, N.F. and Kaveh, M., 1983, *Phil. Mag.* B **47**, 577.
Mott, N.F. and Kaveh, M., 1985, *Adv. Phys.* (in press).
Mott, N.F. and Twose, W.D., 1961, *Adv. Phys.* **10**, 107.
Mott, N.F., Pepper, M., Pollitt, S., Wallis, R.H. and Adkins, C.J., 1975, *Proc. R. Soc.* A **345**, 169.
Peierls, R.E., 1937, *Proc. Phys. Soc.* A **49**, 3.
Rosenbaum, T.F., Andres, K., Thomas, G.A. and Bhatt, R.N., 1980, *Phys. Rev. Lett.* **43**, 1723.
Slater, J.C., 1957, *Phys. Rev.* **82**, 538.
Spear, W.E. and Le Comber, P.G., 1975, *Solid St. Commun.* **17**, 1193.
Street, R.A., 1982, *Phys. Rev. Lett.* **49**, 1187.
Thompson, J.C., 1976, *Electrons in Liquid Ammonia* (Oxford: Clarendon Press).
Thouless, D.V., 1977, *Phys. Rev. Lett.* **39**, 1167.
Turkevich, L.H. and Cohen, M.H., 1983, *J. Non-Cryst. Solids* **61–62**, 13.
White, R.M. and Mott, N.F., 1971, *Phil. Mag.* **24**, 845.
Wilson, A.H., 1931, *Proc. R. Soc.* A **133**, 458.

CHAPTER 2

the metal–nonmetal transition

T.V. Ramakrishnan

Department of Physics, Banaras Hindu University, Varanasi-221 005, India

1. Introduction

The transition of a system from metal to insulator is obviously a basic electronic change. In this review, a few of the relevant physical ideas and models will be discussed. Many of the concepts in this field are due to Sir Nevill Mott, whose continued focus on basic questions and experimental consequences has stimulated most of the work in this area. It is therefore a great pleasure to participate in this festschrift.

The metal–nonmetal transition can occur as some external physical parameter such as pressure, density, composition, disorder, temperature or magnetic field is varied. The transition can be discontinuous, that is, marked by a jump in the density, conductivity and so on, or continuous, when these quantities do not change suddenly across the transition (but their derivatives do). The phenomena and the theories have been reviewed by Mott (1974).

In a crystalline solid, if a one-electron band theory description is adequate the transition could be due to a rearrangement of electronic states such that the highest occupied allowed level becomes separated by an energy gap from the lowest unoccupied level. We do not consider such band-crossing transitions here (see Mott 1974 for a discussion of these). Systems which ought to be metallic in an independent-electron approximation can, because of correlation between electrons, becoming insulating. This is the Mott transition. We discuss a very simple model for correlation effects in solids (the Hubbard model, §2.1). Despite considerable work and a good understanding of its qualitative features, there is no detailed picture yet of its predictions for physical properties in the vicinity of the metal–nonmetal transition. We summarize existing theories which fall into two classes, namely, the correlated ground state (§2.2) and the one-body approximation (§2.3). We then discuss briefly (§2.4) some features present in real systems but not in the Hubbard model, for example, long-range coulomb interaction, orbital degeneracy and electron–lattice coupling. The last subsection

(§2.5) is concerned with some commonly used criteria for the metal–insulator transition.

An entirely different class of metal–insulator transitions is related to the phenomenon of electron localization in a random potential, first considered by Anderson (1958). In a sufficiently random potential, electronic states are not extended but are localized. The consequences of such a localization for metal–insulator transitions were first discussed by Mott. In §3, we outline these ideas, a recent scaling approach to localization and the experimental situation.

In actual systems, disorder and interaction are both present. We discuss in §4 the disordered interacting electron gas. Perturbative effects are mentioned first (§4.1); we then review recent contributions to the theory of metal–insulator transitions in disordered systems with interacting electrons, and compare the predictions with what is known experimentally.

An attempt has been made to describe physical ideas and to highlight open problems and promising trends. However, the review is very incomplete: for example, there is very little discussion of lattice or orbital degeneracy effects in pure or disordered systems. The few attempts to do realistic electronic structure calculations of metal–insulator transitions have not been described. References are suggestive rather than exhaustive.

There are many reviews which cover recent progress in the field of disordered electronic systems (Lee & Ramakrishnan 1985, Bishop & Dynes 1985, Altshuler & Aronov 1984, Fukuyama 1984, Nagaoka & Fukuyama 1982). These may be consulted for further experimental or theoretical details.

2. The Mott transition (correlation induced metal–nonmetal transition)

2.1 *The Hubbard model*

The model and its parameters

This model is perhaps the simplest description of electron correlation effects in solids. Though proposed and used earlier by a number of authors, it was discussed extensively, especially in the context of metal–insulator transitions by Hubbard (1963, 1964). The Hamiltonian is characterized by a kinetic energy term t_{ij} denoting the hopping of an electron from a site i to its nearest neighbour site j, and by the extra energy cost U of putting two electrons (spins up and down) on the same site i.

$$H = \sum_{i,j,\sigma} t_{ij} a_{i\sigma}^{+} a_{j\sigma} + U \sum_{i} n_{i\uparrow} n_{i\downarrow} \tag{1}$$

Here t_{ij} is zero unless i and j are nearest neighbours and is the same for all of these. In the absence of U, this is just a tight-binding Hamiltonian with atomic energy levels ε_i at all sites being the same (zero) and with nearest neighbour overlap t_{ij}

leading to a band of one-electron states. The Mott–Hubbard correlation term U is, in the atomic limit, the difference between the ionization energy I and electron affinity A. The model is characterized by one dimensionless parameter (t/U), and by the electron density or average number of electrons per site $\langle n \rangle$. This can range from zero to two.

The model perhaps does not describe in realistic detail any of the commonly studied systems, for the following reasons. It has no long-range coulomb repulsion between electrons, no orbital degeneracy or many-band effects, no electron–lattice coupling, and assumes a perfectly ordered lattice. We briefly discuss the nature and some of the consequences of these relevant perturbations in §2.4.

The parameters of the model, t_{ij} and U, are poorly known in general, except possibly for the hydrogenic doped semiconductor system. U is most directly determined experimentally by combining photoemission and bremsstrahlung isochromat spectroscopy. The former measures, roughly, energies of the singly ionized electronic states while the latter measures energies of configurations with one electron added. The difference in the energies d^{n-1} and d^{n+1} for a particular ground d electron configuration d^n is just the quantity $I - A = U$ measured *in situ*. A recent example of such a measurement is that of Sawatzky & Allen (1984) on NiO. These authors find that $U \simeq 9$ eV for the d^8 ground state orbital of Ni. The bandwidth and crystal field splitting energies are much smaller, of the order of 1 or 2 eV. However, the energy required to remove an electron from the top of the filled oxygen 2p band and put it in the d shell of a nearby Ni ion is the observed band gap 4·3 eV, much less than the correlation gap of 9 eV. Thus, in this system there are important effects due to other states (p states) not present in a one-band model such as equation (1).

The Hubbard U can also be obtained theoretically by calculating the configurational energies of the appropriate small cluster. Calculations by Fujimori & Minami (1984) for the $(NiO_6)^{10-}$ cluster yield energy separations including U in good agreement with the spectroscopic results of Sawatzky & Allen (1984). For metals, a spin-density functional calculation for the energies of appropriate Wigner Seitz cell configurations leads to an estimate for U. (See for example, Herbst *et al.* 1976). The values of U are in general considerably less than their atomic values $(I - A)$, especially in metals. This point was emphasized quite early by Herring (1966) in his review of magnetism in transition metals. The reduction is due to screening by other electrons physically present in a unit cell of the crystalline solid. U is then a phenomenological constant dependent on density, on electron filling and on the proximity in energy as well as overlap in space of other orbitals. For example, for a lattice of hydrogen atoms it depends on the lattice constant, just as t_{ij} does.

The hopping term t_{ij} is not determinable directly by experiment, but electron energy dispersion curves ($E(k)$ vs. k) have been obtained from angle-resolved photoemission for several transition metals (Eastman & Himpsel 1981). By fitting to a tight-binding type model, the hopping integrals t_{ij} can be inferred. First-principles band-structure calculations also lead to estimates of t_{ij} for simple systems.

Limiting cases

We now discuss known results for the Hubbard model in some limiting cases, especially those relevant for the metal–insulator transition.

(i) For small U, i.e. $t_{ij} \gg U$, the ground state in three dimensions is metallic irrespective of the band filling. This is an expected perturbative result. In one dimension, however, the weakest interaction U destabilizes the Fermi surface.

(ii) For large U, i.e. $U \gg t_{ij}$ *and* with one electron per site (half filled band) the ground state is an antiferromagnetic insulator. Since a configuration with two electrons at one site and a hole at another site costs an extra energy U with respect to the one-electron-per-site configuration, the latter is strongly favoured. The moments (electrons) at nearest neighbour sites have an antiferromagnetic coupling (t_{ij}^2/U). Thus the Néel temperature is $kT_N \simeq (zt^2/U)$ where z is the number of nearest neighbours. Above T_N, the moment-carrying electrons are still localized, one on each lattice site, but the moments are disordered. The system is insulating, since the electron excitation energy or gap is of order $U \gg kT_N$.

There is a clear separation in energy between charge fluctuations and spin fluctuations (U and (t_{ij}^2/U) respectively). The very existence of Mott insulators, that is insulating solid-state systems with local magnetic moments (reflected in the Curie–Weiss-like χ) but without long-range magnetic order, indicates a breakdown of a one-electron description. A magnetic moment at a site means that there are at least two magnetically distinct states of equal energy, one occupied and the other not. It is only because of correlation that the other state is not occupied.

(iii) The system that is not half filled is always metallic, no matter how strong or weak the correlation. With respect to the half filled band, there are doubly occupied or unoccupied sites ('electrons' and 'holes') which are mobile. The system is thus metallic. The insulating phase is unstable against the slightest change in electron density from half filling. This prediction has not been explored in detail either theoretically or experimentally.

(iv) In one dimension, the ground state with one electron per site is an antiferromagnetic insulator for all U. The problem can be solved exactly using the Bethe ansatz (Lieb & Wu 1968). The normal metallic phase is unstable with respect to interactions because of well known logarithmic singularities associated with a one-dimensional Fermi surface. This result suggests that the lower critical dimension for the metal–insulator transition is one.

We now briefly discuss the theory of the metal–insulator transition for the half-filled Hubbard band. The approaches fall into two distinct types of approximations which have not been seriously put together. One of them emphasizes local dynamic correlations through an approximate ground state wave function (§2.2) and the other is a one-body approximation in which one attempts to find the self-consistent symmetry breaking static potential (§2.3).

2.2. *Correlated ground state method* (*Gutzwiller Approximation*)

The effect of the correlation term U in the Hubbard Hamiltonian is to disfavour the double occupancy of any site relative to the non-interacting limit, where this probability amplitude is simply the product of those for the separate single occupancy of up and down spin states. The ground state wave function ψ_G is thus approximated as

$$\psi_G = \prod_i (1-(1-g)n_{i\uparrow}n_{i\downarrow})\psi_F \tag{2}$$

where ψ_F is the free fermion ground state wave function. Since $n_{i\uparrow}n_{i\downarrow}$ has a value of one when both up and down states at site i are occupied, and a value of zero otherwise, a weight of g where $0 \leqslant g \leqslant 1$ is assigned to a doubly occupied site *vis-à-vis* other occupancies. In the limit $g=1$, there is no correlation, and in the limit $g=0$, there are no doubly occupied sites in the ground state. The quantity g is thus an inverse measure of correlation, and is the sole variational parameter with respect to which the ground state energy is minimized. Using the fact that $D_i = n_{i\uparrow}n_{i\downarrow}$ has eigenvalues one and zero only, equation (2) can be written compactly as

$$\psi_G = g^{\sum_i D_i}\psi_F \tag{3}$$

It is not possible to evaluate expectation values exactly using ψ_G, so that further approximations are made. In the very simplest one, due to Gutzwiller (1965), spatial correlations between configurations are neglected wherever they can be. Concretely, for example in calculating the wave function normalization $\langle \psi_G|\psi_G\rangle$, terms like $\langle \psi_F|g^{D_i}g^{D_j}|\psi_F\rangle$ are neglected if sites i and j are different. Similarly, in calculating the expectation value of the hopping term, the nearest neighbour configurations are specified, and all other spatial correlations are ignored. The energy ε_g per site is then found to be

$$\varepsilon_g(d) = 16d(1-2d)\bar{\varepsilon} + Ud \tag{4}$$

where d is the fraction of doubly occupied sites and is related to g by $g = d(\frac{1}{2}-d)^{-1}$. $\bar{\varepsilon}$ is the kinetic energy of the uncorrelated system per spin, i.e. $\bar{\varepsilon} = \sum_{\vec{k}} \varepsilon(\vec{k},\sigma)$ where the summation is over all occupied states $\vec{k}$. For a symmetric band centred around zero energy, $\bar{\varepsilon}$ is negative. The ground state energy is obtained by minimizing $\varepsilon_g(d)$ with respect to g or equivalently in equation (4) as a function of d. For $U=0$, clearly $d=(1/2).(1/2)=(1/4)$ and the ground state energy is $2\bar{\varepsilon}$. As U increases, d decreases smoothly and vanishes at $U_c = 8\bar{\varepsilon}$. Beyond U_c the

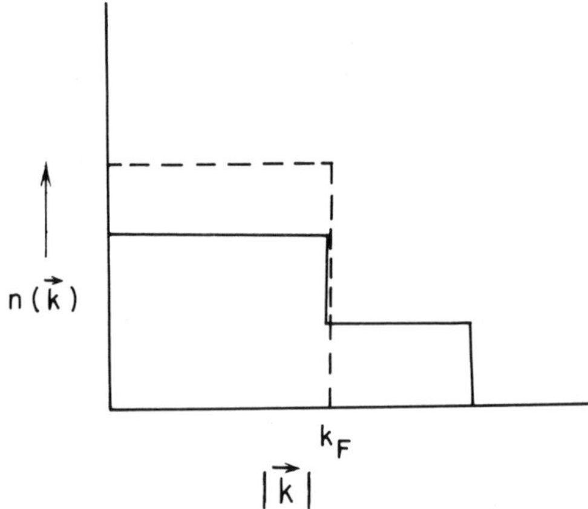

Figure 1. Ground state momentum distribution in the Gutzwiller correlated wavefunction (zero range approximation); full lines. The discontinuity at k_F decreases with increasing correlation. The free Fermi gas distribution is shown by dotted lines.

only sensible solution is $d=0$, at least with the above approximate estimate of the ground state energy.

We now discuss the properties of the system for U close to U_c (Brinkman & Rice 1970, Rice & Brinkman 1971, Vollhardt 1984). The ground state momentum distribution $n(k)$ has the form shown in Figure 1, the free Fermi gas distribution being shown by the dotted line. As particles become correlated in real space, the distribution spreads out in momentum space. Since in the approximation used, all intersite spatial correlations are neglected, $n(k)$ consists of two flat pieces with a discontinuity at the Fermi surface. Since the total number of electrons, $\sum_{\vec{k},\sigma} n(k)=N$, is fixed, this discontinuity decreases as U increases, going to zero as $U \rightarrow U_c$. This vanishing of the Fermi surface jump q indicates an instability of the normal metallic phase. Now q is well known in many-body theory to be related to the renormalization of electron energy due to interactions. (See, for example, Fetter & Walecka 1971.) If the energy shift of the electron with frequency ω and bare energy $\varepsilon(\vec{k})$ is $\Sigma(\omega)$, one has $q=[1-\partial\Sigma(\omega)/\partial\omega]_{\omega=\mu}$ where μ is the Fermi energy. The density of states is $N(\omega)=\sum_{\vec{k},\sigma} \delta(\omega-\Sigma(\omega)-\varepsilon(\vec{k}))$ and is therefore proportional to q^{-1} at the Fermi level. Close to U_c, $q=[1-(U/U_c)^2]$, that is, it goes to zero linearly. The density of states at the Fermi level, and thus the effective mass, diverge as $[1-(U/U_c)]^{-1}$ for $U \rightarrow U_c$. The spin susceptibility is obtained by calculating the change in ground state energy in a small magnetic field, and is enhanced with respect to the Pauli value by the same singular factor as the effective mass. Calculating the change of ground state energy with respect to a change in electron density, that is, a small deviation from half filling, one finds that the compressibility goes to zero as $[1-(U/U_c)]^2$.

All these results imply a localization of fermions due to correlation. Because electrons are nearly locked into their places, the low-lying excitations are very heavy, the spins are also nearly localized and the system is virtually incompressible. The screening length for long wavelength charge or density fluctuations depends on compressibility, varying inversely as its square root. Thus the screening length diverges as $[1-(U/U_c)]^{-1}$. This is another rather direct indication that a metal–insulator transition is at hand. The susceptibility and specific heat (or effective mass) are enhanced identically; this is a sign of short-range correlation effects. Two other conditions under which this happens are the correlated impurity and the disordered metal with spin fluctuation or paramagnon effects. In the former case, one imagines an electron gas with a Hubbard type of correlation only at a small fraction c of sites. In the Hartree–Fock approximation, it turns out (Lederer & Mills 1968) that the terms that are linear in c for susceptibility and the specific heat coefficient are singularly enhanced as $U \rightarrow U_c^{\text{loc.}}$. The ratio is not. Something similar happens to a system of electrons coupled to paramagnons in a disordered metal. On the other hand, for a pure fermion system coupled to paramagnons, the susceptibility is enhanced by the Stoner factor s ($s \gg 1$) but the specific heat only by a factor $\ln s$. Closer examination (see, for example, Vollhardt 1984 for a detailed review) shows that the susceptibility enhancement is due to the density of states or effective mass increase, and *not* due to an incipient singularity in the spin triplet scattering channel (the appropriate Landau parameter or scattering amplitude is non-singular). The Fermi surface volume is unchanged up to $U = U_c$ where the Fermi surface discontinuity vanishes. This result is in agreement with Luttinger's theorem (Luttinger 1960) that interactions do not change the volume of the Fermi surface so long as the ground state is adiabatically continuous with the unperturbed configuration.

This attractive picture for the almost localized Fermi liquid has been applied most successfully and extensively to liquid ^{3}He (Anderson & Brinkman 1978, Vollhardt 1984) by assuming U to be close to U_c and assuming a simple density or pressure dependence for it. The properties of metallic doped V_2O_3 near the metal–insulator critical point, for example the small effective Fermi temperature, approximate proportionality of succeptibility and specific heat enhancement, have also been described in this model (Rice & Brinkman 1971, Mott 1974).

One of the major shortcomings of the approximation used is its inability to describe the phase for $U > U_c$. Closely related is the result that $d \rightarrow 0$ as $U \rightarrow U_c$. This is not correct, since from perturbation theory one can show that there is an amount $d = (t/U)$ of the doubly occupied configuration mixed quantum mechanically into the ground state for large U ($U \gg U_c$). This deficiency arises from the complete neglect of spatial correlations.

There have been several attempts to calculate physical quantities using the Gutzwiller wavefunction without making further approximations, to improve the wavefunction itself and to generalize the approach to non-zero temperatures. We mention a few recent attempts here. Ogawa & Kanda (1978) show that the effect of spatial correlations present in the Gutzwiller ground state can be included for larger and larger clusters, somewhat like the Kikuchi cluster method in lattice gas or Ising models. This leads for example to non-zero range spin correlations, and a quantitative change in U_c, but to no qualitative departure from the Brinkman–Rice behaviour for the highly correlated gas. A non-zero temperature generalization in which a simple non-interacting quasiparticle term for the entropy is added has been developed recently by Rice *et al.* (1985) and applied to the heavy fermion superconductor UBe_{13}, assuming that it can be modelled by a nearly half filled band with small deviation δ from half filling ($\delta \ll 1$), and strong correlation ($U \gg U_c$). Under these conditions, the $T=0$ susceptibility and density of states are enhanced by a factor δ^{-1} and the effective temperature scale is $\delta T_F \ll T_F$. The rather rapid change of specific heat and susceptibility with temperature is fairly well reproduced.

Perhaps the most interesting sidelight on the Gutzwiller wavefunction is provided by the recent work of Kaplan *et al.* (1982) which compares predictions for a one-dimensional chain in the regime of large U with the known results for a one-dimensional Heisenberg antiferromagnet. In the limit of large U or small bandwidth, the latter is a very good approximation for the ground state of the one-dimensional Hubbard model. The correlation functions for finite sized systems in the Gutzwiller ground state (but with no further approximations) are in excellent agreement with those of the antiferromagnetically coupled Heisenberg spins. The ground state energy is also accurately reproduced if an additional term is included in the ground state wavefunction requiring the doubly occupied site and the hole site positions ('electrons' and 'holes') to be next to each other. Without this term, the ground state energy has the right analytical form, i.e. $(-t^2/U)$, but a coefficient which is only 50% of the correct value. These results strongly indicate that the Gutzwiller wavefunction has antiferromagnetic correlations built into it (though no long range antiferromagnetic order or broken symmetry); these are masked by approximate methods of evaluation, such as that outlined above. It is thus possible that a more accurate calculation using the correlated wavefunction of the Gutzwiller type will describe both temporal *and* spatial correlations as one approaches the metal–insulator transition; with an appropriate site symmetry broken one-electron wavefunction, the antiferromagnetically ordered insulating phase can also be discussed. In the past, a theory for the latter has been sought for (and found) within the Gutzwiller scheme by starting with an antiferromagnetic (AF) Hartree–Fock ground state rather than a free fermion state (Ogawa *et al.* 1975). The above type of zero-range approximation is used in calculating expectation values, etc. It turns out for alternant lattices that the *uncorrelated* ($g=0$) AF state is more stable than the Gutzwiller paramagnetic or the correlated AF state for $U \geqslant 0{\cdot}4U_c$, that is, well before the Brinkman–Rice transition. It is not clear whether this result is an artefact of the approximations made in calculating the expectation values with the Gutzwiller wavefunction.

It has been realized recently that the Gutzwiller wavefunction contains p-wave superconductive correlations (Anderson 1984). For U close to U_c, the up and down spin electrons are correlated away from each other. This suggests a node at zero separation in their relative wavefunction, that is, a p-wave type of distribution. The effective interaction parameters of the electrons (Landau parameters) can be calculated, assuming only that these vanish for relative angular momentum $l \geqslant 2$. The superconducting transition temperature for such an interaction can be obtained from standard many-body theory, and is found to be about (1/10) T_F (Valls & Tesanovic 1984). Thus the ground state of the heavy, nearly localized, strongly correlated Fermi liquid may be superconducting, a possibility not considered before.

2.3. *One-body approximation*

Perhaps the approach most commonly used in discussing the Hubbard model is to look for the one-body Hamiltonian that best approximates it. This was the method followed by Hubbard in his early work (1963, 1964) and has been used for example by Cyrot and co-workers (Cyrot 1972, Cyrot & Lacour-Gayet 1972). The most extensive recent work of this kind is due to Economou and co-workers (Economou *et al.* 1978, de Marco *et al.* 1978, White & Economou 1978). The Hartree–Fock approximations discussed by Brandow (1977) are similar.

The approximation is basically that

$$U n_{i\downarrow} n_{i\uparrow} \simeq U\langle n_{i\uparrow}\rangle n_{i\downarrow} + U\langle n_{i\downarrow}\rangle n_{i\uparrow} - U\langle n_{i\downarrow}\rangle\langle n_{i\uparrow}\rangle \tag{5}$$

where $\langle n_{i\sigma}\rangle$ is the thermally averaged electron number at site i with spin σ. For a given set of mean values $\{\langle n_{i\sigma}\rangle\}$, the one-body Hamiltonian is identical with that of the Anderson model for localization (see §3.1), the energy for an electron at site i and with spin σ assuming the random value $V_{i\sigma} = U\langle n_{i-\sigma}\rangle$. If the total number of electrons per site is given, i.e. $\langle n_{i\uparrow}\rangle + \langle n_{i\downarrow}\rangle$ is fixed, only the magnetization per site, $m_i = \langle n_{i\uparrow}\rangle - \langle n_{i\downarrow}\rangle$, is a random variable. Assuming that the Anderson problem for any given random site energy distribution is solved, the physically relaized $\{n_{i\sigma}\}$ is calculated by requiring self consistency, namely by solving the coupled equations $\{n_{i\sigma}\} = f(\{V_{j\sigma}\}) = f(U\{n_{j-\sigma}\})$. One thus has in general N coupled equations, where N is the number of sites. In practice, however, a number of further approximations are made. First, since each site is identical, it is assumed that the magnetic moment m_i is the same in size at each site, and that only the sign is a random variable. The one-body potential then has two values $\pm Um$ distributed with equal probability over the lattice sites. The electronic structure of such a two-component AB alloy has been investigated in several variants of the Coherent Potential Approximation (see, for example, Cyrot 1972; Economou *et al.* 1978). The ground state for all non-zero m is magnetically ordered, the interaction between moments at different sites being due to the hopping motion of electrons. The most commonly investigated order is antiferromagnetic since this is the actual ground state for a half filled band with large U.

The metal–insulator transition in such a one-body picture is due either to gaps in the electronic density of states, or to Anderson localization. We first consider the antiferromagnetically ordered regime, with moments pointing up in one sublattice and down in the other. The one-body potential is similarly periodic, and the original tight-binding band splits into two. In the limit of small bandwidth and large U, the splitting is $Um \simeq U$ corresponding to the correct correlation gap between singly and doubly occupied site energies. As the bandwidth zt increases, the effective gap (between the top of the lower band and the bottom of the upper band) due to antiferromagnetic order decreases, being of order $(U - zt)$. Due to particle–hole symmetry of the original half filled band problem it is easy to show that the number of states in both bands is the same. Thus in the half filled case, the lower band is occupied while the higher one is not. The two bands have been called the lower and upper Hubbard bands by Mott and the energy gap between them is the Mott–Hubbard correlation gap. The system is insulating at $T=0$ if the Mott–Hubbard gap is non-zero, or crudely, if $U > zt$. For some lattices (such as simple cubic or body-centred cubic, which can be partitioned uniquely into alternant cubic lattices containing nearest neighbour sites) the band gap is non-zero no matter how small U (or how large the ratio (t/U)). This means that the metallic state is unstable, just as in one dimension. Fluctuation effects and non-nearest neighbour hopping change this pathological, structure-dependent result. For the tight-binding face-centred cubic solid, Cyrot (1972) finds that antiferromagnetic order sets in at $(U/zt) = 0{\cdot}36$ where moments first appear, and that the band gap is large enough for this system to go insulating at $(U/zt) = 0{\cdot}56$. The moment here is 0·86 of its full value. Thus in this approximation the system goes (at $T=0$) from a paramagnetic metal to antiferromagnetic metal to antiferromagnetic insulator as correlation increases. The transitions are continuous, and are of the band crossing type.

Even when the moments are present on each site but are not ordered, one can have a density of states gap. The hopping in the equivalent equiatomic alloy produces a maximum broadening of zt (i.e. all energy levels of the alloy are in the ranges $(-Um/2 \pm zt/2)$ and $(Um/2 \pm zt/2)$. Therefore for $U > zt$, there is a band gap and the system is a 'disordered' insulator. It is specially clear in the large U limit that the correlation gap U, being due to local correlations, is hardly affected by the disappearance of long-range order (energy scale t^2/U). Using the single-site coherent potential approximation, Cyrot showed (1972) that the density of states goes to zero at the Fermi energy and a gap appears for $(U/zt) > 0{\cdot}67$ (fcc lattice). Thus one can have an insulator which has local moments but no long-range magnetic order. This is in contrast to the ideas of Slater (1951) who argued, from a self-consistent field theory and a band model, that the insulating phase and antiferromagnetism are coterminus. Figure 2 schematically shows the density of states in the Hubbard model for $U > zt$.

Economou and co-workers (Economou *et al.* 1978, White & Economou 1978) have made extensive studies using the one-body approach and a two-site version of the CPA (White & Economou 1977) for dealing with randomness. The effective randomness, i.e. the size of local moments and their mutual correlation,

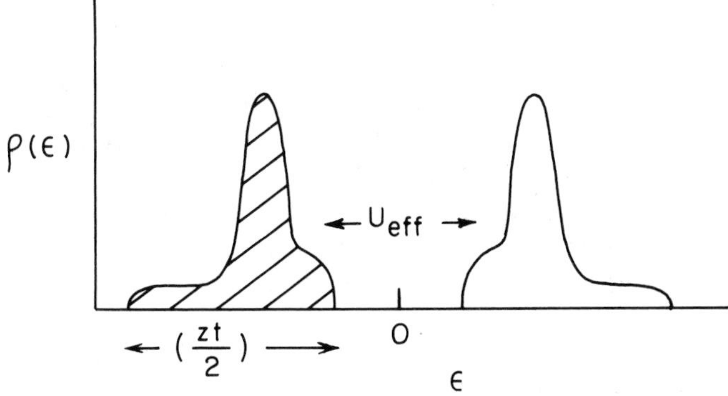

Figure 2. The density of states $\rho(\varepsilon)$ vs. energy ε in a one body approximation to the Hubbard model. The lower and upper Hubbard bands and the occupied states are shown (for the half filled case).

depends on temperature, and influences the electronic structure. A method of taking this into account was devised so that properties of the system could be calculated at non-zero temperatures also and a phase diagram obtained. The calculations show both a metallic and an insulating phase, with the possibility of long-range antiferromagnetic order in each. A schematic phase diagram is shown in Figure 3.

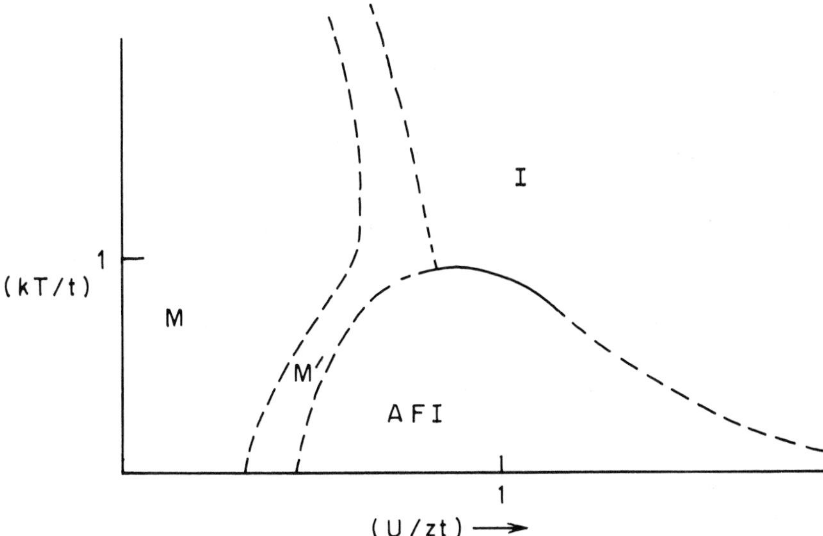

Figure 3. A schematic phase diagram for the Hubbard model in the temperature (k_BT/t) and correlation (U/zt) plane, after de Marco *et al.* (1978). Phases shown are metallic (M), metallic with moments (M′), antiferromagnetic insulator (AFI), and insulating (I). Transitions are continuous except where shown by full lines.

The one-electron-per-site system is metallic for weak correlation (region marked M) and has no local magnetic moments. These develop above a certain critical correlation and in the region marked M′ the system is a metal with moments. At low temperatures in this regime, the moments may order, and one could have an antiferromagnetic metal, for example. At still larger U there is a transition to an antiferromagnetic insulator (labelled AFI). Long range antiferromagnetic order disappears at the Néel temperature (the boundary between AFI and I) above which one has an insulating phase with disordered moments. For $U \gg zt$, the Néel temperature is $T_N \sim k^{-1}(zt^2/U)$, that is, it decreases as U increases. This is shown in the figure. The somewhat arbitrary boundary between metal and insulator, based on a feature in an approximate calculation of the conductivity, is also shown. The transitions are all continuous, except that in an intermediate regime of U, there is a first-order antiferromagnetic to paramagnetic transition which may be an artefact of the approximation used. The Mott transition regime is $(U/zt) \simeq 1$, $(kT/t) \simeq 1$, with $(T_N)_{max} \simeq (t/k)$.

We show for comparison (Figure 4) the experimental phase diagram of V_2O_3 doped with chromium and titanium (McWhan and Remeika 1970). These are supposed to uniformly dilate or compress the system and thus increase or reduce (U/zt). There is a general similarity between the two phase diagrams. However, the observed transitions are first-order except that the metal–nonmetal transition has a critical point. The slope of the M-NM transition line implies from the Clausius–Clapeyron equation that there is a small entropy increase on going from metal to insulator. Also, the metallic phase is unusual, being marked by a very low characteristic (Fermi or degeneracy) temperature. This is reflected in the

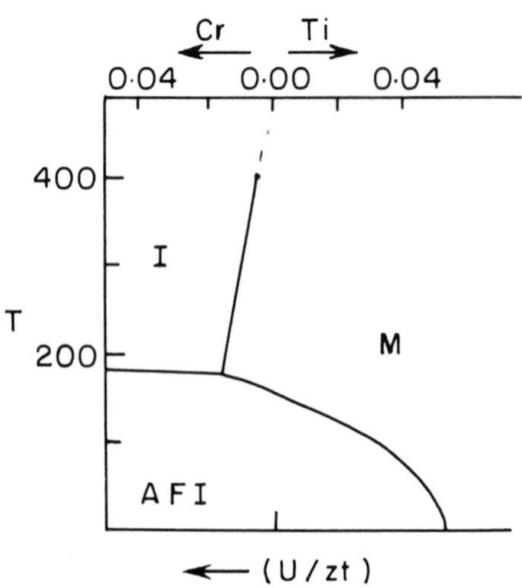

Figure 4. Phase diagram for V_2O_3 alloyed with Cr and Ti (McWhan and Remeika 1970). Metallic, insulating and antiferromagnetic insulating phases are shown. Full and broken lines indicate first-order and continuous transitions respectively.

high specific heat and in the Curie–Weiss-like magnetic susceptibility. The resistivity of the metallic phase is very high and is strongly temperature dependent. A good fit to a T^2 increase is obtained over a wide range of temperatures.

It is clear that the one-body theory, while reasonably accurate for the gross features, does not describe the details well. The failure could be due to one or more of the relevant perturbations discussed below. For example, the first-order transition could be due to electron–lattice coupling. The discontinuous antiferromagnetic insulator to paramagnetic transition is probably a magnetostrictive effect, large because the electronic degrees of freedom are 'soft'. The properties of the metallic state near the transition are, however, similar to those expected for the highly correlated Gutzwiller–Brinkman–Rice many body ground state. That is, the specific heat and susceptibility are critically enhanced and strongly temperature dependent, as is the resistivity on account of the small energy scale and large local fluctuations. Unfortunately, the approximations used so far in the correlated ground state approach are not adequate for describing the antiferromagnetically ordered states and the magnetically disordered insulating regime. The phase diagram of the Hubbard model is not described, nor are the effects of relevant perturbations. We now compare the two approaches and discuss the reasons for their complementary inadequacies.

The one-body approximation is a mean field theory where the local magnetic moment is the mean field. The approximation is accurate, at least regarding energies, when the mean field is large, i.e. for strong correlation ($U \gg zt$) where $m = \langle n_{i\uparrow} \rangle - \langle n_{i\downarrow} \rangle$ is close to its maximum value of unity at each site i. Paradoxically, therefore, the mean field theory is best for both very strong and very weak correlation. In the former case, the $N_i = 2$ configuration is very unfavourable in energy. The most likely configurations have $N_i = 1$ because of the constraint $\langle n_{i\uparrow} \rangle + \langle n_{i\downarrow} \rangle = 1$. At any site i, therefore, there is a slow fluctuation between the two $N_i = 1$ configurations $\langle n_{i\uparrow} \rangle = 1$, $\langle n_{i\downarrow} \rangle = 0$ and the reverse, in the magnetically disordered regime. In the Hartree–Fock or one-body approximation, these symmetry-restoring fluctuations are ignored and a solution with locally broken symmetry assumed. The error in energy is of order (zt^2/U). To this order, one has non-interacting Ising (rather than Heisenberg) spins. By looking at specific nearest neighbour spin configurations, the small residual short-range spin interaction J_{ij} can be calculated. The antiferromagnetically ordered ground state can also be directly addressed in band theory. Thus the low-energy spin excitations and ordering deep in the insulating regime can be accurately described and some qualitative shortcomings of the mean field theory made up. We also note that the one-body approximation reproduces the correct site-energy spectrum. For example, with $\langle n_{i\uparrow} \rangle \simeq 1$, $\varepsilon_{i\uparrow} \simeq 0$ and $\varepsilon_{i\downarrow} = U$. The correct energy levels are $E_i = 0$ for $N_i = 1$ or 0 and $E_i = U$ for $N_i = 2$. However, dynamic degeneracy is replaced by a static one. There is no dynamic interchange between the up and down configurations.

In the weak correlation limit $U \ll zt$, the mean field is zero, but since all configurations are equally and independently accessible, the local fluctuations (both charge and spin) are distributed more or less uniformly over all possible frequency or time scales. The one-body method is unfortunately inadequate in the regime of most interest to us, namely near the metal–insulator transition. The transition occurs in the intermediate coupling regime $U \simeq zt$, $kT \simeq t$, where there is no clear separation in energy scale between charge and spin fluctuations. The mean field is zero or small and fluctuations have low energy, as suggested by results from the Gutzwiller approach. In such a case, mean field theory is poor.

The correlated wavefunction approach, on the other hand, focuses on local correlations. It describes the progressive freezing out of the doubly occupied configurations due to local, temporal quantum fluctuations. The approximations developed so far are basically independent-site models, and do not include spatial correlations, symmetry breaking solutions and non-zero temperature effects in any consistent way, although some attempts have been mentioned earlier. The mean field theory describes neither spatial nor temporal fluctuations accurately, but only averaged, symmetry breaking quantities. A more complete theory including both these effects is necessary for a coherent picture of the metal–insulator transition and the phase diagram in this very simplest of models for the Mott transition.

2.4. Relevant perturbations

Coulomb interaction

For a half filled band, such as a lattice of hydrogen atoms or a doped uncompensated semiconductor, there is one electron per site. In the presence of correlation, a fraction of these sites is doubly occupied, and an equal fraction is empty. The doubly occupied sites can be described as 'electrons' in the upper Hubbard band, while the equal number of unoccupied sites are 'holes' in the lower Hubbard band. In the Hubbard model, 'electrons' and 'holes' have a short-range kinetic correlation. In reality they attract each other by the long-range coulomb force. This leads to mutual binding and favours the insulating phase. Mott (1949) proposed in 1949 that because of this, a situation with a small number of carriers ('electrons' and 'holes') is not possible. For 'electron' and 'hole' densities such that the screened coulomb potential at that density has a bound state, one does not have free carriers and there is no metallic phase. The screened coulomb (Yukawa) potential $V(r) = (e^2/\varepsilon r)\exp(-qr)$ has a bound state for a screening length $q^{-1} \simeq 0{\cdot}83(\varepsilon \hbar^2/m^*e^2) = 0{\cdot}83 a_H$. Using a Fermi–Thomas estimate for q^{-1}, one has

$$(n_{e,h})^{1/3} a_H \gtrsim 0{\cdot}4 \tag{6}$$

for the metallic state to exist, where $n_{e,h}$ is the density of electrons or holes. Thus the free electron or hole density drops discontinuously from this value to zero at the metal–nonmetal transition, due to coulomb interaction. The transition becomes discontinuous or first-order.

Based on this idea, Mott proposed the famous criterion for a metal–nonmetal transition. Since the density of 'electrons' or 'holes' is necessarily less than that of sites or the actual electron density n, the condition (6) is an upper limit on the electron density for which the metallic phase is stable, that is, one has

$$n_c^{1/3} a_H \gtrsim 0{\cdot}4 (n/n_{e,h})^{1/3} \tag{7}$$

The ratio $(n/n_{e,h})^{1/3}$ is of the order of one (somewhat greater than 1) but is not well known. One model calculation of this ratio for doped semiconductors is due to Bhatt & Rice (1981). They used results from the theory of electron–hole drops (see Rice 1977 for a review). These are well studied systems where, as a function of electron–hole density, a transition takes place from an excitonic phase to free electrons and holes. The transition is similar to that from an insulator to a metal. At $T=0$, the transition is strongly first-order. Using the density $n_{e,h}$ at which this occurs, Bhatt and Rice concluded for a single-valley version of Si:P that $(n/n_{e,h})^{1/3} \simeq 4$. The most commonly used value for the Mott criterion is

$$n_c^{1/3} a_H \gtrsim 0{\cdot}25 \tag{8}$$

for a metal. (See §2.5 for a discussion.)

The above line of argument assumes that there are well defined and separated Hubbard bands so that 'electrons' and 'holes' are long-lived excitations (in the absence of coulomb interaction). This is indeed true for the electron–hole drop system. However, it may not be true especially near the metal–insulator transition, where the upper and lower Hubbard bands begin to overlap and $U_{eff} \simeq 0$. How this affects the Mott picture of a discontinuous metal–insulator transition and the Mott criterion is not known.

Electron–lattice interaction

Most systems showing a metal–insulator transition are narrow-band materials, and contain oxygen or chalcogen ions. Both these facts suggest strong electron–lattice coupling. Experimentally, the lattice distortion in systems such as VO_2, V_2O_3 connected with the metal–insulator transition supports this belief. However, there is no serious theory yet of the Hubbard model with a deformable, dynamic lattice.

There has been considerable work on the effect of electron–lattice coupling for $U=0$, reviewed for instance by Toyozawa (1983). For an electron coupled to acoustic phonons it turns out that as the coupling strength increases, there is a fairly rapid transition from a large to a small polaron. The small polaron is a very heavy object strongly localized in the potential well created by the local deformation it causes. A system of such small polarons leads to a specially narrow band, and can undergo a metal–nonmetal transition even if the correlation U is very small, so that lattice deformation aids the Mott transition.

A qualitatively different kind of metal–nonmetal transition is possible if electron–acoustic phonon coupling is strong. The electrons can be paired into bipolarons, a lattice of such bipolarons being an insulating molecular crystal. If the overlap between bipolarons is significant, then the condensed matter system is best described as a covalent insulator. These ideas have not been developed extensively yet. An early realization of this pairing effect is due to Anderson (1975), who proposed their existence in amorphous chalcogenide semiconductors. The idea, roughly, is that two localized electrons deform the local lattice region twice as much, and the deformation energy is therefore four times that for a single electron. Thus a configuration with such a close opposite spin pair is more stable than two electrons separated from each other, provided the relative energy gain is greater than the coulomb repulsion. Such covalent pairs can be described as arising from a negative U. They constitute the states near the Fermi energy in amorphous semiconductors, and thus determine many of their electronic properties.

In several oxide and perovskite systems, there is evidence for bipolarons (Schirmer & Salje 1980) and for a bipolaronic ground state (Lakkis *et al.* 1976). A study of bipolaron formation has been made recently by Hiramoto & Toyozawa (1985). They show that bipolarons are necessarily small since otherwise long-range coulomb repulsion overwhelms deformation gain. Interestingly, there is a regime of electron–phonon coupling strengths for which a small bipolaron is stable, but two distant small polarons are not. In a lattice, such a bipolaron formation could be the driving force for a characteristic metal–insulator transition. At the present level of approximation which neglects phonon dynamics, it is similar to a band-crossing transition into a molecular crystal phase by doubling of the unit cell.

The more realistic situation with both correlation between electrons and electron coupling in a deformable lattice has not been extensively investigated. Results for small one-dimensional chains have been presented by Toyozawa (1983) in a review of electron-induced lattice instabilities. A number of types of insulating phases (CDW, SDW) with different unit cell sizes are possible.

Orbital effects

As mentioned earlier, the Hubbard model ignores orbital degrees of freedom, concentrating on correlation as the dominant physical factor. The former are, however, crucial for such basic atomic properties as Hund's rule coupling, correct atomic moment, configurational splitting and spin orbit interaction. There are also important solid state orbital effects such as crystal field splitting, ligancy etc. The presence of orbitals with other symmetries (other than d), for example, p-like orbitals, has important effects on energy levels as well as on wavefunction localization and symmetry. Thus, the metal–insulator transition is likely to be strongly affected by the extra local degrees of freedom. In addition, new symmetry breaking phases such as orbital superlattices are also possible (Mott & Zinamon 1970, Cyrot & Lyon-Caen, 1975). A detailed analysis of the effects of deformation and correlation has been made by Goodenough (1971) to produce a broad picture of metallic and insulating ground states across classes of homologous compounds. For binary systems, Wilson (1972) has suggested a phase diagram, with the Mott transition regime demarcated.

There has recently been considerable progress in the theory of strongly correlated orbitally degenerate systems in metals. A typical problem is a single such impurity (with orbital degeneracy $N \gg 1$) with $U = \infty$. It turns out that for large N the problem is simpler because quantum (or fluctuation) effects are greatly reduced. Thus there is a $(1/N)$ expansion, first discussed for the mixed valence impurity by Anderson (1981) and by Ramakrishnan (1981). There has been considerable work on the results of such an expansion for Kondo impurity and lattice systems (see for example, Coleman 1984). Special N orbital models for the Mott transition (Oppermann 1982) and for magnetism in transition metals (Samson 1984) have been discussed recently. But the former uses a conduction electron spin-localized spin coupling as the basic interaction, while the latter is designed to mimic Heisenberg spins and has no special behaviour for commensurate band filling. In both, the $(1/N)$ expansion is taken to mean that interaction effects can be treated in systematic perturbation theory. More realistic N orbital models need to be considered.

The valley degeneracy in phosphorus-doped silicon for example also leads to a kind of many-orbital effect. Bhatt & Rice (1981) show by explicit calculations for clusters that the Hubbard U, the shape of the Hubbard bands D^- and D^+, are all strongly affected by valley degeneracy (sixfold in the above case). Correlation effects are reduced, the favoured spin correlations are ferromagnetic rather than antiferromagnetic, and the dominant effect is that of one-electron potential fluctuation or fluctuation in hopping energies (Anderson localization).

Disorder

In most metal–insulator transitions experimentally studied so far, the resistivities just on the metallic side are quite high, of the order of Mott's maximum metallic resistivity. Thus, these systems are electronically quite disordered. The effect of disorder on strongly correlated systems is poorly understood (see §§4.2, 4.3). Alloying in a system with a commensurately filled band introduces disorder, and can also introduce additional electrons or holes. The two effects may be quite distinct. Experimentally (Honig 1982) it seems that isoelectronic impurities have a less drastic effect on the metal–insulator transition in V_2O_3 than impurities with different valence. This is an interesting effect, since the Mott insulating phase in the perfect crystal is unstable with respect to any deviation from commensurate filling. However, the presence of disorder may Anderson localize these carriers and stabilize the insulating phase. Thus, though disorder is always present and its effects are significant, they are only very roughly understood.

2.5. *Criteria for the metal–nonmetal transition*

A number of arguments lead to very similar results for the material condition at which a metal–insulator transition can occur. If the number of outer electrons (charge carriers) per unit volume is n and the effective atomic length scale is $a_H^* = (\varepsilon h^2/m^* e^2)$, the condition is

$$n_c^{1/3} a_H^* \simeq 0{\cdot}25 \tag{9}$$

for the metal–insulator transition to occur. In equation (9), first proposed by Mott (1949), ε is the dielectric constant of the background medium. The criterion is found to be valid over a very wide range of densities n, nearly 8 orders of magnitude (see for example, Edwards & Sienko 1978, 1982, 1983) even for small a_H^* provided the quantities are calculated realistically. For the doped semiconductor system Si:P, ε is the dielectric constant of pure Si, and m^* is the electron band mass. The different arguments for the Mott condition imply different mechanisms for the transition; the similarity of the final criterion is primarily due to physical and dimensional reasons, as we explain below. However, some mechanisms suggest a first-order transition, while the others imply a continuous change. We now discuss the different criteria briefly.

The metal–insulator transition is, crudely, due to competition between potential energy and kinetic energy. The former tends to localize, the latter tends to delocalize. The system is a metal if the latter is larger, and an insulator otherwise. The natural energy scale for atomic systems is $(e^2/\varepsilon a_2^*)$ where a_2^* is a characteristic atomic length $(\varepsilon h^2/m^* e^2)$. For a degenerate electron gas with density n, the kinetic energy is proportional to $(h^2/m^*)n^{2/3}$. However, n and a^* are connected; $n^{-1/3} \sim a^*$. Thus the condition for kinetic energy to be larger than potential energy is, approximately,

$$(h^2/m^*)n^{2/3} \gtrsim A(e^2/a_2^*) \tag{10}$$

or

$$n_c^{1/3} a_2^* \gtrsim A \tag{11}$$

The best estimate for A, obtained from doped semiconductors (basically hydrogenic systems), is $A \simeq 0{\cdot}25$.

The potential energy could be due to short-range correlation U, as in the Hubbard model. Then equation (11) is the condition for the metal–insulator transition in the Hubbard model, that is, $(U/zt) \simeq 1$ (see Figure 3). It has been shown explicitly by Cyrot (1972) and by Edwards & Sienko (1978) that for hydrogenic systems, this criterion also reduces to equation (11). The transition is expected to be continuous.

The criterion (11) was however derived from the argument that coulomb interaction leads to a strongly first-order transition in which the free carrier density jumps from zero to the Mott value (§2.4). This means, as discussed recently by Mott (1984), that if, say, alloying (characterized by the parameter x) changes the atomic or microscopic quantity (U/zt) continuously, there will be a region of x where the system is unstable and phases separate. The cleanest example of this is the electron–hole drop system. If such a phase separation is kinetically impossible, one expects a discontinuous change in electron density or in the Mott–Hubbard gap. An example of this is perhaps the $(V_{1-x}Cr_x)_2O_3$ alloy system. However, in most metal–insulator transitions, there is a considerable amount of randomness. This may make the transition continuous. For the well studied doped semiconductor system. Mott argues (1984) that because of the dominant effect of randomness the transition is of the Anderson localization type, though it occurs at a density close to where the Hubbard bands just split. We return to this question in §4.2.

Another mechanism, the first to be proposed (Herzfeld 1927), is that local field corrections in an insulator drive it metallic when the insulator becomes sufficiently dense. The insulator is supposed to consist of atoms with polarizability α_0, distributed with density n. It follows from electrostatics that the dielectric constant of such a system is

$$\varepsilon = \frac{1+(4\pi n\alpha_0/3)}{1-(4\pi n\alpha_0/3)} \tag{12}$$

Thus for $(4\pi n\alpha_0/3)=1$, the dielectric constant diverges. Dimensionally, it also can be cast in the form of equation (11). The transition is presumably continuous, with the dielectric constant diverging linearly as the density approaches the critical value. It is not clear, however, whether equation (12) survives in that simple form for dense systems. As atoms begin to overlap, other processes contribute to α_0 and the static classical picture underlying the Clausius–Mossotti form is no longer valid (see Castner (1980) for a discussion of some of these effects). In disordered systems, there are large local fluctuations in overlap between states, and thus the whole idea of a constant polarizability well defined on an atomic scale is not appropriate. It is believed that in such systems, the long wavelength *polarizability* $\alpha_{q\to 0}$ diverges at the insulator–metal transition (see §4.2). However, local field corrections have not been explicitly considered for these systems.

In random systems, the metal–nonmetal transition occurs because of electron localization (§3.1). The condition for this is that the fluctuation W in the one-electron potential is of the order of or larger than the electron bandwidth zt or kinetic energy scale. Again, for realistic potentials, the Anderson criterion $W \simeq zt$ (Anderson 1958) works out to be nearly the same as equation (11). Another closely related condition, due to Ioffe & Regel (1960) and to Mott (1972), is that the nature of electron transport changes when the disorder is so strong that the mean-free-path, l, is of the order of the de Broglie wavelength, λ (or equivalently the inverse of the Fermi wavevector k_F). The condition $k_F l \sim 1$ (see detailed discussion in §§3.2 and 3.3) can also be brought into the form of equation (11).

The main conclusion from the above discussion of criteria for the metal–nonmetal transition is that it is not possible to decide on a particular mechanism from agreement of the parameters for the transition with the value predicted. Only the detailed physical behaviour of the system can help distinguish between various possibilities. An example is the prediction of Mott that the transition is discontinuous in a particular way if long-range coulomb interaction effects are dominant. Perhaps the most detailed picture of the metal–nonmetal transition available both theoretically and experimentally is for disordered systems. This is described in the next section.

3. Anderson localization (disorder-induced metal–nonmetal transition)

3.1. Anderson localization

Anderson (1958) considered an electron moving through a medium in which the potential energy varies randomly from point to point. He showed that when the randomness exceeds a certain amount, the electronic states are spatially localized, and not extended. This was in contradiction to the then prevalent belief that no qualitative change occurs in the nature of states with increasing disorder; electrons always diffuse, but more slowly with increasing disorder. After about a decade of neglect, the consequences of this work were developed, principally by Mott (see e.g. Mott & Davis 1979) who introduced the ideas of mobility edge, metal–insulator transition due to disorder (Anderson transition) and minimum metallic conductivity. We discuss these ideas below, and outline recent work on the Anderson transition, stimulated by the scaling theory of localization (Abrahams *et al.* 1979).

Anderson considered a simple tight-binding model

$$H = \sum_{i,j} t_{ij} a_i^+ a_j + \sum_i \varepsilon_i n_i \tag{13}$$

where the sites i are on a lattice, t_{ij} is the nearest-neighbour hopping energy, and ε_i is the energy of the electron at site i (Figure 5). In a perfect solid, ε_i is the same for each site. Here it is assumed to have a value anywhere in the range $\pm W/2$, i.e. $-W/2 \leqslant \varepsilon_i \leqslant W/2$, with equal probability. The energies at different sites are not correlated. The model describes a structurally crystalline but electronically random system. Other possibilities, namely structural randomness or a

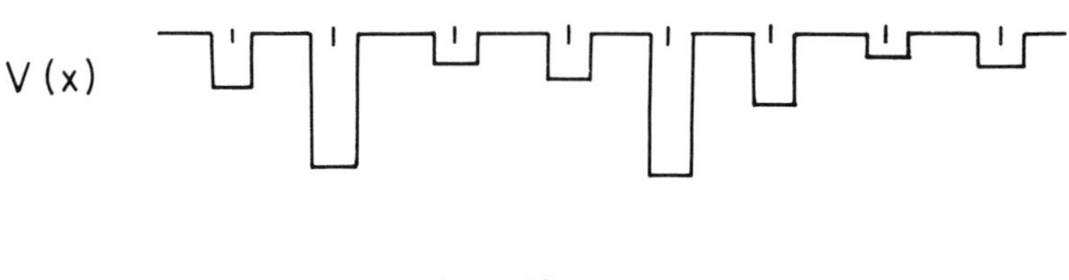

Figure 5. A lattice of random potentials considered in the Anderson model.

combination of the two, have also been discussed. Most of the results mentioned below do not depend on the detailed nature of the disorder, provided only that it is frozen.

For large fluctuations in ε_i and small hopping t_{ij}, electrons will be trapped in regions where the potential is locally attractive. An electron so localized can escape out by admixing its state with that of another. However, states nearby in space (large overlap matrix elements) are unlikely to be close in energy if ε_i fluctuates widely; this admixture is typically of order (t/W) where W is the characteristic difference between site energies. On the other hand, states close in energy are far apart in space so that their overlap is exponentially small. Thus admixture with other tight-binding states does not delocalize an electron trapped in a potential fluctuation so long as perturbation theory in intersite hopping converges for the random medium, that is, so long as $(t/W)<1$. A sophisticated version of this kind of argument was presented by Anderson (1958) who calculated the probability distribution for the on-site residence amplitude as a function of complex energy, and showed that it converges provided $(zt/W)<N_c$. Here N_c is of order unity but depends on electron energy, lattice structure, dimensionality, etc.

When disorder is strong enough for the above condition to be satisfied, states are exponentially localized, that is, the envelope of the wavefunction decreases exponentially with characteristic scale ξ called the localization length. With decreasing disorder, ξ increases till it diverges for critical disorder. In the limit of strong disorder, ξ is of the order of an atomic length.

The criterion for localization has been determined most accurately and extensively by computer simulation. For example, Weaire & Srivastava (1977) find that states at the centre of a tight-binding band of a three-dimensional simple cubic lattice are localized if

$$(zt/W)<0{\cdot}45 \tag{14}$$

The criterion depends on the electron energy in general. States at the band edges arise from a special combination of phases of states at different sites, and are thus most strongly affected by disorder. They are most easily localized. Thus for a given disorder, states in some energy regions are localized and in other regions are extended. The boundary separating extended and localized states was termed

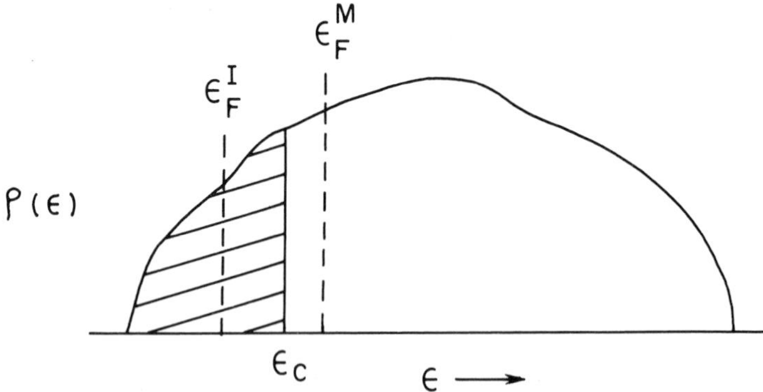

Figure 6. Density of electronic states $\rho(\varepsilon)$ vs. energy ε. The mobility edge ε_C, localized states (shaded), and Fermi energies ε_F^M and ε_F^I for typical metallic and insulating systems are shown.

the mobility edge by Mott (1967), since states on one side of the edge are mobile while states on the other side are not, being spatially localized.

Now at zero temperature, all the available electronic states are occupied in order of increasing energy. If the last occupied state is in the regime of localized states, the low-lying excited states of the system are localized and the system is an insulator. On the other hand, if the Fermi energy is in the region of extended states the system is a disordered metal (see Figure 6 for a schematic illustration). Thus, as disorder increases or as the electron density decreases (or with a combination of the two) the Fermi energy crosses the mobility edge and the system undergoes a transition from a metallic to an insulating state.

This disorder-induced transition was termed the Anderson transition by Mott, who was the first to investigate its characteristics. Much of the experimental activity in the field has been directly due to these contributions. We now outline some of these ideas.

3.2. Disordered metal, and Mott's minimum metallic conductivity

We first consider approaching the Anderson transition from the metallic side. For a disordered metal, the d.c. conductivity σ is given in standard weak scattering transport theory by the formula

$$\sigma = \frac{ne^2\tau}{m} = \left(\frac{e^2}{\hbar}\right)\left(\frac{n}{k_F^2}\right)(k_F l) \tag{15}$$

where n is the electron density, τ the relaxation time and l the mean free path due to collision with random scatterers. Ioffe & Regel (1960) pointed out that mean free paths l smaller than interatomic spacing, a ($\simeq k_F^{-1}$), are impossible. Mott (1972) argued that the condition $k_F l \simeq 1$ sets a lower limit on the mean free path or conductivity. For greater disorder, the mode of transport must change and the system cannot remain metallic. The minimum metallic conductivity has the following values:

$$\sigma_{\min}^{3d} = \left(\frac{1}{3\pi^2}\right)\left(\frac{e^2}{\hbar}\right)k_F \tag{16}$$

$$\sigma_{\min}^{2d} = \left(\frac{1}{2\pi}\right)\frac{e^2}{\hbar} \tag{17}$$

We notice that $\sigma_{\min}^{3d}$ depends on material parameters only through k_F or equivalently the inverse of interatomic spacing a, or the electron density n ($k_F \sim (3\pi^2 n)^{1/3}$). It has a quantum conductance scale ($e^2/\hbar\pi^2$); $\sigma_{\min}^{3d}$ is the zero-temperature conductivity of a degenerate, disordered electron gas. Extending the idea to two-dimensional systems, the minimum conductivity is seen from equation (17) to be universal, independent of material parameters (Licciardello & Thouless 1975). It has the dimensions of conductance. The associated universal resistance is about 30 000 Ω. The maximum metallic resistivity in three dimensions corresponding to $k_F \simeq 10^8\ \text{cm}^{-1}$ is about 1000 μΩ cm.

There is a very large body of experimental evidence from a variety of systems supporting the idea that the conditions (16) and (17) mark a change of transport regime. For example, experiments on disordered systems at not too low temperatures show that with increasing disorder, when σ decreases to $\sigma_{\min}$, the temperature dependence of resistivity changes sign (from positive to negative), and for still greater disorder, the system is an Anderson insulator characterized by diverging resistivity at low temperature. The variation of resistivity with temperature for increasing disorder is shown schematically in Figure 7. We see that a high-temperature conductivity of $\sigma_{\min}$ is the rough boundary between metallic and insulating behaviour. Similarly, measurements on thin films at moderate temperatures show that their resistance per square is either less than 30 000 Ω or much larger. Numerical simulations (Licciardello & Thouless 1975) suggest similar results (see Mott & Davis 1979 for many examples). The Mott argument predicts, however, that across the Anderson transition, there is a discontinuity in the conductivity at zero temperature (Figure 8). The conductivity has the value of equation (16) on the metallic side, and zero on the insulating side. At any non-zero temperature, this behaviour is masked because the insulator also has a non-zero conductivity. This prediction has been the subject of considerable theoretical and experimental attention. The scaling theory predicts a continuous conductivity transition at zero temperature (§3.3) and careful low temperature experiments close to critical disorder confirm this behaviour (see §§3.4 and 4.2 for further discussion).

Approaching the transition from the localized side, the natural presumption is that the localization length diverges. Now the polarizability α_0 depends on the square of the size of the localized wavefunction, so that one expects it to diverge as

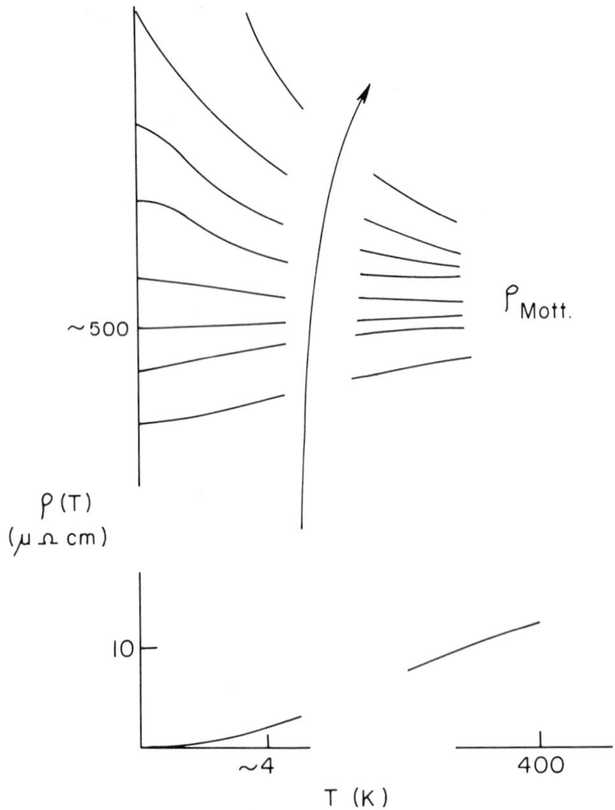

Figure 7. Schematic illustration of resistivity of a metal as a function of temperature. The different curves are for increasing disorder as shown by the arrow. Note the break in temperature and resistivity scales, and convergence to ρ_{Mott} at high temperatures.

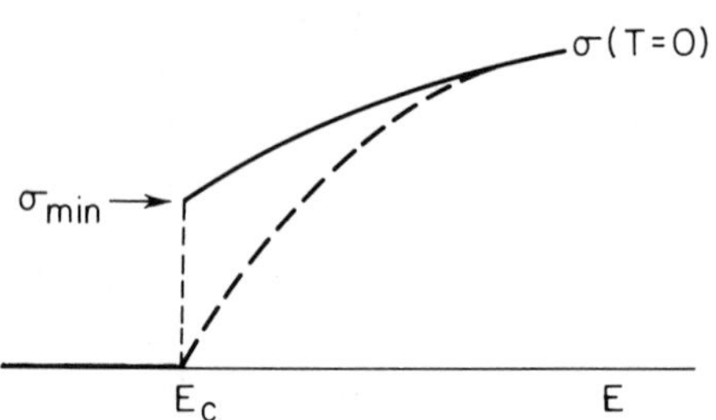

Figure 8. Zero-temperature conductivity of a disordered metal as a function of its Fermi energy E. The minimum metallic conductivity σ_{min} is shown. Dotted line shows the scaling theory prediction.

ξ^2. In a random medium with sizeable fluctuations in wavefunction amplitude over length scales of order ξ, one does not expect a classical local field correction, so that the dielectric constant ε diverges as follows:

$$\varepsilon \sim \xi^2 \tag{18}$$

Considered as a zero temperature phase change, localization appears to be unusual. It is discontinuous as a conductivity transition, but continuous as a dielectric transition with a diverging dielectric constant ε and a related correlation length ξ. We shall see below that this asymmetry is removed in scaling theory. The transition is continuous, and is characterized by a correlation length which diverges in the same way on both the insulating and metallic sides of the transition. The former describes how the dielectric constant diverges, and the latter how the conductivity goes to zero. The exponent is universal.

An interesting question, first addressed by Mott & Twose (1961) (see also Landauer 1970), is the lower critical dimension for the transition. Mott & Twose showed that electronic states are localized in one dimension no matter how weak the randomness. The localization length is of the order of the backscattering mean free path. Thus the metallic ground state is unstable in one dimension against disorder. One of the predictions of scaling theory is that even two-dimensional systems do not have a metallic ground state. The localization length depends exponentially on the mean free path, and not linearly as in one dimension.

3.3. Quantum interference effects and scaling theory

In the picture described above, there is an asymmetric treatment of the effects of quantum interference. The electron wave in a random medium is scattered repeatedly. These scattered components interfere, and for a localized state, interference effects are such that the wave amplitude is large only in a small region of space and falls off exponentially around it. Thus on the insulating side of the Anderson transition, interference effects are strong enough to localize a state on a large length scale ξ. However, for slightly less disorder, on the metallic side this interference is supposed to have no qualitative effect on physical properties. The conductivity, for example, is calculated in a single-collision approximation, or in an approximation where the only effect of multiple collisions is to renormalize the collision time τ.

Abrahams *et al.* (1979) argued that quantum interference effects in a disordered system evolve continuously as a function of scale size. By identifying the leading microscopic contribution to this process, they calculated the functional form of quantum interference, and predicted consequences for observed properties. We now outline this scaling theory. Earlier scaling ideas for localization are due to Wegner (1976) and Thouless (1977, 1979).

The first question is: what is the measure of quantum interference? Abrahams *et al.* argue, following the ideas of Thouless, that conductance itself is such a measure. It is small if quantum interference is large, and vice-versa. Thouless considered putting together blocks of random material of size L. A block can be characterized by two parameters, the strength $t(L)$ of coupling between wavefunctions on one block and on another, and $W(L)$ which is the mismatch between an energy level on one block and the closest level on the neighbouring block. $W(L)$ is the level spacing, which varies as L^{-d}, d being the spatial dimension. If the ratio $[t(L)/W(L)]$ is small, energy levels are stable against perturbation by coupling to other blocks; states in the block L^d are localized on a length scale $\xi < L$. Just as in the Anderson model, the nature of the state depends on the ratio (t/W). More specifically, consider an increase in the scale of the system from L to $L+\delta L$. The ratio (t/W) also changes. The size of this change depends on (t/W) itself. Now (t/W) determines the conductance $g(L)$ of the finite system, as shown by Thouless. The consequence of this argument is therefore that

$$\delta(t/W)=f(t/W)$$

or

$$[L/g(L)]\frac{\mathrm{d}g(L)}{\mathrm{d}L}=\frac{\mathrm{d}\ln g(L)}{\mathrm{d}L}=\beta(g) \tag{19}$$

Now if $L \gg l$, which is the only relevant microscopic length, it is reasonable that $\beta(g)$ does not depend on the details of the system, but only on g which has universal scale $g_c=(e^2/\hbar\pi^2)$, and on dimensionality d. Abrahams *et al.* proposed such a one-parameter theory and constructed $\beta(g)$ from a knowledge of limiting forms and perturbative corrections. We describe these below.

For large conductance the system is a good metal, there are no quantum interference effects, and Ohm's law, which regards current flow as classical, is valid. Thus one has the Ohm's law relation $g=\sigma L^{d-2}$ between conductance g and the conductivity σ which is a material parameter independent of system size L. This leads to

$$\beta(g)=(d-2) \tag{20}$$

for $g \gg 1$ (we measure g in units of $g_c=e^2/\hbar\pi^2$).

For very small conductance, that is, when states near ε_F are localized, d.c. conduction in a large system ($L \gg \xi$) is by hopping on to states which most likely have exponentially small amplitudes at the system boundary. One thus has $g(L) \sim \exp(-L/\xi)$ or

$$\beta(g)=\ln g \tag{21}$$

for $g \ll 1$.

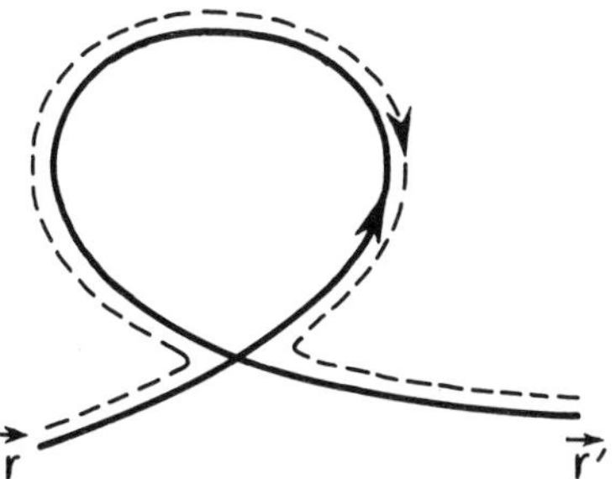

Figure 9. Two paths for an electron propagating from r to r′.

It was shown by Abrahams *et al.* that a particular set of processes contributes a negative $(1/g)$ term to $\beta(g)$ for large g. The process is interference between forward motion and backscattering in the field of the random potential. It is illustrated in Figure 9. An electron can go from r to r′ in many ways. The total amplitude for the process is the sum over that for various paths. Two of these paths are shown in Figure 9. In one (full line) the path returns to the same point (or within a distance of order l) and goes on to r′. In the other (dotted line) the loop path is traversed in the opposite direction. The phases for the two paths clearly reinforce each other by an amount which depends on disorder and on the dimensionality, and this reinforcement increases the probability of return to the same region in space. Phases for nearby paths add more or less coherently. The effect of this quantum interference is to localize the electronic state. In one dimension, interference is perfect and there are no extended states. The perturbative scaling function is

$$\beta(g) = (d-2) - a/g \tag{22}$$

where a is a constant of order unity.

A field theory formulation of the localization problem is due to Wegner (1979). In the field theory the diffusion constant is analogous to spin wave stiffness of a Heisenberg ferromagnet. The field theory, as well as numerical results from computer simulation (Mackinnon & Kramer 1981) all agree with equation (22). There are no terms found to higher order in $(1/g)$, up to $(1/g^3)$. The scaling curve based on these results is shown in Figure 10. We now discuss its consequences.

The function $\beta(g)$ describes how, starting from a value g_0 at a microscopic length scale l, the conductance evolves as a function of scale size. In three dimensions, if g_0 is such that $\beta(g_0)$ is positive, g increases with length scale, till Ohm's law behaviour is attained for large L. The conductivity $\sigma(L=\infty)$ can be calculated from the scaling curve, and one finds

$$\sigma = (g^*/l)\{(g_0 - g^*)/g^*\}^{\nu} \tag{23}$$

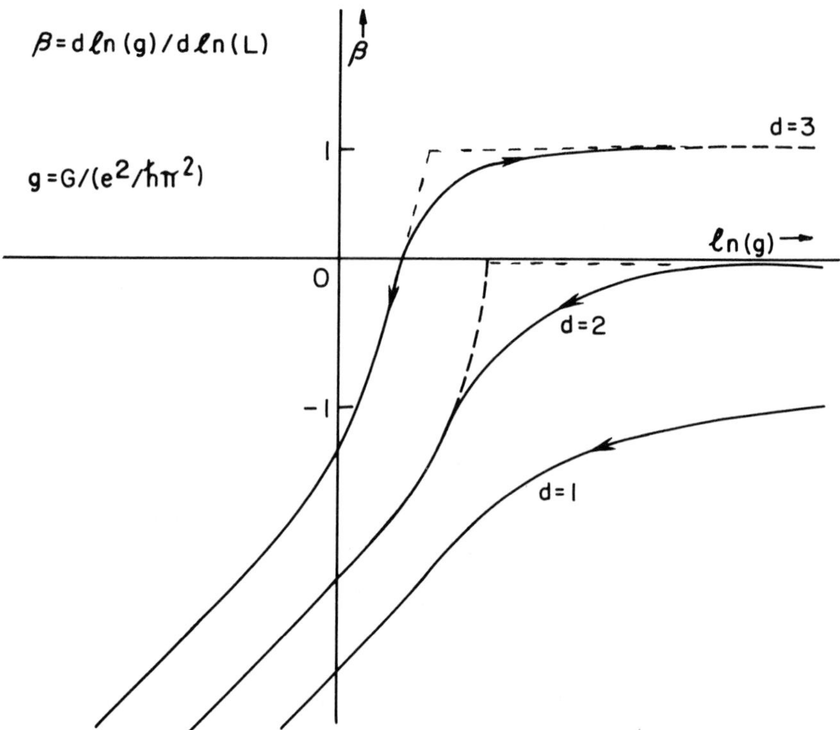

Figure 10. The scaling curve $\beta(g)$ for conductance g of a disordered system as a function of ln g (full lines), for systems of dimensionality $d=3$, 2 and 1. Dotted lines show the behaviour expected from the minimum metallic conductivity concept.

for $g_0 \gtrsim g^*$, where $\beta(g^*)=0$. The first factor in the above equation is the Mott minimum metallic conductivity. The second factor represents the reduction due to quantum interference acting on length scales larger than l. The exponent ν is the inverse slope of the $\beta(g)$ curve near g^*. The perturbative estimate from equation (23) is $\nu=1$. For this value, one has

$$\sigma(L)=\left(\sigma_0-\frac{e^2}{\hbar\pi^3 l}\right)+\frac{e^2}{\hbar\pi^3 L} \tag{24}$$

where σ_0 is the conductivity in the absence of interference effects. Thus the prediction is that $\sigma(L=\infty)\rightarrow 0$ as disorder becomes critical, i.e. as $\sigma_0 \rightarrow (e^2/\hbar\pi^3)=\sigma_{\min}$ or $g_0=g^*$ $(=e^2/\hbar\pi^3)$. The conductivity decreases from the Mott value of $(e^2/\hbar\pi^3 l)$ to zero as L increases from l to infinity. For $\sigma_0 \gtrsim \sigma_{\min}$, diffusion or conductivity is strongly scale dependent, varying roughly as $1/L$ over a sizable range.

A similar calculation for the case $g_0 < g^*$ shows that now, since $\beta(g)<0$, g always decreases with increasing L, and finally goes to the localized regime behaviour $g(L) \sim \exp(-L/\xi)$ with

$$\xi \sim \{(g^* - g_0)/g^*\}^{\nu} \tag{25}$$

Thus the Anderson transition is characterized in scaling theory by a correlation length ξ diverging with the exponent ν. On the insulating side, this is the localization length. On the metallic side, it describes how the conductivity goes to zero, i.e. $\sigma = (e^2/\hbar\pi^3)(1/\xi)$.

In two dimensions, we notice that $\beta(g) < 0$, so that no matter how large g_0 is at the microscopic scale l, one always goes over to the exponentially localized regime for large enough length scales. Explicitly, one finds that perturbatively

$$\sigma(L) = \sigma_0 - (e^2/\hbar\pi^2)\ln(L/l) \tag{26}$$

that is, the conductivity decreases continuously and logarithmically with length scale. From equation (26) one estimates a localization length $\xi \sim l\exp[(\pi/2)k_F l]$ which is exponential in the mean-free-path. There is thus no metallic regime.

In one dimension, $\beta(g)$ is strongly negative and the localization length can be shown to be proportional to the mean free path l.

3.4. Experimental consequences

All the above results are for absolute zero. At any non-zero temperature, inelastic collisions which scatter electrons out of a given energy state cause random differences in the phase changes for the two paths shown in Figure 9. When the phase fluctuation is of order 2π, quantum interference is no longer operative. This happens for distances of order

$$L_{Th} \sim \sqrt{(D\tau_{in})} \tag{27}$$

where D is the diffusion constant for elastic scattering, and τ_{in} is the inelastic scattering time (Thouless 1977). L_{Th} is thus the effective scale size for quantum interference. It increases as T decreases, because τ_{in} becomes longer. The actual dependence varies with the mechanism. For example, at low temperatures and in disordered systems, the electron–electron scattering time goes as $T^{-3/2}$ (Schmid 1974) rather than as T^{-2} which is the clean-system temperature dependence. This effective enhancement of interaction in a disordered system is one of the characteristic effects discussed in the next section. As temperature increases, $\tau_{in}(T)$ decreases, so that the effective scale size for quantum interference decreases, these effects are negligible and one has the Mott behaviour. Using equation (27) in equations (24) and (26) we find that localization leads to non-ohmic behaviour, with the conductivity decreasing as temperature decreases, according to a power law dependence in three dimensions and a logarithmic form in two.

One interesting perturbation of quantum interference is due to a magnetic field H. This introduces a phase difference $\phi = (2e/hc)HA$ between the two paths of Figure 9, where A is the area enclosed by either of the paths. This phase difference reduces quantum interference and thus increases conductance (negative magnetoresistance). The magnetic field provides a length cutoff $L_H \sim (ch/eH)^{1/2}$, the cyclotron radius. In three dimensions the magnetoresistance for large fields H is $\rho(H) \sim \sqrt{H}$ (Kawabata 1981), whereas in two dimensions the magnetoresistance goes as ln H. There are similarly characteristic effects due to magnetic impurity scattering or scattering from spin orbit coupled random impurities.

The observable localization effects can conveniently be divided into two classes. For weak disorder ($k_F l \gg 1$), the effects discussed above lead to characteristic non-classical behaviour. There are other consequences for magnetoresistance and so on. These have all been measured, and the results are in very good agreement with theory.

The second class, which is our main interest here, is the strong disorder regime $k_F l \sim 1$ close to the critical disorder for Anderson localization. In scaling theory, perturbative precursor effects, mentioned above, and effects near the Anderson transition have the same physical origin. The prediction for the transition is that the low-temperature conductivity goes to zero linearly with deviation from critical disorder. This has been confirmed in several experiments. For example, in Nb_xSi_{1-x} alloys (Hertel *et al.* 1983), $T=0$ conductivities nearly twenty times smaller than σ_{min} have been reported; σ goes linearly to zero with $(x-x_c)$ where $x_c \simeq 12\%$ Nb. Similar results have been found for Au_xGe_{1-x}, and for doped compensated semiconductors (see Thomas 1984a,b for a recent summary). However, other results, such as a singular dip in the density of states for Nb_xSi_{1-x}, suggest that interaction effects are also very important (§4) since in a non-interacting system, localization has no singular effect on the density of electronic states. A metal is defined here as a material with low-lying current-carrying excitations, so that at $T=0$, it has a non-zero conductivity, no matter how small.

A new effect, pointed out by Altshuler *et al.* (1981) is that the conductivity is a periodic function of magnetic flux in units of the flux quantum $\phi_0 = (hc/2e)$. This follows from the discussion above regarding the effect of a magnetic field. In a striking experiment on a magnesium film deposited on a quartz fibre along which there is a magnetic field, Sharvin & Sharvin (1981) found a conductivity oscillation with the above periodicity in the magnetic flux. This effect in a *normal* (non-superconducting) disordered metal is a striking example of the reality of quantum interference.

4. The Mott–Anderson transition (metal–nonmetal transition in a disordered interacting electron system)

4.1. *Characteristic perturbative effects*

The localization transition discussed above is for a gas of non-interacting electrons. In reality electrons interact with each other and with the lattice, which is not frozen but deformable and dynamic. The problem of the continuous metal–insulator transition at $T=0$ in an interacting, disordered system has attracted considerable attention in the last two or three years, and it is being realized that several classes of metal–insulator transitions are possible. We first outline the effects of interaction in a disordered system within perturbation theory, and then mention these possibilities as well as experimental results.

The interacting electron gas has been studied for a long time. The results can be described by Landau's Fermi liquid theory where the interaction effects are represented by a small number of interaction parameters. These renormalize physical quantities such as specific heat and susceptibility, but leave the basic structure of the free Fermi gas theory intact, for example, the low-temperature behaviour. It was perhaps because of the strong hold of Fermi liquid theory that the novel effects discovered by Altshuler & Aronov (1979) were not found earlier.

Altshuler & Aronov (1979) treated the disordered Fermi liquid problem by perturbation theory, to the lowest order in interaction strength. They used a diagrammatic many-body method. The result can also be obtained by a calculation in terms of exact eigenstates of the electron in the random potential (Abrahams *et al.* 1981). The shift in energy of the state m, due to an interaction $v(\mathbf{r}-\vec{\mathbf{r}}')$ is Σ_{m}, where

$$\Sigma_{\mathrm{m}} = - \sum_{\mathrm{n\ occupied}} \int \mathrm{d}\vec{\mathbf{r}}\,\mathrm{d}\vec{\mathbf{r}}'\; \psi^*_m(\vec{\mathbf{r}})\psi^*_n(\vec{\mathbf{r}}')\psi_m(\vec{\mathbf{r}}')\psi_n(\vec{\mathbf{r}})v(\vec{\mathbf{r}}-\vec{\mathbf{r}}') \tag{28}$$

In (28), $\psi_m(\vec{\mathbf{r}})$ is the wavefunction of the state m, and the interaction is with all occupied states n. The energy of a particle added to the system with energy E will be shifted by an amount $\Sigma(E)$, i.e.

$$\bar{\Sigma}(E) = \frac{1}{N(0)} \overline{\Sigma_m\, \delta(E-E_m)} \tag{29}$$

$$= -\int \mathrm{d}E'\; \mathrm{F}(E, E'; \vec{\mathbf{r}}, \vec{\mathbf{r}}')\, v(\vec{\mathbf{r}}-\vec{\mathbf{r}}')\, \mathrm{d}\vec{\mathbf{r}}\,\mathrm{d}\vec{\mathbf{r}}' \tag{30}$$

where $\mathrm{F}(E, E'; \vec{\mathbf{r}}, \vec{\mathbf{r}}')$ is, from equations (28) and (29), the probability that a density fluctuation with energy $(E-E')$ propagates from $\vec{\mathbf{r}}$ to $\vec{\mathbf{r}}'$. $N(0)$ is the density of states at Fermi energy. In a disordered system with diffusion constant D and compressibility $(\mathrm{d}n/\mathrm{d}\mu)$, using the Fourier transform of F, one finds that

$$\bar{\Sigma}(E) = \left(\frac{\mathrm{d}n}{\mathrm{d}\mu}\right) \int_{-\infty}^{0} \mathrm{d}E' \cdot \frac{\mathrm{d}q}{(2\pi)^3} \cdot \frac{Dq^2}{(E-E')^2 + (Dq^2)^2}\, v(q) \tag{31}$$

We notice that because of diffusive electron motion, the effective interaction with low-lying excitations ($E-E'$ small) and large distances (q small) is strongly enhanced. Electrons stay around each other for much longer, and the coupling is strongly retarded.

The energy shift, equation (31), leads to a change in the density of electronic states. The change is $[\delta N(E)/N(0)]=(\partial\bar{\Sigma}(E)/\partial E)$ and goes for small E (i.e., near the Fermi energy) as

$$\frac{\delta N(E)}{N(0)}=\frac{v(0)\,N(0)}{4\pi^2(\mathrm{d}n/\mathrm{d}\mu)D}\left(\frac{E}{D}\right)^{1/2} \tag{32}$$

There is thus a square root cusp in the density of states, whose size is larger, the smaller the diffusion constant D. The following extremely simple and physical derivation of equation (32) has been presented by Aronov (1984). The probability that a diffusing electron continues to stay near another electron at time t later is $(1/Dt)^{3/2}\exp(-r^2/Dt)$, where in the present case $r=0$. The two electrons with energy difference E remain coherent with each other for a time $t\lesssim(h/E)$. Thus, for short-range interaction, the probability that these two electrons stay near each other and interfere is

$$P(E)\sim\int_0^{h/E}\mathrm{d}t(1/Dt)^{3/2}\simeq E^{1/2}/D^{3/2} \tag{33}$$

This is the origin of the correction to the density of states calculated above in equation (32).

The square-root dip has been observed in many systems (see for example Lee & Ramakrishnan 1985, or Bishop & Dynes 1985 for a review). An example is shown in Figure 11, taken from the work of Hertel *et al.* (1983) on Nb_xSi_{1-x} alloys. As x decreases towards $x_c\simeq 0{\cdot}12$, the system goes insulating. As the metal–insulator transition is approached, the dip becomes more and more pronounced, the characteristic energy scale Δ in the form $\surd(E/\Delta)$ tending to zero as the square of the zero temperature conductivity. On the insulating side, there seems to be a gap in the density of states or at least a soft gap, with $N(E)\to 0$ as $E\to 0$. In the relatively weakly disordered metal regime, that is, for $x\gg x_c$, the square root behaviour is observed right up to $E=0$, over a considerable range in energy.

In two dimensions, the same effect and argument lead to a logarithmic decrease in density of states with energy E. This has recently been seen in experiments on quench-condensed tin films by White *et al.* (1984).

The lowest-order process discussed above also leads to a decay of the quasi-particle with energy E, provided the interaction $V(q)$ is chosen realistically, for example, if it is the dynamically screened coulomb interaction $V(q,\omega)$. As mentioned earlier, the decay rate due to electron–electron interaction varies as $E^{3/2}$ in three dimensions (Schmid 1974). In two dimensions the rate is $E\ln(E_c/E)$ where E_c is an upper electronic energy cutoff (Abrahams *et al.* 1981). Since the

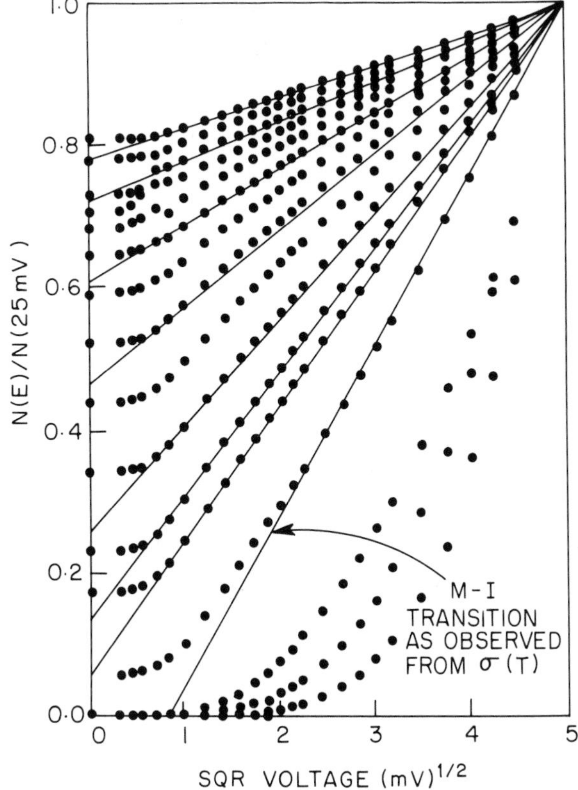

Figure 11. Density of electronic states as a function of energy for Nb_xSi_{1-x} alloy samples (Hertel *et al.* 1982). The curves are for different values of x, approaching the metal–insulator transition ($x_c = 0{\cdot}12$) from the metallic side ($x > x_c$).

quasi-particle energy is E, this result implies that the decay width is larger than the energy. The quasi-particle state is not well defined, and there is no Fermi liquid theory for a disordered system in two dimensions. In three dimensions, disorder has no such qualitative consequences, though the decay probability is enhanced by disorder with respect to the clean limit form E^2.

Altshuler & Aronov (1979) also showed that the diffusive motion of electrons in an interacting system leads to characteristic transport anomalies. There is a term for the d.c. conductivity varying as $\sqrt{T}$ in three dimensions, and as $\ln T$ in two dimensions (Altshuler *et al.* 1980). Thus for thin films, perturbative transport anomalies due to localization and to interaction behave similarly with temperature and are of comparable size. The effect of other perturbations, such as a magnetic field, can help disentangle these contributions. There is a positive spin-splitting contribution to magnetoresistance in an interacting electron gas (see, for example, Kawabata 1981, Lee & Ramakrishnan 1982). In a number of systems, a fairly quantitative analysis of coulomb interaction and localization effects has been made (see, for example, Fukuyama 1984).

A number of long-standing anomalies in transport properties of metallic glasses, doped semiconductors and, generally, disordered metals, are believed to be due to interaction and localization effects (see Kaveh & Mott 1982 for a discussion; see also Lee & Ramakrishnan 1985).

The description of the interaction is rather schematic above. The actual bare interaction, i.e. the coulomb interaction, is not weak, so that effects to all orders in interaction strength have to be included. This has been done by Altshuler & Aronov (1983, 1984) and by Finkelshtein (1983). Since spin is conserved, scattering in electron–hole channels with given total spin can be considered separately. The spin singlet and triplet channels describe density and spin fluctuations respectively. The former is dominated by long wavelength coulomb interaction and its screening, and the latter describes short-range exchange correlations. The effective coupling constant for the former is universal (unity on the scale of $N(0)^{-1}$) while for the latter it is a Landau-like parameter F. The interaction effects mentioned above are characterized by the two parameters 1 and F. If there are additional interactions, both these parameters are appropriately modified. In the next section, we discuss effects beyond the lowest order in these interaction parameters, and the experimental situation.

4.2. Scenarios for the metal–nonmetal transition

It is clear from the form of the interaction effect (equation (32)) that the consequences of interaction and of disorder are strongly coupled. As the system becomes more disordered (D becomes smaller) the density of states is strongly affected. This in turn influences the screening of the coulomb interaction. In addition, there is the effect of interaction on diffusion or on effective disorder. There are at least two electron–hole fluctuation modes (density and spin), and these are mutually coupled in higher order. These effects make the problem of the interacting disordered electron gas extremely difficult.

The pioneering attempt to deal with the coupled effects of localization and interaction is due to McMillan (1981), who attempted to calculate the unscreening of the coulomb interaction self consistently as a function of scale size. Several assumptions in this semi-phenomenological approach do not survive detailed analysis.

A field theoretic formulation of the interacting electron gas problem, due to Finkelshtein (1983) is the basis of much recent work in this area. The electron wave with fluctuating amplitude and given frequency is the basic field. Three basic couplings between these fermion fields which change their wavelength and frequency are shown in Figure 12. They correspond to particle–hole coupling in the spin-singlet channel (Figure 12(a)), short range particle–hole scattering (both singlet and triplet channels) (Figure 12(b)) and particle–particle scattering

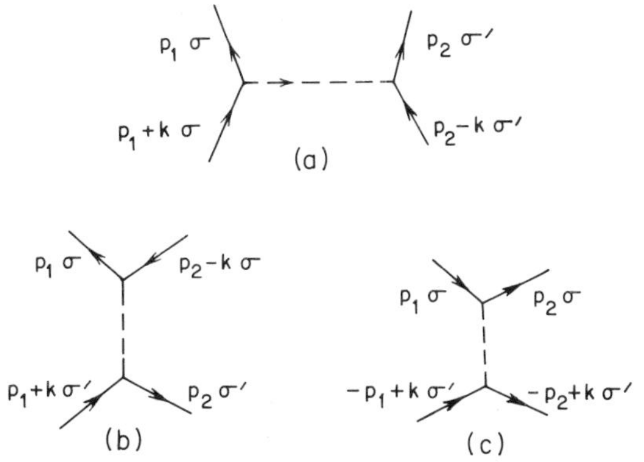

Figure 12. Basic interaction processes in an electron gas.

(Figure 12(*c*)). The aim is to construct a scaling theory for the effective coupling in these channels, starting from given values of interaction strength for short distances and time scales of order l and τ respectively. The coupling at long times and distances, which is affected by interaction between various modes and by diffusive electronic motion, is directly relevant for properties like static susceptibility, d.c. conductivity, specific heat and so on, and is obtained by progressively integrating out effects on length and time scales longer than l and τ respectively.

The general problem is characterized by three coupling constants and a diffusion constant D to begin with and has not been solved. Several special cases have been considered to lowest order in D^{-1}. We mention two here. Assume that triplet interaction is suppressed, as well as that in the particle–particle channel. This can be experimentally realized if there is considerable scattering by magnetic impurities. Such scattering leads to non-zero energy for these two long wavelength modes. One thus has a system of only density fluctuations coupled by coulomb interaction. Altshuler & Aronov (1983) show that for the coulomb case, the metal–insulator transition is continuous, with conductivity going to zero according to the exponent unity. The dielectric constant diverges with the exponent (3/2), the density of states at the Fermi energy varies as $\exp[-1/8(\ln T)^2]$, and the a.c. conductivity as $\omega^{2/5}$. The conductivity exponent is the same as for pure localization, though the other exponents are different (Castellani *et al.* 1984a).

A more general situation corresponds to neglect of particle–particle scattering, but inclusion of both spin-singlet and spin-triplet scattering (Castellani *et al.* 1984b, Finkelshtein 1984). Here it turns out that the effective spin-triplet coupling becomes very large on scaling, and indeed diverges at a finite length scale l_c. At this length scale, density diffusion is non-zero, that is, the conductivity is non-zero, but spin diffusion vanishes. The spin susceptibility also diverges. One thus has a clump of frozen magnetization on a length scale l_c, that is, a local moment. This is reminiscent of the Brinkman–Rice instability. Unfortunately, the perturbative scaling approach (perturbative in the effective coupling strength) breaks down when coupling constants become large and the above results for a zero-temperature spin localization transition preceding a metal–insulator transition are only suggestive.

There is as yet no analysis of interaction effects which also includes the backscattering localization phenomenon, for models exhibiting a metal–insulator transition.

4.3. Experiment

We now briefly summarize the experimental situation. In all the disordered systems studied so far, the $T=0$ metal–insulator transition is continuous. The conductivity goes to zero, with an exponent of unity in all systems except doped, uncompensated semiconductors where the exponent is close to half. (See Thomas 1984b for a succint review; see also Lee & Ramakrishnan 1985, Bishop & Dynes 1985). Stress tuning of the metal–insulator transition clearly establishes this value for Si:P (see for example, Rosenbaum *et al.* 1983). In those cases where it has been measured, such as in Au_xGe_{1-x} alloys and Nb_xSi_{1-x} alloys, there is a pronounced and singular dip in the density of states as the metal–insulator transition is approached. Other transport properties, such as magnetoresistance, also show singular critical behaviour. The special nature of the metal–insulator transition in the uncompensated, doped semiconductor is clear from the fact that with increasing compensation, there is a crossover from an exponent of half to unity for the vanishing of the conductivity (Thomas *et al.* 1982, Zabroskii & Zinoveva 1984). Because of this, as well as extensive experimental work on this class of systems, its central role in the field of metal–insulator transitions and the puzzling nature of results, we discuss these below.

As mentioned above, the conductivity of phosphorus doped silicon and homologous systems goes to zero continuously at the $T=0$ metal–insulator transition. The exponent is half, whereas the exponent is unity in localization theory and in one special version of interacting disordered electron gas theory (see above; Altshuler & Aronov 1983). An exponent of half is possible for a non-interacting disordered electron gas provided singular backscattering effects are suppressed, for example, by a magnetic field or by magnetic impurity scattering. (This diffusion localization mechanism was first discussed by Götze (1979). The static dielectric constant of Si:P diverges (see for example Rosenbaum *et al.* 1983 for a summary of results; see also Thomas 1984a) with an exponent of unity, nearly twice the conductivity exponent.

The a.c. conductivity just on the insulating side has been measured at very low temperatures (Paalanen *et al.* 1983; also Thomas 1984a). It is linear in frequency, and the coefficient diverges as the dielectric constant, that is, linearly with deviation from critical disorder. Bhatt & Ramakrishnan (1984) argue that this behaviour is direct evidence for the importance of long-range coulomb interaction effects of the sort considered by Efros and Shklovskii (Efros 1981,

Shklovskii & Efros 1982). In the absence of such an interaction, the coefficient would diverges as $\varepsilon^{5/2}$ rather than as ε. Secondly, the term linear in frequency would be proportional to $k(T)$. Neither is observed. (See Mott & Davis 1979 for a discussion of the a.c. conductivity of an Anderson insulator.) The linear term in the specific heat does not vanish across the metal–insulator transition. This property is consistent with an Anderson localization transition without interactions, or possibly with the existence of a finite density of electronic two-level systems in the insulating phase implied by the Efros–Shklovskii theory (Bhatt & Ramakrishnan 1984).

There is an increasing body of evidence pointing to strong spin correlations near the metal–insulator transition in doped semiconductors. For example, there is a characteristic enhancement of the specific heat at low temperatures (Thomas *et al.* 1981). By studying spin relaxation in phosphorus doped silicon at very low temperatures, Paalanen *et al.* (1984) found that spin–lattice relaxation rates T_1^{-1} are singularly enhanced near the metal–insulator transition. The rate, instead of being proportional to the temperature as in a metal (Korringa relaxation) is seen to be proportional to the square root of the temperature. From the magnetic field dependence of T_1, it is concluded that spin diffusion is greatly slowed down and the static spin susceptibility is enhanced. The total picture is very different from the first-order transition (Mott transition) expected on neglecting disorder, and the kind of second-order transition (Anderson transition) expected on neglecting interaction. From bits and pieces of experimental information, it seems that disorder, long-range coulomb effects, as well as commensurate band filling (half filling) and associated Hubbard-model spin-correlation effects are all important. The distance between the experimental features of the transition and available theoretical models in this most extensively studied example of metal–insulator transitions is an indication of our present level of understanding.

5. Outlook

Despite the long span of time over which the metal–nonmetal transition has been studied, the large variety of systems explored and the amount of theoretical work done, our understanding of this phenomenon is very incomplete. Efforts to determine the mechanism in a particular transition are broadly of three kinds. The first is to compare the transition criteria predicted by different assumed mechanisms with observation. We have seen in §2.5 that on very general grounds this is unlikely to be a fruitful method. The second is to study the systematics of the transition across classes of chemically related compounds and to parametrize the observed patterns. This has been done very successfully by Goodenough (1971) and for a smaller class of systems by Wilson (1972). However, the success of this approach depends on how realistic the basic models are, and how reliably their consequences have been worked out. The third is to make detailed predictions of physical behaviour near the metal–insulator transition for various kinds of models and compare them with that experimentally observed. This approach has been pioneered by Mott over the years (see Mott 1974, Mott & Davis 1979, Mott 1984). Recently, this approach has been extensively applied to disordered electronic systems near the metal–insulator transition, as summarized above. We mention some of these results again in a broader context.

There is a detailed theory now for the Anderson transition, namely, the disorder-induced metal-insulator transition at $T=0$ in the absence of interactions. There are a number of experimental results supporting many of the conclusions; for example, that both the dielectric transition and the conductivity transition are continuous with related exponents, and that the latter has an exponent of unity. In the presence of interactions and disorder, many kinds of metal–insulator transition appear possible; some of these have been investigated in detail. All of them are continuous zero-temperature transitions. There are no theories as yet for the most commonly observed metal–nonmetal changes, that is, discontinuous changes often with a critical point at $T \neq 0$. There are no calculations of the thermodynamic consequences of the singular dip in the density of electronic states and of poor screening (excitonic effects) in the vicinity of the metal–insulator transition. Interaction effects have been treated in perturbation theory, and the conduction band filling is assumed irrelevant. However, from results for the Hubbard model, and from experimental results in doped semiconductors, it is clear that the random system with one electron per site is special, and that compensation leads to major changes in the nature of the transition. There is no theory yet with this feature, and there have not been enough experiments. In a recent article, Mott (1984) has again pointed out that there is no real understanding of how disorder removes the Mott discontinuity. Existing theories of disordered interacting electron systems do not consider such excitonic effects.

For correlated systems without disorder, we have seen in §2 that there are two complementary approaches, one emphasizing local quantum fluctuations, and the other, symmetry breaking averages. These two need to be put together for a reasonable description of the transition and the phase diagram. There is no theory yet which does this and considers the effect of various relevant perturbations.

It is quite likely that the mechanisms for metal–insulator transitions will be clarified only by making detailed predictions and confronting these with experimental results. In this sense, the field is just beginning to develop.

Acknowledgements

I would like to thank a number of colleagues for sending reprints and preprint copies of their work, in particular R.N. Bhatt, B.H. Brandow, R.C. Dynes, E.N. Economou, J.M. Honig, N.F. Mott, R. Oppermann, T.M. Rice, G.A. Thomas and D. Vollhardt.

References

Abrahams, E., Anderson, P.W., Licciardello, D.C. and Ramakrishnan, T.V., 1979, *Phys. Rev. Lett.* **42**, 673.
Abrahams, E., Anderson, P.W., Lee, P.A. and Ramakrishnan, T.V., 1981, *Phys. Rev.*, B**24**, 6783.
Altshuler, B.L. and Aronov, A.G., 1979, *Sov. Phys.—JETP*, **50**, 968.
Altshuler, B.L. and Aronov, A.G., 1983, *Solid St. Commun.* **46**, 429.
Altshuler, B.L. and Aronov, A.G., 1984, in *Electron–Electron Interactions in Disordered Systems*, edited by M. Pollak and A.L. Efros (Amsterdam: North Holland).
Altshuler, B.L., Aronov, A.G. and Lee, P.A., 1980, *Phys. Rev. Lett.* **44**, 1288.
Altshuler, B.L., Aronov, A.G. and Spivak, B.Z., 1981, *Sov. Phys.—JETP Lett.*, **33**, 94.
Anderson, P.W., 1958, *Phys. Rev.* **109**, 1492.
Anderson, P.W., 1975, *Phys. Rev. Lett.* **34**, 953.
Anderson, P.W., 1981, in *Valence Fluctuations in Solids*, edited by L.M. Falicov, W. Hanke and M.B. Maple (New York: North Holland).
Anderson, P.W., 1984, *Phys. Rev.* B **30**, 1549.
Anderson, P.W. and Brinkman, W.F., 1978, in *Physics of Liquid and Solid Helium*, Part II, edited by K.H. Bennemann and J.B. Ketterson (New York: Wiley).
Anderson, P.W., Abrahams, E. and Ramakrishnan, T.V., 1979, *Phys. Rev. Lett.* **43**, 718.
Aronov, A.G., 1984, *Physica* B+C **126**, 314.
Bhatt, R.N. and Rice, T.M., 1981, *Phys. Rev.* B **23**, 1920.
Bhatt, R.N. and Ramakrishnan, T.V., 1984, *J. Phys. C: Solid St. Phys.* **17**, L639.
Bishop, D.J. and Dynes, R.C., 1985, to be published.
Brandow, B.H., 1977, *Adv. Phys.* **26**, 651.
Brinkman, W.F. and Rice, T.M., 1970, *Phys. Rev.* B **2**, 4302.
Castellani, C.C., di Castro, C., Lee, P.A., and Ma, M., 1984a, *Phys. Rev.* B **30**, 527.
Castellani, C.C., di Castro, C., Lee, P.A., Ma, M., Sorella, S. and Tabet, E., 1984b, *Phys. Rev.* B **30**, 1596.
Castner, T.G., 1980, *Phil. Mag.* **42**, 873.
Coleman, P., 1984, Princeton preprint and to be published.
Cyrot, M., 1972, *Phil. Mag.*, **25**, 1031.
Cyrot, M. and Lacour-Gayet, P., 1972, *Solid St. Commun.* **11**, 1767.
Cyrot, M. and Lyon-Caen, C., 1975, *J. Physique* **36**, 253.
De Marco, R.R., Economou, E.N. and White, C.T., 1978, *Phys. Rev.* B **18**, 3968.
Doniach, S. and Engelsberg, S., 1966, *Phys. Rev. Lett.* **17**, 750.
Eastman, D.E. and Himpsel, F.J., 1981, *Conference on Transition Metals, Leeds, England, 1980* (Inst. Phys. Conf. Ser. 55), p. 115.
Economou, E.N., White, C.T. and de Marco, R.R., 1978, *Phys. Rev.*, B **18**, 3946.
Edwards, P.P. and Sienko, M.J., 1978, *Phys. Rev.* B **17**, 2575.
Edwards, P.P. and Sienko, M.J., 1982, *Accounts of Chemical Research*, **15**, 87.
Edwards, P.P. and Sienko, M.J., 1983, *Int. Rev. Phys. Chem.* **3**, 83.
Efros, A.L., 1981, *Phil. Mag.* B **43**, 829.
Fetter, A.L. and Walecka, J.D., 1971, *Quantum Theory of Many Particle Systems* (New York: McGraw-Hill), p. 309.
Finkelshtein, A.M., 1983, *Sov. Phys.-JETP*, **57**, 97.
Finkelshtein, A.M., 1984, *Z. Phys.* B **56**, 189.
Fujimori, A. and Minami, F., 1984, *Phys. Rev.* B **30**, 957.
Fukuyama, H., 1984, in *Electron Electron Interactions in Disordered Systems*, edited by M. Pollak and A.L. Efros (Amsterdam: North Holland).
Goodenough, J.B., 1971, in *Progress in Solid State Chemistry*, edited by H. Reiss (Oxford: Pergamon Press), **5**, 145.

Götze, W., 1979, *J. Phys. C: Solid St. Phys.* **12**, 1279.
Gutzwiller, M.C., 1965, *Phys. Rev.* A **137**, 1726.
Herbst, J.F., Watson, R.E. and Wilkins, J.W., 1976, *Phys. Rev.* B **13**, 1439.
Herring, C., 1966, in *Magnetism*, edited by G. Rado and H. Suhl (New York: Wiley), Volume IV.
Hertel, G., Bishop, D.J., Spencer, E.G., Rowell, J.M. and Dynes, R.C., 1983, *Phys. Rev. Lett.* **50**, 743.
Herzfeld, K.F., 1927, *Phys. Rev.* **29**, 701.
Hiramoto, H. and Toyozawa, Y., 1985, *J. Phys. Soc. Japan* **54**, 245.
Honig, J.M., 1982, *J. Solid St. Chem.* **45**, 1.
Hubbard, J., 1963, *Proc. R. Soc.* A **276**, 238.
Hubbard, J., 1964, *Proc. R. Soc.* A **281**, 401.
Ioffe, A.F. and Regel, A.R., 1960, *Prog. Semiconductors* **4**, 237.
Kaplan, T.A., Horsch, P. and Fulde, P., 1982, *Phys. Rev. Lett.* **49**, 889.
Kaveh, M. and Mott, N.F., 1982, *J. Phys. C: Solid St. Phys.* **15**, L707.
Kawabata, A., 1981, *J. Phys. Soc. Japan*, **50**, 2461.
Lakkis, S., Schlenker, C., Chakraverty, B. K., Buder, R. and Marezio, M., 1976, *Phys. Rev.* B **14**, 1479.
Landauer, R., 1970, *Phil. Mag.* **21**, 963.
Lederer, P. and Mills, D.L., 1967, *Phys. Rev.* **165**, 837.
Lee, P.A. and Ramakrishnan, T.V., 1982, *Phys. Rev.* B **26**, 4009.
Lee, P.A. and Ramakrishnan, T.V., 1985, *Rev. Mod. Phys.* **57**, 287.
Licciardello, D.C. and Thouless, D.J., 1975, *J. Phys. C: Solid St. Phys.* **8**, 4157.
Lieb, E.H. and Wu, F.Y., 1968, *Phys. Rev. Lett.* **20**, 1445.
Luttinger, J.M., 1960, *Phys. Rev.* **119**, 1153.
Mackinnon, A. and Kramer, B., 1981, *Phys. Rev. Lett.* **47**, 1546.
McMillan, W.L., 1981, *Phys. Rev.* B **24**, 2739.
McWhan, D.B. and Remeika, J.P., 1970, *Phys. Rev.* B **2**, 3734.
Mott, N.F., 1949, *Proc. Phys. Soc.* A **62**, 416.
Mott, N.F. and Twose, W.D., 1961, *Adv. Phys.* **10**, 107.
Mott, N.F., 1967, *Adv. Phys.* **16**, 49.
Mott, N.F. and Zinamon, Z., 1970, *Rep. Prog. Phys.* **33**, 881.
Mott, N.F., 1972, *Phil. Mag.* **26**, 1015.
Mott, N.F., 1974, *Metal–Insulator Transitions* (London: Taylor & Francis).
Mott, N.F. and Davis, E.A., 1979, *Electronic Processes in Noncrystalline Materials* (Oxford: Clarendon Press).
Mott, N.F., 1984, *Phil. Mag.* **50**, 161.
Nagaoka, H. and Fukuyama, H. (eds.), 1982, *Anderson Localization* (Berlin: Springer-Verlag).
Ogawa, T. and Kanda, 1978, *Z. Phys.* B **30**, 355.
Ogawa, T., Kanda, K. and Matsubara, T., 1975, *Prog. Theor. Phys.* **53**, 614.
Oppermann, R., 1982, *Prog. Theor. Phys.* **68**, 1038.
Paalanen, M.A., Rosenbaum, T.F., Thomas, G.A. and Bhatt, R.N., 1983, *Phys. Rev. Lett.* **51**, 1896.
Ramakrishnan, T.V., 1981, in *Valence Fluctuations in Solids*, edited by L.M. Falicov, W. Hanke and M.B. Maple (New York: North Holland), p. 13.
Rice, T.M. and Brinkman, W.F., 1971, in *Alloys, Magnets and Superconductors*, edited by R.E. Mills, E. Ascher and R. Jaffee (New York: McGraw-Hill), p. 593.
Rice, T.M., 1977, in *Solid State Physics*, edited by H. Ehrenreich, F. Seitz and D. Turnbull, Vol. 32, 1.
Rice, T.M., Udea, K., Ott, H.R. and Rudigier, H., 1985, *Phys. Rev.* B **31**, 594.
Rosenbaum, T.F., Milligan, R.F., Paalanen, M.A., Thomas, G.A., Bhatt, R.N. and Lin, W., 1983, *Phys. Rev.* B **27**, 7509.

Samson, J.H., 1984, Cornell University preprint and to be published.
Sawatzky, G.A. and Allen, J.M., 1984, *Phys. Rev. Lett.*, **53**, 2339.
Schirmer, O.F. and Salje, E., 1980, *J. Phys. C: Solid St. Phys.* **13**, L1067.
Schmid, A., 1974, *Z. Phys.* **271**, 251.
Sharvin, D.Yu. and Sharvin, Yu.V., 1981, *JETP Lett.* **34**, 285.
Shklovskii, B.I. and Efros, A.L., 1982, *Sov. Phys.–JETP*, **54**, 218.
Slater, J.C., 1951, *Phys. Rev.* **82**, 538.
Thomas, G.A., Kawabata, A., Ootuka, Y., Katsumoto, S., Kobayashi, S. and Sasaki, W., 1981, *Phys. Rev.* B **24**, 4986.
Thomas, G.A., Ootuka, Y., Katsumoto, S., Kobayashi, S. and Sasaki, W., 1982, *Phys. Rev.* B **25**, 4288.
Thomas, G.A., 1984a, *Phil. Mag.* **50**, 169.
Thomas, G.A., 1984b, *Proc. Conf. Highly Doped Semiconductors*, San Francisco (to be published).
Thouless, D.J., 1977, *Phys. Rev. Lett.* **39**, 1167.
Thouless, D.J., 1979, in *Ill Condensed Matter*, edited by R. Balian, R. Maynard and G. Toulouse (New York: North Holland), p. 1.
Toyozawa, Y., 1983, in *Highlights of Condensed Matter Physics* (Proceedings of Varenna Summer School, 1983) (London: Academic Press).
Valls, O.T. and Tesanovic, Z., 1984, *Phys. Rev. Lett.* **53**, 1497.
Vollhardt, D., 1984, *Rev. Mod. Phys.* **56**, 99.
Weaire, D. and Srivastava, V., 1977, *J. Phys. C: Solid St. Phys.* **10**, 4309.
Wegner, F.J., 1976, *Z. Phys.* B **25**, 327.
Wegner, F.J., 1979, *Z. Phys.* B **35**, 207.
White, A.E., Dynes, R.C. and Garno, J.P., 1985, *Phys. Rev.* B **31**, 1174.
White, C.T. and Economou, E.N., 1977, *Phys. Rev.* B **15**, 3742.
White, C.T. and Economou, E.N., 1978, *Phys. Rev.* B **18**, 3959.
Wilson, J.A., 1972, *Adv. Phys.* **21**, 143.
Zabroskii, A.G. and Zinoveva, K.N., 1984, *Sov. Phys. JETP* **59**, 425.

CHAPTER 3

the periodic system of the elements

D.E. Logan and P.P. Edwards
University Chemical Laboratory, Lensfield Road, Cambridge, CB2 1EW, UK

1. Introduction

On 4 June, 1889, the Russian chemist Mendeleéff (1889) delivered a lecture to the Chemical Society of Great Britain entitled 'The Periodic Law of the Chemical Elements'. Mendeleéff noted ... 'While science is pursuing a steady onward movement, it is convenient from time to time to cast a glance back on the route already traversed, and especially to consider the new conceptions which aim at discovering the general meaning of the stock of facts accumulated from day to day in our laboratories.' The present chapter is an attempt to review the development of the periodic system of the elements (Figure 1). Particular emphasis will be attached to the question as to why elements exist either as metals or nonmetals under ambient conditions.

At each stage in the development of ideas pertaining to a subject of such breadth, there is naturally somewhat of an obsession with the particular, relevant aspect of the problem which is then in vogue. Although the genesis of scientific ideas is rarely systematic, a historical perspective offers the advantage of setting in overall context the role of specific contributions to a subject. In addition, it may enable us to recognize the inception of ideas which were, in a sense, before their time, and which were discarded or neglected in an earlier prevailing climate of thought. In this regard we note that the first semiquantitative attempts to explain the occurrence of metallic or nonmetallic behaviour in the periodic system were made by Goldhammer in 1913 and Herzfeld in 1927. Their simple rationalization, in terms of relevant atomic properties which in a crude sense confer metallic or nonmetallic status upon an element under ambient conditions, leads to what is commonly called the Herzfeld criterion for metallization.

For a variety of undoubtedly complex reasons, the Herzfeld–Goldhammer theory lay dormant for many years (Herzfeld 1966, Cohen 1968, Edwards & Sienko 1983a,b). Recently, however, there has been a renaissance of interest in the approach. The Herzfeld criterion has now been applied to many systems exhibiting induced metal–nonmetal transitions in order to predict the critical

THE PERIODIC TABLE

s-block | d-block | p-block

IA	IIA	IIIB	IVB	VB	VIB	VIIB	VIII			IB	IIB	IIIA	IVA	VA	VIA	VIIA	
H	He																
Li	Be											B	C	N	O	F	Ne
Na	Mg											Al	Si	P	S	Cl	Ar
K	Ca	Sc	Ti	V	Cr	Mn	Fe	Co	Ni	Cu	Zn	Ga	Ge	As	Se	Br	Kr
Rb	Sr	Y	Zr	Nb	Mo	Tc	Ru	Rh	Pd	Ag	Cd	In	Sn	Sb	Te	I	Xe
Cs	Ba	La	Hf	Ta	W	Re	Os	Ir	Pt	Au	Hg	Ti	Pb	Bi	Po	At	Rn
Fr	Ra	Ac															

Ce	Pr	Nd	Pm	Sm	Eu	Gd	Tb	Dy	Ho	Er	Tm	Yb	Lu
Th	Pa	U	Np	Pu	Am	Cm	Bk	Cf	Es	Fm	Md	No	Lw

Figure 1. A representation of the periodic system of the elements.

conditions for metallization (for recent reviews, see Edwards & Sienko 1982, 1983a,b, Ross & McMahan 1981). This resurgence of interest is in part due to the fact that, although the Herzfeld criterion is obtained by simple classical arguments, it is known to give predictions which often agree with experiments as well as theoretical estimates derived from detailed and often difficult quantum mechanical calculations.

This review contains in part a description and assessment of the Herzfeld criterion. In a historical context the resultant emphasis is disproportionate, but the approach may deserve more attention than it has previously received.

The outline of the chapter is as follows. In the remaining paragraphs of this section we outline the experimental basis for the classification of elements as metals or nonmetals. Section 2 deals with certain historical perspectives on the periodic system, while §3 deals with the evolution of what one might term the 'modern' theory of solids dating from the beginning of this century. The final section contains a discussion of the Herzfeld–Goldhammer theory of metallization, and its application to the question of the metallic versus nonmetallic status of elements within the periodic system.

1.1. Metals and nonmetals in the periodic system

Perhaps the most fundamental classification of the elements within the periodic system (Figure 1) is in terms of their metallic or nonmetallic properties.

Landau & Zeldovitch (1941) and Mott (1961, 1974) pointed out that the metallic and nonmetallic states of matter can only be distinguished unambiguously at the absolute zero of temperature. Thus, at $T = 0$ K, electrons in solids are either localized or itinerant; the electrical resistivity of a metal is either finite or zero, whereas in nonmetals it tends to infinity. Clearly, at temperatures above the absolute zero, this distinction may become ambiguous.

With the exception of a fairly small class of materials, for example, doped semiconductors (Rosenbaum *et al.* 1983 and references therein), very little conductivity data is presently available for the low-temperature extreme. Nevertheless, even armed with room-temperature conductivity data (Figure 2), one can fairly easily identify elements of the periodic system for which the appellation metallic or nonmetallic is most appropriate. The difference in conductivities between a metallic element and a nonmetallic element (under ambient conditions) is striking. For example, the measured electrical conductivities at room temperature (Figure 2) range from values of about 10^5 $(\Omega\,\mathrm{cm})^{-1}$ for the 'simple' Group IA metals, to approximately 10^{-18} $(\Omega\,\mathrm{cm})^{-1}$ for nonmetallic elemental sulphur. This range of approximately 10^{23} in conductivities may be the widest range of any known physical property of the elements at room temperature. Obviously, this difference is even more exaggerated at low temperatures (Kittel 1976).

In addition, it was noted some time ago that metallic structures cease at C, Si and Ge in the first three periods (Figure 1) and at Sb and Po in the fourth and fifth periods. Thus, Hume-Rothery & Raynor (1956) identified three types of elemental structures within the periodic system (Figure 3). Class III is representative of nonmetallic elements in which valence electrons are localized in directional chemical bonds. In this instance a formal valency (N) is associated with the atomic constituent and elemental structures are well described in terms of the $(8-N)$ rule (Hume-Rothery & Raynor 1956). For the elements of Class I, the itinerancy of valence electrons ensures a breakdown of the $(8-N)$ rule, and non-directional 'metallic bonding' is reflected in the high-coordination structure types (bcc, fcc, hcp, and so on).

2. The periodic system of the elements: historical perspectives

There are three distinguishable stages in the history of the periodic system, namely those of initiation, phenomenological description and theoretical development (van Spronsen 1969). The first two stages were essentially complete by 1870. The latter stage, still in an active state of development, concerns the fundamental question of metallic versus nonmetallic status of the elements under ambient conditions.

Probably the first attempt to draw a table of chemically similar substances was made in France in 1782 by de Morveau (de Morveau 1782, Mazurs 1974) and repeated some five years later by a Commission on Chemical Nomenclature (de

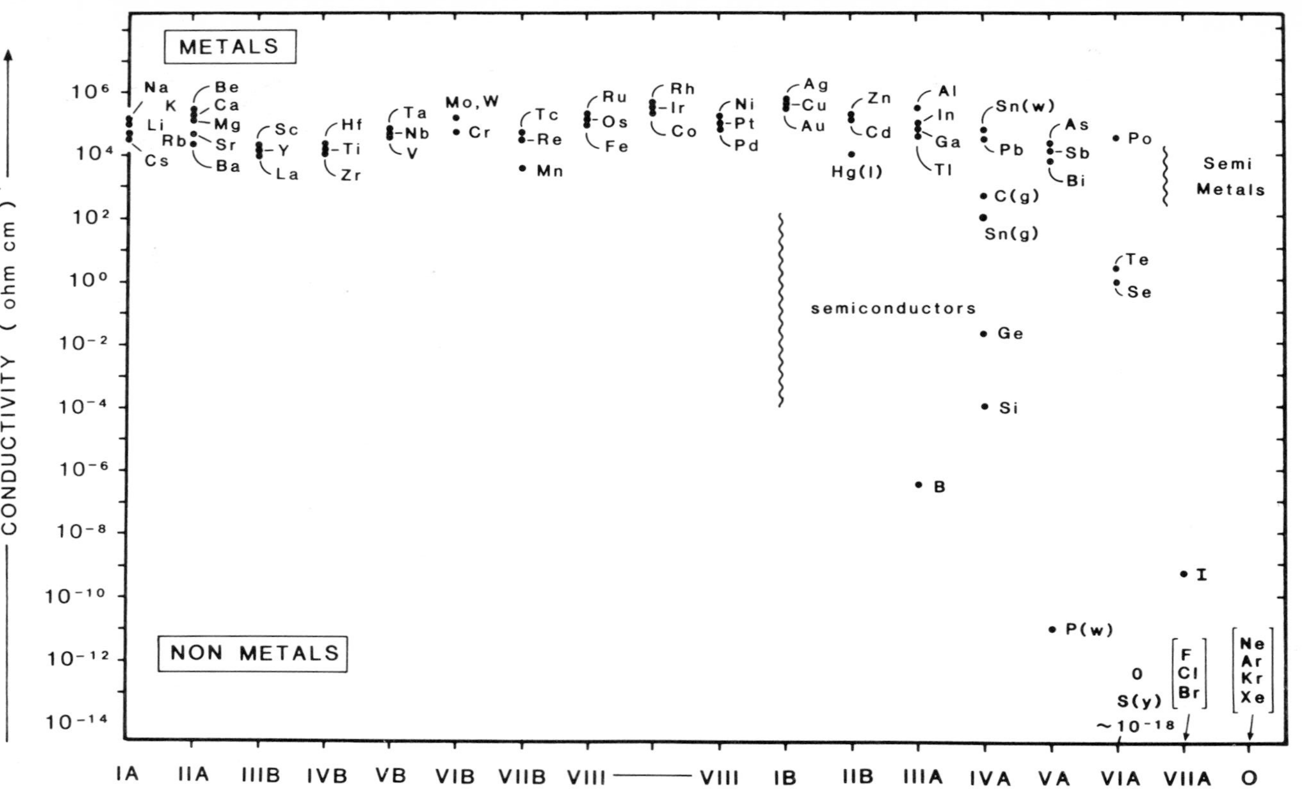

Figure 2. The electrical conductivities at room temperature of the majority of elements of the s, p and d block of the periodic system. The experimental data is taken from Meaden (1966), as cited by Kittel (1976), and from the Handbook of Chemistry and Physics.

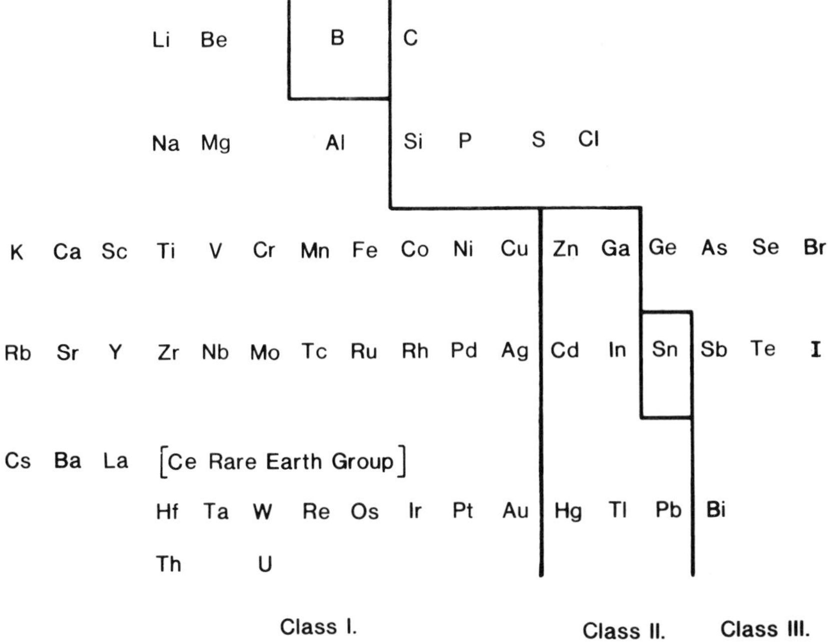

Figure 3. Crystal structures and the periodic system; the elements are divided into three classes by means of heavy lines. The corresponding crystal structures are discussed briefly in the text, and extensively in Hume-Rothery & Raynor (1956) from which this figure is taken.

Morveau *et al.* 1787). The table in question is reproduced from Mazurs (1974) in Figure 4, but the names of the substances have been changed to the present chemical sybols and formulae.

Lavoisier, a member of the 1787 Commission, used this same table in a textbook published in 1789 (Lavoisier 1789). We note that the demarcation between metallic and nonmetallic substances was based exclusively on the (then) known high-density, metallic elements, such as Fe, Ag, Hg, Au, and so on. The alkali and alkaline earth elemental metals (Group IA, IIA, Figure 1) were unknown at that time, and their oxides and hydroxides were judged to be simple, non-decomposable elements. Thus, Lavoisier (1789) defined an element as an entity whose separation or decomposition into other substances had not yet been achieved. He identified, for example, NaOH as an element (Figure 4). The alkali metals themselves were not discovered until 1807, the occasion being the isolation of elemental sodium and potassium by Sir Humphry Davy (Davy 1808). Thus, the pre-1807 Periodic System, as envisaged in Figure 4, is important in that it reminds us that prior to Davy's discovery and isolation of sodium and potassium, *all* of the established metals were high-density elements. As Mendeleéff (1905) noted, the preparation of metallic sodium may be regarded as one of the greatest discoveries in science. Davy's monumental work posed a

NON DECOMPOSABLE SUBSTANCES					
Simple Substances.	Simple Combustible Substances.	Metallic Substances.		Simple Earths.	Alkalies.
Light	N	As	Fe	SiO_2	NaOH
Heat	C	Mo	Sn	Al_2O_3	KOH
O	S	W	Pb	BaO	NH_4OH
H	P	Mn	Cu	CaO	
	Cl	Ni	Hg	MgO	
	B	Co	Ag		
	F	Bi	Pt		
	etc.	Sb	Au		
		Zn			

Figure 4. A representation—perhaps the first—of a table of chemically simple substances. This forerunner of the modern periodic table is attributed to de Morveau *et al.* (1787); modified from the original (Mazurs 1974).

major dilemma to the classification of the periodic system, since both sodium and potassium possessed many of the physical and chemical properties of the known metals, but had exceptionally low densities. Indeed, Erman and Simon in 1808 attempted to resolve this difficulty by proposing that sodium and potassium be termed 'metalloids'—meaning 'those that are like metals'—to indicate that they are similar to metals only in certain respects! In his second Bakerian Lecture, Davy (1808) also addressed the controversial problem of whether these low-density elements could be called metals, and remarked 'The great number of philosophical persons to whom this question has been put, have answered in the affirmative'. Therefore, the initiation of the periodic system was now well and truly founded. The discovery of the alkali metals meant that the conception of elements became broader; for example, in sodium and potassium *chemical* properties (that is, high reactivity) were observed which were but feebly shown by the other metals of the day (Figure 4).

Cannizaro, in a famous lecture delivered at Karlsruhe in 1860, made perhaps the first clear distinction between atoms and molecules, and unambiguously defined such concepts as valence (van Spronsen, 1969). This initiated the second stage of the history proper of the periodic system of the chemical elements. Now, once the time was ripe, the systematic classification of the elements arose almost simultaneously in most leading countries in Europe and in North America (van Spronsen 1969, Mazurs 1974). However, the discovery of the periodic law is (quite rightly) attributed to Mendeleéff (1869a,b, 1889) and Meyer (1870); see also van Spronsen (1969) and Mazurs (1974). Mendeleéff's work was presented at the beginning of 1869 before the Russian Chemical Society, and in a paper published in the Journal of the Russian Chemical Society in the same year (Mendeleéff, 1869a,b). The substance of the paper contained various conclusions: (i) The elements, if arranged according to atomic weights, exhibit an evident periodicity of properties. (ii) Elements which are similar as regards their chemical properties have atomic weights which are either of nearly the same value (e.g. Pt, Ir, Os) or which increase regularly (e.g. K, Rb, Cs). (iii) The magnitude of the atomic weight determines the character of an element. (iv) Certain characteristic properties of the elements can be predicted from their atomic weights.

Mendeleéff's paper (1869a,b), in addition to his main proposed table of the elements, contained suggestions for three other possible representations (Mazurs 1974). It is interesting to note that, in a footnote to this paper, Mendeleéff discussed a representation of the periodic system (Figure 5) that he said would probably not be a convenient one. The modern form of this table (Figure 1) is particularly popular today! For our present discussion, it clearly best reflects the metal–nonmetal transition within the periodic system (Figures 2 and 3). Thus, elemental properties in the p-block change from nonmetallic to metallic from top

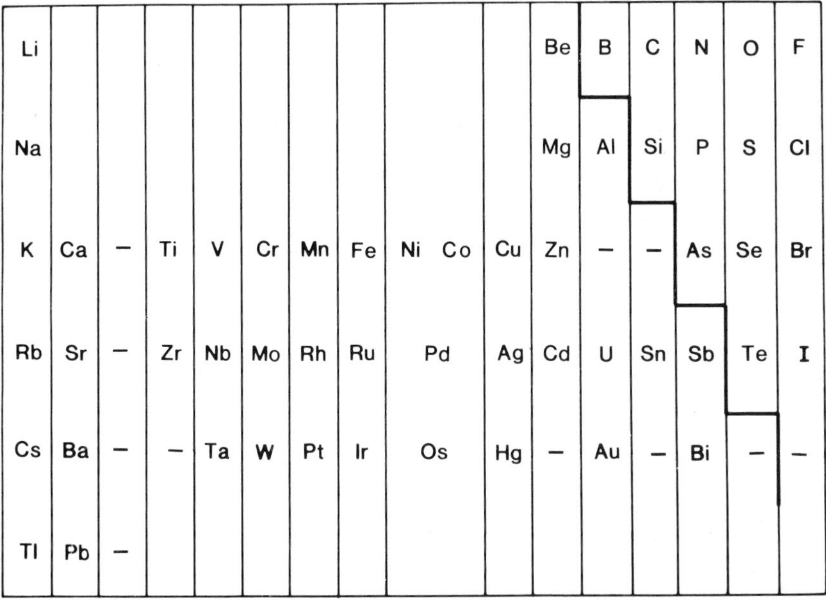

Li										Be	B	C	N	O	F
Na										Mg	Al	Si	P	S	Cl
K	Ca	—	Ti	V	Cr	Mn	Fe	Ni Co	Cu	Zn	—	—	As	Se	Br
Rb	Sr	—	Zr	Nb	Mo	Rh	Ru	Pd	Ag	Cd	U	Sn	Sb	Te	I
Cs	Ba	—	—	Ta	W	Pt	Ir	Os	Hg	—	Au	—	Bi	—	—
Tl	Pb	—													

Figure 5. A periodic table of the elements (Mendeleéff 1869), type IIC–4, taken from Mazurs (1974).

to bottom in vertical groups, and from metallic to nonmetallic behaviour from left to right in a horizontal group.

Meyer's (1870) periodic arrangement of the (then known) elements was suggested by the striking regularity of the curve of atomic volume (that is, atomic weight divided by the elemental density) of the solid elements, using H as the unit of atomic weight and water as the unit of density. The periodic changes in the atomic volume as a function of increasing atomic weight are shown in Figure 6, taken from the original paper of Meyer (1870).

Therefore by 1870, or thereabouts, the periodic system had been accepted as fundamental to chemistry and its future development. It is important to record that the periodic table of the elements reached something of its final form almost 50 years before the theory of the atom revealed the basis for the periodicity of atomic structure. As Mendeleéff (1905) pointed out, '... for the aims of (empirical) chemistry, it is possible to take advantage of the laws discovered by chemistry, without being able to explain their causes'. The periodic system at this stage rested on a solid empirical corpus. The theoretical basis, in terms of the atomic constitution of elements, was achieved after atomic weight (Figure 6) had essentially been replaced by the atomic number (Figure 10), the importance of which was gradually revealed in theoretical attempts to explain atomic structure.

3. Electrons in solids

The 'modern' theory of solids is essentially based on the hypothesis that the electrons in metals are free to move from atom to atom, while those in nonmetals are not (Figure 2). This fundamental idea can be interpreted in terms of models that are qualitatively easy to understand, but often difficult to implement in a quantitative manner. The metallic state has proved to be one of the great fundamental states of matter. As one can see from Figures 2 and 3, the elements definitely favour the metallic state: over two thirds are metals under ambient conditions. Here we present a brief history of the electron theory of metals.

3.1. The independent-electron theory

An early attempt to explain the passage of electricity through a metal was made by Weber (1875) who imagined that a 'molecule' of a metal consisted of a number of electrically charged particles in relative motion to one another. The discovery of the electron by Thompson in 1897, and the all-important inferences for the constitution of matter, resulted in better-founded attempts to explain the mechanism of metallic conduction. The first contribution was made by Riecke in 1898, but this work was quickly superseded by that of Drude (1900) at the turn of the century, who gave the first theory capable of computing observed quantities from quasi-microscopic concepts.

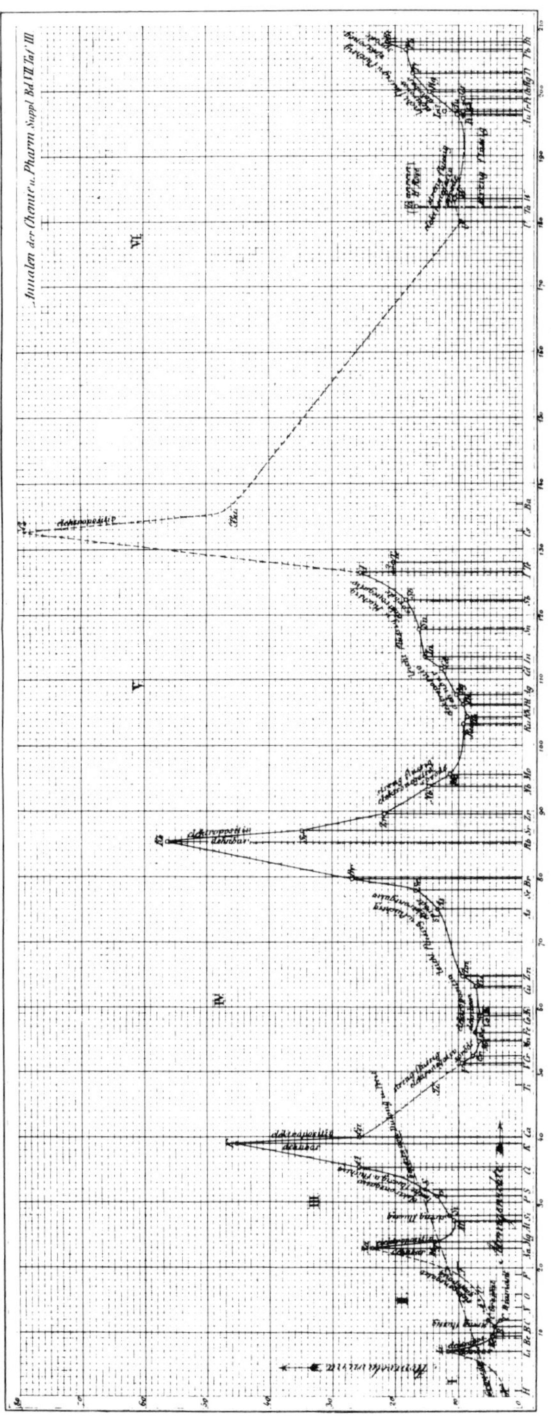

Figure 6. The atomic volume curve by Meyer (1870).

Drude essentially pictured metals as consisting of a 'gas' of negative mobile carriers (electrons) moving against a stabilizing background of positive ions. The electrons have a characteristic average velocity and pursue random and independent motion through the metal due to collisions with metal ions, which, owing to their large mass can be considered to be stationary. In 1904–5, Lorentz, and somewhat later Bohr (1911), gave an improved mathematical formulation of Drude's theory. However, the Drude–Lorentz theory, based on electron motion alone, took only minimal account of the presence of the positive ions (to provide the necessary scattering of conduction electrons).

In many applications, the combined theories gave a satisfactory account of certain electrical and thermal properties, but they could not explain the extraordinary magnetic properties of the metallic elements; that is, the occurrence of weak and temperature-independent paramagnetism for a large number of elements in the periodic system. These difficulties and inconsistencies could not have been resolved until 1925, the occasion being the publication by Pauli (1925) of his paper 'Über den Zusammenhang des Abschlusses der Elektronengruppen im Atom mit der Komplexstruktur der Spektren'. This paper formulated the exclusion principle to explain the closure of atomic shells of electrons. It is interesting to note that, according to Wilson (1980), Pauli did not go on to extend his exclusion principle to the conduction electrons in a metal. Neither, apparently, did Fermi, though he is often credited for having done so. Pauli's 1927 paper 'Über Gasentartung und Paramagnetismus' was largely taken up with a discussion of the difference between Bose–Einstein and Fermi–Dirac statistics. It has even been suggested (Wilson 1980) that, except for the derivation of the paramagnetic susceptibility, the paper contained very little of physical interest. Pauli's contribution, however, was extremely influential in that it enabled a resolution of one of the fundamental stumbling blocks in the classical Drude–Lorentz theory; namely, on the one hand an apparent necessity for the number of conduction electrons in a metallic element to be of the order of one per atom and, on the other, an apparent necessity for the number to be negligibly small! The first requirement was necessary to explain the Hall coefficient data, while the second was the only possible deduction from the magnetic susceptibility data for metals (Wilson 1980). The application of Fermi–Dirac statistics not only gave the correct magnitude for the magnetic susceptibilities of metals, but also explained its remarkable temperature independence. It was Sommerfeld (1927, 1928), already familiar with the failings of the Drude–Lorentz theory, who saw in the work of Pauli (1927) and Fowler (1926) the key to the solution of the difficulties of the earlier theory.

Sommerfeld took over Lorentz's theory in its entirety, but now with the conduction electrons obeying the new Fermi–Dirac statistics instead of classical Maxwellian statistics. The results of Sommerfeld's theory were announced in September 1927 at the Volta congress (Kuhn *et al.* 1967, Hoddeson & Baym 1980), attended by leading physicists of the period. Soon afterwards, he published an outline of his theory in 1927, with a full treatment appearing in two parts early in 1928. The response to Sommerfeld's work was, by and large, extremely favourable. Some indication of this can be gauged by a letter from Bohr to Hume-Rothery in February, 1928, answering an enquiry regarding his 1911 thesis (Hoddeson & Baym 1980): '... Nowadays the old theories based on the classical mechanics can hardly make claim of actual physical interest.... Sommerfeld's work surely means a decisive step as regards the adequate quantum theoretical treatment of the metallic problem'. Despite its rousing success and appeal to a large proportion of the scientific community, Sommerfeld's theory was not without difficulties (Bloch 1980, Peierls 1980). First, it made no mention in general terms of the mutual (repulsive) interaction of conduction electrons. Secondly, as Bethe recalls (Hoddeson and Baym 1980), '... he didn't even care terribly much why the electrons were free, which I thought was a very important thing to know'.

It turned out that Sommerfeld had used just enough quantum mechanics to identify the electronic states and their energies in metals, but he had not utilized the full theoretical machinery already developed by Heisenberg, Schrödinger, Dirac and others, for atomic states (Hoddeson & Baym 1980). The transition to a complete *en principe oui* quantum mechanical theory of electrons in metallic elements was carried out by Bloch, whose thesis project, an investigation into the theory of metals, was suggested by Heisenberg. Bloch's thesis, and subsequent publication in 1928, laid the foundation of the quantum theory of electrons in solids and, with the work of Wilson (1931a,b), the basis for the essential difference between metals and nonmetals within the independent-electron approximation.

Metals versus nonmetals

Bloch (1928, 1929) found that the wavefunction for an electron in a perfect periodic lattice has the form

$$\psi_{\mathbf{k}} = \exp(i\mathbf{k}\,.\,\mathbf{r})U_{\mathbf{k}}(\mathbf{r}) \qquad (1)$$

where $\mathbf{r}$ denotes the electron coordinate, $\mathbf{k}$ is the crystal wavevector and $U_{\mathbf{k}}$ has the periodicity of the lattice. This startling result (Bloch 1980) arose from Bloch's realization that 'the solutions of the Schrödinger equation (for an electron in a periodic potential) differed from the de Broglie wave of a free particle only by a modulation with the period of the potential'. Bloch's paper was followed by a cluster of rapid developments in the theory of solids between 1928 and 1933. However, the story is far from complete. Bloch himself (1980) admits to an uneasy feeling that a model of independent electrons (that is, a model in which electron–electron interactions are suppressed) might represent a rather poor approximation to the actual situation. In addition, recognition of the fundamental qualitative difference between electrons in elemental metals and nonmetals was overlooked (Bloch 1980, see also Cottrell 1966).

As Wilson (1931) pointed out, Bloch had taken up the question of the mechanics of electrons in a metallic lattice and had shown that if the lattice is perfect an electron can travel quite freely through it. Wilson (1931a) also suggested that 'On this view all the electrons in a metal are free'. He proposed (Wilson 1931a,b) that the 'free' electrons in a solid would form open and closed shell configurations in much the same way as do electrons in an atom, and it is only in the former case that the electrons can be viewed as conduction electrons. It is worth outlining this important argument—due to Wilson (1931, 1980)—in greater depth since it represents the basis of the band-theory view of the periodic system of elements.

How does one explain the electronic origins of the difference between metals and nonmetals? Recall that all solutions of the Schrödinger equation for an electron subject to a periodic field of potential $V(x, y, z)$, namely

$$\nabla^2\psi + (2m/\hbar^2)(E - V)\psi = 0 \tag{2}$$

are of the form

$$\psi_\mathbf{k} = \exp(i\mathbf{k}\,.\,\mathbf{r})U_\mathbf{k}(x, y, z) \tag{3}$$

with $U_\mathbf{k}$ being periodic with the period of the potential. Therefore, equation (3) represents an electron moving with wavenumber **k** and *no* scattering. It would seem at first sight (Wilson 1931a,b) that in Bloch's theory '... *all* substances should have infinite conductivity at absolute zero temperature'. As Wilson (1980) recalls, '... Bloch's theory had in fact proved too much. Before his paper appeared it was difficult to explain the existence of metals. Afterwards, it was the existence of insulators that required explanation.'

This fundamental problem had been overlooked because it was generally taken for granted that the difference between metals and nonmetals was a quantitative and not a qualitative one. This concept was challenged by Wilson (1931a): 'It is not possible to maintain that the difference between good and bad conductors is one of degree only, the electrons in poor conductors being more tightly bound than in metals, and giving rise to a smaller current. There is an essential (qualitative) difference between a semiconductor, such as germanium, and a good conductor, such as silver, which must be accounted for by any theory ...'. He further suggested (Wilson 1931a,b) that the Bloch theory could be enormously simplified and made much more intuitive if one assumed that the electrons in crystalline solids, like valence electrons in single atoms, could form either open or closed shell configurations. In solids these open or closed shells overlap to form bands. A completely filled band cannot show a preponderance of electron waves in any direction, that is, there can be no electric current even in the presence of an electric field. An external electric field can only induce a current by exciting some electrons into higher energy states. In a completely filled zone, by virtue of the Pauli principle, there are obviously no directly accessible vacant states into which excitation of electrons can occur.

In contrast, an electric current flow can occur in partially filled bands in the presence of an electric field. Here the electron energy can be raised via the excitation of the electron by the external electric field into vacant and immediately accessible energy states.

It therefore followed on the Wilson scheme (1931, 1980) that an elemental solid with an odd valency *had to be a metal*, whereas elements with an even valency might produce either a metal or a nonmetal. Wilson's model seemed at first sight to predict that divalent elements such as calcium could be nonmetals. However, it was early realized (Wilson 1980) that a filled and an empty band could overlap to give metallic character. It is perhaps typical of the almost instantaneous acceptance of the 'new' quantum mechanics of the period that few doubted this view (see, also, the earlier comments by Bohr to Hume-Rothery). Moreover, the approximation in which each electron moved independently of all others was also fairly readily accepted.

Thus, the idea of partially filled energy bands imparting metallic character to solids had, according to Wilson (1931), universal applicability. However, even a cursory glance at the atomic constitution of the elements (Figure 1) reveals some disturbing features. For example, why is solid hydrogen a nonmetal while the rest of the Group IA elements are prototypical highly conducting metals? The Group VIIB elements are made up of atomic states with an incompletely filled *n*p shell (e.g. Cl^0, $1s^2 2s^2 3p^5$). In the solid state one would imagine a broadened p band with a positive hole which, on the basis of the arguments above, should lead to metallic character in solid chlorine. Anderson (1971) also notes this dichotomy in his work *Concepts in Solids*.

The first major indication of a possible inadequacy in the independent-electron band theory came in 1937. At a conference chaired by Mott on electrical conductivity processes in solids, de Boer & Verwey (1937) presented electrical conductivity data for a variety of transition metal oxides, including NiO, CoO, MnO, Fe_2O_3, Fe_3O_4, Mn_3O_4 and Co_3O_4. With the exception of Fe_3O_4, all of these compounds showed conductivities in the range 10^{-10} to 10^{-7} $(\Omega\,cm)^{-1}$, indicating they are nonmetals. de Boer and Verwey pointed out that electronic configurations for the monoxides, such as NiO($t_{2g}^6 e_g^2$); CoO($t_{2g}^5 e_g^2$); MnO($t_{2g}^3 e_g^2$), would lead to materials with partially filled energy bands which, on the independent-electron band theory view, should therefore be metallic. Obviously, with the exception of magnetite, all of these 'open shell' oxides are insulators. This is even more intriguing in view of the relatively small differences between the metal–metal distances ($d_{m\text{-}m}$) in the nonmetallic transition metal monoxides and the corresponding (highly conducting) elemental metals:

System	Co (metal)	CoO	Mn (metal)	MnO
$d_{m\text{-}m}/Å$	2·50	3·01	2·74	3·14

Furthermore, it now transpires that TiO, a superconducting oxide below 1·5 K, possesses itinerant 3d electrons and is clearly metallic. In the discussion following de Boer and Verwey's paper, Peierls (1937) built upon the Dutch authors' insightful comments to suggest that a rather drastic modification of the (then) present electron theory of metals was required in order to account for these observations. In particular, Peierls suggested that the coulomb repulsion between d electrons was ensuring electron localization at individual lattice sites, rather than itinerant electron mobility throughout the entire lattice. De Boer and Verwey (1937) also suggested that the nonmetallic status of these oxides is explained by the fact that any itinerant electrons would have a large probability of being localized in the presence of a large energy barrier to intersite electronic transport.

This identification of the breakdown of the conventional Bloch–Wilson view of electrons in solids lay dormant for more than a decade. As Mott (1980) recalls '... this problem worried me throughout the war period.' In a determined campaign beginning in 1949, Mott, however, as Brandow (1976, 1977) has put it '... urged that this problem be recognized for what it was—a fundamental challenge to the existing conceptions of solid-state theory.'

3.2. Beyond the independent-electron view

The qualitative difference between the properties of a low-density, non-metallic form of an element or compound (for example, the monovalent alkali elements at very low densities) and those of its high-density, metallic form are not easily describable in the Bloch–Wilson scheme (Mott 1961, Krumhansl 1965). Consider an array of (neutral) alkali atoms, e.g. Cs, at various densities. We expect for this case that a considerable overlap of the 6s valence electron wavefunctions occurs at high liquid or solid densities. If one follows the conventional Bloch–Wilson picture, then in the ground state we have a half filled 6s electron band, itinerant electrons and metallic status. Thus, the valence electrons in high-density metallic Cs are Bloch-like extended states (equation 1). Within the independent-electron approximation, the width of the 6s band (Δ) would be expected to decrease with decreasing density, but the low-lying excited states would, on this view, be infinitesimally close and metallic status would be predicted for *all* elemental densities (Krumhansl 1965). Mott (1961) pointed out, '... this is against common experience and, one might say, common sense.' From the experimental standpoint, the transition to a nonmetallic state as Cs is expanded is unquestionable (see, for example Hensel, 1984, Freyland & Hensel this volume, p. 93). Therefore, electron–electron interactions, which come into particular prominence at low densities, fundamentally modify the description. The electronic wavefunctions appropriate to the low-density regime reflect a situation in which electrons are localized, not itinerant (Hubbard 1964a, b, Mott, 1974). Thus the system must undergo a metal–nonmetal transition as a function of density on account of electron correlation. A formulation of this state of affairs is due to Hubbard (Hubbard 1964a,b, Mott 1974, Friedel 1984), who included the electron–electron correlation term for intra-atomic electron interactions (that is, for two electrons on the same atom), as embodied in the so-called Hubbard U parameter. At low densities, the coulomb intra-site repulsions dominate and the system is nonmetallic. At high elemental densities considerable orbital overlap occurs and intersite electron transfer can occur with relative ease. Within this model the criterion for a metal–nonmetal transition is

$$\Delta \simeq U \tag{4}$$

and detailed estimates give $\Delta/U = 1{\cdot}2$ as the critical condition for the metallization process (Hubbard 1963, 1964a,b, Caron & Kemeny 1971; Mott 1974; Ramakrishnan this volume, p. 23). The use of equation (4), together with certain approximations, leads to a simple expression for the critical density for the metal–nonmetal transition (Caron & Kemeny 1971, Berggren 1973, Mott 1974), namely

$$n_c^{1/3} a_H^* = 0{\cdot}2 \tag{5}$$

for $\Delta/U = 1{\cdot}0$, where a_H^* is the effective Bohr radius of the localized-electron state in the low-density regime.

In the atomic limit, the Mott–Hubbard correlation energy U is the difference between the ionization energy (I) and the electron affinity (A) of an atom; the energy ($I - A$) then represents the extra energy cost of putting two electrons on the same site. Friedel (1984) has recently used such 'atomic' values of U to suggest a natural demarcation of a large number of elements in the periodic system between metals and nonmetals (Figure 7). On this basis, the metallic elements arise from assemblies of atoms for which the parameter ($I - A$) is relatively small (about 7 eV and below), whereas atomic states for which U is in excess of about 8–9 eV tend to give rise to nonmetallic elements in the condensed phase. This is perhaps a good indication of the importance of atomic properties in dictating metallic versus nonmetallic status.

The influence of the 1937 conference on conduction processes was far reaching; for example, it served as an impetus for a paper published by Mott (1949), who once again addressed another fundamental aspect of the problem of metals versus insulators. Mott's model (1961) is based on the dielectric screening properties of an electron gas. Consider again the array of caesium atoms at various densities. For an isolated caesium atom, a valence electron and the caesium cation attract each other via a potential which at large distances is of the form:

$$V(r) = (-e^2/r) \tag{6}$$

Therefore, a non-conducting atomic state exists when the valence electron is essentially trapped by this long-range (attractive) coulomb field, and a series of

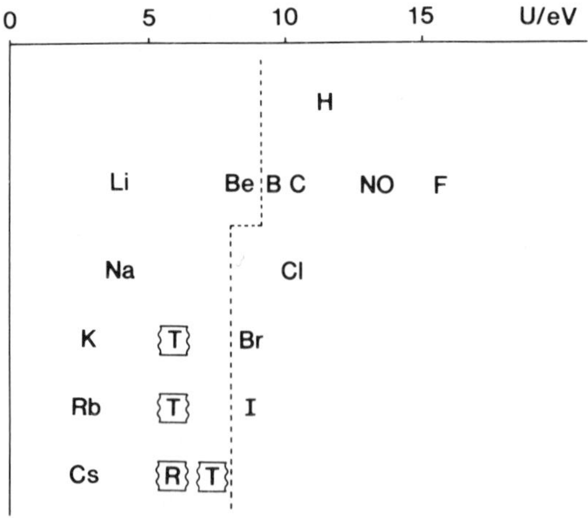

Figure 7. A possible demarcation of the periodic system of the elements in terms of the (atomic) Mott–Hubbard U. Here T represents the elements of the first, second and third transition series, after Friedel (1984).

bound, localized states will result. Under these circumstances, an ionization process requiring some 3·89 eV is necessary for electron excitation. This state of affairs cannot exist in a metal, as was first discussed by Mott and Littleton (1938). Here the attractive interaction between the two oppositely charged particles is screened, via the sea of itinerant electrons. According to the Thomas–Fermi prescription, this screened potential may be written in Yukawa form,

$$V(r) = (-e^2/r)\exp(-qr) \tag{7}$$

where q^{-1} is a characteristic screening length, given by

$$q^2 = [4me^2(3n/\pi)^{1/3}]/\hbar^2 \tag{8}$$

In a metal, the screening length is sufficiently small (usually of the order of an atomic spacing) that the screened potential is too weak to produce a bound state. Thus, metallic status is achieved by having a sufficiently high electron concentration so that the ionization energy of any particular 'atom' in the system then vanishes. Coming from the metallic side, the model predicts a metal–insulator transition (the Mott transition) when the screened potential in equation (7) starts to have localized states. Such states appear when

$$qa_{\mathrm{H}}^* \simeq 1$$

giving

$$n_c^{1/3} a_{\mathrm{H}}^* \simeq 0{\cdot}25$$

as the Mott criterion for metallic character. Although the Mott picture oversimplifies the details of the screening process, there is a wealth of experimental information supporting the basic physical approach (for a review see Edwards & Sienko 1983a,b).

For both the Mott and Hubbard schemes, we note the emergence of two characteristic parameters, the effective Bohr radius, a_{H}^{*}, and the density, n, which play a central role in consideration of metallic versus insulating behaviour; the former is a parameter characteristic of the atomic or localized electron state. In this sense, density and atomic parameters play an integral part in determining the metallic versus the nonmetallic states of matter.

4. The Herzfeld–Goldhammer theory of metallization

As mentioned in the introduction, the first semiquantitative attempts to explain the occurrence of metallic or insulating behaviour in the periodic system were made by Goldhammer in 1913, and by Herzfeld (1927) in a paper entitled 'On atomic properties which make an element a metal'. For many years the Herzfeld–Goldhammer theory lay dormant. Recently, however, there has been a renaissance of interest in the approach, and the so-called Herzfeld criterion has now been applied to many systems exhibiting induced metal–insulator transitions in order to predict the critical experimental conditions for metallization. This resurgence of interest is in part due to the underlying simplicity of the approach, and also to the fact that the Herzfeld criterion often yields predictions which agree well with experiments and with theoretical estimates based on detailed and often difficult quantum mechanical calculations.

The Herzfeld–Goldhammer approach is based on what is known as a polarization or dielectric catastrophe, and the idea is that the static dielectric constant, ε, of the condensed phase system (liquid or solid) should diverge to infinity as the insulator–metal transition density is approached from the insulating side. If, then, one can describe the dielectric constant of the system as the transition is approached, one can determine the insulator–metal transition density at which ε diverges. First, why might one expect such a divergence in the static dielectric constant? To appreciate this intuitively, recall from electrostatics that the macroscopic polarization, $\mathbf{P}(\mathbf{R})$, at a point $\mathbf{R}$ in the insulator, is related to the electric field in the medium—the Maxwell field, $\mathbf{E}(\mathbf{R})$—by

$$\mathbf{P}(\mathbf{R}) = \frac{1}{4\pi}(\varepsilon - 1)\mathbf{E}(\mathbf{R}) \tag{9}$$

(note of course, that the R-dependence vanishes for a homogeneously polarized dielectric.) In physical terms, however, the macroscopic polarization, $\mathbf{P}$, is simply the mean dipole moment per unit volume of the system, i.e.

$$\mathbf{P}(\mathbf{R}) = n\langle \mathbf{p}(\mathbf{R}) \rangle \tag{10}$$

where n is the (number) density of atoms and $\langle \mathbf{p}(\mathbf{R}) \rangle$ is the mean dipole moment of an atom immersed at point **R** in the medium. As the density of the system is increased towards the insulator–metal transition density, the valence electrons become progressively less tightly bound to their parent atoms; at the transition, the valence electrons become itinerant and the system acquires metallic status. As the transition is approached, therefore, one might anticipate a divergence in the mean dipole moment of an atom, as this is essentially a measure of the mean distance of a valence electron from its parent atom. The macroscopic polarization, and thus the dielectric constant, would in consequence diverge at the insulator–metal transition.

A simple description of the dielectric behaviour of a non-polar insulator is embodied in the familiar Clausius–Mossotti (CM) relation which, for a one-component system, takes the form

$$\frac{(\varepsilon - 1)}{(\varepsilon + 2)} = \frac{4\pi}{3} n\alpha_0 \tag{11}$$

where α_0 is the static scalar polarizability of an isolated atom. Before we discuss the CM equation in more detail, note that as $n \rightarrow n_c = 3/(4\pi\alpha_0)$, $\varepsilon \rightarrow \infty$ and a second-order dielectric catastrophe is thus predicted at the critical density n_c : this is the essence of the Herzfeld metallization criterion and may be related to the onset of electron itinerancy as described below.

The CM equation is commonly derived using continuum dielectric theory via the introduction of a Lorentz virtual cavity (Fröhlich 1958); here instead we outline a simple derivation, illustrated for a one-component isotropic system and based on classical statistical mechanics, in order to illustrate the microscopic nature of the approximations made in deriving the result. The macroscopic polarization is given formally by

$$\mathbf{P}(\mathbf{R}) = \left\langle \sum_{\mathrm{i}} \mathbf{p}_{\mathrm{i}}\, \delta(\mathbf{R}_{\mathrm{i}} - \mathbf{R}) \right\rangle \tag{12}$$

where $\mathbf{R}_{\mathrm{i}}$ denotes the centre of mass position of atom i and the sum is over all atoms in the system. $\mathbf{p}_{\mathrm{i}}$ denotes the (induced) dipole moment of atom i for a fixed centre-of-mass configuration of the system, and $\langle \ldots \rangle$ denotes the appropriate ensemble average. To proceed further we use the customary microscopic dipole approximation model for $\mathbf{p}_{\mathrm{i}}$ (see, for example, Böttcher 1973),

$$\mathbf{p}_{\mathrm{i}} = \alpha_0 \mathbf{e}(\mathbf{R}_{\mathrm{i}}) \tag{13}$$

where $\mathbf{e}(\mathbf{R}_i)$ is the electric field acting to polarize atom i for a fixed centre-of-mass configuration of the system. A given atom is polarized not only by the externally applied electrostatic field, $\mathbf{E}^0$, but also by the electric fields due to the induced dipoles of all other atoms. In consequence of this mutual polarization, $\mathbf{e}(\mathbf{R}_i)$ is a superposition of the external electric field plus the fields due to all other induced atomic dipoles, namely

$$\mathbf{e}(\mathbf{R}_i) = \mathbf{E}^0(\mathbf{R}_i) + \sum_{j(\neq i)} \mathbf{T}(\mathbf{R}_i, \mathbf{R}_j) \cdot \mathbf{p}_j \tag{14}$$

where $\mathbf{T}(\mathbf{R}) = \nabla\nabla R^{-1}$. Equations (13) and (14) constitute the familiar Yvon–Kirkwood model (Kirkwood 1936, Yvon 1937). To obtain the CM equation as a first approximation to the dielectric behaviour of the insulator, we make the approximation of neglecting fluctuations in $\mathbf{p}_i$ away from its mean value, $\langle \mathbf{p}(\mathbf{R}_i) \rangle$. We thus replace $\mathbf{p}_i$ in (12) by $\langle \mathbf{p}(\mathbf{R}_i) \rangle = \mathbf{P}(\mathbf{R}_i)/n$, and with the further assumption that the macroscopic polarization does not vary appreciably over distances comparable to atomic dimensions (Logan 1981), equations (12)–(14) yield:

$$\mathbf{P}(\mathbf{R}) = n\alpha_0 \mathbf{E}^{\mathrm{LO}}(\mathbf{R}) \tag{15}$$

$\mathbf{E}^{\mathrm{LO}}(\mathbf{R})$ is the familiar Lorentz local field, or effective field acting on an atom in the condensed phase, and is simply related to the Maxwell field, $\mathbf{E}(\mathbf{R})$, by (Böttcher 1973):

$$\mathbf{E}^{\mathrm{LO}}(\mathbf{R}) = \mathbf{E}(\mathbf{R}) + \frac{4\pi}{3}\mathbf{P}(\mathbf{R}) \tag{16}$$

Equations (15), (16) and (9) immediately yield the CM relation, (11). It is perhaps worth noting that the CM equation is essentially exact for a cubic arrangement of identical point-polarizable particles (Lorentz 1952). It should further be stressed that $\mathbf{E}^{\mathrm{LO}}$ is a valid effective field only for a medium consisting of atoms or non-dipolar molecules; for a system consisting of dipolar molecules there is an additional contribution to $\mathbf{P}$ arising from orientational polarization of the permanent electric dipoles (see, for example, Fröhlich 1958).

To relate explicitly the polarization catastrophe predicted by the CM equation with the onset of electron itinerancy, Herzfeld (1927) considered a simple mechanical model (the so-called Drude oscillator) used in the classical theory of electrons (Lorentz 1952, Rosenfeld 1951), in which the valence electrons of an atom are regarded as harmonically bound to their parent atoms. Consider for simplicity an isolated atom with a single valence electron of mass m, in an external electrostatic field, $\mathbf{E}^0$, and let $\mathbf{p} = e\mathbf{r}$ be the instantaneous dipole associated with the instantaneous displacement, $\mathbf{r}$, of the harmonically bound electron from its parent atom. The classical equation of motion for $\mathbf{p}$ is then

$$m\ddot{\mathbf{p}} + m\omega_0^2\mathbf{p} = e^2\mathbf{E}^0 \tag{17}$$

where ω_0 is the characteristic angular frequency of the valence electron. If now the atom is immersed in a medium, $\mathbf{E}^0$ must be replaced by the effective field acting to polarize the atom in the condensed phase, which we take to be the Lorentz local field, $\mathbf{E}^{LO} = \mathbf{E} + (4\pi/3)n\langle\mathbf{p}\rangle$, where $\langle\mathbf{p}\rangle$ is the mean dipole moment of an atom. The equation of motion for $\mathbf{p}$ is then

$$m\ddot{\mathbf{p}} = e^2\mathbf{E} - m\omega_0^2\mathbf{p} + (4\pi/3)ne^2\langle\mathbf{p}\rangle \tag{18}$$

where $(m/e)\ddot{\mathbf{p}} = \mathbf{F}$ is the instantaneous force acting on the valence electron. As the valence electron is bound to its parent atom, the mean force, $\langle\mathbf{F}\rangle$, acting on it must be zero, and (10) therefore yields

$$\langle\mathbf{p}\rangle = \frac{\alpha_0}{1 - (4\pi/3)n\alpha_0}\mathbf{E} \tag{19}$$

where $\alpha_0 = e^2/(m\omega_0^2)$ is the polarizability of the harmonically bound electron. Since $\mathbf{P} = n\langle\mathbf{p}\rangle$, equations (19) and (9) yield the CM equation, (11). From (19), $\langle\mathbf{p}\rangle \to \infty$ as $n \to n_c = 3/(4\pi\alpha_0)$, and the dielectric catastrophe occurring as $n \to n_c$ in the CM equation, (11), is associated with the onset of itinerancy, as manifest in the resultant mechanical instability of the harmonically bound system.

The CM equation is commonly cast in the form

$$(\varepsilon - 1)/(\varepsilon + 2) = R/V \tag{20}$$

where V is the molar volume. $R = 4\pi N_A\alpha_0/3$ (where N_A is Avagadro's number) is the long-wavelength limit of the *low density* molar refractivity, by virtue of the assumption that α_0 is the polarizability of an isolated atom: we return to this point below. The Herzfeld criterion, then, predicts that a material will show metallic behaviour if the molar volume is reduced below the value $4\pi N_A\alpha_0/3$. This simple criterion, requiring solely a knowledge of the free-atom polarizability, was used by Herzfeld to predict which elements in the condensed phase are metallic, which are insulating and which are borderline. As discussed by Edwards & Sienko (1983b), the overall features of the periodic classification of the elements conform rather well to this simple criterion. For example the group VIIA elements (the halogens) and the rare gases all have $R/V < 1$ under ambient conditions and nonmetallic behaviour is predicted, and observed, for these elements. Conversely, the group IA elements (alkali metals) have $R/V > 1$ and metallic character is predicted for all the members. In contrast the elements of groups IIIA–VIA of the periodic table effectively straddle the dividing line between localized and itinerant electron behaviour.

As representative examples of metallic and insulating behaviour, we focus attention here on the alkali metals and the rare gases. As one expects physically, the free-atom polarizabilities increase progressively as we descend either of the two groups. Of particular importance, however, is the relative magnitude of the polarizabilities of counterpart atoms in the two groups; the polarizability of an alkali metal atom with atomic number $Z+1$ is between one and two orders of magnitude larger than its rare gas counterpart with atomic number Z. For example, α_0 for Li(Cs) is 24 Å^3 (55 Å^3), whereas that for He (Xe) is 0·205 Å^3 (4·05 Å^3) (Teachout & Pack 1971). The relative behaviour of α_0 as we descend the two groups is illustrated schematically in Figure 8, and in Figure 9 we plot the resultant R/V values for the Group IA and 0 elements. A wide variety of experimental data is available for the molar volumes of the elements under

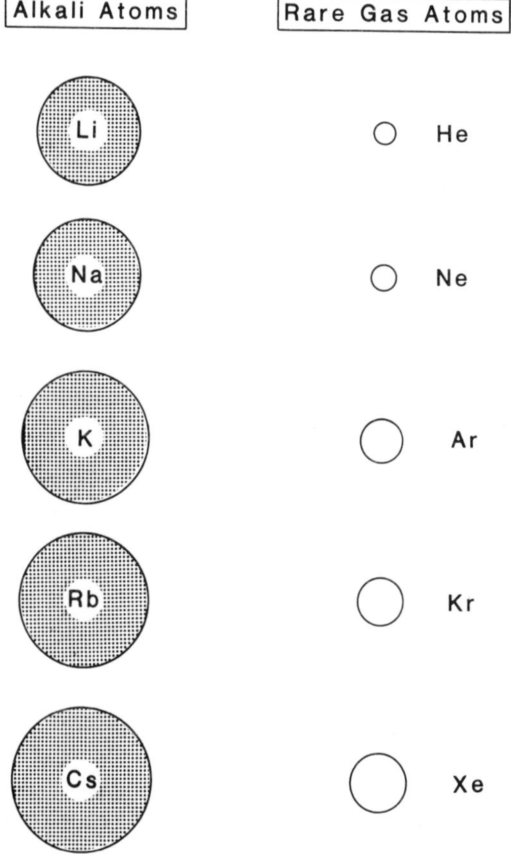

Figure 8. A schematic representation of the free-atom polarizabilities of the alkali and rare gas atoms.

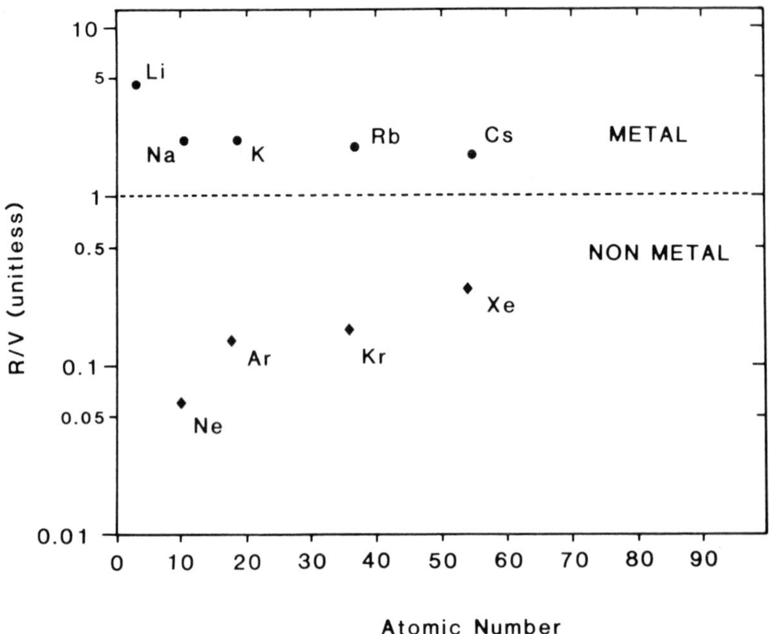

Figure 9. The ratio (R/V) for the elements of Group 1A and Group 0 of the periodic system of the elements. Here R is the low-density molar refractivity and V is the molar volume.

ambient conditions, and these are summarized in Figure 10 for the entire periodic table (Greenwood and Earnshaw 1984). As is seen from Figure 9, the molar volumes of the alkali metals are somewhat larger than their rare gas counterparts, but the ratio R/V is nonetheless always considerably greater than unity for the alkali metals and always substantially less than unity for the rare gases.

The Herzfeld criterion has now been applied to a wide variety of systems exhibiting induced metal–nonmetal transitions in order to estimate the critical experimental conditions for the onset of metallic character. For example, Ross (1972) has successfully employed the criterion to predict the required volume reduction for the metallization of elemental and molecular hydrogen, the solid rare gases and the alkali halides. Berggren (1974) has similarly used the metallization criterion to predict n_c for n-type Si and Ge, electron–hole droplets, metal atoms in rare gas matrices and the supercritical fluid alkali metals. There has also recently been considerable interest in the direct experimental detection of an impending polarization or dielectric catastrophe, in both single- and multicomponent systems which undergo a metal–insulator transition (Castner 1980a,b, Ortuno 1980, Capizzi *et al.* 1980, Paalanen *et al.* 1983, Freyland *et al.* 1983, Thompson 1984). Capacitance and other measurements on the Si:P system at various donor concentrations reveal the onset of a polarization catastrophe as n_c is approached from the insulating side. Similar results have also been reported

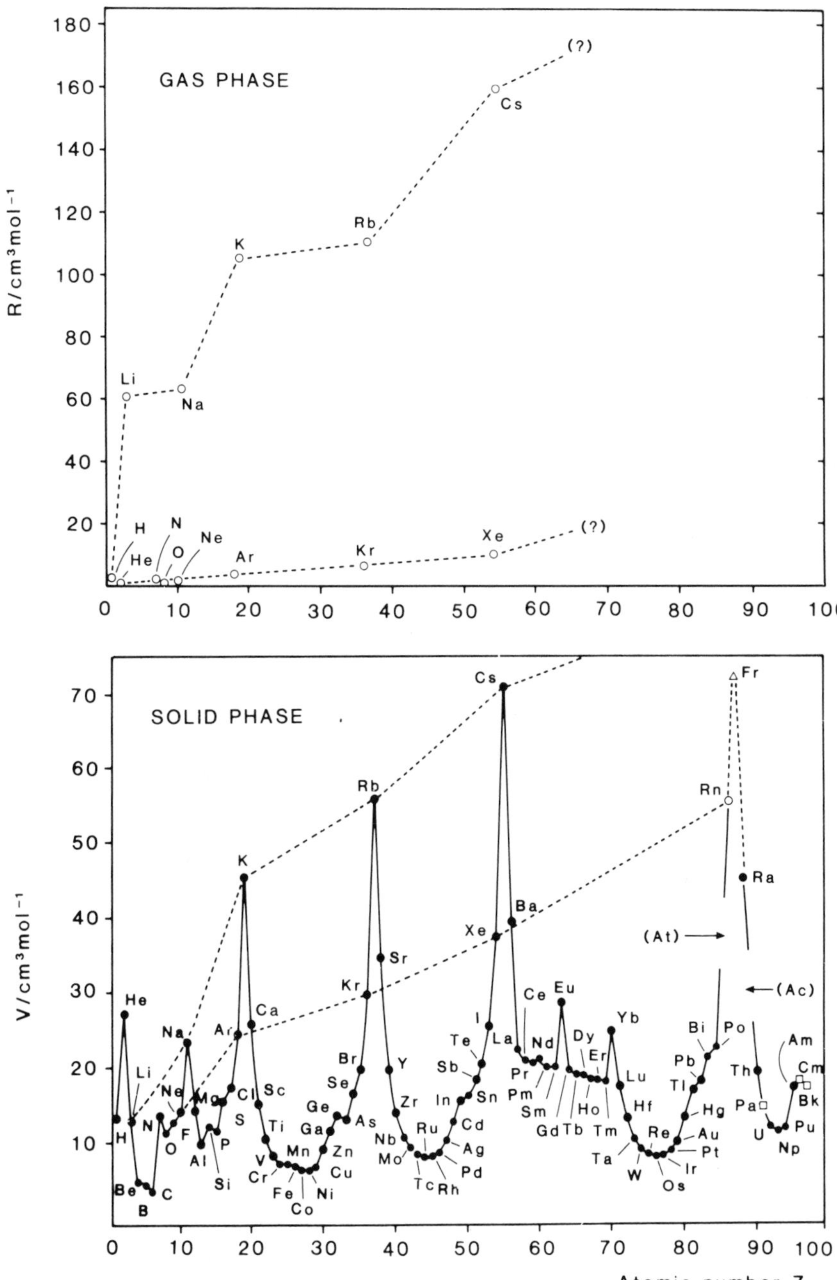

Figure 10. The molar refractivities (R, top portion of figure) and molar volumes (V, bottom) for a wide variety of elements within the periodic system. The molar volumes are appropriate to ambient conditions. See Figure 6 for the earlier atomic volume curve of Meyer.

for amorphous Si:Au alloys, n-type Ge, sodium–ammonia solutions, molten K–KCl mixtures and Hg–Xe thin films (for detailed references see Edwards & Sienko 1982, Thompson 1984).

Although the Herzfeld criterion is extremely simple, and is arrived at by classical arguments, it often yields predictions which are in quantitative accord with experiments, as well as with estimates derived from detailed quantum mechanical calculations. For example, the reported M–NM transition densities of He, Ar and Xe (experimental and/or calculated), and of TlI, I_2, NaCl, KCl, KBr and CsI (experimental), and those calculated via the Herzfeld criterion generally differ by at most 15% (Vaišnys & Žmuidzinas 1978, Young *et al.* 1981, Aidun *et al.* 1984). Similarly, Ross (1972) used the criterion to predict a metallization onset in condensed hydrogen at a density of 1·02 g cm^{-3}; experimental work by Hawke *et al.* (1978) yields a density of 1·06 g cm^{-3}. Of course for many systems, indeed a majority, the Herzfeld criterion is not quantitatively accurate. For example, the predicted critical densities for metallization of the fluid alkali metals are typically in disagreement with experiment by a factor of 2–3 (Berggren 1974, Logan and Edwards 1985), while for n-type Si the discrepancy is larger (Castner 1980a). Although even order-of-magnitude disagreement between theory and experiment is not uncommon in condensed matter chemistry and physics, these examples serve to indicate that, despite its appealing simplicity, the Herzfeld criterion is a substantial over-simplification of 'reality'. We conclude this section, therefore, by discussing further the limits of applicability of the criterion.

A commonly advanced objection to use of the CM equation as a means of estimating the critical density for metallization, is that 'such a critical value n_c $(=3/4\pi\alpha_0)$ is never reached' (Böttcher 1973). Consider, however, the alkali metals. Under normal (ambient) conditions, elemental densities range from $4{\cdot}6\times10^{22}$ cm^{-3} in the case of Li to $8{\cdot}61\times10^{21}$ cm^{-3} for Cs, values which in all cases are in excess of $n_c=3/4\pi\alpha_0$, which ranges from $9{\cdot}9\times10^{21}$ cm^{-3} in the case of Li to $4{\cdot}3\times10^{21}$ cm^{-3} for Cs. A weaker form of the above criticism is based on the observation that the predicted n_c is often in excess of the maximum density which is physically permitted if the constituent atoms are regarded as hard spheres of diameter σ. For the hard-sphere system we can define a reduced (dimensionless) density $n^*=n\sigma^3$. Due to the hard-core condition the density is limited by its maximum value at close packing, n_M^*, given by $n_M^*=\sqrt{2}\,(n_M^*=3\sqrt{3}/4)$ in the case of fcc (bcc) close packing and by $n_M^*=1$ for simple cubic packing. From (11), the Herzfeld criterion predicts the onset of metallization at a reduced critical density n_c^* such that $4\pi n_c^*\alpha_0^*/3=1$, where $\alpha_0^*=\alpha_0/\sigma^3$ is the reduced atomic polarizability. Using the appropriate values for α_0 and σ (Teachout & Pack 1971, Hirschfelder *et al.* 1954), the Herzfeld criterion applied to the rare gases predicts that n_c^* ranges from $\sim$4 for Xe to $\sim$10 for Ne, values clearly far in excess of n_M^* for any of the close packed lattices. In contrast, estimates of n_c^* for the alkali metals range from $\sim$0·6 to $\sim$0·8, values which are lower than n_M^* for any of the cubic close packed lattices, and which essentially reflect the large polarizabilities of the alkali atoms. Further, even in the somewhat extreme case of the rare gases, recall that the reported

M–NM transition densities for He, Ar and Xe agree to within 15% with those calculated from the Herzfeld criterion. This is contrary to what one would expect from the above hard-sphere arguments, and in part reflects the obvious fact that the repulsive part of a real intermolecular potential is always 'softer' than the characterization embodied in a hard core.

A further, justified criticism of the simple Herzfeld criterion (Bradley 1966, Samara 1975, Castner 1980a,b, Holz & Bennemann 1976, Logan & Edwards 1985) is the implicit assumption that the polarizability of an atom in the condensed phase is equal to its gas phase value, α_0. For most simple gases, up to moderate pressures, the CM equation does show good agreement with experiment; indeed, for the rare gases the maximum deviation from pure CM behaviour is rarely more than a few percent throughout the entire fluid density regime. In general, however, the assumption of constant polarizability will not be tenable, as the internal electronic degrees of freedom of an atom may be substantially modified by interatomic interactions in the condensed phase. Thus α_0 in the CM equation, (11), should be replaced by an effective polarizability, A, which is in general a function of density and temperature.

In broad terms, two essentially competing physical effects may be identified which cause A to differ from its gas phase value α_0. First, positive deviations from pure CM behaviour are observed almost ubiquitously in the dilute limit as n is increased from zero: this is true of most simple non-polar gases (Böttcher 1973), and also for more complex systems such as n-type Si, where n corresponds to the donor impurity density (Castner *et al.* 1975). The principal intermolecular interactions which give rise to these positive deviations, and thus cause A to be larger than the isolated atom α_0, are the long-range dipole–induced dipole interactions embodied in the classical Yvon–Kirkwood model described earlier. Further, the Yvon–Kirkwood model, which incorporates solely these long-ranged interactions, necessarily gives the simple CM relation, (11), as a lower bound on ε (Prager *et al.* 1970). A second effect contributing to the effective polarizability of an atom in the condensed phase may be distinguished at higher densities where overlap effects become important, and which, over a given density range, may cause A to decrease with increasing density. This is to be expected intuitively: the ability of an atom to distort in an electric field, and thus its polarizability, may be expected to diminish as the space available for distortion (that is, the volume) becomes smaller. Samara (1975) has elegantly demonstrated the importance of this contribution by analysis of experimental data pertaining to a variety of cubic crystal systems, and its relevance has long been appreciated in attempts to explain the dielectric behaviour of the rare gases (Mazur 1958, and references therein).

For a given system, then, two essentially competing influences contribute to the effective polarizability, A. At relatively high densities, where both effects are important, the balance between them may be delicate: this may in part account for the apparent quantitative success of the simple Herzfeld criterion in predicting metallization densities for the rare gases.

Several papers have been devoted directly to means whereby the simple

Herzfeld criterion may be extended (Holz & Bennemann 1976, Castner 1980a,b, Logan & Edwards 1985). The subject is, however, still in relative infancy, and we shall not comment on it further here.

Despite the caveats which must always be borne in mind when applying the criterion, it is clear that the simple Herzfeld–Goldhammer approach to metallization is a valuable one which has qualitative, and occasionally even quantitative, validity. Future theoretical work will undoubtedly lead to substantial refinement of the basic model, but the criterion in its simplest form is likely to survive, at least as a qualitative source of guidance for the design of experiments on metal–nonmetal transitions.

References

Aidun, J., Bukowinski, M.S.T. and Ross, M., 1984, *Phys. Rev.* B **29**, 2611.
Anderson, P.W., 1963, *Concepts in Solids* (New York: W. A. Benjamin).
Ashcroft, N.W. and Mermin, N., 1976, *Solid State Physics* (New York: Rinehart and Winston).
Berggren, K.-F., 1973, *Phil. Mag.*, **27**, 1027.
Berggren, K.-F., 1974, *J. Chem. Phys.*, **60**, 3399.
Bethe, H.A., 1980, *Proc. R. Soc.* A, **371**, 49.
Bloch, F., 1928, *Z. Phys.*, **52**, 555.
Bloch, F., 1929, *Z. Phys.*, **57**, 545.
Bloch, F., 1980, *Proc. R. Soc.* A, **371**, 24.
de Boer, J.H. and Verwey, E.J.W., 1937, *Proc. R. Soc.*, **49** (extra), 59.
Bohr, N., 1911, 'Studier over Metallernes Elektronteori', *Ph.D. thesis* (Thaning and Appel, Copenhagen); reprinted with English translation in *Niels Bohr, Collected Works*, Vol. 1, edited by L. Rosenfeld and J.R. Neilsen (Amsterdam: North Holland).
Böttcher, C.J.F., 1973, *Theory of Electric Polarization*, (Amsterdam: Elsevier).
Bradley, R.S., 1966, *Phys. Stat. Solidi*, **15**, K35.
Brandow, B.H., 1976, *Int. J. Quant. Chem. Symp. No. 10*, p. 417.
Brandow, B.H., 1977, *Adv. Phys.*, **26**, 651.
Capizzi, M., Thomas, G.A., De Rosa, F., Bhatt, R.N. and Rice, T.M., 1980, *Phys. Rev. Lett.*, **44**, 1019.
Caron, L.G. and Kemeny, G., 1971, *Phys. Rev.* B, **3**, 3007.
Castner, T.G., 1980a, *Phil. Mag.* B, **42**, 873.
Castner, T.G., 1980b, *Phys. Rev.* B, **21**, 3523.
Castner, T.G., Lee, N.K., Cieloszyk, G.S. and Salinger, G.L., 1975, *Phys. Rev. Lett.*, **34**, 1627.
Cohen, M.H., 1968, *Rev. Mod. Phys.*, **40**, 839.
Cottrell, A.H., 1960, *Contemp. Phys.*, **1**, 417.
Davy, H., 1808, *Phil. Trans. R. Soc.*, **98**, 1.
Drude, P., 1900, *Ann. Phys., Lpz.*, **1**, 566; ibid, **3**, 369.
Edwards, P.P. and Sienko, M.J., 1982, *Acc. Chem. Res.*, **15**, 87.
Edwards, P.P. and Sienko, M.J., 1983a, *Int. Rev. Phys. Chem.*, **3**, 83.
Edwards, P.P. and Sienko, M.J., 1983b, *J. Chem. Ed.*, **60**, 691.
Erman, P. and Simon, P.L., 1808, *Gilbert's Annalen*, **28**, 131.
Fowler, R.H., 1926, *Mon. Not. R. Astr. Soc.*, **87**, 114.
Freyland, W., 1981, *Comm. Solid St. Phys.*, **10**, 1.
Friedel, J., 1984, in *Physics and Chemistry of Electrons and Ions in Condensed Matter*, edited by J.V. Acrivos, N.F. Mott and A.D. Yoffe (Dordrecht: Reidel), p. 45.

Freyland, W., Garbade, K. and Pfeiffer, E., 1983, *Phys. Rev. Lett.*, **51**, 2144.
Fröhlich, H., 1958, *Theory of Dielectrics* (Oxford: Oxford University Press).
Goldhammer, D.A., 1913, *Dispersion und Absorption des Lichtes* (Leipzig: Teubner).
Greenwood, N.N. and Earnshaw, A., 1984, *Chemistry of the Elements* (Oxford: Pergamon Press).
Hawke, P.S., Burgess, T.J., Duerre, D.E., Huebel, J.G., Keeler, R.N., Klapper, H. and Wallace, W.C., 1978, *Phys. Rev. Lett.*, **41**, 994.
Hensel, F., 1984, in *Physics and Chemistry of Electrons and Ions in Condensed Matter*, edited by J.V. Acrivos, N.F. Mott and A.D. Yoffe (Dordrecht: Reidel), p. 401.
Herzfeld, K.F., 1927, *Phys. Rev.*, **29**, 701.
Herzfeld, K.F., 1966, *J. Chem. Phys.*, **44**, 429.
Hirschfelder, J.O., Curtiss, C.F. and Bird, R.B., 1954, *Molecular Theory of Gases and Liquids* (New York: Wiley).
Hoddeson, L.H. and Baym, G., 1980, *Proc. R. Soc.* **A**, **371**, 8.
Holz, G.A. and Bennemann, K.H., 1976, *Phys. Rev. Lett.*, **37**, 1507.
Hubbard, J., 1963, *Proc. R. Soc.* A, **276**, 238.
Hubbard, J., 1964a, *Proc. R. Soc.* A, **277**, 237.
Hubbard, J., 1964b, *Proc. R. Soc.* A, **281**, 401.
Hume-Rothery, W. and Raynor, G.V., 1956, *The Structure of Metals and Alloys*, Institute of Metals, Monograph and Report Series No. 1 (London: The Institute of Metals).
Hurd, C.M., 1975, *Electrons in Metals: An Introduction to Modern Topics* (New York: Wiley).
Kirkwood, J.G., 1936, *J. Chem. Phys.*, **4**, 592.
Klapper, H. and Wallace, W.C., 1978, *Phys. Rev. Lett.*, **41**, 994.
Kittel, C., 1976, *Introduction to Solid State Physics* (New York: Wiley).
Kohn, W., 1957, *Solid St. Phys.*, **5**, 258.
Krumhansl, J.A., 1965, in *Physics of Solids at High Pressures*, edited by C.T. Tomizuka and R.M. Emrick (New York: Academic), p. 425.
Kuhn, T.S., 1967, *Sources of History for Quantum Physics: an Inventory and Report* (Philadelphia: American Philosophical Society), cited in Hoddeson and Baym (1980).
Landau, L. and Zeldovitch, J., 1943, *Acta Physicochimica URSS*, **18**, 194.
Lavoisier, A., 1789, *Traité élémentaire de Chimie* (Paris: Cuchet).
Logan, D.E., 1981, *Molec. Phys.*, **44**, 1271.
Logan, D.E. and Edwards, P.P., 1985, submitted for publication.
Lorentz, H.A., 1952, *Theory of Electrons* (New York: Dover).
Mazur, P., 1958, *Adv. Chem. Phys.*, **1**, 309.
Mazurs, E.G., 1974, *Graphic Representations of the Periodic System During One Hundred Years* (Alabama: The University of Alabama Press).
Meaden, G.T., 1966, *Electrical Resistance of Metals* (London: Heywood).
Mendeleéff, D., 1869a, *Zh. Russk. Khim. Obshch.*, **1**, 60; cited by B.N. Menshutkin, 1934, *Nature*, **133**, 946.
Mendeleéff, D., 1869b, *Z. Chem.*, **12** [2] 5, 405; cited by Mazurs, 1974.
Mendeleéff, D., 1889, *J. Chem. Soc.*, **LV**, 634.
Mendeleéff, D., 1905, *The Principles of Chemistry*, 3rd English edn., Vol. II (London: Longmans, Green & Co).
Meyer, L., 1870, *Ann. Chem., Justus Liebigs*, Suppl. 7, 354.
de Morveau, A., 1782, *J. phys., hist. nat. et arts*, **19**, 370; cited by Mazurs, 1974.
de Morveau, Lavoisier, Berthollet and de Fourcroy, 1787, *Méthode de nomenclature chimique* (Paris: Cuchet, Hassenfratz et Adet); cited by Mazurs, 1974.
Mott, N.F., 1937, *Proc. R. Soc.*, **49** (extra), 72.
Mott, N.F., 1949, *Proc. Phys. Soc.* A, **62**, 416.
Mott, N.F., 1961, *Phil. Mag.*, **6**, 287.
Mott, N.F., 1974, *Metal–Insulator Transitions* (London: Taylor & Francis).

Mott, N.F., 1980, *Proc. R. Soc.* A, **371**, 56.
Mott, N.F., 1984, *Rep. Prog. Phys.*, **47**, 909.
Mott, N.F. and Littleton, M.J., 1938, *Trans. Faraday Soc.*, **34**, 485.
Ortuno, M., 1980, *J. Phys. C: Solid St. Phys.*, **13**, 6279.
Pauli, W., 1925, *Z. Phys.*, **31**, 765.
Pauli, W., 1927, *Z. Phys.*, **41**, 81.
Peierls, R.E., 1937, *Proc. R. Soc.*, **49** (extra), 72.
Peierls, R.E., 1980, *Proc. R. Soc.* A, **371**, 28.
Prager, S., Kunken, W. and Frisch, H.L., 1970, *J. Chem. Phys.*, **52**, 4925.
Riecke, E., 1898, *Ann. Phys. Lpz.*, (3), **66**, 353.
Rosenbaum, T.F., Milligan, R.F., Paalanen, M.A., Thomas, G.A., Bhatt, R.N. and Lin, W., 1983, *Phys. Rev.* B, **27**, 7509.
Rosenfeld, L., 1951, *Theory of Electrons* (Amsterdam: North Holland).
Ross, M., 1972, *J. Chem. Phys.*, **56**, 4651.
Ross, M. and McMahan, 1981, in *The Physics of Solids Under High Pressure*, edited by J.S. Schilling and R.N. Shelton (Amsterdam: North Holland).
Samara, G.A., 1975, *Chem. Phys. Lett.*, **33**, 319.
Sommerfeld, A., 1927, *Naturwissenschaften*, **15**, 825.
Sommerfeld, A., 1928, *Z. Phys.*, **47**, 1.
van Spronsen, J.W., 1969, *The Periodic System of Chemical Elements. A History of the First Hundred Years* (Amsterdam: Elsevier).
Teachout, R.R. and Pack, R.T., 1971, *Atomic Data*, **3**, 195.
Thompson, J.C., 1984, *J. Solid St. Chem.*, **54**, 308.
Vaišnys, J. and Žmuidzinas. J.S., 1978, *Appl. Phys. Lett.*, **32**, 152.
Weber, W., 1875, *Ann. Phys. Lpz.* (2), **156**, 1.
Wilson, A. H., 1931a, *Proc. R. Soc.* A, **133**, 458.
Wilson, A. H., 1931b, *Proc. R. Soc.* A, **134**, 277.
Wilson, A. H., 1954, *The Theory of Metals* (Cambridge University Press).
Wilson, A.H., 1980, *Proc. R. Soc.* A, **371**, 39.
Young, D.A., McMahan, A.K. and Ross, M., 1981, *Phys. Rev.* B, **24**, 5119.
Yvon, J., 1937, *Actual. Scient. et Ind.*, Nos. 542, 543 (Paris: Hermann & Cie).

CHAPTER 4

the metal–nonmetal transition in expanded metals

W. Freyland and F. Hensel
Universität Marburg, West Germany

1. Introduction

The behaviour of fluid metals near the liquid–vapour critical point has attracted considerable experimental as well as theoretical attention for many decades. This interest originates in part from the fact that fluid metals expanded to critical conditions exhibit remarkable changes in their electronic properties. A continuous metal–nonmetal (M–NM) transition occurs with decreasing density of the liquid or increasing density of the vapour.

A comprehensive theoretical discussion of this type of M–NM transition was first given by Mott (1961). A common effect of increasing density is to broaden the bands of allowed energy and to decrease the gaps of forbidden energy irrespective of whether the material is crystalline or fluid. In the latter case, however, the density of states is expected to tail into the gap due to the lack of long-range order. The broadening of the bands can ultimately lead to a continuous transformation from an insulating to a metallic state resulting from an overlap of the highest occupied and the first excited band. Besides this band-crossing transition, a second kind of density induced M–NM transition has been discussed by Mott (1956, 1961) for systems of one-electron atoms in which the coulomb interaction between electrons and holes is important. However, as was first shown by Hubbard (1964), the intra-atomic coulomb repulsion alone leads to a transition. For large interatomic distances an antiferromagnetic insulator is predicted and the 'Mott transition' can be described by the crossing of two Hubbard bands (see, for example, Mott 1974). For disordered systems including fluids another form of M–NM transition can occur according to whether the Fermi energy of a degenerate electron gas lies above or below a mobility edge. This Anderson transition has been extensively discussed in numerous papers by Mott (e.g., Mott 1974). Expanded fluid metals offer the rare possibility for an experimental verification of these transitions in one-component systems. Their advantage is that the complexity of the concentration-induced M–NM transition in binary systems can be avoided. For the latter it is necessary to allow for the

Table 1. Experimental critical-point data for fluid metals.

Metal	T_c/K	p_c/bar	ρ_c/g cm^{-3}	References
Hg	1750	1671	5·75	Götzlaff 1983
Cs	1925	92	0·37	Jüngst 1985
Rb	2020	125	0·29	Knuth 1985
Mo	14300	5700	2·9	Seydel & Fucke 1978

influence of the dominant solvent on the behaviour of the minority component which exhibits the transition.

However, expanded metals suffer from some limitations for the investigation of the M–NM transition. The very high temperatures and pressures necessary for the continuous expansion of the liquid metal up to the critical point (see Table 1) will smear out the M–NM transition by considerable thermal excitation of electrons. In addition they are technologically attainable only for a restricted number of metals. For most metals the critical region lies at higher temperatures and pressures than are accessible to conventional static experimental techniques. Only transient experiments such as shock waves or exploding wires reach temperatures and pressures high enough to explore, for example, the critical point of molybdenum (Seydel & Fucke 1978). Static experiments have been carried out only for Hg, Cs, Rb and K, but even for these metals with the lowest critical temperatures not all measurements have been in agreement. This is mainly due to two causes. The severe problems of temperature and pressure measurement and control together with the highly reactive nature of fluid metals have limited the accuracy of these experiments. The second serious difficulty is that the well established rules and methods for the determination of critical-point properties of normal insulating fluids cannot be applied to fluid metals just as they are. The difference when dealing with the critical-point phase behaviour of metals lies in the existence of competing interactions. Regardless of the way in which the effective interactions in a metal are described, the M–NM transition implies that the description must change with density: from metallic cohesion in the dense liquid to van der Waals' interaction between neutral atoms and molecules in the dilute vapour. In contrast to normal fluids, such as argon, for which to a first approximation the behaviour can be described by reference to a single pair-potential at all densities, in fluid metals the interactions are manifold and complicated. It is not surprising, therefore, that the liquid–vapour phase behaviour of metals differs strongly from that of normal fluids as shown by the reduced diagram in Figure 1.

During the past two decades various reviews on expanded metals have been published which emphasize different topics (Alekseev & Iakubov 1983, Cusack 1978, Endo 1982, Freyland 1981, Hensel 1980, 1984, Yonezawa & Ogawa 1982). Therefore we may concentrate here on a review of more recent experimental results and give a brief account of some new theoretical calculations on the

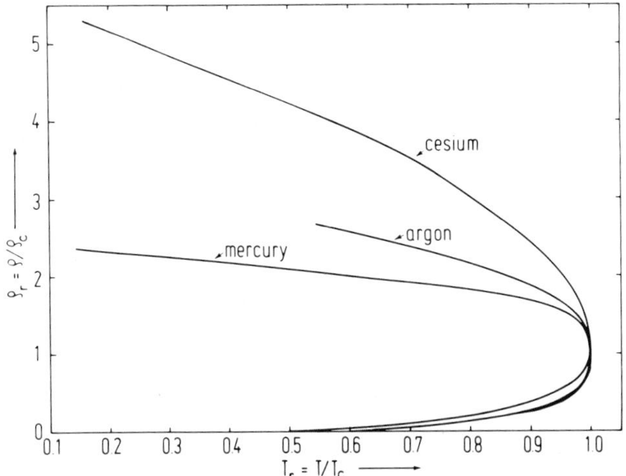

Figure 1. Density variation along the liquid–vapour coexistence curve for metals in comparison with argon; plotted are reduced densities ρ/ρ_c versus reduced temperature T/T_c.

M–NM transition problem of expanded metals. In a few places we include results from older work for reasons of clarity, so that the reader can follow the development of the subject.

2. Physico-chemical properties of expanded fluid metals

2.1. Electronic transport

Usually, the M–NM transition is marked by clear changes in the electronic transport properties from nearly-free-electron behaviour to thermally activated processes. This is the case for crystalline systems and some disordered solid systems—for reviews see the Proceedings of the Scottish University Summer School of Physics (1978) and Mott (1974). In fluid metals the transformation in the electronic structure is smeared out by the relatively high thermal energy kT dictated by the high critical temperatures. So no abrupt variations in the electron transport can be expected for a continuous expansion. Only a detailed investigation of both temperature and density dependence gives some information on the transformation of the transport characteristics. There is general agreement among experimenters that this requirement can best be achieved by simultaneous measurements of transport and pVT data, so that errors resulting from separate determinations of these quantities are minimized. In the following we present some recent results on the electrical conductivity of expanded Hg and Cs which have been obtained by this technique. Measurements of the transport properties of expanded metals have been reviewed several times—see the reviews cited in the introduction—so that we can concentrate here on a brief summary of the main characteristics derived from conductivity, thermopower, and Hall mobility measurements.

For fluid mercury, measurements of the d.c. conductivity, σ, and absolute thermoelectric power, S, have been reproduced independently by different groups, so that here data with sufficient accuracy have been established which allow a detailed discussion of temperature and density effects during the M–NM transition. Figure 2 represents a selection of such results as a function of density ρ

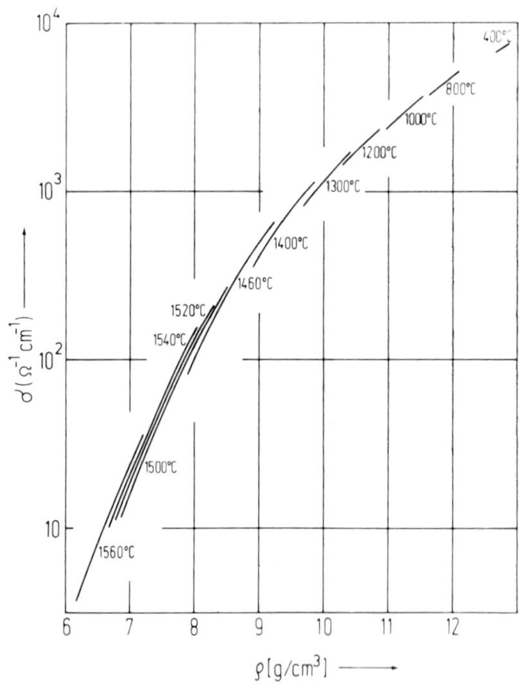

Figure 2. Density dependence of the electrical conductivity σ of expanded Hg for different sub- and supercritical temperatures (Schönherr *et al.* 1979).

at different constant sub- and supercritical temperatures taken from the work of Schönherr *et al.* (1979). During an expansion between the triple point and the critical point three density regimes and corresponding transport characteristics can be distinguished. Above about 11 g cm^{-3} ($\approx 2\rho_c$) or $\sigma \gtrsim 2\times 10^3$ $(\Omega\,\text{cm})^{-1}$, both the temperature and volume coefficients of σ, $(\partial \ln \sigma/\partial T)_V$ and $(\partial \ln \sigma/\partial \ln V)_T$, show only a weak density dependence (see, for example, Schmutzler & Hensel 1972), and the Hall constant equals the free electron value (Even & Jortner 1972). Around $2\rho_c$ the electron mean-free-path becomes comparable with the mean interatomic distance, that is, the Ioffe–Regel limit is reached. With further expansion, clear deviations from the nearly-free-electron model occur.

Between 11 and 9 g cm^{-3} the Hall coefficient strongly increases by a factor of 3 above its free-electron value (Even & Jortner 1972). In this range where σ drops by an order of magnitude, the transport behaviour is consistent with the predictions of the random-phase model (see, for example, Cusack 1982). As was first suggested by Mott (1966) a pseudo-gap may form, that is, a minimum in the density of states $N(E)$, which precedes the separation of the 6s valence and the 6p conduction band in expanded Hg. If the gap is deep enough, states may become localized in the Anderson sense (Mott 1966). A third transport regime sets in below 9 g cm^{-3} (for comparison, the critical density ρ_c is 5·75 g cm^{-3}). The temperature dependence of σ and the correlation of σ and the absolute thermoelectric power S show semiconducting characteristics:

$$\sigma = \sigma_0 \exp(-\Delta E/kT)$$

$$S = \frac{k}{e}\left(\frac{\Delta E}{kT} + \text{constant}\right)$$

The corresponding experimental evidence is plotted in Figure 3. In order to allow for the temperature dependence of ΔE this has been taken as $\Delta E = E_0 - \gamma T$ (see Schönherr *et al.* 1979). The σ_0 values obtained from this evaluation are of order 10^2 $(\Omega\,\text{cm})^{-1}$ and are comparable with those found for transport at a mobility edge $(E_c - E_F)$. In summary, the transport properties of expanded Hg strongly indicate that a transition from metallic to semiconductor-like behaviour sets in near 9 g cm^{-3}, well above the critical density ρ_c. Approaching the critical-point region, strong anomalies have been reported by several authors for the thermoelectric power S (Duckers & Ross 1972, Alekseev *et al.* 1980a and Yao & Endo 1982). Near the critical isochor where σ is of order 1 $(\Omega\,\text{cm})^{-1}$, the thermopower seems to change from negative to large positive values. A demonstration of this effect is given in Figure 4, taken from the work of Yao & Endo (1982). This change in sign of S possibly may be related to a transition to plasma states in the supercritical region (Alekseev *et al.* 1980a).

In fluid alkalis metallic transport behaviour may prevail for an expansion over a wider range of densities (see Figure 1). Results of the d.c.-conductivity of Cs from a simultaneous measurement of σ and pVT data are shown in Figure 5 (Franz 1980). Metallic characteristics consistent with the nearly free electron model seem to exist down to about $2\rho_c$ and $\sigma \gtrsim 10^3$ $(\Omega\,\text{cm})^{-1}$. This is indicated by the negative temperature coefficient of σ (see Figure 5), the behaviour of the Hall coefficient of Cs which remains free-electron-like down to 1 g cm^{-3} (Even & Freyland 1975), and the Ioffe–Regel limit which for expanded alkali metals is reached near 10^3 $(\Omega\,\text{cm})^{-1}$. In this density range the conductivity agrees well with the predictions calculated according to the Faber–Ziman theory (Block *et al.* 1977, Kahl & Hafner 1984). Below $2\rho_c$ the temperature coefficient of σ changes

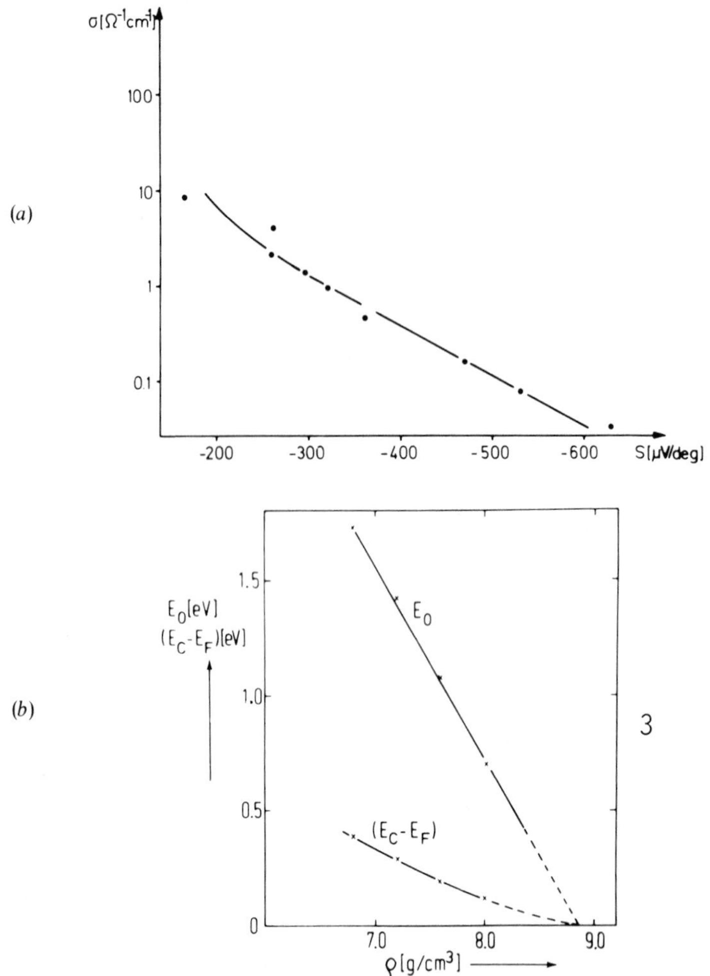

Figure 3. (*a*) Correlation of the logarithm of electrical conductivity σ and absolute thermoelectric power S for fluid Hg (Schmutzler & Hensel 1972); (*b*) activation energy of σ versus density for fluid Hg (Schönherr *et al.* 1979).

sign and $\sigma(\rho)$ exhibits a close similarity in Cs and Rb (Franz *et al.* 1980a). Around ρ_c the conductivity is of order 10^2 $(\Omega\,\mathrm{cm})^{-1}$ and the thermopower drops to large negative values (Freyland *et al.* 1974). Again a plot of $\ln\sigma$ vs. S indicates semiconducting behaviour in fluid Cs for $500 > \sigma > 10$ $(\Omega\,\mathrm{cm})^{-1}$ (Freyland *et al.* 1974, Alekseev *et al.* 1975).

For the electronic structure of dense metal vapours different ionization equilibria and the interaction of charged and neutral particles play a central role—for a recent review see Alekseev & Iakubov (1983). Focusing on the charge–neutral interaction and the density dependence of the generation of positive

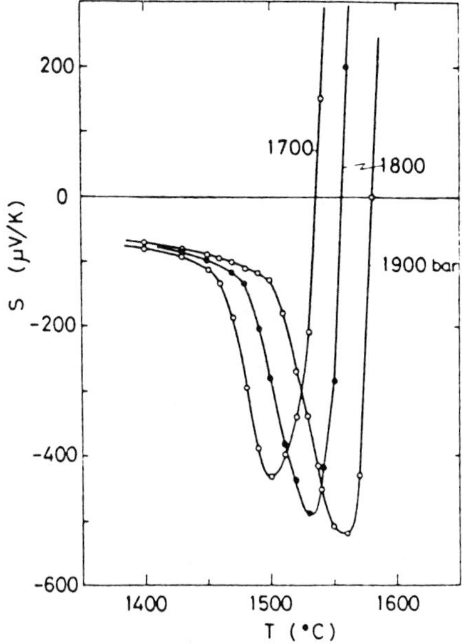

Figure 4. Thermopower anomaly in fluid Hg near the critical point (From Yao & Endo 1982).

molecular ions, Hernandez and others (Hernandez *et al.* 1984, Hernandez 1984, 1985) have recently shown that the electronic conduction in dense vapours such as Cs, Rb and Hg can be described for conditions up to the critical point. A comparison of these model calculations with the observed conductivities of dense Hg vapour at a constant supercritical temperature is given in Figure 6. The experimental conductivities have been obtained with two separate techniques, d.c. and a.c. measurements in the microwave range (Götzlaff 1983, Schönherr 1984). The unusually high conductivities ($>10^{-3}$ $(\Omega\,\mathrm{cm})^{-1}$) observed above 1 g cm^{-3} are comparable with those measured by Hubbard & Ross (1983) in the coexisting vapour phase. As is seen in Figure 6, the model accounts for the observed density dependence of σ up to ρ_c (Hernandez *et al.* 1984).

2.2. Magnetic properties and electronic correlation

Among the different mechanisms which have been discussed for the M–NM transition in expanded fluid metals, electronic correlation effects play a central role (see Mott 1974). The Mott–Hubbard transition is a direct consequence of intra-atomic Coulomb interaction. One of the first theoretical studies which dealt with the magnetic properties of a highly correlated electron gas was that of Brinkman & Rice (1970). They predicted a strong enhancement of the spin susceptibility and the electronic specific heat, which should be determined by the

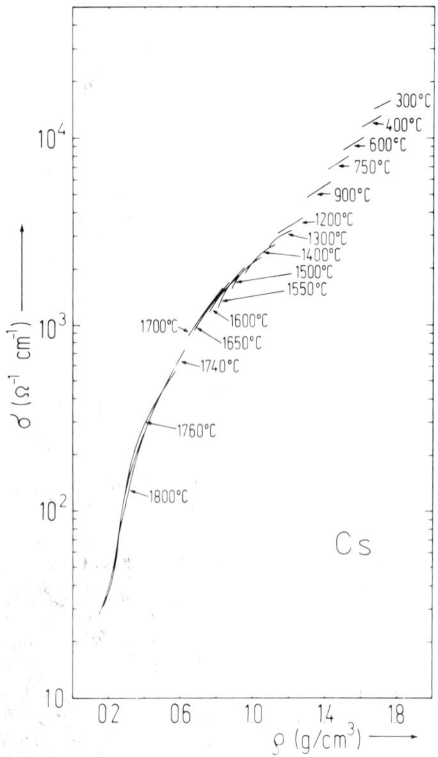

Figure 5. Variation of the d.c. conductivity in expanded Cs with density and temperature (Franz 1980).

ratio of the intra-atomic coulomb repulsion, the Hubbard $U=\langle e^2/r_{12}\rangle$ and the average electronic energy without correlation. Near the transition the electron gas should have nearly antiferromagnetic order, that is, a small fraction of the atomic sites should be doubly occupied, the spin on the other sites resonating between the two possible positions.

In recent years, experimental data on the magnetic properties of expanded metals have been obtained, which give some insight into the change of electronic correlation on expansion, in particular for fluid alkalis. These comprise measurements of the static magnetic susceptibility along the liquid–gas coexistence curve (Freyland 1979, 1980), n.m.r. experiments on expanded Cs and Na (El-Hanany *et al.* 1983, Bottyan *et al.* 1983), and recent conduction electron spin resonance (c.e.s.r.) results for liquid Na over a smaller range of expansion (Nicoloso & Freyland 1985). For expanded Hg the Knight shift has been measured through the M–NM transition range by El-Hanany & Warren (1975). In the following we will describe these results in some detail and will discuss them in relation to recent theoretical investigations of the magnetic characteristics of expanded alkalis near the M–NM transition. We will focus on the alkali metals as

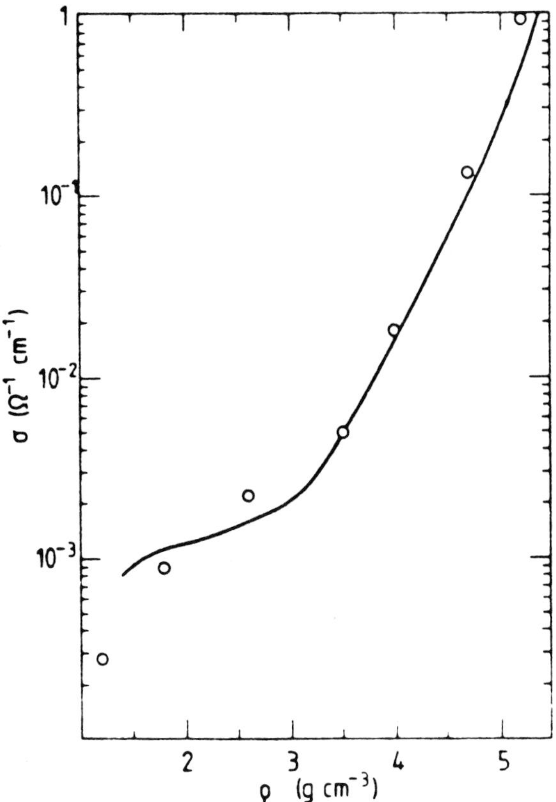

Figure 6. Change of the electronic conduction in dense Hg vapour for densities up to the critical density at constant temperature. Solid line: experimental behaviour; circles: theoretical points (Hernandez *et al.* 1984).

the n.m.r. results of expanded Hg have been reviewed before (see, for example, Warren 1977).

Figure 7 gives a general view for the variation of the electronic magnetic susceptibility χ^e of Cs for an expansion from the dense metallic liquid at the triple point to the low-density insulating vapour phase. For experimental reasons these measurements have been performed along the liquid–gas saturation curve so that temperature is an implicit variable. In order to compare the change in χ^e from metallic to nonmetallic states the gram susceptibility χ^e_g is plotted in Figure 7. The electronic part χ^e has been determined from the total measured susceptibility χ^{tot} by subtraction of the ion core contribution χ^{ion} which is assumed to be constant. The different contributions are related by

$$\chi^{tot} = \chi^{ion} + \chi^e = \chi^{ion} + \chi^{pe} + \chi^{de}$$

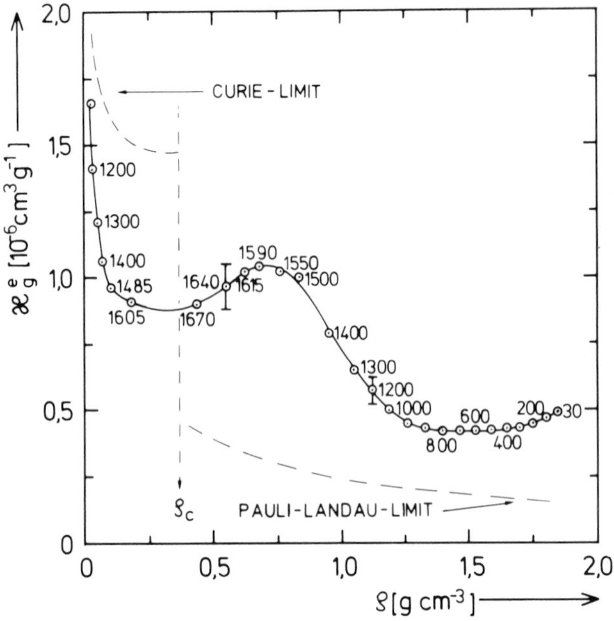

Figure 7. Electronic part of the magnetic susceptibility of expanded Cs. The gram susceptibility is plotted versus density for different temperatures along the liquid–vapour coexistence curve (Freyland 1979).

where χ^{pe} and χ^{de} are the paramagnetic and diamagnetic valence electron contributions to χ^{e}. For the density scale, ρ, in Figure 7 the improved Cs data (see Figure 15) have been used, which yields a slightly different behaviour of χ around the critical point in comparison with the previously published results (Freyland 1979). From the behaviour of $\chi^{e}(\rho)$ the following conclusions can be drawn. Relative to the calculated free-electron limit, the Pauli–Landau susceptibility of conduction electrons, a clear paramagnetic enhancement is observed which gets more pronounced approaching the critical point. A shallow maximum appears around $2\rho_c$. It is interesting to point out that the Ioffe–Regel limit is reached for the expanded alkalis at these densities. On the nonmetallic side of the transition below ρ_c the vapour susceptibility shows a diamagnetic deviation from the Curie limit for localized $s=1/2$ electrons. So we have to conclude that spin pairing processes which lead to aggregated species like molecules are an important feature when approaching the transition from the nonmetallic side. A quantitative discussion of these species on the basis of the χ^{e} data alone is not possible, as different equilibria, such as dissociation and ionization reactions, have to be considered. If one assumes only the monomer–dimer equilibrium, both species in the electronic ground state, one estimates from the χ data near $\rho_c/2$ a mole fraction of 0·25 for Cs_2 and 0·4 for Rb_2 molecules (for the Rb data see Freyland 1980).

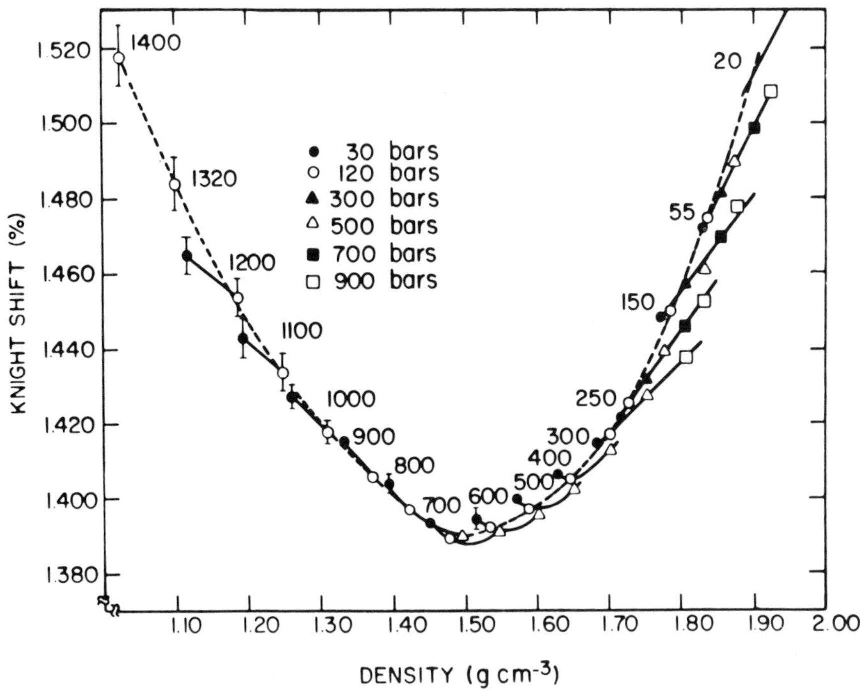

Figure 8. Knight shift of expanded liquid Cs for conditions near the liquid–vapour coexistence curve; the numbers along the dashed curve indicate temperatures in °C (El-Hanany *et al.* 1983).

Figure 8 presents the results of El-Hanany *et al.* (1983) on the ^{133}Cs Knight shift $\Delta H/H$ as a function of density for various temperatures. It is apparent that the density dependence of the n.m.r. shift shows the same trend as the magnetic susceptibility and so confirms the low-density spin enhancement. It is also clear that this enhancement is predominantly due to volume expansion and that temperature plays a minor role. The n.m.r. relaxation rate gives a valuable microscopic insight into the nature of the interaction which is responsible for the spin enhancement in the expanded alkalis. If exchange correlation effects—which lead to the Stoner type enhancement in metals at normal densities (Moruzzi *et al.* 1978)—are important, then the spin-lattice relaxation rate should decrease relative to the Korringa rate upon expansion. However, the experimental trend is opposite in both expanded Cs and Na—for further details see the original papers on the n.m.r. properties. This behaviour is consistent with fluctuations towards a spin density wave or incipient antiferromagnetic coupling of the type expected in the Brinkman–Rice model. In this interpretation of the relaxation behaviour it is assumed that electron–ion scattering effects are not dominant for densities well above $2\rho_c$.

For the interpretation of the spin enhancement and, more generally, of the change in the electronic structure of expanded alkalis, various theoretical studies have been published which emphasize different mechanisms. Franz (1984) has developed a theoretical model which emphasizes disorder due to the randomness in the number of nearst neighbours. The density of the expanded metal is related to the mean coordination number N_1 of the atoms, the nearest neighbour distance R_1 is kept constant (see the neutron diffraction results for expanded Rb below). She uses a simple tight-binding description and considers the expanded liquid like a random alloy where the atomic sites are either occupied or vacant. The susceptibility enhancement in this model is mainly ascribed to an increase in the density of states at the Fermi energy E_F, $N(E_F)$, due to narrowing of the conduction band with expansion and reduced coordination number. Thus the enhancement factor should scale with N_1 and should be the same for all alkali metals. As a consequence of disorder this model predicts the coexistence of localized and extended states at the same energy for densities approaching the M–NM transition region. According to this calculation the transformation to nonmetallic behaviour may set in around $2\rho_c$ where a relatively large fraction of localized states appears. These pile up at E_F and thus yield a rapid reduction in the metallic cohesive energy.

A different theoretical approach has been proposed by Rose (1980) who neglects the randomness of the electronic potentials (that is, the Anderson localization mechanism), and concentrates on the role of electronic correlation. Within the framework of a spin-split self-consistent band-structure calculation, the ground-state energy is determined as a function of average density and spin moment. In this model a first-order transition from the paramagnetic metal to the spin-ordered insulator is predicted for expanded alkalis, where the transition densities n_c are given by $r_c = 1{\cdot}04 r_0 + 2{\cdot}8$; r_0 roughly correlates with the atomic radius and $r_c = (3/4\pi n_c)^{1/3}$. A striking feature of this calculation is that the spin enhancement should become systematically smaller for the heavier alkalis. This trend is indicated in the experimental results shown in Figure 9. Here the volume spin susceptibilities of Na, Rb and Cs are compared as a function of reduced volume (V_m = volume at the melting point). For the data in Figure 9 the conduction electron diamagnetism has been corrected according to the theory of Kanazawa & Matsudawa (1960).

Figure 9 has two further interesting aspects. As is observed for Cs, the spin susceptibility approaches the free spin Curie value for an expansion above the peak in χ_V^{ep}. The Curie law plotted in Figure 9 has been calculated for the appropriate densities and temperatures along the coexistence curve of Cs. An important implication of this Curie law limitation for the interpretation of the susceptibility enhancement has recently been pointed out by Warren (1984). Whereas the exchange-enhanced susceptibility of the Stoner form is not restricted to values below the Curie limit, the Brinkman–Rice theory demands $\chi/\chi_{\text{Curie}} \rightarrow 1$ for sufficiently large values of $N(E_F)$.

A second striking observation of the volume dependence of χ_V^{ep} concerns the pronounced minimum which is apparent for all the alkalis at slight volume

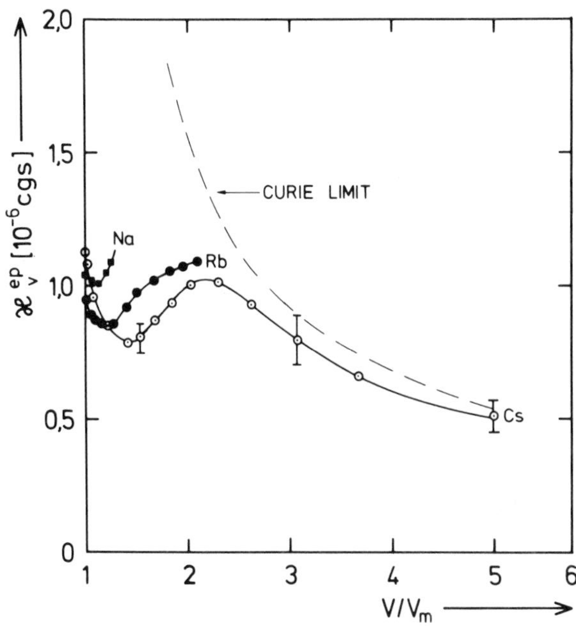

Figure 9. Spin susceptibility of expanded alkali metals versus reduced volume (V_m = volume at melting point); the volume susceptibility is plotted.

expansion. As was suggested in the first interpretation of the Cs results (Freyland 1979), two competing interactions should be considered. Starting from the triple point, exchange correlation effects first prevail; with further expansion the Brinkman–Rice enhancement takes over. A quantitative basis for this explanation was recently given by a band-structure calculation of Warren & Mattheiss (1984). They performed a self-consistent, relativistic, augmented plane wave (APW) calculation for expanded Cs, with fixed nearest neighbour distance but varying coordination numbers from 8 for body-centred cubic to 4 for diamond structure. Within this one-electron model they find a continuously decreasing volume susceptibility for the Stoner-type correlation. This again indicates the importance of many-electron correlation effects in the understanding of the spin enhancement for an expansion beyond the minimum in χ_V^{ep}.

Beyond this density range, significant changes occur in the conduction electron wave function. The calculation of Warren and Mattheiss yields an increasing charge density at the nucleus with expansion. The experimental answer is opposite. From the relation of the Knight shift and the spin susceptibility, which may be written as

$$\frac{\Delta H}{H} = \frac{8\pi}{3} \langle |\Psi(0)|^2 \rangle_{E_F} \, \Omega \chi_V^{ep}$$

the average probability density at the nucleus, $\langle |\Psi(0)|^2 \rangle_{E_F}$, may be obtained ($\Omega$ is the atomic volume). For Cs, $\langle |\Psi(0)|^2 \rangle_{E_F}$ remains almost constant for an expansion of about 40% (near the minimum in χ_V^{ep}) and then decreases with further expansion. It is unlikely that conduction electrons become less s-like with expansion, and so the reduction in $\langle |\Psi(0)|^2 \rangle_{E_F}$ implies an increased charge density in the interstitial volume between neighbouring atomic sites. This is consistent with antiferromagnetic coupling and is expected if, in the course of the metal–nonmetal transformation, local aggregations such as molecular species become important.

Another interesting aspect of the comparison in Figure 9 is the shift of the susceptibility minimum towards smaller reduced volumes for the lighter alkali metals. Possibly for liquid Na the Brinkman–Rice enhancement sets in for only 10% expansion. This effect influences all the magnetic properties, as is more clearly demonstrated in Figure 10. For expanded Na above 500°C the electron

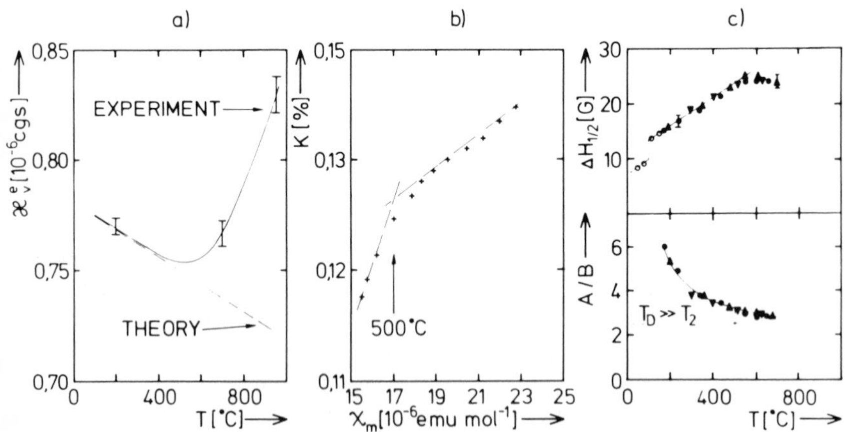

Figure 10. Magnetic properties of expanded liquid Na: (*a*) electronic volume susceptibility (Bottyan *et al.* 1983); (*b*) Knight shift versus molar susceptibility (Bottyan *et al.* 1983); (*c*) c.e.s.r. results of half width $\Delta H_{1/2}$ Dysonian asymmetry ratio A/B versus temperature (Nicoloso & Freyland 1984).

susceptibility deviates from the theoretical nearly-free-electron value (Figure 10(*a*)), a break occurs in the relation of Knight shift and magnetic susceptibility (Figure 10(*b*)), which again indicates decreasing $\langle |\Psi(0)|^2 \rangle_{E_F}$, and the c.e.s.r. properties deviate from the nearly-free-electron predictions, that is, the line width no longer follows the Elliot relation ($\Delta H_{1/2} \propto T$) and the Dyson asymmetry ratio approaches the limit where the electron diffusion time gets much larger than the spin–spin relaxation time (Figure 10(*c*)).

In conclusion, we find that the behaviour of the magnetic properties of expanded alkali metals strongly supports the predictions of the Brinkman–Rice theory for a highly correlated electron gas. The n.m.r. results give clear indications that antiferromagnetic interactions play an important role. The susceptibility peak observed for expanded Cs around $2\rho_c$ probably defines the transition from a paramagnetic metal to antiferromagnetic nonmetallic states. This is particularly remarkable in the light of the Curie law limitation of χ. In a previous discussion of the M–NM transition in expanded alkalis it was speculated (Freyland 1981) that possibly two transitions may be considered, a magnetic transition near $2\rho_c$ and a conductivity transition near ρ_c, where σ approaches Mott's minimum metallic conductivity limit of order 10^2 $(\Omega\,\mathrm{cm})^{-1}$. In view of the present more detailed knowledge of the magnetic properties, the possibility that nonmetallic states dominate already below $2\rho_c$ cannot be excluded. In the fluid phase atomic centres are not fixed and so a likely analogue to the antiferromagnetic ground state is the singlet state of aggregated species like molecules. The density dependence of χ in expanded Cs around the critical point may be explained in this way without invoking the formation of a pseudo-gap as a precursor of a conductivity transition. In this context it is interesting to point out that Redmer & Röpke (1985), in a theoretical calculation of the equation of state of dense Cs plasma, predict a high concentration of dimers for conditions near the critical point.

2.3. *Optical and dielectric properties*

The only optical experiments bearing on the M–NM transition in expanded liquid metals are reflectivity measurements on mercury by Ikezi *et al.* (1978) and by Hefner *et al.* (1980). Figures 11 and 12 show the results of Hefner *et al.* (1980) expressed as the real part of the dielectric constant $\varepsilon_1(\omega)$ (Figure 11) and the a.c. conductivity $\sigma(\omega) = \omega\varepsilon_2(\omega)/4\pi$ (Figure 12) along the coexistence line of mercury. The change in the shape of the $\varepsilon_1(\omega)$ and $\sigma(\omega)$ curves nicely illustrates the gradual diminution of metallic properties with decreasing density. In the nearly-free-electron range for densities ρ larger than twice the critical density ρ_c, the low-frequency optical conductivity is Drude-like. A gradual change from metallic to nonmetallic behaviour occurs in the density range between 11 and 9 $\mathrm{g\,cm^{-3}}$. For still smaller densities $\sigma(\omega)$ indicates that a gap opens. A similar conclusion has been reached in a direct analysis (Krohn & Thompson 1980) of the reflectivity data using a particular model of the density of states. The density for opening of the gap is in good agreement with the onset of characteristic nonmetallic transport properties (Figure 3) and a sharp drop in the ^{199}Hg Knight shift (Warren & Hensel 1982). The exact interpretation of the observations is not certain yet, but there remains no doubt that mercury changes macroscopically to a nonmetallic, effectively 'semiconducting' state when it approaches the critical region, that is, $\rho \lesssim 1{\cdot}5\rho_c$ and $T \gtrsim 0{\cdot}96T_c$.

We turn next to the behaviour of the optical absorption spectrum on the

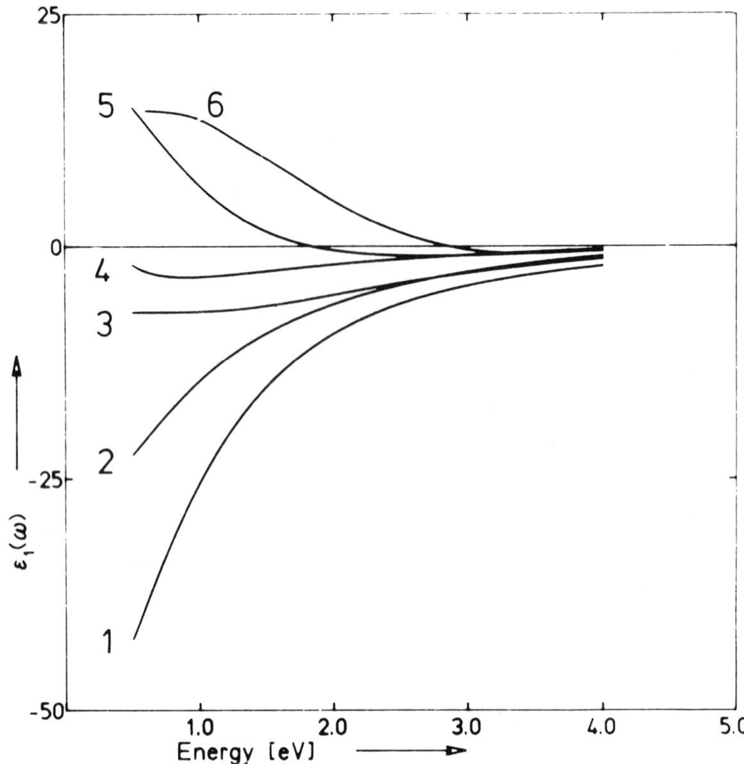

Figure 11. Real part of the dielectric constant ε_1 of expanded Hg along the coexistence curve; for the (ρ, T) conditions see Figure 12 (Hefner *et al.* 1980).

insulating vapour side (Overhof *et al.* 1976, Ikezi *et al.* 1978). At very low densities a line spectrum is observed with the main absorption lines at 4·89 eV and 6·7 eV corresponding to transitions between the 6s ground state and the 6p triplet and singlet state of the Hg atom. As the density is increased, the sharp lines broaden due to interactions with neighbouring atoms, resulting in a relatively steep absorption edge which moves rapidly to lower energies with increasing density. Figure 13 gives a few selected data for the density dependence of the edge at a constant temperature of 1550°C, far above the critical temperature (Uchtmann 1978). Bhatt & Rice (1979) and Uchtmann *et al.* (1981) have shown that a uniform density increase is insufficient to explain a line broadening as large as the observed shift in Figure 13 and that one must take density fluctuations into account. The absorption edge is then lowered by the environment of the atom being excited, and the edge is thus explained in terms of absorption by excitonic states of large randomly distributed clusters. From the large values of the absorption coefficient it can be concluded that the singlet exciton (6^1p_1) with large oscillator strength broadens faster than the triplet exciton (6^3p_1) with small

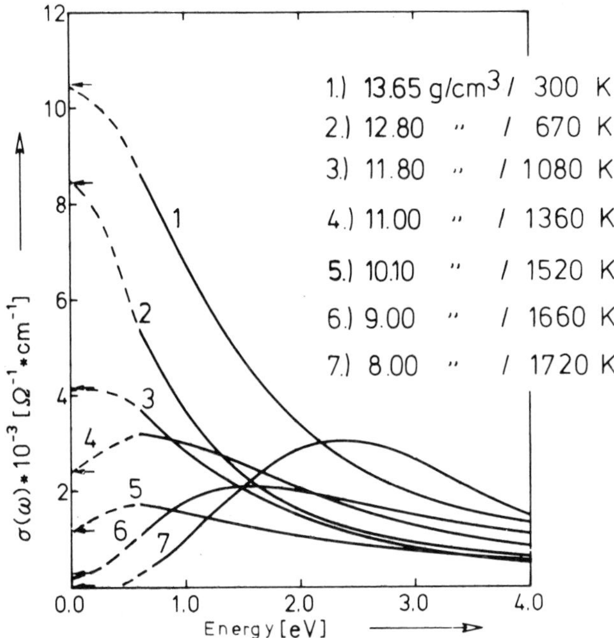

Figure 12. Optical conductivity $\sigma(\omega)$ of expanded Hg along the coexistence curve; the density temperature values are given in the figure and the arrows indicate the d.c. conductivity results (Hefner *et al.* 1980).

oscillator strength. A detailed analysis (Uchtmann *et al.* 1981) of the density dependence of the absorption edge shows that the singlet contribution dominates for densities larger than 1 g cm^{-3} whereas for $\rho < 1$ g cm^{-3} the shape of the edge is dominated by the triplet transition.

Recent experimental works on the near-infrared (Hefner & Hensel 1982, Yao *et al.* 1985) and low-frequency (Schönherr 1984) dielectric constant in mercury vapour indicate that the liquid–vapour critical point phase transition is preceded by a 'dielectric transition' which is signalled by a relatively steep anomaly in the real part of the dielectric constant ε_1. Selected experimental results for ε_1 at a constant photon energy $\hbar\omega = 1{\cdot}27$ eV (Yao *et al.* 1985) as a function of pressure are shown in Figure 14 for different sub- and supercritical temperatures. As the pressure is increased at a constant temperature T larger than about 1400°C (i.e. about $0{\cdot}96T_c$) ε_1 initially follows Clausius–Mosotti behaviour up to the dielectric transition pressure p_0. At p_0 the dielectric constant quickly rises. The most surprising feature of the data is the abrupt enhancement (dashed part of the curves) of ε_1 which is hard to measure. At subcritical temperatures, it looks like a first-order transition. It is obvious that at subcritical temperatures the dielectric transition precedes the vapour-to-liquid transition which is indicated in Figure 14 by the open circles. For supercritical temperatures the enhancement of ε_1 is

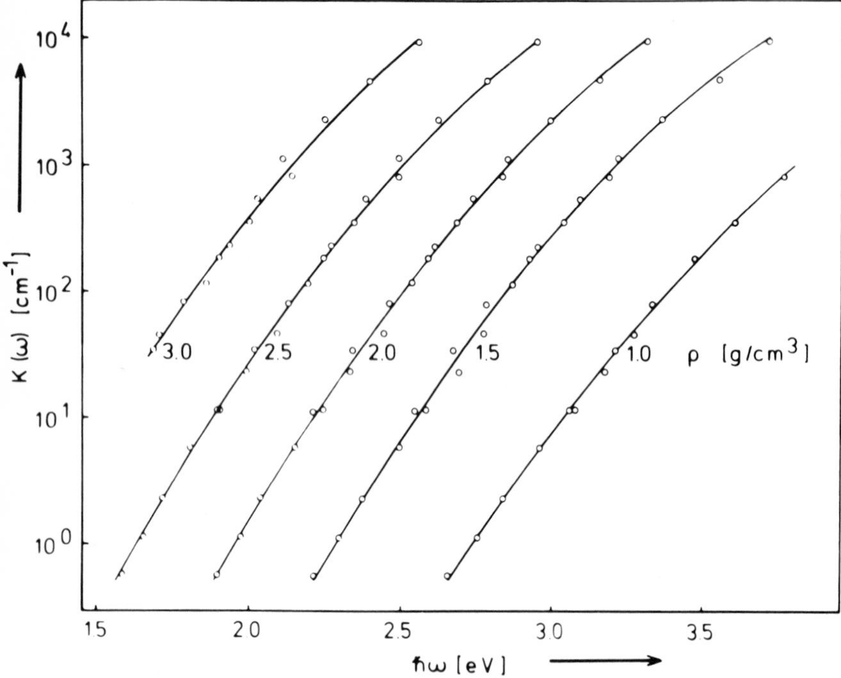

Figure 13. Optical absorption edge of dense Hg vapour at different densities for a constant supercritical temperature of 1550°C; the absorption constant is plotted against photon energy (Uchtmann 1978).

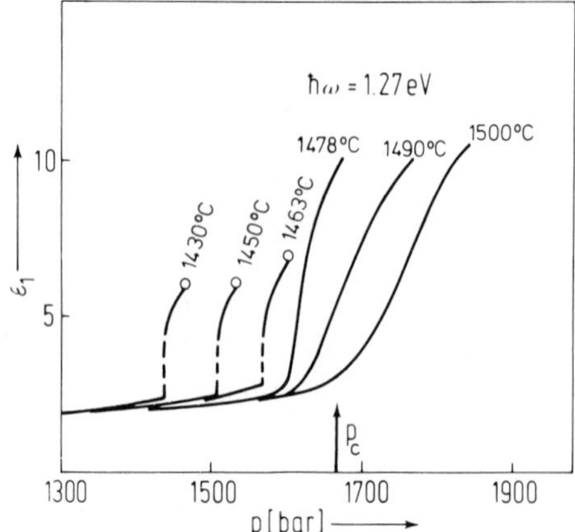

Figure 14. Dielectric constant of dense Hg vapour as a function of pressure for different sub- and supercritical temperatures; the dashed lines indicate the dielectric transition for subcritical temperatures (Yao *et al.* 1985).

smeared out. The pattern of the ε_1 curves in Figure 14 immediately suggests that the dielectric anomaly is intimately related to the gas–liquid critical-point phase transition in mercury.

This becomes especially obvious when ε_1 is studied in the vicinity of the critical point as a function of temperature at constant density. Far above the critical temperature ($T \gtrsim 1{\cdot}04 T_c$) ε_1 is nearly temperature independent at constant density and shows only slight positive deviations from the simple Clausius–Mosotti values ε_{1CM}. This effect changes for temperatures $T \lesssim 1{\cdot}04 T_c$. Here ε_1 shows a strongly temperature dependent anomalous part $\Delta\varepsilon_1 = (\varepsilon_1 - \varepsilon_{1CM})$. The functional form of the temperature dependence of $\Delta\varepsilon_1$ along isochores has a cusplike anomaly near the critical point. The striking increase in the dielectric constant $\Delta\varepsilon_1/\varepsilon_1$ reaches about 0·7 for fluid mercury at densities close to the critical values. This value is much higher than that of normal nonmetallic fluids for which a comparatively moderate dielectric anomaly occurs very close to the critical point. The experimentally observed $\Delta\varepsilon_1/\varepsilon_1$ for CO is 10^{-3} (Pestak & Chang 1981).

Hefner & Hensel (1982) have speculated that the enhanced dielectric anomaly in mercury is due to a transition from a normal dielectric gas phase to an equilibrium dispersed system containing charged dense droplets. The underlying idea of this proposal is that a charge present in a vapour of neutral atoms can cause a local density increase, which can lead in the vicinity of the vapour–liquid critical temperature to an analogue of a vapour–liquid transition in the compressed region, that is, dense charged clusters are formed which approach metallic conditions while the macroscopic system is still in the nonmetallic region (Hefner *et al.* 1982). This interpretation is based on a paper by Lifshitz & Gredescul (1970), who showed by an analysis of the free energy of a system consisting of an electron and a medium of classical particles that the appearance of an indirect particle interaction via the electron can cause a transition from a normal gas distribution to a dense cluster in a narrow interval of temperature or density, respectively. The hypothesis that charges tend to stabilize dense clusters is also supported by a computer experiment by Lagarkov & Sarychev (1978) who examined electrons obeying Newton's laws with a semiclassical density of states, interacting with a set of neutral mercury atoms. An analysis of the spatial correlation function for the electron–neutral Hg interaction showed that the density of the cluster can approach values characteristic of the liquid.

Hernandez (1982) has performed a quantum mechanical variational calculation to probe the requisite conditions for electron stabilization of high density fluctuations in fluid Hg. The onset of stabilization has been found to occur when electrons are self-consistently bound to small volumes with an excess density at the cluster centre of at least 50% more than average. The calculated density temperature threshold for the onset of the electron–cluster stability is in qualitative agreement with the experimentally observed dielectric anomaly. The picture of charged dense droplets in mercury vapour has been questioned by Turkevich & Cohen (1984a). Alternatively, they make the interesting proposal that mercury for densities smaller than the M–NM transition density and larger than the dielectric transition density constitutes a disordered, inhomogeneous excitonic insulator phase. The underlying suggestion is that improved coordination in the fluid, that is, clustering, can reduce the bottom of the Frenkel exciton band, leading ultimately to an exciton condensation instability. The permanent dipole moments of the condensed excitons are estimated to order as a ferroelectric phase (Turkevich & Cohen 1984b). The advantage of the proposed phase diagram is that it explains both the M–NM transition in liquid mercury and the dielectric transition in mercury vapour.

2.4. Thermodynamic and structural properties

A theoretical description of the thermodynamics of liquid metals is particularly complicated, due to the density dependence of the effective interatomic interaction and the different contributions, such as the exchange correlation and the bandstructure term, to the cohesive energy—for reviews see Shimoji (1977), Ashcroft & Stroud (1978). For expanded metals approaching the M–NM transition further crucial problems arise. Possible variations in the nature of electronic correlation have to be considered as indicated by the magnetic properties. Variations in the electronic structure, or, more generally, in the nature of the interatomic forces, have to be taken into account. First attempts to approach these problems theoretically for fluid monovalent metals were made by Nara *et al.* (1977) who calculated an equation of state treating the ionic system by a lattice gas model and including the Hubbard correlation energy explicitly. Their main objective was the interrelation between the liquid–vapour phase transition and the M–NM transition, a problem first formulated by Landau & Zeldovitch (1943). For a recent extended review see Yonezawa & Ogawa (1982). However, at present the theoretical attempts to model the statistical mechanics of the M–NM transition in fluids are still insufficient to provide a clear-cut answer. At present no theory exists which incorporates both the fluid aspects and the variation in the electronic structure from extended to localized states.

Experimentally, the key to the thermodynamic properties of a fluid system lies in accurate pVT measurements. For the problem of M–NM transition in expanded metals these experiments have to be extended to the critical-point region, which is a crucial task in itself. Determinations of critical-point data, especially for fluid Hg and Cs, have been reproduced several times by different groups with different techniques; some used the steepest descent of electronic properties like conductivity as a measure for critical-point characteristics. Over the years these results from different sources showed quite considerable scattering (see the general reviews in the introduction). For example, the critical temperature of Cs from five separate determinations was given between 1740°C and 1785°C. Only recently in an exact isochoric determination of the pVT data of Cs was it realized that T_c of Cs is much lower, and we think is now best given as

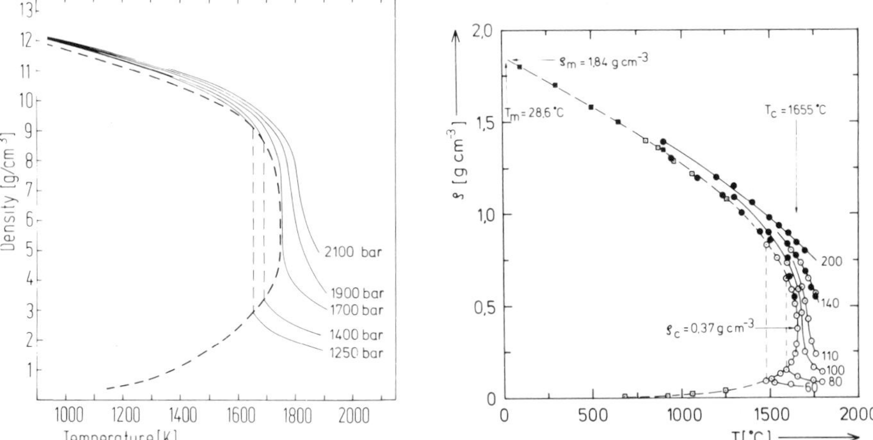

Figure 15. Density–temperature–pressure data of expanded fluid Hg (left) and Cs (right). The results of Hg have been compiled from several independent experiments, those of Cs are mainly from Franz (1980) (full circles) and Jüngst (1985) (open circles).

1652°C (Jüngst 1985—see also Table 1). In Figure 15 we present pVT data for fluid Hg and Cs at sub- and supercritical conditions which mainly stem from isochoric measurements and we believe that at present these results have the highest accuracy up to the critical-point region.

A detailed discussion of the thermophysical properties of metallic fluids has recently been given by Hensel (1982) and so we will concentrate here on some basic observations relevant to the M–NM transition problem. In principle, one is lead to expect that the M–NM transition will show up in the thermodynamic properties as the nature of the interatomic forces changes. An indication of this type seems to appear in the volume dependence of the internal pressure $p_i = (\partial U/\partial V)_T$, which mainly reflects the cohesive part of the internal energy U. An exact relation of p_i with the measured quantities $p = p(V, T)$ and the derivative $(\partial p/\partial T)_V$, the isochoric thermal pressure coefficient, is given by the thermodynamic equation of state

$$\begin{aligned}(\partial U/\partial V)_T &= T(\partial S/\partial V)_T - p \\ &= T(\partial p/\partial T)_V - p\end{aligned}$$

As for both the alkalis and mercury the isochores, within experimental errors, are straight lines, $(\partial U/\partial V)_T$ is obtained directly from a (p, T) plot by extrapolation to $T = 0$. Figure 16 shows the volume dependence of the internal pressure for Hg along the liquid–vapour coexistence curve (Schönherr *et al.* 1979). A rather prominent change in the curvature of $U(V)_T$ occurs near a density of $2\rho_c$ and has been confirmed separately by Endo (1982). A similar observation is obtained from the pVT data for Cs and Rb (Franz 1980).

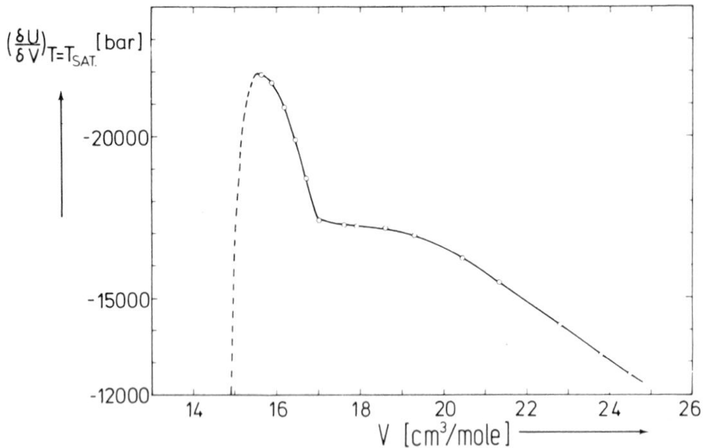

Figure 16. Internal pressure $(\partial U/\partial V)_T$ for expanded Hg at conditions near the liquid–gas saturation curve (Schönherr *et al.* 1979).

For the second derivatives of the free energy, like the thermal expansion coefficient $\alpha_p = V^{-1}(\partial V/\partial T)_p$ or the isothermal compressibility $\chi_T = -V^{-1}(\partial V/\partial p)_T$, no clear breaks are visible in the behaviour approaching the M–NM transition region. An example of the density dependence for different temperatures is given for χ_T of Hg in Figure 17, which also contains the pressure coefficient of conductivity demonstrating the close correlation of the derivatives approaching the critical point (Götzlaff 1983). This is not surprising, as for any

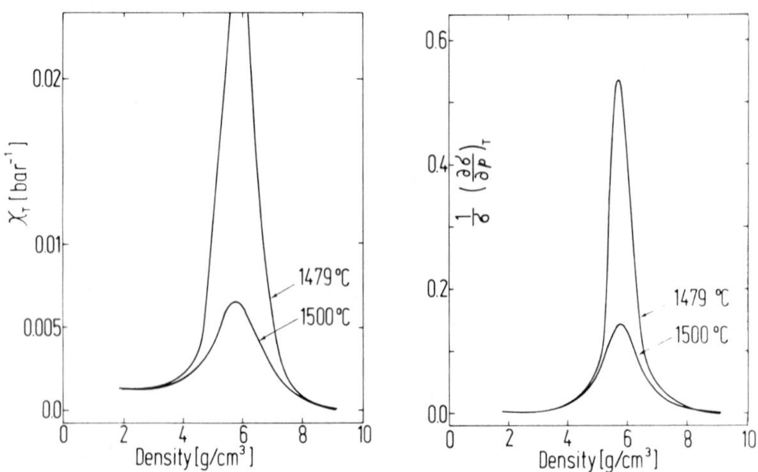

Figure 17. Isothermal compressibility, $\chi_T = -V^{-1}(\partial V/\partial p)_T$, and isothermal pressure coefficient of the d.c. conductivity of fluid Hg for densities and temperatures near the critical point (Götzlaff 1983).

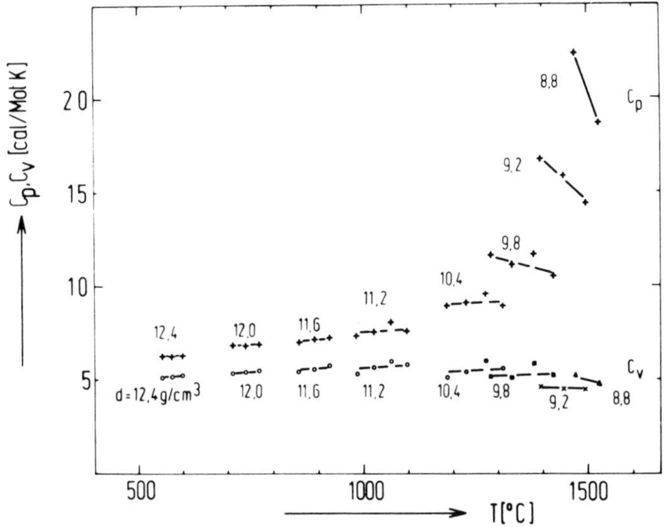

Figure 18. Heat capacity, C_p and C_V, respectively, of expanded liquid Hg (Levin & Schmutzler 1984).

fluid these quantities diverge as the critical point is approached. No matter what the form of the molecular interactions is, the correlation length of density fluctuations becomes long-range approaching the critical point. As for simple fluids like Ar this may set in around $2\rho_c$ for metals, too, and thus strongly determines these coefficients. This seems to apply also to the caloric properties such as the heat capacity C_V. The specific heat at constant pressure C_p has been determined from direct calorimetric measurements both for the alkali metals (Alekseev *et al.* 1980a) and for Hg (Levin & Schmutzler 1984) for a wide range of expansion below $2\rho_c$. Figure 18 represents the Hg results as a function of temperature for different densities, together with C_V data which have been derived from the relation

$$C_V = C_p - T \cdot V\alpha_p \cdot (\partial p/\partial T)_V$$

From a comparison of Hg and Ar in a reduced density plot Schmutzler and Levin conclude that within experimental errors these two fluids show the same trend in C_V for densities near the metal–semiconductor transition of Hg. Concerning a possible enhancement of the electronic contribution to the specific heat in expanded alkalis no clear conclusion can yet be reached.

Coming back to the change in the metallic cohesion approaching the M–NM transition, some progress has recently been achieved in the theoretical interpretation of the structural results, which have been obtained by neutron diffraction for expanded Rb (Frantz *et al.* 1980b) and are represented in Figure 19. The main experimental observations are:

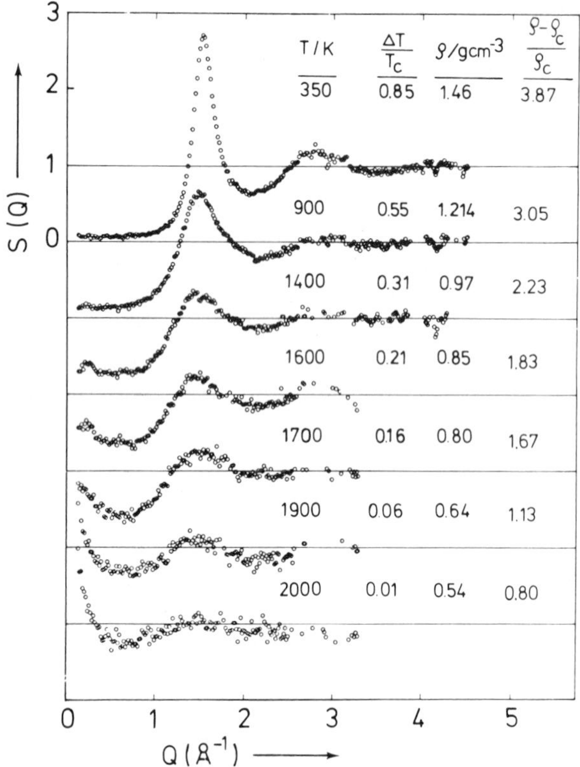

Figure 19. Static structure factor of expanded liquid Rb determined by neutron diffraction (Franz *et al.* 1980).

1. With reducing density, the nearest-neighbour coordination number decreases linearly with density, whereas the mean interatomic distance does not change appreciably. In this respect expanded metals again show a close similarity to other simple fluids like Ar.

2. Well before the end of the metallic transport regime near $2\rho_c$ is reached the structure factor $S(Q)$ exhibits a strong increase for small Q-values. Qualitatively this reflects a clear change in the attractive part of the interatomic interaction towards long-range forces, which is what would be expected if screening is reduced with expansion. Quantitative agreement with experiment has recently been reported on the basis of calculations which use thermodynamic perturbation theories like the Weeks–Chandler–Andersen approach (McLaughlin & Young 1984, Kahl & Hafner 1985). McLaughlin and Young point out that below $2\rho_c$ a nearly-free-electron description of the long wavelength behaviour fails completely.

3. The metal–nonmetal transition and critical phenomena

It can be expected that the gross changes in electronic structure at the M–NM transition, which on the basis of the results of the foregoing sections are presumably relevant in the vicinity of the critical point, may noticeably influence the gas–liquid critical point phenomena of metals. Experimental studies relevant to this question have received less attention and are rather new. As the experimental difficulties in the critical-point region of metals are rather severe this is not surprising. There are only two recent experiments on the equation of state of Hg (Götzlaff 1983, Seyer *et al.* 1985) and Cs (Jüngst 1985) relatively close to the respective critical points for which comparatively accurate temperature control and optimal elimination of temperature gradients are claimed. These data seem to be accurate enough to permit the first determination of the asymptotic behaviour of the thermophysical properties of metals near the gas–liquid critical point.

Figure 20 shows as an example a plot of coexisting vapour (ρ_V) and liquid (ρ_L) densities of Cs together with the curve of average densities $\bar{\rho}=\frac{1}{2}(\rho_L+\rho_V)$ versus temperature (Jüngst 1985). The form of the coexisting curve is clearly asymmetric compared to those of simple non-conducting fluids. The asymmetry, however, is very similar to that observed for the metal–ammonia (Chieux *et al.* 1980) and electron–hole liquid phase diagrams (see, for example, Thomas 1984).

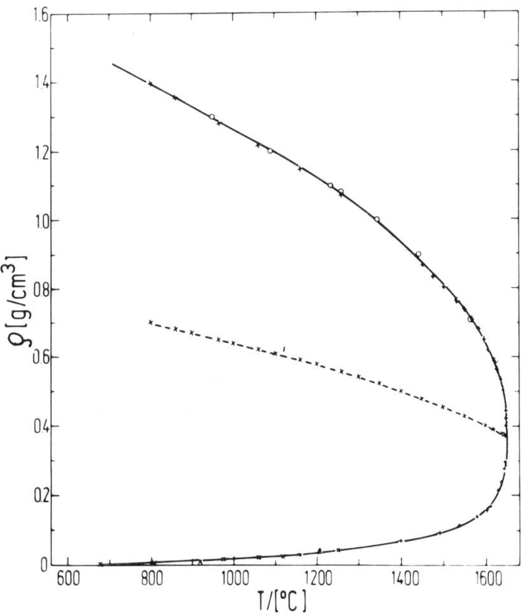

Figure 20. Coexisting liquid and vapour densities of fluid Cs together with curve of rectilinear diameter (dashed line) (Jüngst 1985).

The most striking feature of Figure 20 is the breakdown of the law of rectilinear diameter. The fact that this law does not hold indicates already an unusual behaviour of the critical exponents β_L and β_V which are used in the power law analysis of coexistence curves in the asymptotic expression

$$\frac{\rho_{L,V} - \rho_c}{\rho_c} \sim \frac{|T - T_c|^{\beta_{L,V}}}{T_c}$$

Here β_L is the exponent of the liquid (ρ_L) branch, β_V of the vapour (ρ_V) branch of the coexistence curve. Experimentally it is found that the liquid branch can be fitted with $\beta_L \approx 0{\cdot}46$ over a relatively large temperature range, whereas the value β_V for the vapour branch depends strongly on the relative temperature distance from the critical point. Close to T_c, β_V approaches a value close to that observed for β_L. These values are comparable with $\beta = 0{\cdot}5$ obtained for classical mean field theory. Similar behaviour is shown by the isothermal compressibility χ_T of caesium (Jüngst 1985). Figure 21 shows the power law analysis for the divergence of χ_T along the critical isochore of caesium according to the asymptotic expression

$$\chi_T \approx \frac{|T_c - T|^{-\gamma}}{T_c}$$

The critical exponent $\gamma \approx 1$ is again consistent with mean-field behaviour.

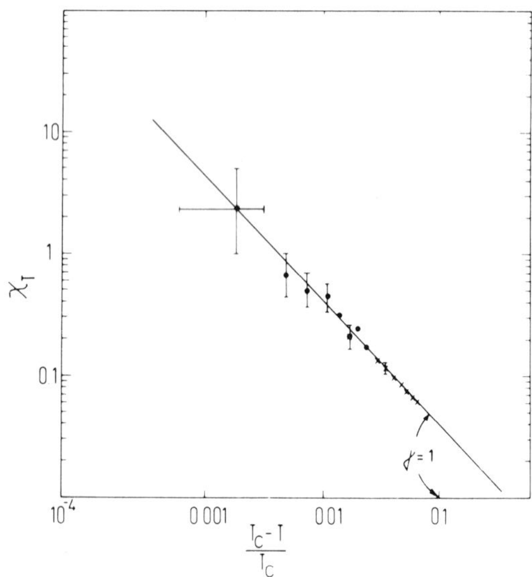

Figure 21. Power law of the isothermal compressibility χ_T along the critical isochore of Cs (Jüngst 1985).

Similar observations have been made for mercury (Götzlaff 1983, Hensel 1984) for which close to the critical point the experimentally deduced exponents are also consistent with the classical mean-field theories. To our knowledge this has never before been observed for other one-component liquids.

According to modern theory of critical phenomena, systems can be grouped in universality classes. Systems within a universality class have the same critical exponents and the same scaling functions characterizing the thermodynamic behaviour in the vicinity of the critical point (see, for example, Domb & Green 1976). In particular, it is commonly believed that all fluids with short range intermolecular forces, including polar fluids, belong to the same static universality class as a uni-axial ferromagnet represented by the three-dimensional Ising model or the Landau–Ginzburg–Wilson model with a one-component order parameter. This suggestion has been confirmed experimentally with considerable certainty for non-conducting normal fluids (see, for example, Sengers 1982). In contrast to this are the experimental observations for mercury and caesium. It seems that long-range interatomic interactions or the presence of competing interactions close to the critical point of fluid metals reduces the upper critical dimensionality from 4 to 3 resulting in critical exponents very close to mean-field values (see, for example, Als-Nielsen & Birgenau 1977).

It must be pointed out, however, that at present, because of the severe experimental problems connected with the high critical temperatures of metals, reliable experimental data can only be obtained for the region $|T_c - T|/T_c \gtrsim 10^{-3}$, so that it may be argued that the indices could change at lower values of $|T_c - T|/T_c$ to the non-classical values normally observed for fluids.

References

Alekseev, V.A., Vedenov, A.A., Ovcharenko, V.G., Krasitskaya, L.S., Ryzhkov, Yu.F. and Starostin, A.I., 1975, *High Temperature–High Pressure*, **7**, 677.

Alekseev, V.A. and Iakubov, I.T., 1983, *Phys. Rep.*, **96**, 1.

Alekseev, V.A., Blagouranov, L.A. and Philippov, L.P., 1980a, *J. Physique*, **41**, Colloque C-8, 607.

Alekseev, V.A., Ovcharenko, V.G. and Ryzhkov, Yu.F., 1980b, *J. Physique*, **41**, Colloque C-8, 91.

Als-Nielsen, J. and Birgenau, R.J., 1977, *Am. J. Phys.*, **45**, 554.

Ashcroft, N.W. and Stroud, D., 1978, *Solid St. Physics*, **33**, 1.

Bhatt, R.N. and Rice, T.M., 1979, *Phys. Rev.* B, **20**, 466.

Block, R., Suck, J.B., Freyland, W., Hensel, F. and Gläser, W., 1977, *Liquid Metals 1976* (*Inst. Phys. Conf. Ser. 30*), edited by R. Evans and D.A. Greenwood, p. 126.

Bottyan, L., Dupree, R. and Freyland, W., 1983, *J. Phys. F: Metal Phys.*, **13**, L173.

Brinkman, W.F. and Rice, T.M., 1970, *Phys. Rev.* B, **2**, 4302, ibid B **2**, 1324.

Chieux, P., Damay, P., Dupuy, J. and Jal, J.F., 1980, *J. Phys. Chem.*, **84**, 1211.

Cusack, N.E., 1982, *Prog. Theor. Phys.* (*Japan*), *Supplement*, **72**, 81.

Domb, C. and Green, M.S. (eds), 1976, *Phase Transitions and Critical Phenomena*, Vol. 6 (New York: Academic Press).

Duckers, L.J. and Ross, R.G., 1972, *Phys. Lett.* A, **30**, 715.

El-Hanany, U. and Warren, W.W., Jr., 1975, *Phys. Rev. Lett.*, **34**, 1276.
El-Hanany, U., Brennert, G.F. and Warren, W.W., Jr., 1983, *Phys. Rev. Lett.*, **50**, 540.
Endo, H., 1982, *Prog. Theor. Phys.* (*Japan*), *Suppl.*, **72**, 100.
Even, U. and Freyland, W., 1975, *J. Phys. F: Metal Phys.*, **5**, L104.
Even, U. and Jortner, J., 1972, *Phil. Mag.*, **25**, 715.
Franz, G., 1980, *PhD Thesis*, Universität Marburg.
Franz, G., Freyland, W. and Hensel, F., 1980a, *J. Physique*, **41**, Colloque C-8, 70.
Franz, G., Freyland, W., Gläser, W., Hensel, F. and Schneider, E., 1980b, *J. Physique*, **41**, Colloque C-8, 194.
Franz, J.R., 1984, *Phys. Rev.* B, **29**, 1565.
Freyland, W., 1979, *Phys. Rev.* B, **20**, 5104.
Freyland, W., 1980, *J. Physique*, **41**, Colloque C-8, 74.
Freyland, W., 1981, *Commun. Solid St. Phys.*, **10**, 1.
Freyland, W., Pfeifer, H.P. and Hensel, F., 1974, *Proceedings of the 5th International Conference on Amorphous and Liquid Semiconductors*, edited by J. Stuke and W. Bredig (London: Taylor & Francis), p. 1327.
Friedman, L.R. and Tunstall, D.P. (eds), 1978, *Metal–Nonmetal Transition in Disordered Systems* (Edinburgh: Scottish Universities Summer School).
Götzlaff, W., 1983, *Diplom-Thesis*, Universität Marburg.
Hefner, W. and Hensel, F., 1982, *Phys. Rev. Lett.*, **48**, 1026.
Hefner, W., Schmutzler, R.W. and Hensel, F., 1980, *J. Physique*, **41**, Colloque C-8, 62.
Hefner, W., Sonneborn-Schmick, B. and Hensel, F., 1982, *Ber. Bunsenges. Phys. Chem.*, **86**, 844.
Hensel, F., 1980, *Angewandte Chemie*, **92**, 598; *Angewandte Chemie, International Edition in English*, **19**, 593.
Hensel, F., 1982, *Proceedings of 8th Symposium on Thermophysical Properties*, edited by J.v. Sengers (New York: ASME), p. 151.
Hensel, F., 1984, in *Nato ASI Series C*, Vol. 30, edited by J.V. Acrivos, N.F. Mott and A.D. Yoffe (Dordrecht: Reidel), p. 401.
Hernandez, J.P., 1982, *Phys. Rev. Lett.*, **48**, 1682.
Hernandez, J.P., 1984, *Phys. Rev. Lett.*, **53**, 2320.
Hernandez, J.P., 1985, *Phys. Rev.* B, to be published.
Hernandez, J.P., Schönherr, G., Götzlaff, W. and Hensel, F., 1984, *J. Phys. C.*, **17**, 442.
Hubbard, J., 1964, *Proc. R. Soc.* A, **277**, 237; ibid, **281**, 401.
Hubbard, S.R. and Ross, R.G., 1983, *J. Phys. C.*, **16**, 6921.
Ikezi, H., Schwarzenegger, K., Simons, A.L., Passaner, A.L. and McCall, S.L., 1978, *Phys. Rev.* B, **18**, 2494.
Jüngst, S., 1985, *PhD Thesis*, Universität Marburg.
Kahl, G. and Hafner, J., 1984, *Phys. Rev.* A, **29**, 3310.
Kanazawa, H. and Matsudawa, N., 1960, *Prog. Theor. Phys.*, **23**, 433.
Knuth, B., 1985, *Diplom-Thesis*, Universität Marburg.
Krohn, C.E. and Thompson, J.C., 1980, *Phys. Rev.* B, **21**, 2619.
Largarkov, A.N. and Sarychev, A.K., 1978, *Teplofizika Vysokikh Temperatur*, **16**, 903.
Landau, L. and Zeldovitch, G., 1943, *Acta Physica Chimica USSR*, **18**, 194.
Levin, M. and Schmutzler, R.W., 1984, *J. Non. Cryst. Solids*, **61–62**, 83.
Lifshitz, I.M. and Gredescul, S.A., 1970, *Sov. Phys.—JETP*, **30**, 1197.
McLaughlin, I.L. and Young, W.H., 1984, in *Liquid and Amorphous Metals V*, edited by C.N.J. Wagner and W.L. Johnson (Amsterdam: North Holland Physics Publishing).
Moruzzi, V.L., Janak, J.F. and Williams, A.R., 1978, *Calculated Electronic Properties of Metals* (New York: Pergamon).
Mott, N.F., 1956, *Can. J. Phys.*, **34**, 1356.
Mott, N.F., 1961, *Phil. Mag.*, **6**, 287.

Mott, N.F., 1966, *Phil. Mag.*, **13**, 989.
Mott, N.F., 1974, *Metal–Insulator Transitions* (London: Taylor and Francis).
Nara, S., Ogawa, T. and Matsubara, T., 1977, *Prog. Theor. Phys.*, **57**, 1474.
Nicoloso, N. and Freyland, W., 1985, to be published.
Overhof, H., Uchtmann, H. and Hensel, F., 1976, *J. Phys. F: Metal. Phys.*, **6**, 523.
Pestak, M.W. and Chang, M.H.W., 1981, *Phys. Rev. Lett.*, **46**, 939.
Redmer, R. and Röpke, G., 1985, *Physica* A, to be published (see also further references therein).
Rose, J.H., 1980, *Phys. Rev.* B, **23**, 552.
Schmutzler, R.W. and Hensel, F., 1972, *J. Non. Cryst. Solids*, **8–10**, 718.
Schönherr, G., 1984, in *Nato ASI Series C*, Vol. 30, edited by J.V. Acrivos, N.F. Mott and A.D. Yoffe (Dordrecht: Reidel).
Schönherr, G., Schmutzler, R.W. and Hensel, F., 1979, *Phil. Mag.* B, **40**, 411.
Sengers, J.V., 1982, in *Phase Transitions*, edited by M. Levy, J.C. LeGuillon and J. Zinn-Justin (Plenum Publishing Corporation).
Seydel, U. and Fucke, W., 1978, *J. Phys. F: Metal Phys.*, **8**, L157.
Seyer, P., Schmutzler, R.W. and Hensel, F., 1985, to be published.
Shimoji, M., 1977, *Liquid Metals* (London: Academic Press).
Thomas, G.A., 1984, *Journal of Physical Chemistry*, **88**, 3749.
Turkevich, L.A. and Cohen, M.H., 1984a, *Ber. Bunsenges. Phys. Chem.*, **88**, 297.
Turkevich, L.A. and Cohen, M.H., 1984b, *Phys. Rev. Lett.*, **53**, 2323.
Uchtmann, H., 1978, *PhD Thesis*, Universität Marburg.
Uchtmann, H., Popielawki, J. and Hensel, F., 1981, *Ber. Bunsenges. Phys. Chem.*, **85**, 555.
Warren, W.W., Jr., 1977, *Liquid Metals 1976* (*Inst. Phys. Conf. Ser. 30*), edited by R. Evans and D.A. Greenwood.
Warren, W.W., Jr., 1984, *Phys. Rev.* B, **29**, 7012.
Warren, W.W., Jr. and Hensel, F., 1982, *Phys. Rev.* B, **26**, 5980.
Warren, W.W., Jr. and Mattheiss, L.F., 1984, *Phys. Rev.* B, **30**, 3103.
Yao, M. and Endo, H., 1982, *J. Phys. Soc., Japan*, **51**, 1504.
Yao, M., Uchtmann, H. and Hensel, F., 1985, *Phil. Mag.*, to be published.
Yonezawa, F. and Ogawa, T., 1982, *Prog. Theor. Phys.* (*Japan*) *Suppl.*, **72**, 1.

CHAPTER 5

the metal–nonmetal transition in ammonia and methylamine solutions

J. C. Thompson
The University of Texas at Austin, USA

1. Introduction

Solutions of alkali metals in liquid ammonia are ionically conducting and blue coloured when dilute. Concentrated solutions are electronically conducting and bronze coloured. One of the earliest, perhaps *the* earliest, comments on the metal–nonmetal (M–NM) transition, as a transition, was made in 1921 by C.A. Kraus (1921) in discussing his conductivity data for metal–ammonia solutions. He had determined the primary carrier to be negative, and massless by chemical standards. He then stated:

> In dilute solutions the process is electrolytic. The negative carrier is chemically uncombined, but is associated with one or more molecules of the solvent... As the concentration of the solution increases the nature of the phenomenon changes only insofar as the combination of the negative carrier with NH_3 is affected. At the higher concentrations, the negative carriers are free from association with the NH_3 molecules... It may be surmised, therefore, that conduction in metals is effected by the same carrier.

That such a conclusion should be reached between the times of Drude and Sommerfeld is quite remarkable. Kraus later added (Kraus 1931),

> I should point out that the property of metallicity is not inherent in the atoms themselves, but, rather, depends on their state or condition... Hg in the solid or liquid state is a metal while, in the gaseous state, it is as much a nonmetal as is Ar.

Shortly thereafter, K.F. Herzfeld (1927) also noted that metal–ammonia solutions undergo a M–MN transition. He took the change in conductivity as an example of the release of the metal valence electrons following a polarization catastrophe.

So it was natural that N.F. Mott (1961) should include metal–ammonia solutions as examples of systems exhibiting M–NM transitions in his extensive review. He not only remarked upon the conductivity transition but upon the apparent association of that change with the phase separation. A great deal of effort has gone into the study of these unique materials and into the relation of their properties to the various theories of the M–MN transition.

With the assault of the 'gang-of-four' in 1979 (Abrahams *et al.* 1979) attention turned to quantum localization at the $T=0$ limit and to doped, stressed Si (Rosenbaum *et al.* 1983). The application of the ideas of the gang-of-four to the hot metal–ammonia solutions is not as straightforward as at a few millikelvin. Even in the nonmetallic state there are mobile yet localized carriers so that one cannot fix the transition, x_c, by a vanishing conductivity. In the absence of x_c the evaluation of critical exponents, etc., is impossible. Similarly, experiments to determine the dielectric constant, ε_0, are difficult in a system of finite conductivity so that the data required to establish the divergence of that quantity are limited. So then, one is restricted to qualitative observations not dissimilar to those made by Mott over 20 years ago.

2. The transition

The rapid rise in conductivity (Kraus 1921, Cohen & Thompson 1968) seen in the region marked B–C in Figure 1 is the major reason one speaks of a M–NM transition in metal–ammonia solutions. The most rapid rise is near 4 mole% metal. The thousandfold increase in σ resulting from a fourfold increase in the metal concentration (or a 40% decrease in the ion–ion separation) is consistent with observations of the M–NM transition in other systems. Since T/T_F is below 0·05, these are 'low temperature' experiments in the sense, say, of those on Si:P at 4 K when $T_F=120$ K (Rosenbaum *et al.* 1983). However, they do not reproduce mK conditions nor the high mobility in the metallic state and frozen-out carriers in the nonmetallic state found in Si:P. Metal–ammonia mobilities never exceed a few tens of $cm^2\ V^{-1}\ s^{-1}$. Solvated electron mobilities are 10^{-2}–$10^{-3}\ cm^2\ V^{-1}\ s^{-1}$.

The phase separation between metallic and nonmetallic phases shown in Figure 2 (Chieux & Sienko 1970, Hsu *et al.* 1984) has long been recognized as an important aspect of the mobility transition (Pitzer 1958, Mott 1961, Krumhansl 1965, Edwards & Sienko 1981, Thomas 1984). The consolute point in sodium–ammonia is near 4·1 mole % metal. The shape of the phase transition curve is, in the main, given by mean field theory with $\beta=1/2$ (Chieux & Sienko 1970). Such behaviour is generally observed for phase separation in conducting fluids where coulomb interactions are important (Ichikawa & Thompson 1973). Very close to the consolute point there is a crossover to the non-classical exponent $\beta=0{\cdot}31$. The region of the M–NM transition in the homogeneous alkali system is marked by cross-hatching above the consolute point (the colour change also occurs at the same composition). It is generally assumed that when the Mott criteria for a

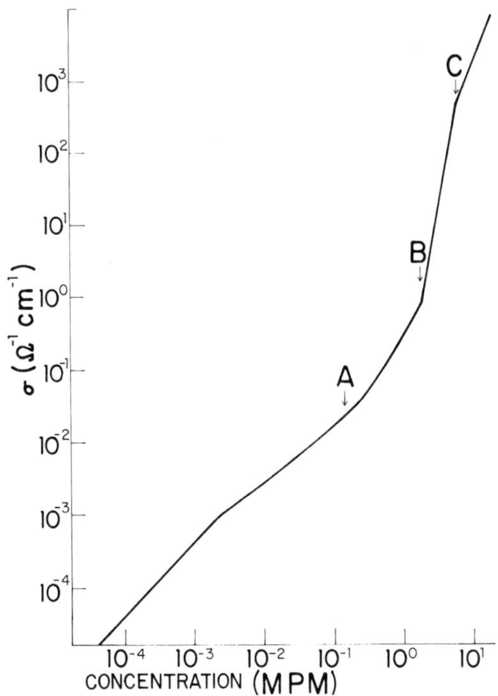

Figure 1. Conductivity of NH_3:Na solutions (Kraus 1921). A marks the range of association processes, while the M–NM transition is in the range B–C. MPM is mole % metal.

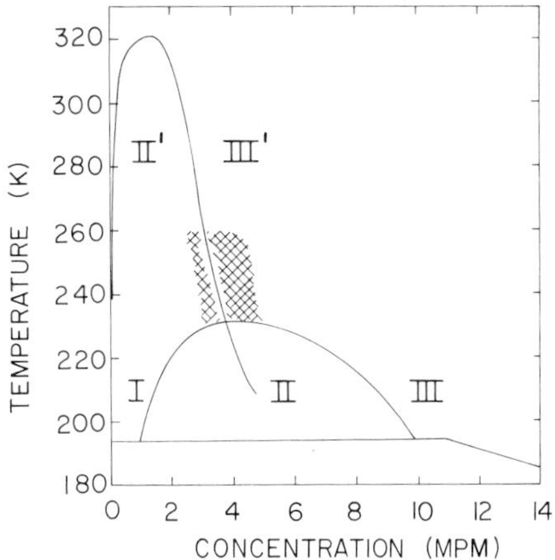

Figure 2. Phase diagram for NH_3:Na and NH_3:Eu solutions. Region I is nonmetallic, regions II and II′ are inhomogeneous, and regions III and III′ are metallic. The crosshatching marks the M–NM transition for homogeneous NH_3:Na solutions.

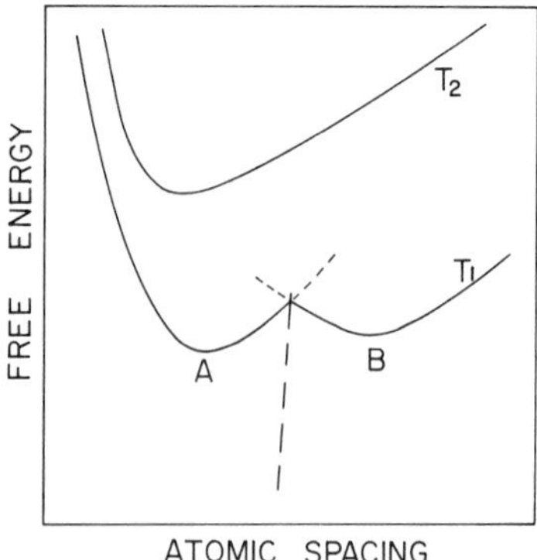

Figure 3. Free energy as a function of atomic spacing. At T_1 there is a phase separation between phases with the spacings marked with A and B. There is no separation for T_2.

M–NM transition (Mott 1974, Edwards & Sienko 1981):

$$n_0^{1/3} a_{\mathrm{H}}^* = 0{\cdot}25 \tag{1}$$

is satisfied, the electron delocalizes and, in a fluid, there is a phase separation. Free-energy curves similar to those sketched in Figure 3 occur when cohesive energies change (Mott 1961), as from metallic to van der Waals. At low temperatures, T_1, there is a phase separation, and at higher temperatures, T_2, a continuous transition. The differences in x_c between the Na and Eu solutions is at least in part the consequence of the two valence electrons per atom donated by the Eu. On this basis the critical compositions should differ by a factor of two. That the factor is closer to three is probably a reflection (at least in part) of the same ionic size and polarizability effects that scramble the order of the consolute temperatures among the alkalis. While the competition between metallic and other cohesive forces clearly brings the metal–nonmetal transition and phase separation together in europium–ammonia or sodium–ammonia solutions, such juxtaposition is not necessary. There is no phase separation in caesium–ammonia solutions at the M–NM transition (Swenumson *et al.* 1978) or elsewhere, while the phase separation (or critical fluctuations) are at significantly lower densities than the conductivity change in fluid mercury or in lithium–methylamine (Edwards 1980).

3. Transport coefficients

The conductivity for metal–ammonia solutions in Figure 1 rises most rapidly in the neighbourhood of 3 mole % metal (Swenumson *et al.* 1978) just below the consolute point at 4 mole % metal. The Mott–Ioffe–Regel criterion

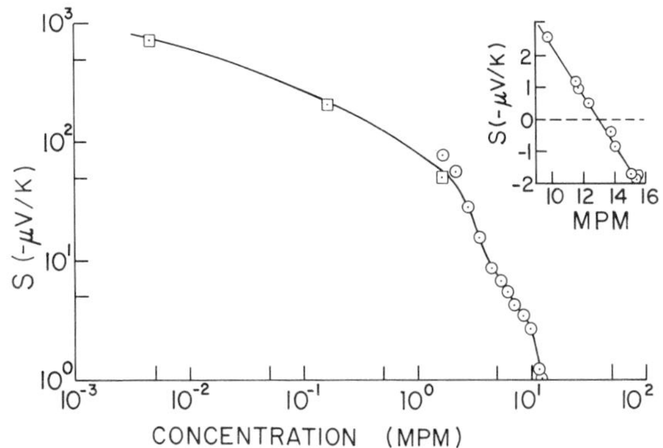

Figure 4. Thermoelectric power S for NH_3:Na solutions as a function of concentration. Note that S is negative except above 13 mole % metal (MPM).

(Ioffe & Regel 1960, Mott 1974) for the metal–nonmetal transition is that

$$k_F l \sim 1 \tag{2}$$

where l is the mean free path. This criterion is satisfied near 5 mole % metal, just above the consolute point. All other charge transport parameters show some sign of change in the same range for NH_3 (Thompson 1976). In lithium–methylamine solutions the transition occurs at 15 mole % metal or higher. This trend is consistent throughout all the observations. The maximum conductivity is about 400 $(\Omega\ \mathrm{cm})^{-1}$ in lithium–methylamine at 22 mole % metal, which is the saturation composition. Rather than the phase separation which marks the M–NM transition in NH_3 solutions, there is an end to the solubility in CH_3NH_2 at the point of electron delocalization. The two do not seem to be related, as a fourfold solvation for Li in NH_3 is also observed. No true metallic phase is observed for lithium–methylamine (Toma *et al.* 1976), yet most of the precursors are observed.

The thermopower S rises (Dewald & Lepoutre 1954, 1956, Damay *et al.* 1971, Hahne *et al.* 1976, Hirasawa *et al.* 1978, Niibe *et al.* 1982) in magnitude rapidly as the concentration drops below 4 mole % metal, see Figure 4, yet reaches the standard (Mott & Davis 1979) nonmetallic value of $k/e = 86\ \mu\mathrm{V\ K}^{-1}$, only below 1 mole % metal. Similarly, the expected relation

$$\ln \sigma = -(e/k)S \tag{3}$$

is only satisfied well below 1 mole % metal. Similar results are found (Toma *et al.* 1976) in lithium–methylamine at 6 mole % metal.

The Hall coefficient R_H and mobility $\mu_H = \sigma R_H$ also show the effects of localization as shown in Figure 5. R_H begins its drop below the free-electron value already at 6 mole % metal.

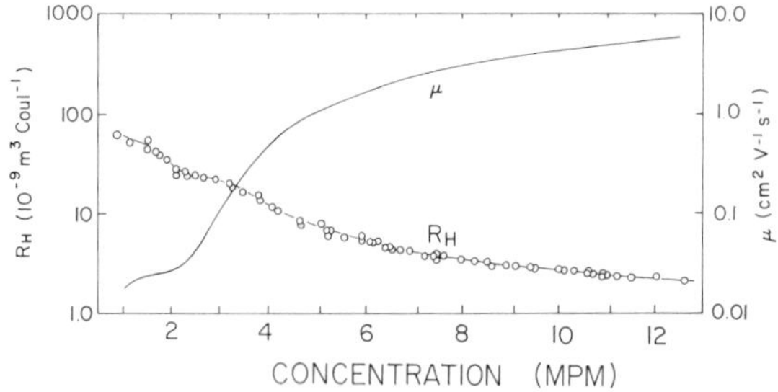

Figure 5. Hall coefficient R_H and mobility μ for NH_3:Cs solutions.

Mott (1974) introduced the g-factor given by

$$g = N(E_F)/N(E_F)_{free} \tag{4}$$

where $N(E)$ is the density of states and $N(E_F)_{free}$ is the free electron density of states. He suggested as a criterion for the localization of carriers that g should drop below 1/3. Friedman (1971) showed in a strong scattering regime that g entered R_H and σ in such a way that $\mu_H \propto \sigma^2$. In the metal–ammonia solutions the exponent is found to be 1·5 rather than 2·0 (Even *et al.* 1977). However, the values of g derived from R_H and σ reach 1/3 in the neighbourhood of 3 mole % metal. Nakamura *et al.* (1974) have derived a relation for S and g which yields essentially the same transition composition (Hirasawa *et al.* 1978).

Much recent attention has been focused on the dielectric constant ε near the M–NM transition in Si–P and fluid Hg (Herzfeld 1927, Hefner & Hensel 1982, Rosenbaum *et al.* 1983, Bhatt 1984, Thompson 1984a,b, Logan & Edwards 1985). From the elementary arguments by Herzfeld (1927) based on the Clausius–Mossotti relation, as well as more sophisticated scaling theories of the gang-of-four, ε is expected to diverge at the M–NM transition. Elegant measurements (Rosenbaum *et al.* 1983) on P-doped, stressed Si have found the expected divergence, as discussed elsewhere in this volume. Microwave and optical measurement in metal–ammonia solutions also show a strong enhancement of ε, though experimental problems limit the quality and range of the current data. Figure 6 shows some of the results (Thompson 1984a,b, and references therein). The pure NH_3 value is near 20, and a twenty-fold rise has been reported by Radscheit & Breitschwerdt (1973) at 1·2 mole % metal. However, as might be expected, the peak is rounded and shifted at the high frequencies of these experiments. Though small, the peak observed in the visible range at 0·6 eV is consistent (Koehler & Lagowski 1969, Somoano & Thompson 1970) with data taken at higher energies and is real.

Goetze (1981) has developed a theory of the change of $\varepsilon(\omega) = \varepsilon_1(\omega) + i\varepsilon_2(\omega)$ in

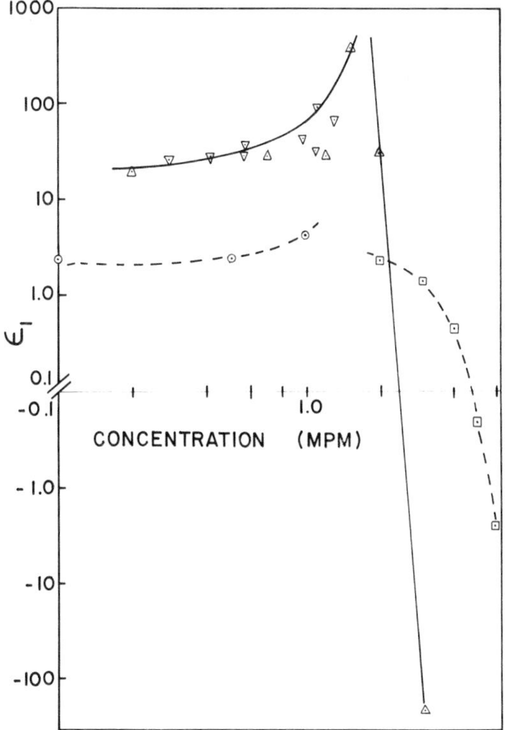

Figure 6. Real part of the dielectric constant ε_1. The triangles are for microwave data and the circles and squares for optical data. Note the break in the scale between positive and negative values.

the region of a M–NM transition. He predicts that the zero-frequency divergence weakens and converts to a peak at non-zero frequencies as the system moves into the metallic range. Eventually Drude behaviour takes over. Goetze points out that the usual Drude sum rule

$$ne^2/m = \int \omega\varepsilon_2(\omega)\,\mathrm{d}\omega \tag{4}$$

must be obeyed. Somoano & Thompson (1970) and also Koehler & Lagowski (1969) have observed such behaviour in metal–ammonia solutions. However, even in the most metallic solutions (Thompson & Cronenwett 1967) the sum rule yields half the expected value. This has been attributed to an effective mass differing from unity. The results of the sum rule calculation are shown in Figure 7 and continued into the nonmetallic state using the oscillator strengths of the solvated electron absorption. Goetze's sum rule would seem to be obeyed.

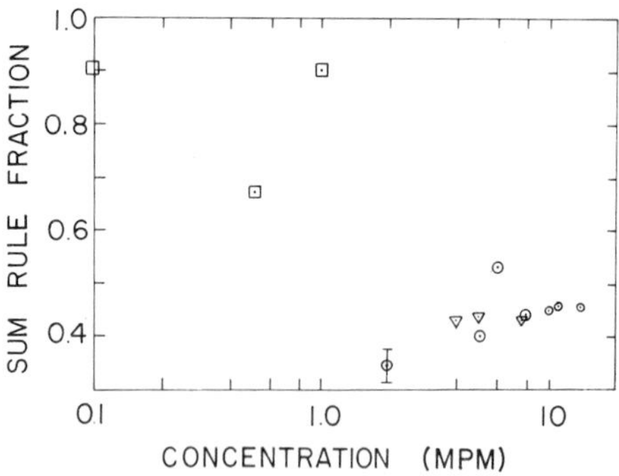

Figure 7. The ratio of the sum rule observed for $\mathrm{Im}(\varepsilon)=\varepsilon_2$ (squares) and for $\mathrm{Im}(1/\varepsilon)$ (circles and triangles) to the calculated values. Substantial extrapolation is required for the ε_2 integral, but not for the other.

4. Magnetic phenomena

Edwards (1984) has recently given an extensive review of the magnetic effects observed at the M–NM transition. This includes magnetic resonance studies as well as susceptibility. A systematic shift of the 'transition' to lower conformations is observed and has been lucidly explained by D. F. Holcomb (1978). There is extensive new data by Niibe *et al.* (1984) and Nakamura *et al.* (1984).

The susceptibility rises from temperature-independent Pauli paramagnetism toward Curie–Weiss behaviour as NH_3 is added to the concentrated solutions. There is, however, a significant spin pairing in the nonmetallic state. Therefore, the magnetic transition is from a set of itinerant spins to an assembly of spin-paired, molecular-type entities (Edwards 1984). Such antiferromagnetic coupling is the normal consequence of the exchange coupling of two spins. And such coupling is to be expected as the localized electrons become more numerous as one approaches the M–NM transition from *below*. The problem is that the spin pairing begins far below the M–NM transition, and that the lifetime of the singlet state is near 1 μs (Edwards 1984). The description of a 'species' with the correct properties in this fluid system is still controversial (Schindewolf 1984).

Warren (1971) has pointed out, in studies of Ga_2Te_3, that there is an enhancement of the nuclear spin-lattice relaxation rate over that given by the Korringa relation. This enhancement is shown in Figure 8 for several systems (Nakamura *et al.* 1984). Also shown is a corresponding reduction seen in the electron relaxation rate computed by Edwards (1984). A similar reduction, but by a smaller factor, for the electron relaxation in caesium–ammonia solutions can be derived from the work of Catterall and Burton reported at Colloque Weyl VI (Burton 1984). The more modest modifications of the rates in NH_3 over CH_3NH_2 solutions is consistent for both nuclear and electron relaxation. Holcomb (1978) has emphasized that the use of the metallic relaxation formulae

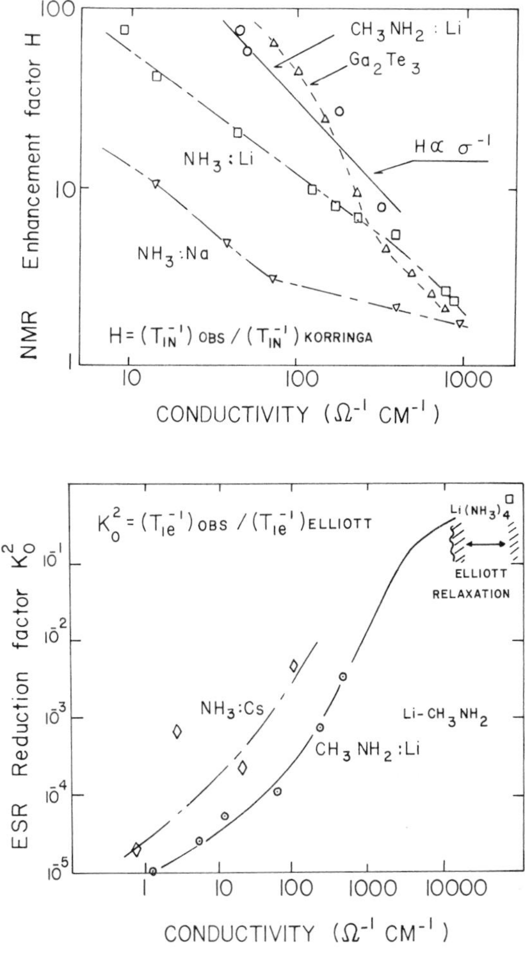

Figure 8. The n.m.r. enhancement factor H and the e.s.r. reduction factor K_0^2 as a function of conductivity for several materials. H is taken from Nakamura *et al.* (1984). K_0^2 is taken from Edwards (1984) and from Burton (1984).

in this low conductivity regime is questionable. In the absence of a better theory for nuclear and electronic relaxation in the region of electron localization, such imperfect guides are the only ones available. The differences they suggest between ammonia and methylamine are currently unexplained.

5. Other results

The change in cohesive forces which occurs upon the electron localization–delocalization process has already been cited as the basis of the phase separation.

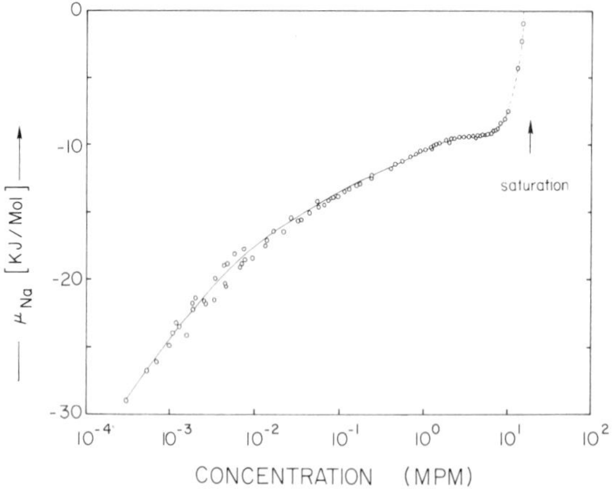

Figure 9. The chemical potential of Na in NH_3.

Other thermodynamic parameters are also expected to change, though there has been relatively little systematic study in this or any other system.

Variations in the solute chemical potential, μ, or activity coefficient, are directly related to the phase separation. Figure 9 shows the flattening indicative of critical fluctuations just above the consolute point (Schindewolf & Werner 1980). This is, of course, just the range of the M–NM transition. The isothermal compressibility contains the second derivative of μ and also reflects the transition. So also does the adiabatic compressibility which may be calculated from sound speed measurements. Bowen and co-workers (Bowen 1970, Fenner & Bowen 1980) have made extensive studies of both the sound speed and attenuation. The latter gives independent evidence of the presence or absence of the critical fluctuations associated with the phase separation. However, at this stage it reveals nothing new about the nature of the transition. The sound speeds, Figure 10, are solute-independent in the dilute solutions where the structure is dominated by the solvated electron and the solvent. As the metallic state is entered, the sound speeds become solute-dependent because the solvation number varies with solute (4 for Li, 6 for K etc.). A simple hard-sphere model has been successful in describing the observations (Thompson 1971).

6. Discussion

A model of both the metallic and nonmetallic states is required for an understanding of the data reviewed above and of the M–NM transition. The natures of the localized and extended states must be described if the delocalization process is to be determined.

Concentrated NH_3:M solutions behave as simple liquid metals and have

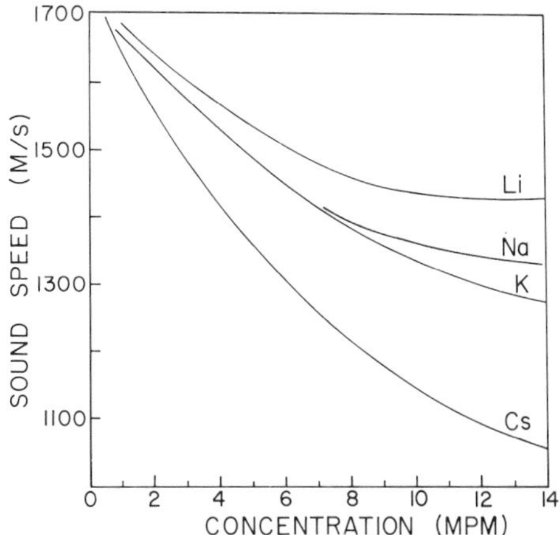

Figure 10. The speed of sound in NH_3 solutions at 240 K.

been described within the Mott-inspired Faber–Ziman theory of binary metal alloys (Ashcroft & Russakoff 1970, Schroeder & Thompson 1968). The electrons are taken as free and as scattered by solvated metal ions and by ammonia molecules. Even at the most elementary level (Pohler & Thompson 1964) the right order-of-magnitude of the conductivity is obtained. In the absence of extensive structural data the basis of the small *positive* temperature coefficient of conductivity cannot be assigned but it is not unusual when the conductivity is in the range of 10^3–10^4 $(\Omega\ \mathrm{cm})^{-1}$, as here (Allgaier 1969). Consistent interpretation of the conductivity and compressibility can be made with cation solvation numbers of 4, 4, 6 and 8 for Li, Na, K and Cs solutions respectively.

The localized electron state is the solvated electron. The interaction of the charge and molecular dipoles, together with long-range polarization, stabilizes the electron in a polaronic state. Extensive theoretical effort (for reviews see Kevan & Webster 1976, Thompson 1984a,b, Brodsky & Tsarevsky 1984) leads to a picture of the solvated electron as an electron in a molecular-sized cavity surrounded by several oriented solvent molecules, though this view is not universally accepted (Tuttle *et al.* 1984). The conventional view places the ground-state electron in a 1s level about 1·3 eV below the vacuum in NH_3. There is a 2p excited state and an optical absorption associated with excitation from the ground state. There is still controversy on the role of bound-to-free transitions in the absorption. A variety of data (Schindewolf 1984) support the association of solvated electrons and cations into a species of the stoichiometry $e^-M^+e^-$ well below the M–NM transition. This entity has, in effect, a negative Hubbard U similar to that found in the D^- state in an amorphous chalcogenide (Mott & Davis 1979).

Krohn *et al.* (1980) have interpreted recent electrochemical and photoemission data in terms of a conduction band *intrinsic* to NH_3 with a band of *intrinsic* localized states below it. Current estimates place the bottom of the conduction band, V_0, at 0·8 eV below the vacuum (Thompson *et al.* 1985). The tail states fall off exponentially below V_0 with a characteristic energy of 0·2 eV so that there is little of the intrinsic level remaining at the solvated electron level: 1·3 eV below the vacuum. The total density of states (Henglein 1974) then is as shown in Figure 11. The placement of the vacant M^+ level above e_s^- reflects the dissolution reaction: $M \rightarrow M^+ + e^-$. There is some auto-ionization of NH_3 so that NH_4^+ and NH_2^- should be added as well.

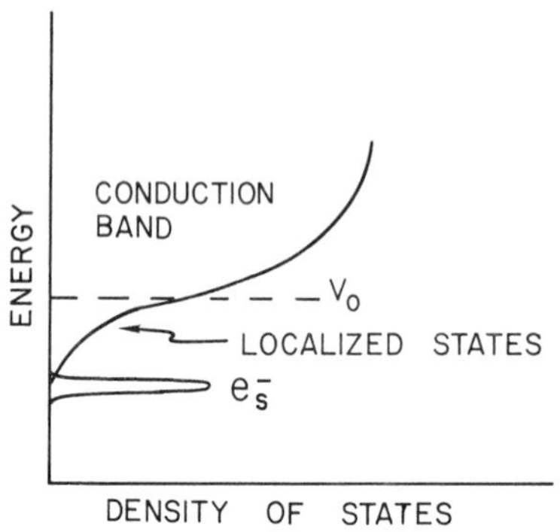

Figure 11. The density of electronic states in NH_3.

These are the states which evolve into a metallic band as metal (and electrons) are added to NH_3 or CH_3NH_2. Both e_s^- and M^+ levels may be expected to broaden. The competition of disorder and correlation then begins. As already noted, little of the simplification of $T \rightarrow 0$, which permits analysis of Si:P, is available here.

The major problem is in identifying the transitional composition, x_c. If the consolute point, 4 mole % metal, is chosen then the Mott conductivity $\sigma_{\min} = 0{\cdot}05e^2/a = 0{\cdot}05(e^2/\hbar)n_c^{1/3}$ turns out to be 120 $(\Omega\,\text{cm})^{-1}$. The conductivity reaches that value at about 4·5 mole % metal, close enough. In Si:P the rise in σ is given by $\sigma = \sigma_0[(n/n_c) - 1]^{0\cdot5}$ with σ_0 a factor of 10 greater than $\sigma_{\min}$ (Rosenbaumm *et al.* 1983). Using $x_c = 4$ mole % metal and $\sigma_0 = 84$ $(\Omega\text{ cm})^{-1}$ (close to $\sigma_{\min}$), the power law for σ is found to be close to 1·0 in potassium–ammonia solutions (Swenumson *et al.* 1978) (see Figure 12). Though this exponent does agree with that found in amorphous $Si_{1-x}Nb_x$, the present data only cover the range for $x/x_c > 1{\cdot}10$. The agreement is of questionable significance. The determination of an exponent for CH_3NH_2–Li is more difficult as there is no clear indication of a phase separation (Nakamura *et al.* 1974, Buntaine & Sienko 1980, Hagedorn & Lelieur 1980, Odom & Bowen 1984). Using the Ioffe–Regel criteria, equation (1), and densities extrapolated from 10 mole % metal (Yamamoto *et al.* 1971) a critical composition near 20 mole % metal is indicated. Too little data is available before saturation to

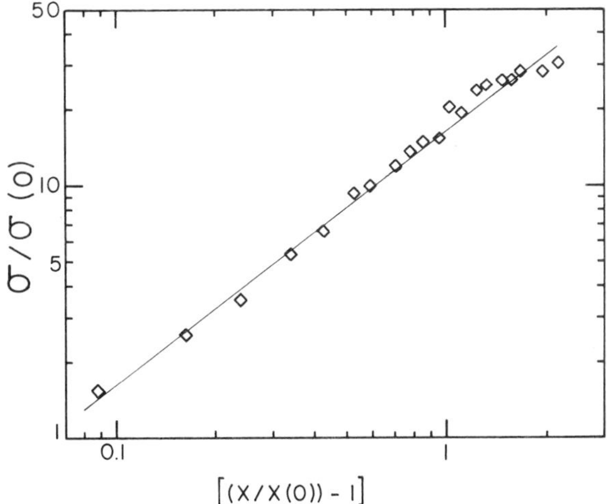

Figure 12. The ratio of the conductivity in NH_3:K solutions to the Mott conductivity as a function of the reduced concentration. $X(0)$ is the concentration at which the Mott conductivity is observed. The line has unit slope.

fix any dependencies. The range of accessible temperatures in metal–ammonia solutions is limited by the phase separation. No distinctions can be made among possible forms of $\sigma(T)$ near the M–NM transition. At lower concentrations an exponential in $1/T$ is required, while a linear dependence of σ on T works at high concentrations (Cohen & Thompson 1968).

Thomas (1984) reports the suggestion that the difference in critical exponent for σ in Si:P and in Si–Nb alloys may be attributed to the excess of scattering sites over carriers in the latter system. If this is the case, then the unit exponent in potassium–ammonia is to be expected. Electron–cation and electron–solvent dipole scattering are known to be effective from the metallic side, at least (Ashcroft & Russakoff 1970).

7. Summary

Solutions of alkali metals in liquid NH_3 show a metal–nonmetal transition in close juxtaposition to the liquid–liquid phase separation. Alkaline earths show the phase separation at *about* the same valence electron density and the electronic transition is again correlated with it. The same mechanism is responsible for both, as Mott suggested many years ago. Yet, as Ziman (1979) aptly observed, metal–ammonia solutions do not provide the idealization of a disordered conducting system. Quantitative comparison with recent quantum localization theories is difficult for these hot fluids. Determination of the relative roles of disorder and of correlation cannot be made. Nevertheless, there are tantalizing similarities between what is seen here and what has been theorized.

Acknowledgements

My work in this field has benefited from the insight and imagination of Sir N.F. Mott and from financial aid from the US NSF and DOE and the R.A. Welch Foundation.

References

Abrahams, E., Anderson, P.W., Licciardello, D.C. and Ramakrishnan, T.V., 1979, *Phys. Rev. Lett.*, **42**, 693.
Allgaier, R.S., 1969, *Phys. Rev.*, **185**, 227.
Ashcroft, N.W. and Russakoff, G., 1970, *Phys. Rev.* A, **1**, 39.
Bhatt, R.N., 1984, *Phil. Mag.* B, **50**, 189.
Bowen, D.E., 1970, in *Metal–Ammonia Solutions*, edited by J.J. Lagowski and M.J. Sienko (London: Butterworth), p. 355.
Brodsky, A.M. and Tsarevsky, A.V., 1984, *J. Phys. Chem.*, **88**, 3790.
Buntaine, J.R. and Sienko, M.J., 1980, *J. Physique*, **41C**, 8–36.
Burton, K., 1984, *PhD Thesis*, Salford University.
Chieux, P. and Sienko, M.J., 1970, *J. Chem. Phys.*, **53**, 566.
Cohen, M.H. and Thompson, J.C., 1968, *Adv. Phys.*, **17**, 857.
Damay, P., Chieux, P. and Lepoutre, G., 1971, *Ber. Bunsenges. Phys. Chem.*, **75**, 642.
Dewald, J.F. and Lepoutre, G., 1954, *J. Am. Chem. Soc.*, **76**, 3369.
Dewald, J.F. and Lepoutre, G., 1956, *J. Am. Chem. Soc.*, **78**, 2956.
Edwards, P.P., 1980, *J. Phys. Chem.*, **84**, 1215.
Edwards, P.P., 1984, in *Physics and Chemistry of Electrons and Ions in Condensed Matter*, edited by J.V. Acrivos, N.F. Mott and A.D. Yoffe (Dordrecht: Reidel), p. 297.
Edwards, P.P. and Sienko, M.J., 1981, *J. Am. Chem. Soc.*, **103**, 2967.
Even, U., Swenumson, R.D. and Thompson, J.C., 1977, in *Liquid Metals 1976* (Inst. Phys. Conf. Ser. 30), edited by R. Evans and D.A. Greenwood (London: Institute of Physics), p. 424.
Fenner, D.B. and Bowen, D.E., 1980, *J. Phys. Chem.*, **84**, 1190.
Friedman, L., 1971, *J. Non. Cryst. Solids*, **6**, 329.
Goetze, W., 1981, in *Recent Developments in Condensed Matter Physics*, Vol. 1, edited by J.T. Devreese (New York: Plenum), p. 133.
Hagedorn, R. and Lelieur, J.-P., 1980, *J. Phys. Chem.*, **84**, 3652.
Hahne, D., Krebs, P. and Schindewolf, U., 1976, *Ber. Bunsenges. Phys. Chem.*, **80**, 804.
Hefner, W. and Hensel, F., 1982, *Phys. Rev. Lett.*, **48**, 1026.
Henglein, A., 1974, *Ber. Bunsenges. Phys. Chem.*, **78**, 1078.
Herzfeld, K.F., 1927, *Physical Review*, **39**, 701.
Hirasawa, M., Nakamura, Y. and Shimoji, M., 1978, *Ber. Bunsenges. Phys. Chem.*, **82**, 815.
Holcomb, D.F., 1978, in *Metal–Nonmetal Transition in Disordered Systems*, edited by L.R. Friedman and D.P. Tunstall (Edinburgh: Scottish Universities Summer School), p. 251.
Hsu, P.S., Zimm, C.B. and Glaunsinger, W.S., 1984, *J. Solid St. Chem.*, **54**, 346.
Ichikawa, K. and Thompson, J.C., 1973, *J. Chem. Phys.*, **59**, 1680.
Ioffe, A.F. and Regel, A.R., 1960, *Progress in Semiconductors*, **4**, 237.
Kevan, L. and Webster, B., 1976, *Electron–Solvent and Anion–Solvent Interactions* (Amsterdam: Elsevier).
Koehler, W.H. and Lagowski, J.J., 1969, *J. Phys. Chem.*, **73**, 2329.

Kraus, C.A., 1921, *J. Am. Chem. Soc.*, **43**, 749.
Kraus, C.A., 1931, *J. Franklin Institute*, **212**, 537.
Krohn, C.E., Antoniewicz, P.R. and Thompson, J.C., 1980, *Surface Science*, **101**, 241.
Krumhansl, J.A., 1965, in *Physics of Solids at High Pressure*, edited by C.T. Tomizuka and R.M. Emrick (New York: Academic), p. 425.
Logan, D.E. and Edwards, P.P., 1985, submitted for publication.
Mott, N.F., 1961, *Phil. Mag.*, **6**, 287.
Mott, N.F., 1974, *Metal–Insulator Transitions* (London: Taylor & Francis).
Mott, N.F. and Davis, E.A., 1979, *Electronic Processes in Non-Crystalline Materials* (Oxford: Clarendon).
Nakamura, Y., Matsumura, K. and Shimoji, M., 1974, *J. Chem. Soc. Faraday Trans. I*, **70**, 273.
Nakamura, Y., Niibe, M. and Shimoji, M., 1984, *J. Phys. Chem.*, **88**, 3555.
Niibe, M. and Nakamura, Y., 1984, *J. Phys. Chem.*, **88**, 5608.
Niibe, M., Nakamura, Y. and Shimoji, M., 1982, *J. Phys. Chem.*, **86**, 4513.
Odom, B.M. and Bowen, D.E., 1984, *J. Phys. Chem.*, **88**, 3904.
Pitzer, K.S., 1958, *J. Am. Chem. Soc.*, **80**, 5046.
Pohler, R.F. and Thompson, J.C., 1964, *J. Chem. Phys.*, **40**, 1449.
Radscheit, H. and Breitschwerdt, K., 1973, in *Electrons in Fluids*, edited by J. Jortner and N.R. Kestner (Berlin: Springer), p. 316.
Rosenbaum, T.F., Milligan, R.F., Paalanen, M.A., Thomas, G.A., Bhatt, R.N. and Lin, W., 1983, *Phys. Rev.* B, **27**, 7509.
Schroeder, R.L. and Thompson, J.C., 1968, *Bull. Am. Phys. Soc.*, **13**, 397.
Schindewolf, U., 1984, in *Physics and Chemistry of Electrons and Ions in Condensed Matter*, edited by J.V. Acrivos, N.F. Mott and A.D. Yoffe (Dordrecht: Reidel), p. 361.
Schindewolf, U. and Werner, M., 1980, *J. Phys. Chem.*, **84**, 1123.
Somoano, R.B. and Thompson, J.C., 1970, *Phys. Rev.* A, **1**, 376.
Swenumson, R.D., Even, U. and Thompson, J.C., 1977, *Phil. Mag.*, **37**, 311.
Thomas, G.A., 1984, *Phil. Mag.*, **50**, 169.
Thompson, J.C., 1971, *Phys. Rev.* A, **4**, 802.
Thompson, J.C., 1976, *Electrons in Liquid Ammonia* (Oxford: Clarendon).
Thompson, J.C., 1984a, in *Physics and Chemistry of Electrons and Ions in Condensed Matter*, edited by J.V. Acrivos, N.F. Mott and A.D. Yoffe (Dordrecht: Reidel), p. 385.
Thompson, J.C., 1984b, *J. Solid. St. Chem.*, **54**, 308.
Thompson, J.C. and Cronenwett, W.T., 1967, *Adv. Phys.*, **16**, 439.
Thompson, J.C., Antoniewicz, P.R., Bennett, G.T. and Coffman, R.B., 1985, to be published.
Toma, Nakamura, Y. and Shimoji, M., 1976, *Phil. Mag.*, **33**, 181.
Tuttle, T.R., Golden, S., Lwenje, S. and Stupak, C.M., 1984, *J. Phys. Chem.*, **88**, 3811.
Yamamoto, M., Nakamura, Y. and Shimoji, M., 1971, *Trans. Faraday Soc.*, **67**, 2292.
Warren, W.W., 1971, *Phys. Rev.* B, **3**, 3708.
Ziman, J.M., 1979, *Models of Disorder* (Cambridge: University Press).

CHAPTER 6

metal–metal halide melts

W.W. Warren, Jr
AT&T Bell Laboratories, Murray Hill, New Jersey 07974, USA

1. Introduction

In a series of pioneering investigations three quarters of a century ago, Aten (1909, 1910a, 1910b) showed that metals can form liquid solutions with their molten halides. The view of many up to that time had been that such 'solutions' are actually colloidal suspensions of metal in the molten salt. If there remained any doubts that these are true solutions, however, they were resolved definitively by the extensive studies of several groups, especially that of Bredig and co-workers, in roughly the decade 1955–65. The equilibrium phase diagrams of a large number of metal–metal halide systems were determined during this period together with basic physical properties of the liquid solutions. Most of this work is summarized in reviews by Bredig (1964) and Corbett (1964). It was firmly established by this time that liquid metal–metal halide mixtures form solutions which, under some conditions, link the metallic and nonmetallic states by continuous metal–nonmetal transitions at intermediate concentrations.

Beneath obvious general similarities lies a rich diversity of more detailed properties of the metal–metal halide melts. One aspect of this variety is illustrated by the selection of temperature–composition phase diagrams shown in Figure 1. Even within a single family, the alkali metal–alkali halide solutions, there is found widely differing solubility behaviour. Solutions of the light alkali metals such as Na–NaCl separate into metal- and salt-rich phases over a broad composition range. These solutions exhibit continuous solubility and metal–nonmetal transitions only above relatively high consolute temperatures T_c (Figure 1(*a*)). Solutions of the heavier alkali metals have progressively lower values of T_c until, for the caesium solutions, the 'miscibility gap' is altogether absent (Figure 1(*b*)).

For polyvalent metals, the formation of lower oxidation states broadens the nonmetallic range and shifts the miscibility gap to more metal-rich concentrations. The system Hg–$HgCl_2$ shown in Figure 1(*c*), for example, can be divided approximately into two ranges: a nonmetallic solution of $HgCl_2$ with the 'subhalide' Hg_2Cl_2 and a metal–salt solution Hg–Hg_2Cl_2. The latter has a wide

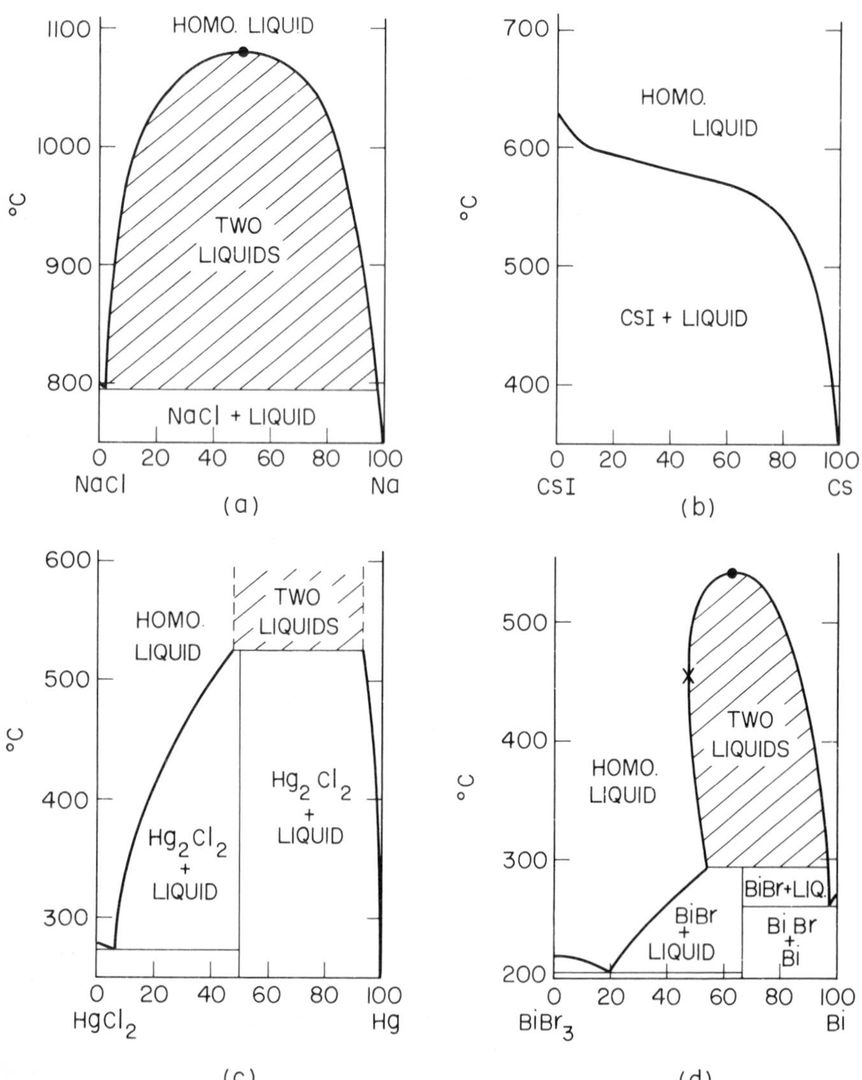

Figure 1. Temperature–composition phase diagrams of some selected metal–metal halide melts showing regions of homogeneous liquid solution and liquid–liquid phase separation: (*a*) Na–NaCl (Bredig 1964); (*b*) Cs–CsCl (Bredig *et al.* 1955); (*c*) Hg–$HgCl_2$ (Yosim & Mayer 1960); (*d*) Bi–$BiBr_3$ (Yosim *et al.* 1962). Closed circles (●) in (*a*) and (*d*) denote consolute points and cross (X) in (*d*) indicates point of maximum retrograde solubility.

range of insolubility and a very high value of T_c. The phase diagram of Bi–$BiBr_3$ (Figure 1(*d*)) is similar but contains the interesting feature of 'retrograde solubility' on the $BiBr_3$-rich side of the miscibility gap. Within a narrow composition range around 50 mole % Bi, a homogeneous solution will actually phase-separate on *heating*, then become homogeneous again at a still higher temperature.

Despite the long history of research on metal–metal halide melts, vigorous activity has continued to grow up to the present. Much of the most recent work has been motivated by strong interest in the metal–nonmetal transition and related problems concerning the states of electrons introduced into nonmetallic media. In the last few years we have seen not only further studies of basic physical properties such as electrical conductivity, density and magnetic susceptibility, but also increasing use of spectroscopic techniques such as optical absorption, nuclear magnetic resonance (n.m.r.), electron spin resonance (e.s.r.) and neutron scattering. These new approaches have resulted in the resolution of some long-standing problems concerning the electronic properties of the solutions and have uncovered new features of the transition of a metal into a nonmetallic ionic state.

It is the purpose of this chapter to summarize current understanding of the electronic structure and properties of metal–metal halide melts with strong emphasis on examples of recent work. Attention is focused mainly on two families of solutions for which sufficiently low T_c values permit study of the complete transition from the metallic to the nonmetallic state. These are the alkali metal–alkali halide and bismuth–bismuth halide solutions. So-called 'ionic alloys', which resemble metal–metal halide solutions because of strong charge-transfer effects, are also discussed briefly. For treatments of other systems and more complete surveys of the older literature, the reader is referred to the reviews of Bredig (1964), Corbett (1964) and Warren (1981).

2. The metallic and nonmetallic states in liquids—some basic concepts

The continuous transition from the metallic to the nonmetallic state is illustrated for some alkali metal and bismuth halide solutions in Figure 2. As the concentration† of metal decreases, the d.c. electrical conductivity in each case drops from $\sigma \sim 10^4$ $(\Omega\ \mathrm{cm})^{-1}$ in the pure liquid metal to $\sigma \sim 1$ $(\Omega\ \mathrm{cm})^{-1}$ in the pure molten halide. There are ranges of metallic and nonmetallic behaviour at the respective extremes of metal content and, at intermediate compositions, metal–nonmetal transitions. The most rapid change in conductivity falls in quite different ranges of composition for the two families of solutions. The reasons for this are discussed later in this chapter. Before considering the solutions in detail, however, we summarize briefly some essential physical ideas which are relevant to the three regimes.

† The compositions of solutions are denoted in this chapter by the mole fraction of metal x for solution $M_x(MX_n)_{1-x}$ or the mole % metal $100x$.

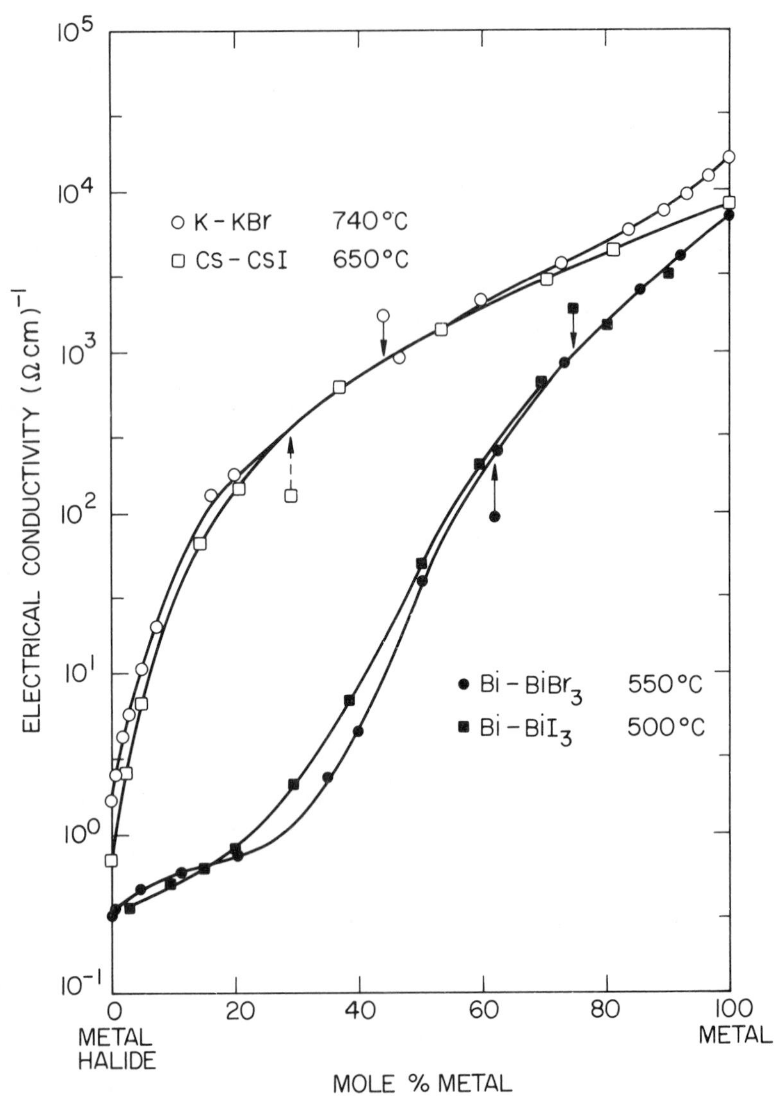

Figure 2. D.c. electrical conductivity versus composition for some metal–metal halide melts: (○) K–KBr at 740°C (Bronstein & Bredig 1958; Bronstein *et al.* 1962); (□) Cs–CsI at 650°C (Sotier, Ehm and Maidl 1984); (●) Bi–BiBr$_3$ at 550°C (Grantham 1965); (■) Bi–BiI$_3$ at 500°C (Grantham & Yosim 1963). Arrows indicate critical concentrations at consolute point or, for Cs–CsI, composition of peak in $S_{cc}(0)$.

2.1. Metallic state

Pure liquid metals are random structures in which the ionic potential is strongly screened by the high density of conduction electron charge. Strong screening leads to weak scattering of electrons in plane-wave states and the electronic mean-free-paths l are greater than the average interionic separation a. Sodium ($l/a \simeq 40$) is the 'best' liquid metal in this sense. Bismuth is a marginal case with $l/a \simeq 1{\cdot}1$. In the range $l/a \gtrsim 1$, the nearly-free-electron theory of Ziman (1961) permits calculation of the electrical resistivity of most liquid metals to within about a factor of two (Shimoji 1977).

Solution of a metal halide in an elemental liquid metal introduces additional scattering centres. These shorten the mean-free-path and reduce the conductivity. The effect can be treated in principle by the Faber–Ziman theory (Faber & Ziman 1965), a generalization of the nearly-free-electron theory to alloys. However, practical application of the Faber–Ziman theory is hindered in most cases by the absence of knowledge of the density and composition correlations in the liquid structure. These can be determined by neutron diffraction using the method of isotopic substitution (Jal *et al.* 1985b), but such data are not available for most metal–metal halide solutions. It then becomes necessary to use theoretical liquid structure factors or severe approximations whose validity is uncertain.

The nearly-free-electron model remains a valid description until the scattering becomes so strong that $l \sim a$. This is the Ioffe–Regel limit (Ioffe & Regel 1960). For many liquid systems, this occurs when the conductivity is 1000–2000 $(\Omega\,\mathrm{cm})^{-1}$.

2.2. Metal–nonmetal transition

A number of physical mechanisms for electron localization at the metal–nonmetal transition have been proposed. It is undoubtedly correct that different mechanisms dominate in different physical systems and it may be that more than one effect is at work in an individual case. The transitions which we list below are those which may be important for metal–metal halide solutions. For more complete discussions see Mott (1974) and Mott & Davis (1979).

(*a*) **Mott–Hubbard transition**

Electron–electron interactions favour localization to reduce double occupancy of individual sites. The localized states are characteristically singly occupied hydrogenic states.

(*b*) **Anderson transition**

Electrons are localized by structural disorder (Anderson 1958). The localized states are described in general as having amplitude on several sites, but the amplitude decays exponentially with distance away from a particular central site. Mott (1969) proposed a minimum metallic conductivity σ_{min} at the onset of the Anderson transition. For liquids, σ_{min} typically has a value $\sim 300\ (\Omega\,\mathrm{cm})^{-1}$. Recent careful studies of heavily doped semiconductors at low temperatures, however, showed a continuous decrease of σ through σ_{min} rather than the predicted abrupt drop to $\sigma = 0$ (Rosenbaum *et al.* 1983).

(*c*) **Polaron (self-trapping) transition**

The electron–lattice interaction causes the electron to 'dig a hole' for itself, that is, to form a potential well in which it becomes trapped. The form of polaron which we will consider for metal–metal halide melts is the F-centre analogue—an electron solvated by metal cations.

(*d*) **Percolation transition**

This is the topological transition at the point where the metallic constituent just forms a connecting path through the solution. On the nonmetallic side, localization takes the form of isolated metallic clusters.

The relation of the metal–nonmetal transition to the liquid–liquid phase separation is a topic of continuing interest. This subject was first discussed by Landau & Zeldovich (1943) in the context of the liquid–gas equilibrium in metals. They suggested the possibility of separate first-order electronic and density (liquid–gas) transitions in fluid metals. While it is clear from Figure 2 that there is no sharp (first-order) electronic transition in the homogeneous metal–metal halide solutions, it is characteristic that the phase separation tends to separate the nonmetallic and metallic liquids. There is a rough correlation in the concentrations of the metal–nonmetal transition and the phase separation—values of σ at the critical concentration x_c fall in the range $10^2 < \sigma_c < 10^3\ (\Omega\,\mathrm{cm})^{-1}$ for both families of solution represented in Figure 2. Thus there is also an approximate equivalence between σ_c and σ_{min}. To evaluate the significance of this, however, one must first understand the underlying mechanism of the metal–nonmetal transition.

2.3. Nonmetallic state

The structure of pure molten salts is closely related to the degree of dissociation. Alkali halide melts are fully dissociated ionic liquids in which the inter-ionic coulomb forces impose a large measure of compositional order. This means that a metal cation M^+ has a high probability of being surrounded by a number (4–6) of halide anions X^- and vice versa (Enderby 1982). Many polyvalent metal halides, in contrast, retain substantial molecular character in the melt. This is revealed by relatively low values of the electrical conductivity in such cases. For example, $HgCl_2$ retains triatomic molecules in the melt which has a conductivity of only 3×10^{-5} $(\Omega\,cm)^{-1}$ (Grantham & Yosim 1966) compared with the ionic conductivity $\sigma_i \sim 1$ $(\Omega\,cm)^{-1}$ of a typical dissociated melt.

Whatever the structure of the pure molten salt, the central question raised by solution of small quantities of metal concerns the fate of the 'excess' electrons—the valence electrons of the dissolved metal. There are many possible states that will be discussed: retention of the electrons by the ions to form dissolved neutral atoms; delocalized dilute metallic states; Mott–Anderson multi-site localized states; polaronic states (F-centres); metal dimers or polymers; and metal ions of lower oxidation state (in polyvalent metals). Determination of the electronic state and transport mechanism for low metal concentrations in molten halides is crucial for understanding the subsequent transition to the metallic state at higher concentrations. We shall see, after consideration of some specific cases, that there is no universal description of the excess electron and that several of the above possibilities occur in different solutions.

3. Alkali metal–alkali halide solutions

The alkali metal–alkali halide solutions are the prototypical metal–salt solutions. They are probably the family most amenable to theoretical understanding. One reason is their relative simplicity and the advanced state of our knowledge of the pure components. The alkali metals are the best examples of nearly-free-electron metals. The halides, for their part, are fully dissociated ionic liquids whose structures have been determined in some detail (see, for example, Biggin & Enderby 1982). A second simplifying factor is the existence of only one stable (monovalent) oxidation state. As objects for research, the principal drawback of the alkali metal solutions is an experimental consideration, namely, the extreme chemical reactivity of the alkali metals.

3.1. Electron localization at low metal concentrations

If one adds a small amount ($x \lesssim 0{\cdot}01$) of alkali metal to one of its molten halides, the otherwise colourless melt becomes strongly coloured. This effect is due to a strong absorption band in the visible or near infrared (Figure 3). These bands are remarkably similar to those associated with F-centres in additively coloured alkali halide crystals, although the liquid-state spectra are typically shifted to the red by a few tenths of an electron-volt. Absorption bands essentially identical to those in liquid alkali metal–halide solutions were obtained by Schmitt & Schindewolf (1977) using direct injection of electrons into pure molten alkali halides. The transition energies ε_m for various molten alkali halides containing excess metal or injected electrons are summarized in Table 1.

Arguing primarily on thermodynamic grounds, Pitzer (1962) proposed that

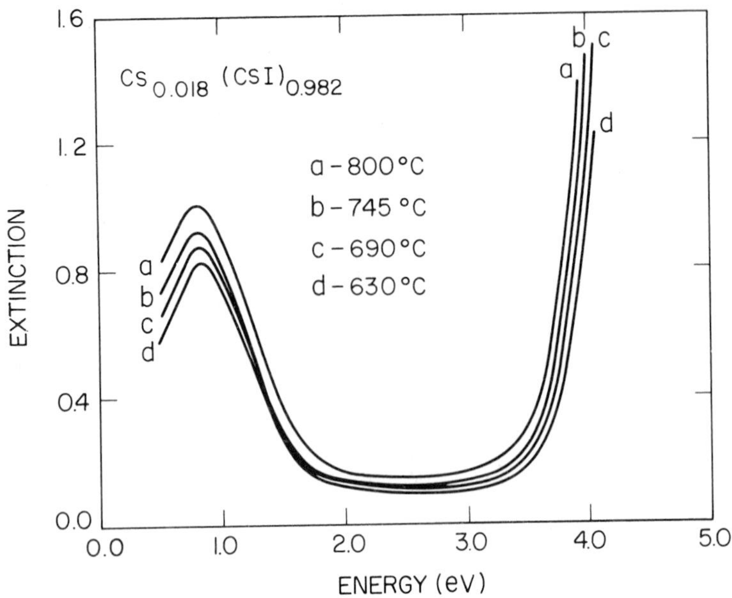

Figure 3. Optical absorption spectra of Cs–CsI ($x=0.018$) at various temperatures (Freyland *et al.* 1984). Spectra show fundamental absorption edge near 4.0 eV and absorption at 0·8 eV due to excess metal.

Table 1. Optical transition energies ε_m for dilute alkali metals and injected electrons in molten alkali halides.

	Alkali metal–halide solution		Injected electrons[d]	
Host	ε_m eV	T °C	ε_m eV	T °C
NaCl	1·61[c]	809	1·68	810
NaBr	1·42[a]	830	1·42 ± 0·06	875
NaI	1·5[b]	750	1·38 ± 0·06	801
KCl	1·30[a]; 1·32[b]	853; 800	1·24	853
KBr	1·18[b]	775	1·18	775
KI	1·12[b]	750	—	—
CsCl	1·01[a]; 1·06[b]	718; 700	1·05	718
CsBr	—	—	0·94	720
CsI	0·86[a]	698	—	—

[a] Freyland *et al.* (1984).
[b] Rounsaville & Lagowski (1968).
[c] Gruen *et al.* (1969).
[d] Schmitt & Schindewolf (1977).

excess electrons in liquid alkali halides are solvated by metal ions to form F-centre analogues. The optical absorption would then be due to electronic transitions to the first excited state. These centres resemble their crystalline counterparts but they may differ in the cavity size, number of coordinating metal ions, lack of well-defined symmetry, etc. Freyland *et al.* (1984) have shown, in fact, that there occurs at the melting point a clear deviation from the Mollwo–Ivey Law, $\varepsilon_m \propto (\text{volume})^{-2/3}$, as it generally applies for F-centres in alkali halide crystals. The essence of the model for liquids is simply that an electron replaces a negative ion in the coulomb-ordered structure.

Until recently there was no direct confirmation of the F-centre model nor was it even definitely known that the excess electrons are localized. Various other models were introduced involving low densities of delocalized electrons or localization by disorder. These models have been summarized by Nachtrieb (1975) and by Warren (1981). There is now, however, little doubt that the excess electrons are indeed localized and in the form of F-centre analogues. Let us now review the evidence for this conclusion.

The nuclear spin relaxation rate in alkali metal–alkali halide melts is governed by a characteristic time τ for fluctuations of the local hyperfine field produced by nearby unpaired electrons. Nuclear magnetic relaxation studies of Cs–CsI by Warren *et al.* (1984) revealed striking peaks in the concentration dependence of the ^{133}Cs and ^{127}I relaxation rates around $x = 0{\cdot}05$ (Figure 4). The observed peaks reflect a strong increase in τ at low metal concentrations. From a value $\tau = 10^{-15}$ s in the pure metal, τ increases monotonically to $\sim 10^{-12}$ s for $x \simeq 0{\cdot}01$. Characteristic times in the picosecond range have also been measured in Cs–CsCl by Warren & Sotier (1981) and in Na–NaBr by Warren *et al.* (1983). Nicoloso & Freyland (1983a) have shown that the electrons contribute Curie-type paramagnetism, as expected for localized states. The experiments show, therefore, that excess electrons are localized in the dilute limit and the localization times are comparable with the diffusion-limited lifetime of a local configuration of ions.

One indication of the structure of the localized states is provided by the magnitudes of the hyperfine fields. Warren & Sotier (1981) and Warren *et al.* (1984) found that the fields are too low to be explained by atomic or multi-site localized states. The hyperfine field is a measure of the charge on the sites of the resonant nuclei, so these results indicate that the charge is distributed mainly off the sites as should be the case for F-centres. Corrected for the volume difference between solid and liquid, the hyperfine fields agree well with first- and second-neighbour hyperfine fields measured around F-centres in crystals.

Electron spin resonance (e.s.r.) has been observed for low metal concentrations in various alkali halide eutectic mixtures by Nicoloso & Freyland (1983b). The measured shifts of the electronic g-factor are consistent with the F-centre model. However, the experimental necessity to use eutectic hosts with their lower melting temperatures complicates the interpretation compared with a single halide host. More local ionic configurations are possible in the eutectics

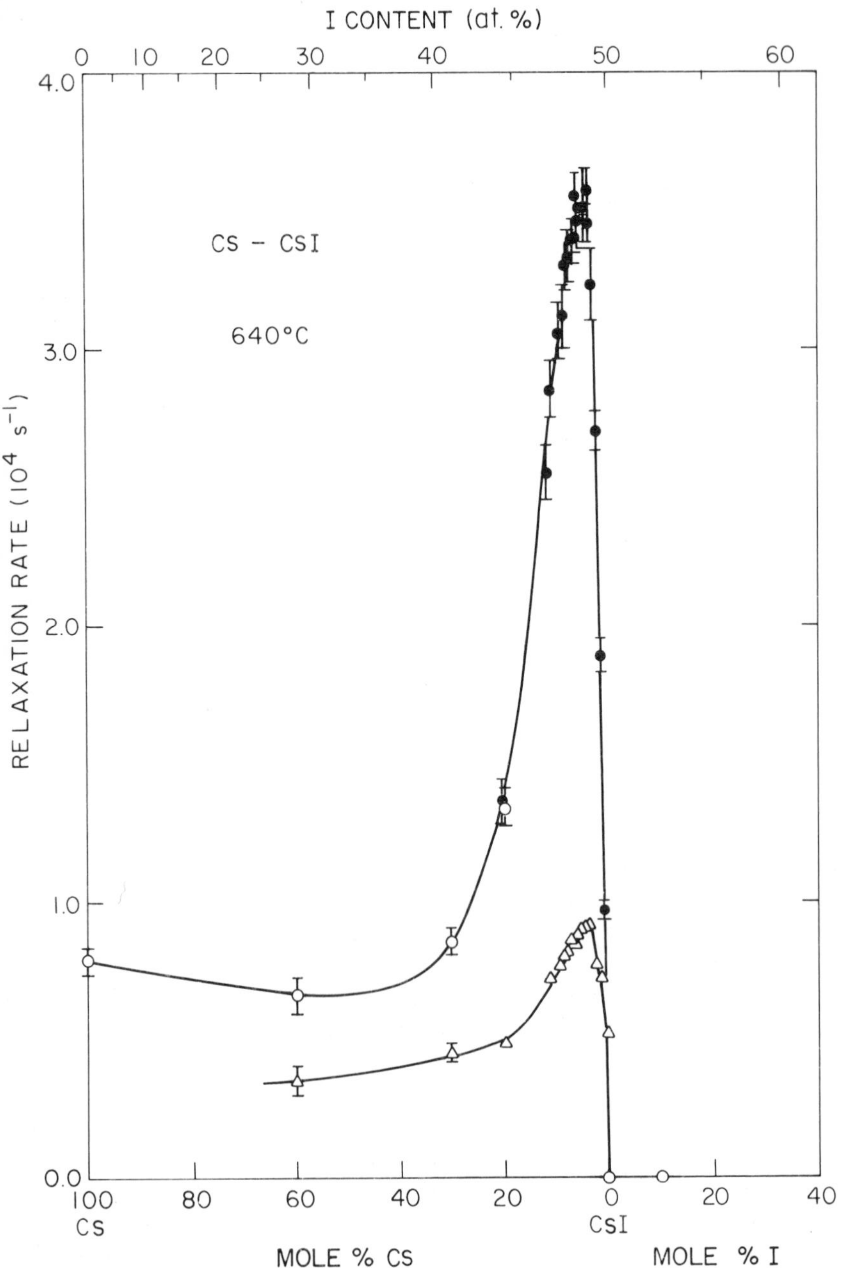

Figure 4. Nuclear relaxation rates for ^{133}Cs (○, ●) and ^{127}I (△) in Cs–CsI–I solutions at 640°C (Warren *et al.* 1984). Peaks are due to electron localization at low metal concentrations.

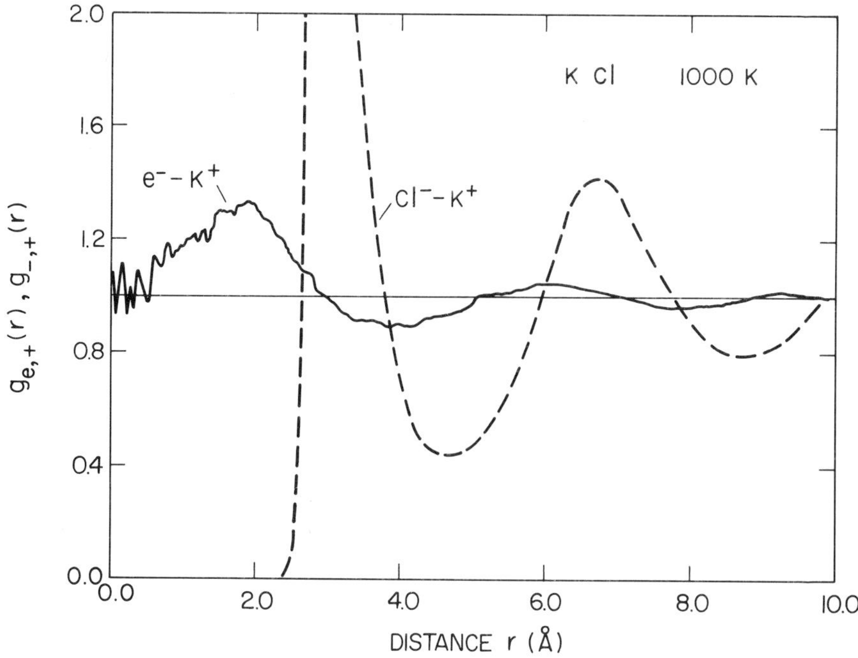

Figure 5. Computed radial pair correlations between electrons and cations (solid line) and between anions and cations (broken line) for K–KCl (Parrinello & Rahman 1984).

leading to the possibility of more than one type of localized centre.

Parrinello & Rahman (1984) have shown how the dilute metal–salt solution can be modelled using a combination of molecular dynamics and Feynman path integral techniques. Their results for the electron–cation partial pair distribution function $g_{e,+}(r)$ and the cation–anion function $g_{-,+}(r)$ are shown in Figure 5. The solvated character of the electronic state is clearly demonstrated by the peak in $g_{e,+}(r)$ at finite r. The coordination number for the electron is approximately 4 metal ions. The mean electron–cation separation is substantially less than the anion–cation distance given by the main peak in $g_{-,+}(r)$. This indicates a strong 'polaron effect'. Such self-trapping explains the difference between the optical excitation energy (~ 1 eV) and the total binding energy ($\sim kT$) implied by the 10^{-12} s lifetime of the state. The optical excitation, according to the Franck–Condon principle, occurs in the unrelaxed configuration, while in thermal dissociation the metal ions must relax outward, recovering coulomb energy. The strong polaron effect implies a relatively small or negative excess volume of mixing. This has, in fact, been observed in recent studies of K–KBr using neutron transmission to determine the density (Jal *et al.* 1985a).

3.2. Conduction in nonmetallic solutions: the 'conductivity dilemma'

Addition of only a few mole % of alkali metal to a molten halide introduces an electronic contribution to the electrical conductivity which completely dominates the ionic contribution (Figure 2). Achieving an adequate understanding of the electronic transport mechanism has been complicated by a longstanding dilemma posed by a strikingly different property of Na solutions compared with the heavier alkali metals K, Rb and Cs. Whereas the conductivity of the latter systems increases rapidly, roughly exponentially with x, the increase in Na solutions is sublinear at temperatures well below T_c (Figure 6). This behaviour, sometimes represented as a decrease in the equivalent molar

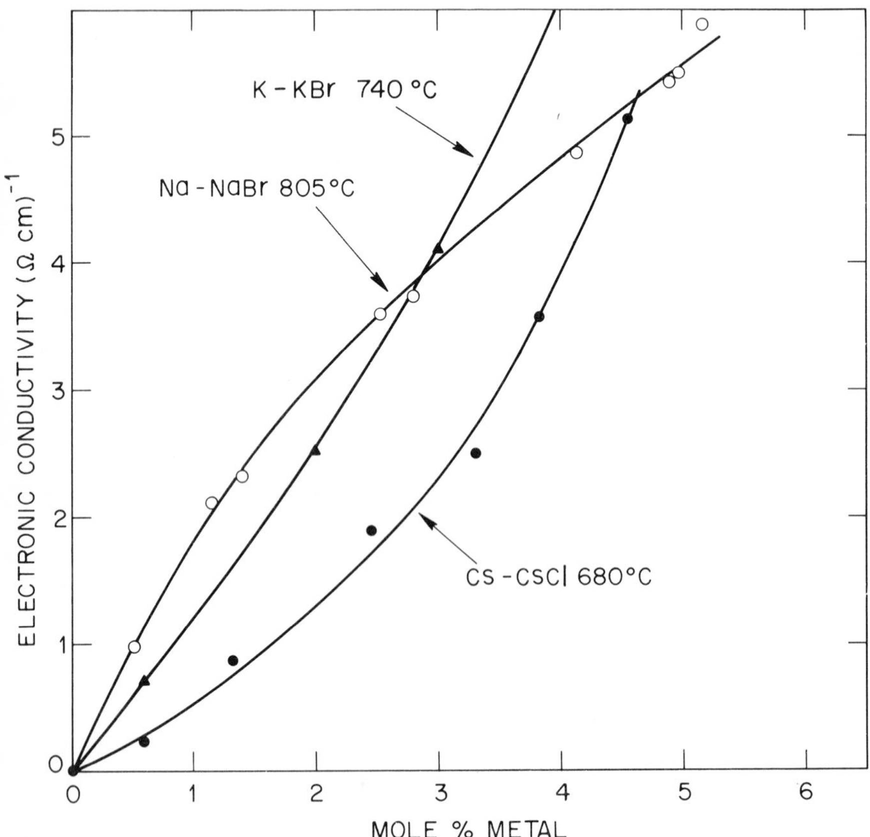

Figure 6. Electronic conductivity versus excess metal content in dilute alkali metal–alkali halide melts: (○) Na–NaBr at 805°C (Bronstein & Bredig 1958); (▲) K–KBr at 740°C (Bronstein & Bredig 1958; Bronstein *et al.* 1962); (●) Cs–CsCl at 680°C (Nachtrieb *et al.* 1976). Negative curvature is typical of sodium–sodium halide solutions at temperatures well below T_c.

conductance (Bredig 1964), persists to the maximum concentration possible before liquid–liquid phase separation occurs. At much higher temperatures, where the solubility is higher (Figure 1(*a*)), the conductivity variation becomes comparable with those of the other metal solutions. In their early work, Bronstein & Bredig (1958) suggested that formation of Na_2 dimers at low temperature might effectively reduce the *concentration* of conducting electrons and thus lower the conductivity at higher metal concentrations. Later, Katz & Rice (1972) and Durham & Greenwood (1976) offered alternative explanations based on reduced electronic *mobilities* at higher metal concentrations.

A comparison of n.m.r. properties of Na–NaBr, Cs–CsCl and Cs–CsI by Warren *et al.* (1983) separates the effects of electron concentration and mobility and supports the original ideas of Bronstein and Bredig. The method exploits the fact that the n.m.r. shift and relaxation rate are proportional to the electron concentration while the relaxation rate also depends on the characteristic time τ. For diffusive transport, τ is related to the mobility of the localized electrons

$$\mu = e\langle a^2 \rangle / 6kT\tau$$

where $\langle a^2 \rangle$ is the mean-square jump distance. Now the n.m.r. shift $\Delta H/H$ and relaxation rate $1/T_1$ are related through

$$1/T_1 = c(T)\tau(\Delta H/H)$$

where $c(T)$ depends only on the temperature and certain coupling constants. Thus the curvature of a plot of $1/T_1$ versus $\Delta H/H$ for various concentrations at constant temperature reveals the variation of τ, and hence μ, without any assumptions about the electron concentration. Such a plot is shown in Figure 7. It may be clearly seen that τ remains essentially constant in Na–NaBr whereas it decreases rapidly with increasing Cs content in the caesium halides. Both $1/T_1$ and $\Delta H/H$ are proportional to σ_{elec} for Na–NaBr. The electron concentration in Na–NaBr is thus reduced at higher metal concentrations by formation of states which are both non-magnetic and non-conducting. The dimer Na_2 suggested by Bronstein & Bredig (1958) would satisfy this criterion, but there are also other possibilities. One such is Na^-, reported in Na-hexamethylphosphoramide by Edwards *et al.* (1981).

The following picture of transport at low metal concentrations has emerged from the foregoing experiments. In Na solutions at low temperature, a certain fraction of excess electrons is removed from the conductivity process by formation of species such as Na_2 or Na^-. These are stable for times long compared with the jump times of the remaining electrons. The mobility of the remainder is determined by the lifetime of a favourable site for localization and is independent of concentration. When such a site is disrupted by ionic diffusion, the electron enters an extended state at the conduction band edge (Littlewood 1981) and moves rapidly a distance a to a new localization site. Warren *et al.* (1983)

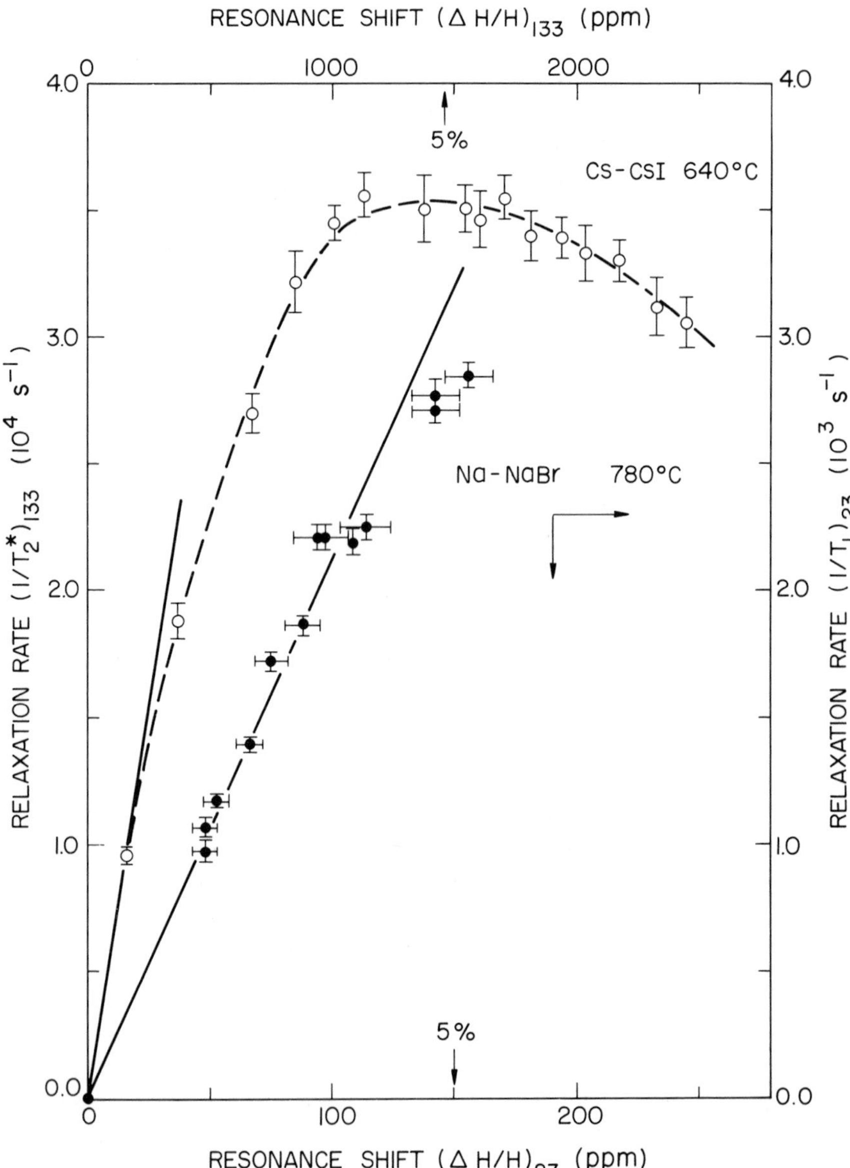

Figure 7. Nuclear relaxation rates vs. resonance shifts: (○) ^{133}Cs in Cs–CsI (left-hand and upper scales); (●) ^{23}Na in Na–NaBr (right-hand and lower scales) (Warren *et al.* 1983). Arrows indicate shift values corresponding to 5 mole % metal concentration. Nearly linear plot of Na–NaBr implies constant mobility, contrasting with Cs–CsI.

found a value $\langle a^2 \rangle^{1/2} \sim 20$ Å from the correlation times and conductivity values. A similar distance is found for the caesium solutions, but here the mobility is strongly dependent on concentration. A possibility suggested by the authors is that species such as Cs_2 also tend to form in the caesium solutions, but they are in rapid equilibrium with localized states so that their lifetimes are less than or comparable with the solvation time. The mobility would be enhanced by this equilibrium since an electron can be localized at a new site after formation and dissociation of the species. Littlewood (1981) proposed a similar mechanism involving rapid exchange between F-centre analogues and occupied conduction states.

3.3. Intermediate concentrations: the metal–nonmetal transition

Let us consider now what happens as we add increasing amounts of metal to a molten alkali halide and pass over eventually to the metallic state. An immediate difficulty, evident from a glance at Figure 2, is the smooth, continuous character of the conductivity through the metal–nonmetal transition. Indeed the concept of a 'transition' at finite temperatures is only an approximate one. Landau & Zeldovich (1943), Mott (1974) and others have pointed out that a metal and nonmetal are rigorously distinguishable only at 0 K. Nevertheless, one can identify ranges in liquids for which the metallic or nonmetallic models are clearly appropriate and useful. We must ask, therefore, over what composition range are these two regimes jointed and how does the microscopic electronic structure evolve from the nonmetallic to the metallic limit.

One criterion often invoked for the onset of metallic behaviour is the minimum metallic conductivity $\sigma_{\min}$ suggested by Mott (1969) to characterize the boundary between delocalized and disorder-localized electronic states. We have already pointed out that $\sigma_{\min}$ correlates roughly with the critical concentration in metal–salt solutions. However, there is currently little evidence that electrons remain localized by disorder up to $x \simeq x_c$. In fact, the structure of the states at small values of x resemble point defects in crystals more than the multi-state, finite-correlation-length localized states usually associated with strong disorder.

Another approach to identifying the transition, independent of the details of electronic structure, is the so-called polarization catastrophe—the divergence of the refractive index on approaching the metal–nonmetal transition from the insulating side (Herzfeld 1927, Castner *et al.* 1975). The onset of this effect has been observed by Freyland *et al.* (1983) in K–KCl (Figure 8). A power-law fit of the dielectric susceptibility

$$\chi = \text{const}\,(x_t/x - 1)^{-\zeta}$$

yielded $x_t = 0{\cdot}1$ for the metal concentration at the transition and $\zeta = 1{\cdot}4 \pm 0{\cdot}2$ for the critical exponent. These data are still rather far from the transition, that is,

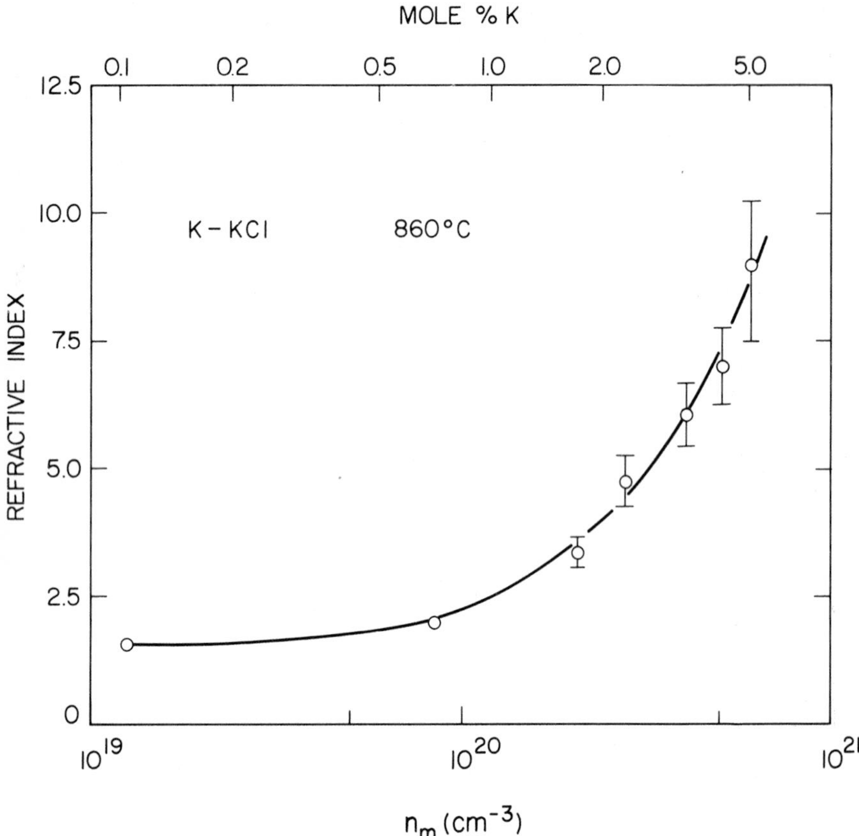

Figure 8. Refractive index vs. excess metal density (n_m) for K–KCl at 860°C and a wavelength of 2·4 μm (Freyland *et al.* 1983). Upper scale indicates metal concentration in mole % K.

$(x_t/x-1) \geqslant 1$, but they show that the dielectric transition occurs at a substantially smaller value of x than either σ_{min} or the critical point.

In the preceding section we discussed the importance of 'spin-paired' species such as M_2 or M^- in the range of a few mole percent excess metal. Such species would reduce the paramagnetic susceptibility. Nicoloso & Freyland (1983a) have measured the magnetic susceptibility of several alkali–alkali halide systems. With increasing x, they found that the paramagnetic susceptibility indeed drops rapidly from the value expected for isolated F-centres. They propose that larger spin-paired aggregates and small metallic clusters form as the metal–nonmetal transition is approached. The scenario is plausible, but is not yet supported by structural evidence. It leads essentially to a percolation transition around $x = 0{\cdot}1$.

Littlewood (1981) has emphasized the role of electron correlation effects and proposed that the transition is of the Mott–Hubbard type. Thus the large coulomb repulsion in a doubly occupied F-centre would prevent delocalization until a critical excess electron concentration is reached. Littlewood estimated that this would occur at $x \simeq 0{\cdot}06$ in Cs–CsI. Spin pairing in this model would take the form of doubly occupied extended states at the edge of the conduction band. If the structural inferences of Nicoloso and Freyland are correct, this Mott–Hubbard description would have to be elaborated to include the effects of polyatomic species.

The liquid–liquid phase separation represents still another aspect of the transition to the metallic state. It is clear from the phase diagrams (Figure 1) that concentrated solutions of delocalized electrons and negative ions are inherently unstable. There is a strong tendency to separate into a metallic phase and a nonmetallic (or barely metallic) one. The tendency toward separation is present even in the caesium solutions where there is no actual phase separation. Yokokawa *et al.* (1979) and Yokokawa & Kleppa (1982) found strong peaks in $S_{cc}(0)$, the long-wavelength-limit concentration–concentration correlation function. For Cs–CsCl they estimated $x_c = 0{\cdot}29$ and $T_c = 881$ K, a point just below the liquidus line.

Concentration fluctuations in the region above T_c may be studied by small-angle neutron scattering. The scattering function for concentration fluctuations is expressed in terms of a correlation length ξ by the Ornstein–Zernicke law

$$S_{cc}(k) = \text{const}\ \xi^2/(k^2 + \xi^2)$$

Results of Chieux *et al.* (1980) given in Figure 9 show contours of constant ξ plotted in the temperature–composition plane for K–KBr. It may be seen that fluctuations exceeding 20 Å extend over a wide composition range from roughly $x = 0{\cdot}30$ to $x = 0{\cdot}55$. The critical exponent for the divergence of ξ at constant x_c is $\gamma = 1{\cdot}21 \pm 0{\cdot}02$. The interplay between these fluctuations and the electronic properties presents an interesting problem that has yet to be investigated in detail.

3.4. *Metallic state*

Dilute halogen impurities in liquid alkali metals ionize to form well defined X^- states at energies several electron-volts below the conduction band states. This was established in a series of investigations by Flynn & Rigert (1973), Flynn & Lipari (1971) and Flynn (1974). One of the interesting properties of these states is the unusually large impurity diamagnetism caused by screening charge expelled from the region of the negative impurity ion.

At higher concentrations of halide, there is now clear evidence for an ionic configuration obtained by neutron diffraction studies of K–KCl by Jal *et al.* (1980, 1985b) and Rb–RbBr by Chabrier *et al.* (1982). The total radial distribution

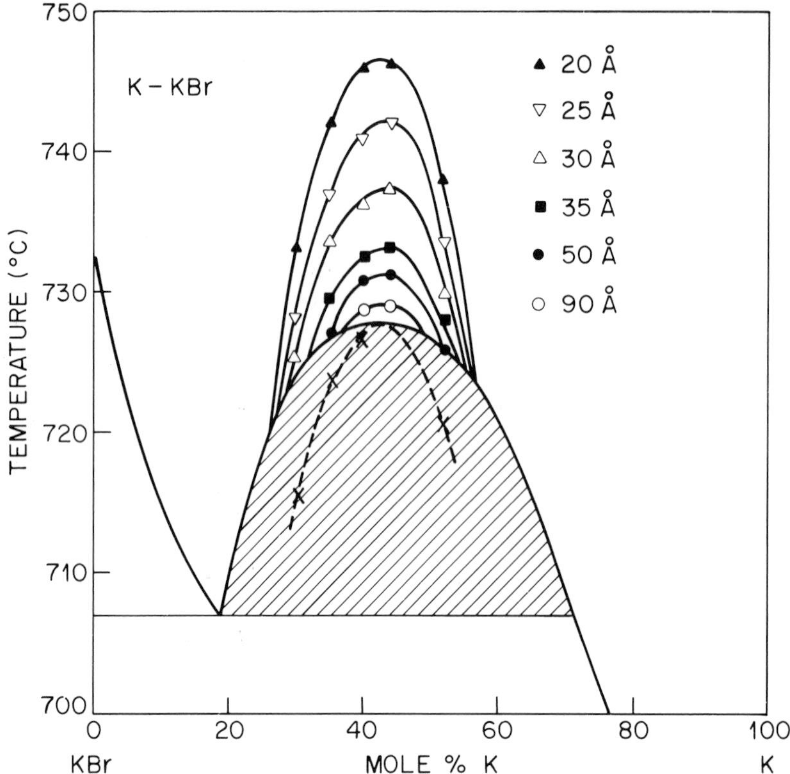

Figure 9. Contours of constant coherence length (ξ) for concentration fluctuations superimposed on temperature–composition phase diagram of K–KBr (Chieux *et al.* 1980). Broken line (– – × – –) within the two-phase region indicates spinodal estimated by extrapolation of the neutron scattering intensity.

function g(r) is shown in Figure 10 for the pure components Rb and RbBr and a solution of 80 mole % Rb in RbBr. The essential feature of the latter is a peak at the Rb^+–Br^- separation of RbBr which is evident even in this solution of relatively high metal content. Fainchstein *et al.* (1984) observed the onset of nonmetallic character in the optical reflectivity of Cs–CsCl around 85 mole% Cs. This rather surprising effect in an otherwise metallic solution is the result of increased absorption. It is unclear how it is related to the ionic configurations found by neutron diffraction.

The negative ions are strong scatterers of conduction electrons and make a major contribution to the electrical resistivity in metallic solutions. Using a model of charged hard spheres to calculate the partial structure factors, Senatore *et al.* (1982) applied the Faber–Ziman theory to compute the change of resistivity $\Delta\rho/(1-x)$ on adding halides to pure alkali metals. In most cases, their results agree with experiment within a few percent.

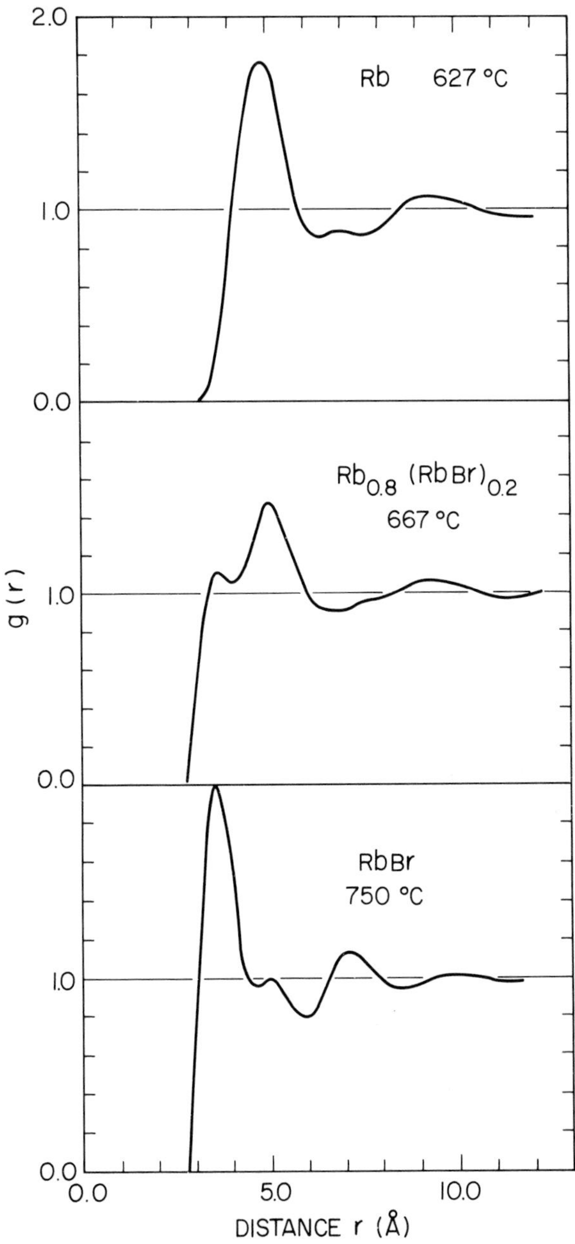

Figure 10. Total pair-correlation function g(r) for pure Rb, pure RbBr and Rb–RbBr ($x = 0{\cdot}8$) obtained by neutron diffraction (Chabrier *et al.* 1982).

The nearly-free-electron picture embodied in the Faber–Ziman theory should be valid in the range between the pure metal and the Ioffe–Regel limit. The mean separation between halide ions in alkali-metal solutions implies that this limit is reached for x in the range 0·50–0·60 and $\sigma \sim 1500\,(\Omega\,\mathrm{cm})^{-1}$ (Figure 2). This agrees with the ^{133}Cs nuclear relaxation rates measured in Cs–CsI by Warren *et al.* (1984). For $x \leqslant 0{\cdot}60$, the observed rates begin to deviate strongly from the Korringa rate characteristic of simple, nearly-free-electron metals. This effect reflects the change in hyperfine field dynamics when the electron mean-free-path no longer exceeds the average separation between nuclei. Between the Ioffe–Regel limit and the onset of localization lies the so-called strong-scattering or diffusive transport regime. Franz (1984) has calculated the conductivity of K–KBr using a quantum percolation model and finds reasonable agreement with experiment in this range.

4. Bismuth–bismuth halide solutions

The bismuth–bismuth halide solutions are relatively easy to study experimentally because of their low melting and consolute temperatures (Figure 1) and their chemical stability. They are the most thoroughly investigated family of liquid metal-halide solutions. Yet despite an abundance of data, their structural and electronic properties remain poorly understood when compared with the alkali–alkali halide solutions. The principal reason, as we shall see, is the rich chemistry of Bi in solution.

4.1. Mixed valence and polyatomic species

Bismuth dissolved in its molten halides may form monomers of differing valence such as Bi^+ or polymers with the general form Bi_n^{m+}. At low metal concentrations, the dominant species is known to be the monomer Bi^+ (Topol *et al.* 1961, Topol & Osteryoung 1962). This suggests a simple mixed-valence model for the range $0 \leqslant x \leqslant 0{\cdot}67$ in which Bi is present as Bi^{3+} and Bi^+. Magnetic susceptibility data are roughly consistent with the presence of just these two diamagnetic species (Topol & Ransom 1963, Topol & Lieu 1964, Hosokawa *et al.* 1984). Raleigh (1963) proposed an electronic conduction mechanism for this model based on Bi^+–B^{3+} intervalence exchange, that is, thermally activated two-electron hopping. Applied to Bi–BiI_3, the Raleigh theory provides an adequate explanation of the slow increase in conductivity with added bismuth, and the model can account for the temperature dependence of the conductivity.

Closer examination reveals, however, that the actual situation in bismuth–bismuth halide solutions is considerably more complex than the Bi^+–B^{3+} model. In Bi–$BiCl_3$ and Bi–$BiBr_3$ at temperatures well below T_c the conductivity *decreases* with increasing bismuth concentration (Grantham 1965). Further, optical absorption spectra in these solutions indicate the presence of more than one absorbing species associated with low concentrations of added metal (Boston & Smith 1962, Boston *et al.* 1963, Boston 1970). These effects are attributed to polyatomic species which decrease the ionic conductivity by virtue of their large size. They also inhibit electronic conductivity by raising the energy for intervalence exchange. Such species can be expected to dissociate at higher temperatures, resulting in higher electronic and ionic conductivity contributions.

The exact form of the polyatomic species has not been definitely established. Topol *et al.* (1961) and Topol & Osteryoung (1962) originally suggested Bi_4^{4+} in the form of the polymeric 'subhalide' $(BiX)_4$. This has been questioned by Corbett (1976) who favours Bi_3^+ as being more compatible with chemical bonding requirements. In fact, it is quite likely that more than one such species is present. Studies of the solid-state phases of Bi–$BiCl_3$ have revealed several species extending up to Bi_9^{5+} (Hershaft & Corbett 1963). These are manifestations of the well known tendency of bismuth to form polyatmoic clusters.

The nuclear magnetic resonance properties of Bi–BiI_3 and Bi–$BiBr_3$ suggest a clear difference in the formation of polyatomic units in these two solutions. The ^{209}Bi nuclear relaxation rates in Bi–$BiBr_3$ (Figure 11) increase very rapidly as the salt is added to liquid bismuth (Dupree & Gardner 1980). The effect is so strong, in fact, that n.m.r. has not been observed in solutions of less than 70 mole % metal. The relaxation behaviour is characteristic of electric quadrupolar relaxation by large, slowly tumbling species. The relaxation rates are lower at higher temperatures, as expected, since these species should then tumble more rapidly and dissociate to smaller units. It should be noted that the nuclear relaxation increase in Bi–$BiBr_3$ occurs well within the metallic range where the conductivity is roughly 10^3 $(\Omega\ cm)^{-1}$.

A rapid increase in nuclear relaxation also occurs in Bi–BiI_3 (Dupree & Warren 1977) but in this case the effect falls in the nonmetallic range around 20 mole % Bi (Figure 11). At these compositions the conductivity is less than 1 $(\Omega\ cm)^{-1}$. The relaxation rates in Bi–BiI_3 *increase* at higher temperatures in contrast with Bi–$BiBr_3$. The Knight-shift data (Figure 12) together with the relaxation behaviour suggest the presence in Bi–BiI_3 of polyatomic Bi structures sufficiently large as to look locally metallic to ^{209}Bi nuclei. Thus, even in macroscopically nonmetallic solutions, large shift values and relatively weak electric quadrupole relaxation are observed. These anomalous properties disappear on heating as we should expect if clusters dissociate to yield a more homogeneous ionic melt. Neutron diffraction peaks (Ichikawa & Matsumoto 1981) are consistent with the Bi–Bi distances expected for Bi_9^{5+} in Bi–BiI_3. But questions of uniqueness are raised by such data whose analysis requires fitting several possible lines to one or two experimental peaks.

In summary, a picture has emerged of the bismuth–bismuth halide solutions in which the monomer Bi^+ is in equilibrium with one or more polyatomic species. The latter are present even in bismuth-rich solutions in Bi–$BiBr_3$ and they effectively suppress electronic conductivity in salt-rich Bi–$BiCl_3$ and Bi–$BiBr_3$ at

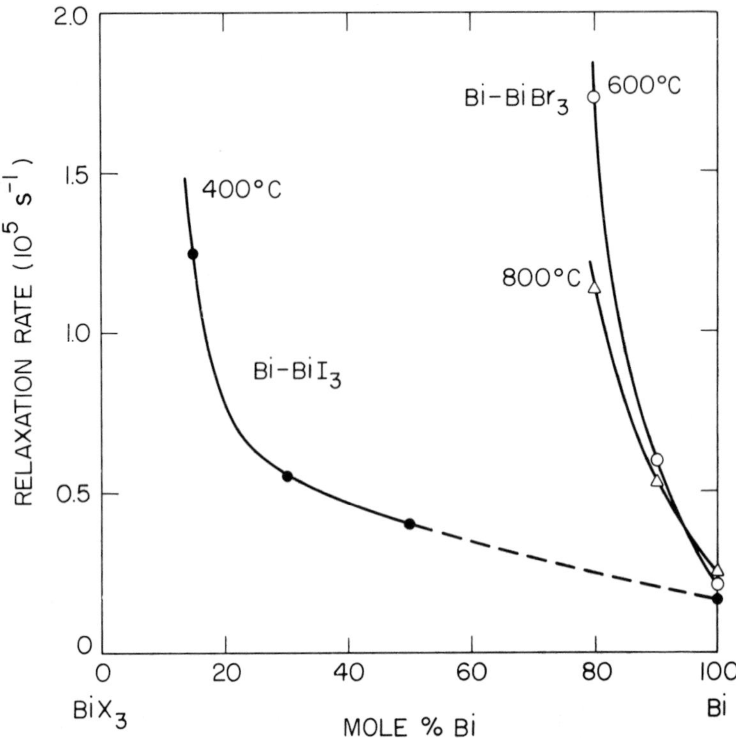

Figure 11. [209]Bi nuclear spin relaxation rates at constant temperature vs. composition for Bi–BiI_3 (closed circles), (Dupree & Warren 1977) and Bi–$BiBr_3$ (open circles and triangles), (Dupree & Gardner 1980).

low temperatures. These species are present in much lower concentrations, if at all, in metal-rich Bi–BiI_3. In that system, polyatomic units reveal themselves in the n.m.r. properties only in the salt-rich solutions. It remains a major challenge to identify clearly the species present and to determine their concentrations at various temperatures and compositions.

4.2. Liquid–liquid phase separation

Bismuth-bismuth halide solutions Bi–$BiCl_3$, Bi–$BiBr_3$ and Bi–BiI_3 all exhibit regions of liquid–liquid immisibility in the range $x \gtrsim 0{\cdot}50$. The $BiCl_3$ and $BiBr_3$ solutions have the additional interesting property of retrograde solubility near $x = 0{\cdot}50$. This strongly suggests a change in liquid structure between the low temperature homogeneous liquid and the high temperature liquid. Dissociation of the polyatomic species discussed in the preceding section is one possibility.

Careful measurements by Hoshino *et al.* (1979) revealed a sigmoidal

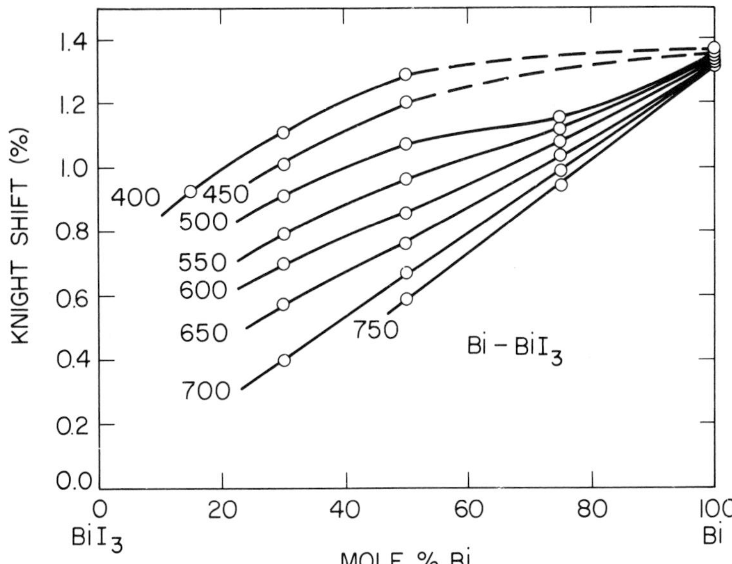

Figure 12. ^{209}Bi n.m.r. shift vs. composition at various temperatures (°C) for Bi–BiI$_3$ (Dupree & Warren 1977). At the lowest temperatures, shifts retain metallic values in nonmetallic solutions (15–30 mole % Bi).

anomaly in the temperature dependence of the electrical conductivity of Bi–BiBr$_3$ near the point of maximum retrograde solubility (Figure 13). Furthermore, extrapolations through the two-phase region of data for the low temperature homogeneous liquid do not 'connect' with the measured conductivity in the high-temperature phase. Thus, again we have evidence of the presence in the low-temperature phase of species which inhibit conduction relative to the high-temperature phase. Corbett *et al.* (1968) suggested that thermal dissociation of large polybismuth cations may, in fact, be responsible for the retrograde solubility effect.

Application of pressure has a strong effect on the region of liquid–liquid immiscibility. The effect on the consolute temperature is given by the thermodynamic relation

$$dT_c/dP = (\partial^2 V_m/\partial x^2)_{x_c}/(\partial^2 S_m/\partial x^2)_{x_c}$$

where V_m and S_m are the molar volume and entropy, respectively (see, for example, Schneider 1978). Since $\partial^2 S_m/\partial x^2 < 0$ and $\partial^2 V_m/\partial x^2 > 0$ (Keneshea & Cubicciotti 1958, 1959a, 1959b), we should expect T_c to decrease with pressure in the bismuth–bismuth halides. This was observed by Tamura *et al.* (1980) who studied the phase separation of Bi–BiBr$_3$ under pressures up to 17 kbar. The two-phase region shrinks under pressure and disappears completely at about 16 kbar (Figure 14).

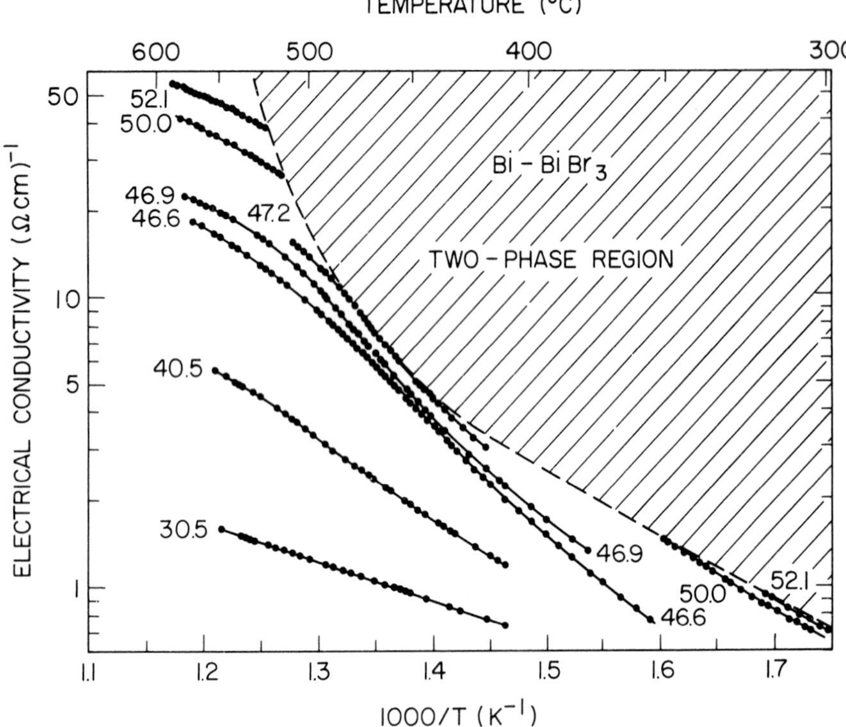

Figure 13. Semi-logarithmic plot of d.c. electrical conductivity vs. inverse temperature for various compositions (mole % Bi) of Bi–$BiBr_3$ near the retrograde solubility region. Broken line denotes conductivity value along phase boundary (Hoshino *et al.* 1979).

With the suppression of the phase separation by high pressure, it is possible to measure physical properties of the solution in a range which would not be possible under normal conditions. Measurements of the conductivity of Bi–$BiBr_3$ under pressure revealed anomalies in both the temperature and composition dependence for $x \sim 0{\cdot}50$, the region of retrograde solubility (Tamura *et al.* 1980). In the same range, Hoshino *et al.* (1980) observed a sharp change in the composition dependence of the velocity of sound in Bi–$BiBr_3$ at 12 kbar. Thus, even in the absence of the phase separation there are structural changes around 50 mole % Bi.

4.3. Metal–nonmetal transition

The available evidence indicates that salt-rich bismuth–bismuth halide solutions consist of Bi^{3+} and Bi^{+} in equilibrium with larger species. The role of polyatomic species in suppressing electronic conductivity is particularly evident

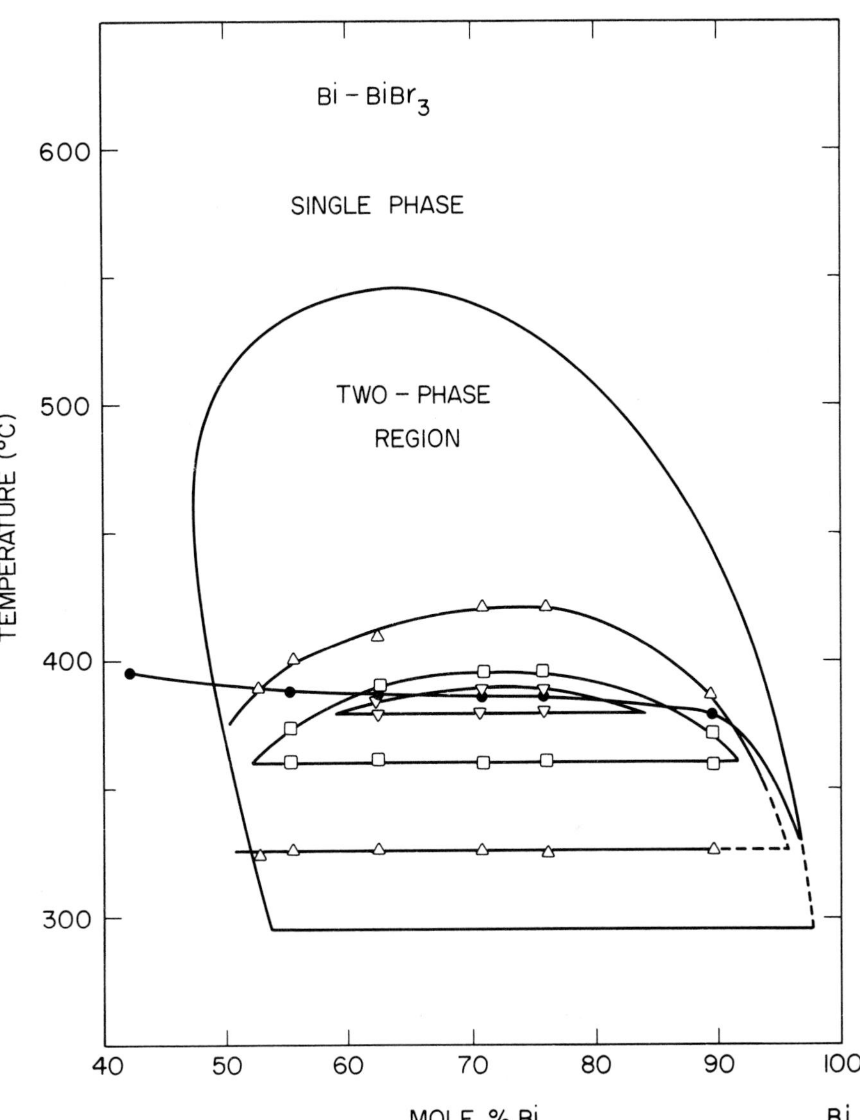

Figure 14. Pressure variation of the two-phase region of the temperature–composition phase diagram of Bi–$BiBr_3$ (Tamura *et al.* 1980): solid line, 0 kbar; (△) 7 kbar; (□) 12 kbar; (▽) 15 kbar; (●) liquids at 17 kbar.

in Bi–$BiCl_3$ and Bi–$BiBr_3$ at low temperatures close to the boundary of the liquid–liquid phase separation—the retrograde region.

At temperatures above T_c, the continuous metal–nonmetal transition may be observed. The most rapid increase in conductivity occurs in the range $0{\cdot}4 \lesssim x \lesssim 0{\cdot}6$ (Figure 2). At lower temperatures, with depression of T_c under pressure, the 'transition' becomes somewhat sharper (Hosokawa *et al.* 1984). The value of the conductivity in the middle of the transition range is roughly $30\ (\Omega\ \mathrm{cm})^{-1}$. This is about the same as the conductivity of the alkali metal–alkali halide solutions at $x \simeq 0{\cdot}1$, the composition of the polarization catastrophe in K–KCl.

At the present time there is no adequate theory or model of the metal–nonmetal transition in the bismuth–bismuth halide solutions. It has frequently been remarked that the onset of the transition on the metallic side ($x \sim 0{\cdot}6$) coincides roughly with the value of σ_{min} proposed by Mott. However, we have no evidence that the conditions considered in the derivation of σ_{min} (localization by disorder) actually apply in bismuth–bismuth halide solutions. In fact, there is considerable evidence for significant changes in liquid structure in the transition range, although the details remain obscure. Electron localization takes the form of discrete species such as Bi^{+} and the polyatomic units. These species are present to some degree even at metal concentrations well above the metal–nonmetal transition. Because of their importance for the electronic properties, it is unlikely that substantial progress will be made toward understanding the metal–nonmetal transition until the various species have been delineated.

4.4. Metallic state

Bismuth is a borderline nearly-free-electron metal. If we make the usual assumption that there are five conduction electrons per atom, the conductivity implies $l/a \simeq 1{\cdot}1$. (For three electrons per atom this value increases to $1{\cdot}5$.) Addition of molten halides causes a rapid decrease in conductivity (Figure 2). The effect can be treated, in principle, using the Faber–Ziman theory (Ichikawa & Shimoji 1969). But such calculations require a number of approximations necessitated mainly by a lack of liquid structure information for the solutions. A further difficulty is the possibility, illustrated by the data of Dupree & Gardner (Figure 11), that large species form at quite high metal concentrations. These would reduce the number of available conduction electrons and invalidate the nearly-free-electron model. Because of these ambiguities it is not possible to state with any precision where the Ioffe–Regel limit occurs in bismuth–bismuth halide solutions. Available evidence suggests only that it occurs at relatively high metal concentrations in comparison with the alkali metal–alkali halide solutions.

5. Ionic alloys

Many binary alloys of liquid metals exhibit pronounced semiconducting or nonmetallic properties at compositions of simple stoichiometry. The chemical bonding responsible for this effect may be covalent or ionic and is, in general, of intermediate character. In a few cases, there is good evidence for nearly complete charge transfer to the more electronegative metal. These are the so-called 'ionic alloys' or 'charge-transfer insulators'. The relative simplicity of ionic bonding makes them particularly attractive candidates for theoretical interpretation with simple models (Holzhey *et al.* 1982, Franz *et al.* 1982). Viewed as pseudo-binary solutions of an elemental metal with a binary ionic liquid composition, ionic alloys can be seen to be closely analogous to the metal–metal halide solutions.

The clearest example yet found of an ionic liquid alloy is Cs–Au. Both pure components are good nearly-free-electron metals. Yet their equiatomic alloy CsAu is essentially a molten salt, analogous to CsI. The electrical conductivity value is typical of a dissociated molten salt (Schmutzler *et al.* 1976) and electromigration studies established that the conduction mechanism is, in fact, ionic (Kruger & Schmutzler 1976). Solid CsAu forms in the CsCl crystal structure and is distinguished from the caesium halides mainly by its smaller energy gap (2·6 eV at 300 K).

Excess electrons are localized in dilute solutions of Cs in CsAu. The ^{133}Cs nuclear relaxation rates measured by Dupree *et al.* (1980) exhibit a peak similar to that found for Cs–CsI (Figure 4). However, the correlation times are shorter, indicating weaker localization in Cs–CsAu, and the localization–delocalization transition occurs at lower caesium concentration ($x \simeq 0{\cdot}07$). The structure of the localized states is similar to the F-centre analogues of the halide solutions, but there is more electronic charge on the cations in Cs–CsAu.

Neutron diffraction studies of Cs–CsAu by Martin *et al.* (1980) show that the Au^- ion is formed in Cs-rich, metallic solutions. For $x = 0{\cdot}75$, the main peak in the total pair-correlation function occurs at the position found in CsAu. The authors attributed this to 'microscopic' segregation, that is, the presence of only two near-neighbour distances—the metallic Cs^+–Cs^+ value and the Cs^+–Au^- separation of the salt. These observations in Cs–Au are very similar to the results discussed in §3.4 for alkali metal–alkali halide solutions.

Cs–CsAu most closely resembles the metal–metal halide solutions and is a useful standard of comparison for ionic character in liquid alloy systems. A number of other alloys (Rb–Au, Mg–Bi, Li–Pb, Li–Bi, Na–Bi, Cs–Sb, etc.) span a spectrum of lower ionic character at the appropriate stoichiometries (Robertson 1983). In some cases there is still considerable disagreement as to the structure and bonding type. An example is Tl–Te for which Cutler (1977) has made extensive use of models which are essentially molecular. Gay *et al.* (1982), in contrast, argued that their neutron diffraction data are best explained by a model of dissociated Tl^+ and Te^{2-} ions. Similar neutron experiments point to a structure for Ni–Te which is ionic but modified by the effects of d-electron overlap (Enderby 1982). The chemical bonding and properties of semiconducting liquid alloys are sources of continuing research interest. A complete discussion of these issues, however, would lead us well beyond the intended scope of this chapter.

Acknowledgements

The author is pleased to acknowledge the valuable contributions of R. Dupree, J. Franz, W. Freyland and S. Sotier, with whom he has discussed various aspects of this subject. Figures 3 and 9 are reproduced with permission from the *Journal of Physical Chemistry*, Copyright 1984 and 1980, respectively, American Chemical Society.

References

Anderson, P.W., 1958, *Phys. Rev.*, **109**, 1492.
Aten, A.H.W., 1909, *Z. Phys. Chem.* (*Leipzig*), **66**, 641.
Aten, A.H.W., 1910a, *Z. Phys. Chem.* (*Leipzig*), **73**, 578.
Aten, A.H.W., 1910b, *Z. Phys. Chem.* (*Leipzig*), **73**, 624.
Biggin, S. and Enderby, J.E., 1982, *J. Phys. C: Solid St. Phys.*, **15**, L305.
Boston, C.R., 1970, *Inorg. Chem.*, **9**, 389.
Boston, C.R. and Smith, G.P., 1962, *J. Phys. Chem.*, **66**, 1178.
Boston, C.R., Smith, G.P. and Howick, L.C., 1963, *J. Phys. Chem.*, **67**, 1849.
Bredig, M.A., 1964, in *Molten Salt Chemistry*, edited by M. Blander (New York: Interscience), p. 367.
Bredig, M.A., Bronstein, H.R. and Smith, Jr., W.T., 1955, *J. Am. Chem. Soc.*, **77**, 307.
Bronstein, H.R. and Bredig, M.A., 1958, *J. Am. Chem. Soc.*, **80**, 2077.
Bronstein, H.R., Dworkin, A.S. and Bredig, M.A., 1962, *J. Chem. Phys.*, **37**, 677.
Castner, T.G., Lee, N.K., Cieloszyk, G.S. and Salinger, G.L., 1975, *Phys. Rev. Lett.*, **34**, 1627.
Chabrier, G., Jal, J.F., Chieux, P. and Dupuy, J., 1982, *Phys. Lett.*, **93A**, 47.
Chieux, P., Demay, P., Dupuy, J. and Jal, J.F., 1980, *J. Phys. Chem.*, **84**, 1211.
Corbett, J.D., 1964, in *Fused Salts*, edited by B. Sundheim (New York: McGraw-Hill), Chap. 6.
Corbett, J.D., 1976, in *Progress in Inorganic Chemistry*, edited by S.J. Lippard (New York: Wiley), Vol. 21, p. 129.
Corbett, J.D., Albers, F.C. and Sallach, R.A., 1968, *Inorg. Chimica Acta*, **3**, 22.
Cutler, M., 1977, *Liquid Semiconductors* (New York: Academic).
Dupree, R. and Gardner, J.A., 1980, *J. Physique*, *Coll.* C8, **41**, C8–20.
Dupree, R. and Warren, W.W., 1977, in *Liquid Metals 1976*, (Inst. Phys. Conf. Ser. 30) edited by R. Evans and D.A. Greenwood (Bristol and London: Institute of Physics), p. 454.
Dupree, R., Kirby, D.J., Freyland, W. and Warren, Jr., W.W., 1980, *Phys. Rev. Lett.*, **45**, 130.
Durham, P.J. and Greenwood, D.A., 1976, *Phil. Mag.*, **33**, 427.
Edwards, P.P., Guy, S.C., Holton, D.M. and McFarlane, W., 1981, *J. Chem. Soc. Chem. Commun.*, p. 1185.
Edwards, P.P., Guy, S.C., Holton, D.M. and McFarlane, W., 1983, *Phil. Mag.*, **47**, 367.
Enderby, J.E., 1982, *J. Phys. C: Solid St. Phys.*, **15**, 4609.
Faber, T.E. and Ziman, J.M., 1965, *Phil. Mag.*, **11**, 153.
Fainchtein, R., Even, U. and Thompson, J.C., 1984, *J. Non. Cryst. Solids*, **61–62**, 47.
Flynn, C.P., 1974, *Phys. Rev.* B, **9**, 1984.
Flynn, C.P. and Lipari, N.O., 1971, *Phys. Rev. Lett.*, **27**, 1365.
Flynn, C.P. and Rigert, J.A., 1973, *Phys. Rev.* B, **7**, 3656.
Franz, J., 1984, *J. Non. Cryst. Solids*, **61–62**, 41.

Franz, J., Brouers, F. and Holzhey, C., 1982, *J. Phys. F: Metal Phys.*, **12**, 2611.
Freyland, W., Garbade, K. and Pfeiffer, E., 1983, *Phys. Rev. Lett.*, **51**, 1304.
Freyland, W., Garbade, K., Heyer, H. and Pfeiffer, E., 1984, *J. Phys. Chem.* , **88**, 3745.
Gay, M., Enderby, J.E. and Copestake, A.P., 1982, *J. Phys. C: Solid St. Phys.*, 4641.
Grantham, L.F., 1965, *J. Chem. Phys.*, **43**, 1415.
Grantham, L.F. and Yosim, S.J., 1963, *J. Chem. Phys.*, **38**, 1671.
Grantham, L.F. and Yosim, S.K., 1966, *J. Chem. Phys.*, **45**, 1192.
Gruen, D.M., Krumpelt, M. and Johnson, I., 1969, in *Molten Salts*, edited by G. Mamantov (New York: Dekker), p. 169.
Hershaft, A. and Corbett, J.D., 1963, *Inorg. Chem.*, **3**, 979.
Herzfeld, K.F., 1927, *Phys. Rev.*, **29**, 701.
Holzhey, C., Brouers, F., Franz, J.R. and Schirmacher, W., 1982, *J. Phys. F: Metal Phys.*, **12**, 2601.
Hoshino, H., Tamura, K. and Endo, H., 1979, *Solid St. Commun.*, **31**, 687.
Hoshino, H., Tamura, K., Hosokawa, S., Suzuki, K., Misonou, M. and Endo, H., 1980, *J. Physique, Coll.*, C8, **41**, C8–52.
Hosokawa, S., Endo, H. and Hoshino, H., 1984, *J. Non. Cryst. Solids*, **61–62**, 77.
Ichikawa, K. and Matsumoto, T., 1981, *Phys. Lett.*, **83A**, 35.
Ichikawa, K. and Shimoji, M., 1969, *Phil. Mag.*, **19**, 33.
Ioffe, A.F. and Regel, A.R., 1960, *Prog. Semiconductors*, **4**, 237.
Jal, J.F., Dupuy, J. and Chieux, P., 1980, *J. Physique, Coll.*, C8, **41**, C8–257.
Jal, J.F., Dupuy, J. and Chieux, P., 1985a, *J. Phys. C: Solid St. Phys.*, **18**, 1347.
Jal. J.F., Dupuy, J. and Chieux, P., 1985b, *J. Phys. C: Solid St. Phys.* (in the press).
Katz, I. and Rice, S.A., 1972, *J. Am. Chem. Soc.*, **94**, 4824.
Keneshea, Jr., F.J. and Cubicciotti, D., 1958, *J. Phys. Chem.*, **62**, 843.
Keneshea, Jr., F.J. and Cubicciotti, D., 1959a, *J. Phys. Chem.*, **63**, 1112.
Keneshea, Jr., F.J. and Cubiciotti, D., 1959b, *J. Phys. Chem.*, **63**, 1472.
Kruger, K.D. and Schmutzler, R.W., 1976, *Ber. Bunsenges. Phys. Chem.*, **80**, 816.
Landau, L. and Zeldovich, J., 1943, *Acta Physicochimica URSS*, **18**, 194.
Littlewood, P., 1981, *Phys. Rev.* B, **24**, 1710.
Martin, W., Freyland, W., Lamparter, P. and Steeb, S., 1980, *Phys. Chem. Liquids*, **10**, 61.
Mott, N.F., 1969, *Phil. Mag.*, **19**, 835.
Mott, N.F., 1974, *Metal–Insulator Transitions* (London: Taylor & Francis).
Mott, N.F. and Davis, E.A., 1979, *Electronic Processes in Non-Crystalline Materials* (Oxford: Clarendon).
Nachtrieb, N.H., 1975a, in *Non-Simple Liquids*, edited by I. Prigogine and S.A. Rice (New York: Wiley Interscience), Advances in Chemical Physics, Vol. 31, p. 465.
Nachtrieb, N.H., Hsu, C., Sosis, M. and Bertrand, P.A., 1976, in *Proceedings of the International Symposium on Molten Salts*, edited by J.P. Pemsler, J. Braunstein and K. Nobe (Pennington: Electrochemical Society), p. 506.
Nicoloso, M. and Freyland, W., 1983a, *Z. Phys. Chem.*, **135**, 39.
Nicoloso, M. and Freyland, W., 1983b, *J. Phys. Chem.*, **87**, 1997.
Parrinello, M. and Rahman, A., 1984, *J. Chem. Phys.*, **80**, 860.
Pitzer, K.S. 1962, *J. Am. Chem. Soc.*, **84**, 2025.
Raleigh, D.O., 1963, *J. Chem. Phys.*, **38**, 1677.
Robertson, J., 1983, *Phys. Rev.* B, **27**, 6322.
Rosenbaum, T.F., Milligan, R.F., Paalanen, M.A., Thomas, G.A., Bhatt, R.N. and Lin, W., 1983, *Phys. Rev.* B, **27**, 7509.
Rounsaville, J.F. and Lagowski, J.J., 1968, *J. Phys. Chem.*, **72**, 1111.
Schmitt, W. and Schindewolf, U., 1977, *Ber. Bunsenges. Phys. Chem.*, **81**, 584.
Schmultzler, R.W., Hoshino, H., Fischer, R. and Hensel, F., 1976, *Ber. Bunsenges. Phys. Chem.*, **80**, 197.

Schneider, G.M., 1978, in *Chemical Thermodynamics*, edited by M.L. McGlashan (London: The Chemical Society), Vol. 2, p. 105.
Senatore, G., Giaquinta, P.V. and Tosi, M.P., 1982, *Physica*, **112B**, 360.
Shimoji, M., 1977, *Liquid Metals* (London: Academic), p. 254.
Sotier, S., Ehm, H. and Maidl, F., 1984, *J. Non. Cryst. Solids*, **61–62**, 95.
Tamura, K., Hoshino, H. and Endo, H., 1980, *Ber. Bunsenges. Phys. Chem.*, **84**, 235.
Topol, L.E. and Lieu, F.Y., 1964, *J. Phys. Chem.*, **68**, 851.
Topol, L.E. and Osteryoung, R.A., 1962, *J. Phys. Chem.*, **66**, 1587.
Topol, L.E. and Ransom, L.D., 1963, *J. Chem. Phys.*, **38**, 1663.
Topol, L.E., Yosim, S.J. and Osteryoung, R.A., 1961, *J. Phys. Chem.*, **65**, 1511.
Warren, Jr., W.W., 1981, in *Advances in Molten Salt Chemistry*, edited by G. Mamantov and J. Braunstein (New York: Plenum), Vol. 4, p. 1.
Warren, Jr., W.W. and Sotier, S., 1981, in *Proceedings of the Third International Symposium on Molten Salts*, edited by G. Mamantov, M. Blander and G.P. Smith (Pennington: Electrochemical Society), p. 95.
Warren, Jr., W.W., Sotier, S. and Brennert, G.F., 1984, *Phys. Rev.* B, **30**, 65.
Warren, Jr., W.W., Sotier, S. and Brennert, G.F., 1983, *Phys. Rev. Lett.*, **50**, 1505.
Yokokawa, H. and Kleppa, O.J., 1982, *J. Chem. Phys.*, **76**, 5574.
Yokokawa, H., Kleppa, O.J. and Nachtrieb, N.H., 1979, *J. Chem. Phys.*, **71**, 4099.
Yosim, S.J. and Mayer, S.W., 1960, *J. Phys. Chem.*, **64**, 909.
Yosim, S.J., Ransom, L.D., Sallach, R.W. and Topol, L.E., 1962, *J. Phys. Chem.*, **66**, 28.
Ziman, J.M., 1961, *Phil. Mag.*, **6**, 1013.

CHAPTER 7

doped semiconductors

Robert F. Milligan

Department of Physics, Muhlenberg College, Allentown, PA 18104, USA

1. Introduction

Semiconducting materials, such as silicon and germanium, provide an ideal medium for studying the metal–nonmetal (M–NM) transition. Because they are crucial to the electronics industry, the technology for preparing very pure samples and samples with carefully controlled amounts of selected impurities is well established. When an impurity from the fifth column of the periodic table (such as phosphorus, arsenic or antimony) is introduced into Si or Ge it replaces a host atom at a tetrahedrally bonded site. Since only four of the five valence electrons of the impurity are required for bonding, the fifth electron is very weakly bound. This weakly bound electron moves in hydrogen-like orbitals with an effective Bohr radius, a_{H}^{*}, of some tens of Ångströms. Only small energies ($\sim 30\,\mathrm{meV}$) are required to move this electron up into the bottom of the conduction band, thus these 'donor' impurities are generally ionized at moderate temperatures and the material conducts.

An analogous situation exists if 'acceptor' impurities from the third column (such as boron, aluminium, gallium or indium) are substituted for Si or Ge atoms. Since these impurities have only three valence electrons the fourth is supplied by an electron from the tightly bound electrons of a neighbouring Si atom. This electron removal produces a positively charged hole. At low temperatures we can view the acceptor impurity as consisting of a singly charged negative core (because of the extra electron) with a weakly bound hole moving in a large hydrogenic orbital. Again, moderate temperatures can ionize the impurity, with the hole gaining sufficient energy to move into the top of the valence band, becoming a mobile charge carrier.

Whether we have mobile negative charge carriers (n-type material) due to excess donor impurities or mobile positive charge carriers (p-type material) due to excess acceptor impurities, the criterion used by experimentalists to distinguish these materials from metallic conductors is that at sufficiently low temperatures the conductivity goes to zero. As the temperature drops the capture of the charge carriers by the charged impurities (carrier freeze out) localizes the previously itinerant charges.

What we have just described is the situation for a lightly doped semiconductor with an impurity concentration, n, less than about 10^{17} elec. cm^{-3} for Ge or 10^{18} elec. cm^{-3} for Si. For these materials the average spacing between impurities ($\sim n^{-1/3}$) is large compared to the size of the effective Bohr radius, that is, there is negligible overlap of the extended wave functions with those of neighbouring impurity sites. As n increases there is eventually significant wave-function overlap and the material has finite conductivity even as the temperature T goes to zero. The critical concentration, n_c, where this occurs, is very well defined and is related to a_H^* by the Mott criterion (Mott 1974):

$$n_c^{1/3} a_H^* \simeq 0{\cdot}26 \tag{1}$$

This criterion for the M–NM transition is valid for a wide variety of materials with n_c varying over eight orders of magnitude as discussed by Edwards & Sienko (1978).

The low temperature behaviour of the conductivity σ, or alternatively resistivity, $\rho \equiv 1/\sigma$, for the Si:P system is shown in Figure 1 where we plot log ρ

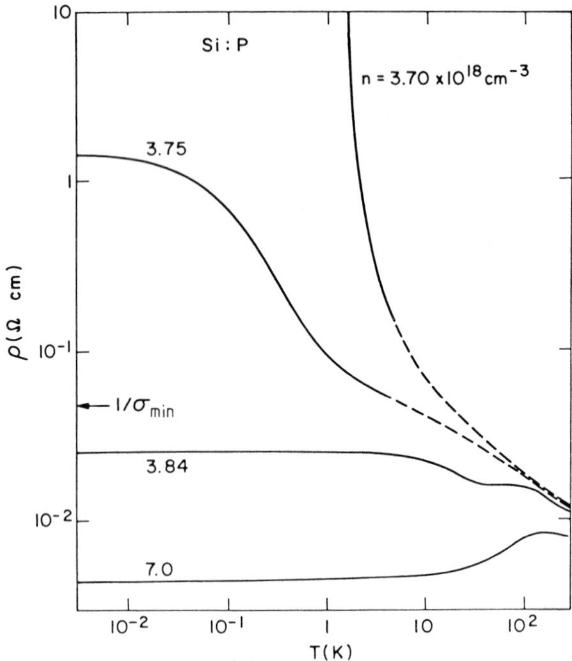

Figure 1. Logarithm of resistivity, ρ, vs. logarithm of temperature, T, for several samples of Si:P with donor density n above and below the critical density $n_c = 3{\cdot}74 \times 10^{18}$ elec. cm^{-3}. Near n_c a small change in n has a large effect on ρ at low T. (Rosenbaum *et al.* 1983)

against log T for several samples near $n_c = 3{\cdot}74 \times 10^{18}$ elec. cm^{-3}. The bottom curve represents a sample which is clearly metallic; σ remains finite down to $T=0$. The other three samples are much closer to n_c and have nearly identical values of ρ at room temperature although vastly different values at low T. This behaviour is to be expected, since at 300 K these samples all have roughly the same carrier concentration and carrier mobility, but at low T carrier freeze out causes the sample with $n < n_c$ to behave like an insulator. This figure illustrates an important point: the M–NM transition in the doped semiconductors is a zero T phenomenon unlike some other systems discussed in this book which manifest the transition at higher temperatures.

2. Nature of the transition

Interest in the M–NM transition goes back to the classic papers of Mott (1949, 1956) and Anderson (1958) which emphasize, respectively, the role of electron correlation and disorder in the phenomenon of electron localization. In the Mott picture a regular array of hydrogen atoms whose spacing is small compared to a_H^* will be a metal, since band calculations would yield a half-filled electron band made up of the perturbed 1s states. At low densities where there is negligible wave-function overlap, coulomb repulsion inhibits electrons from hopping from one host nucleus to another and the system is insulating at $T=0$. For this case there is a lower filled band separated from an upper empty band by an energy gap. Increasing the density broadens the bands and narrows the gap, but the system remains insulating at $T=0$ until the gap vanishes at a density n_c.

Anderson showed that even without electron interactions, localized electron states can result from the application of a sufficiently large random potential. In such non-interacting systems, the M–NM transition is viewed not as a closing of a true one-electron energy gap, but as that of a mobility gap. The mobility gap is the distance of the Fermi level from a critical energy (mobility edge) at which the one-electron states become delocalized. The relevance of the Anderson idea to the doped semiconductors has to do with the random distribution of the impurity sites which leads to an effective random potential in which the electrons move.

The relative importance of electron correlations and disorder to the M–NM transition in doped semiconductors has been debated for several years. As we shall see in this chapter, both aspects seem to be required for an accurate description.

3. The conductivity at $T=0$

The M–NM transition in doped semiconductors affects the transport, optical and magnetic properties of the material. Since most of the theoretical and experimental work has focused on the effect of carrier concentration on the zero temperature d.c. conductivity, $\sigma(0)$, that property will be emphasized in this chapter. The review by Milligan *et al.* (1985) discusses many of the other aspects of the transition in more detail.

As suggested by Figure 1, on the insulating side of the transition ($n<n_c$) $\sigma(0)=0$ while on the metallic side ($n>n_c$) $\sigma(0)$ is finite. For $n \gg n_c$ we can use Boltzmann transport theory to find $\sigma(0)$. The result (Ziman 1964) is

$$\sigma(0)=\sigma_B=e^2 S_F l/12\pi^3\hbar \tag{2}$$

where S_F is the Fermi surface area and l the elastic mean free path. For a spherical Fermi surface $S_F=4\pi(3\pi^2 n)^{2/3}$ and equation (2) reduces to $\sigma_B=ne^2\tau/m$, where $\tau=l/v_F$ is the mean scattering time. Using $a\equiv n^{-1/3}$ as the average spacing between electrons, we obtain for a spherical Fermi surface

$$\sigma(0)=\left(\frac{\pi}{3}\right)^{1/3}\frac{e^2}{\pi\hbar a}\cdot\frac{l}{a} \tag{3}$$

In 1960 Ioffe and Regel showed that the Boltzmann theory holds only when l is greater than the mean separation between scattering centres. In an otherwise perfect crystal at $T=0$ the scattering centres for a doped semiconductor are the impurity atoms. For an uncompensated sample the impurity concentration equals the electron concentration, n, which makes the Ioffe–Regel criterion $l \geqslant a$. This suggests a lower limit to the conductivity given by

$$\sigma_{IR} \sim e^2/3\hbar a \tag{4}$$

In 1972 Mott extended the ideas of Ioffe and Regel and formally introduced the idea of a 'minimum metallic conductivity', σ_{min}. In Mott's view, as n is reduced towards n_c the metallic conductivity $\sigma(0)$ falls with l until it reaches a minimum value, σ_{min}, at $l \simeq a$. For $n<n_c$ the conductivity drops discontinuously to zero as seen in Figure 2(*a*).

Mott finds

$$\sigma_{min}=C\, e^2/\hbar a \tag{5}$$

where C has a value of $\sim$0·025 to 0·1, significantly less than the 1/3 factor in σ_{IR}. Mott attributes this small value to a reduction in the density of states at the Fermi level, compared to the free-electron estimates.

An alternative description of the transition was offered by Cohen and co-workers (Webman *et al.* 1975) who viewed it not as a quantum phenomenon but rather as classical percolation. They consider the barely metallic material as an insulating medium permeated by small metallic regions whose size is still large compared to a or l. This model predicts a soft n-dependent onset to the transition

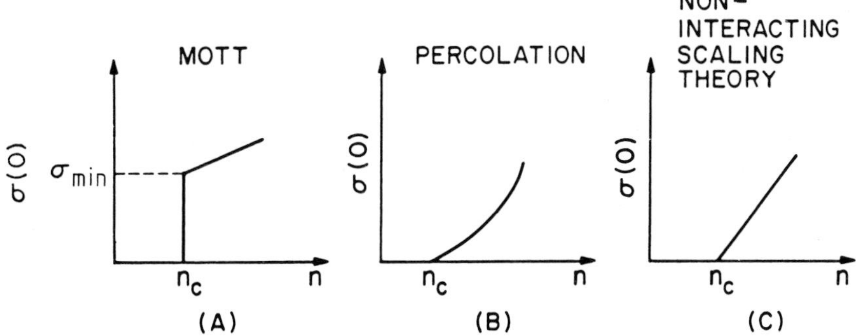

Figure 2. Some proposed variations of $\sigma(0)$ near the M–NM transition: (*a*) Mott's model showing a minimum metallic conductivity, (*b*) classical percolation and (*c*) scaling theory. (Milligan *et al.* 1985)

given by

$$\sigma(0) \propto (n-n_c)^t \tag{6}$$

with $t \sim 1{\cdot}8$ as shown in Figure 2(*b*). Most experiments contradict such a picture of the transition.

A third view of the transition was suggested by Abrahams *et al.* (1979). Their scaling theory for non-interacting electrons in a disordered system assumes that near the transition the conductance is the only relevant parameter describing the localization. For $n < n_c$, $\sigma(0)=0$, but just above n_c a localization length ξ replaces a in equation (5) and yields

$$\sigma(0) = C'\, e^2/\hbar\xi \tag{7}$$

where C' is a constant. The localization length diverges at the transition, having the critical form

$$\xi = \xi_0(n/n_c - 1)^{-\nu} \tag{8}$$

As a function of n, $\sigma(0)$ can be written as

$$\sigma(0) = \sigma_0(n/n_c - 1)^{\nu} \tag{9}$$

where $\sigma_0 \sim \sigma_{min}$ and ν is estimated to be 1. This critical, but continuous, behaviour is shown in Figure 2(*c*).

4. Temperature corrections to σ

Until recently the primary question being debated was whether the transition in doped semiconductors was continuous, as suggested by scaling theory, or was characterized by a minimum metallic conductivity, σ_{min}. The answer to the question was frustrated by the difficulty of accurately knowing the relative carrier concentrations of a series of samples with values of n within a percent or so of n_c. Another problem was how to estimate $\sigma(0)$ from conductivity data taken at finite temperatures. Experiments on Si:P by Rosenbaum and co-workers (1981a, 1983) suggested that the transition was continuous, as predicted by equation 9, although they found $\nu \sim 0{\cdot}5$ and $\sigma_0 \sim 20\sigma_{min}$ in contrast to the values arising from scaling theory. In addition, these workers found that the temperature dependence of the conductivity was of the form (Rosenbaum *et al.* 1981a, 1981b, 1983)

$$\sigma(T) = \sigma(0) + AT^{1/2} \tag{10}$$

where A took on a positive or negative value depending on how close n was to n_c. This form of $\sigma(T)$ was predicted by theories which included the coulomb interactions between charges (Altshuler & Aronov 1979a,b, 1981, Altshuler *et al.* 1980a,b, Lee & Ramakrishnan 1982). The $T^{1/2}$ dependence of $\sigma(T)$ at low temperature is now a generally observed phenomenon for metallic samples of many systems near the transition. Localization theory (Gorkov *et al.* 1979) predicts a linear temperature-correction term for $\sigma(T)$. Including both interaction and localization effects yields

$$\sigma(T) = \sigma(0) + AT^{1/2} + BT \tag{11}$$

This form of $\sigma(T)$, first used by Thomas *et al.* (1982b) to fit the experimental results of Ge:Sb up to 500 mK, often provides a better description of $\sigma(T)$ over an extended T range than does the $T^{1/2}$ term alone. This is particularly true for systems such as Ge which have small characteristic energies.

5. Tuning through the transition

Experiments designed to study the M–NM transition must vary the parameter n/n_c. Initial experiments utilized many samples, each with a different n. More recently, experimenters have chosen to use a single sample and drive it through the transition by applying either stress or magnetic field. In both cases the effective Bohr radius, a_H^*, is altered, hence so is n_c (see equation (1)).

5.1. *Stress tuning*

Early experimental work by Fritzsche (1962) and Cuevas & Fritzsche (1965a,b) (see also the review by Fritzsche, 1978) demonstrated the importance of stress to the conductivity of silicon and germanium at temperatures above 1·2 K.

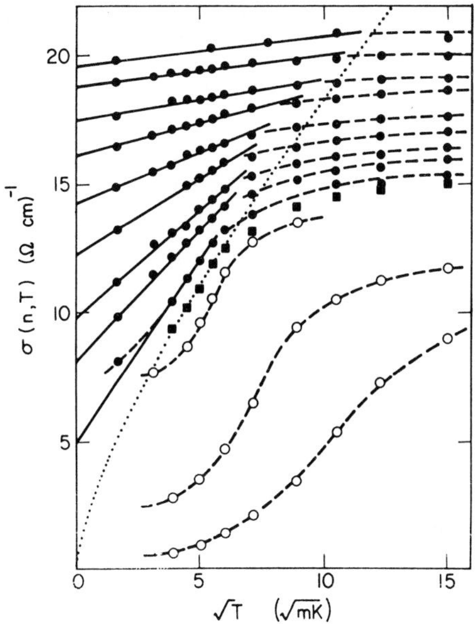

Figure 3. Conductivity plotted against $T^{1/2}$ for a Si:P sample under uniaxial stress $\mathscr{S}$ with a P concentration below n_c. Each curve is for a different stress (high $\mathscr{S}$ at top) with the values of $\mathscr{S}$ being 8·03, 7·82, 7·56, 7·36, 7·15, 7·00, 6·83, 6·71, 6·59, 6·33, 5·73, 5·12 kbar. On the metallic side, the $T^{1/2}$ behaviour (solid lines) is restricted to lower and lower temperatures (dotted line) as the transition is approached. The solid squares are at an interpolated value of $\mathscr{S}=\mathscr{S}_c$ where $n=n_c$. The open circles are not included in the analysis because they are affected by rounding. (Thomas *et al.* 1983)

For many valley semiconductors the application of stress mixes the ground state with excited state wave functions, thus changing a_H^* (Bhatt 1982, Milligan *et al.* 1985). For a uniaxial stress of ~2 kbar Bhatt predicts a 3% decrease in n_c for Si:P, allowing a barely insulating sample to be driven across the transition.

Paalanen and co-workers at AT&T Bell Labs have utilized this principle to study Si:P samples within 0·1% of the transition for temperatures down to 3 mK (Paalanen *et al.* 1982). Figure 3 shows a plot of the conductivity as a function of $T^{1/2}$ as found by Thomas *et al.* (1983). Each series of data points is for a different stress, with the highest stress at the top. At the lowest temperatures the metallic samples extrapolate to finite σ as $T^{1/2}\to 0$. As the stress, $\mathscr{S}$, approaches the critical value $\mathscr{S}_c$, considerable rounding occurs in the data, shown here as open circles. The dotted line is an estimate of the region below which deviations from $T^{1/2}$ dependence are expected and emphasizes the importance of working at the lowest possible temperatures when near the transition.

It is possible to make a linear conversion of $\mathscr{S}/\mathscr{S}_c$ into n/n_c (Paalanen *et al.* 1982). The resulting plot is shown in Figure 4 (Thomas *et al.* 1983) for two finite temperatures and for the extrapolated value of σ as $T^{1/2}\to 0$. There are two important features of this plot. The first is the obvious continuous nature of the

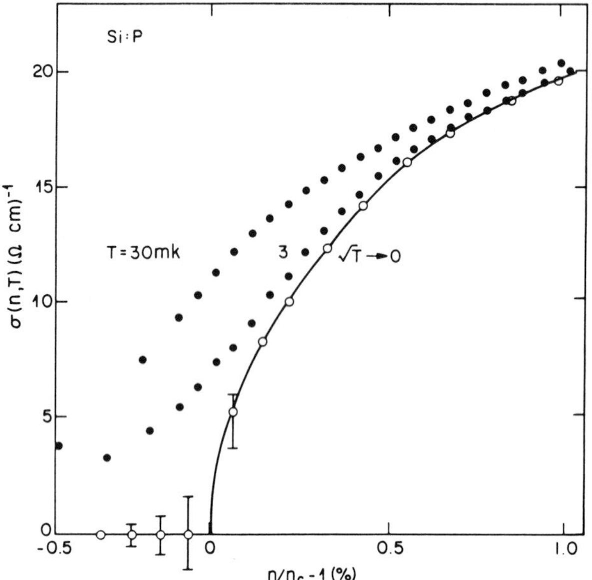

Figure 4. Conductivity of Si:P plotted against (n/n_c-1) using the data from Figure 3, where n/n_c is determined from $\mathcal{S}/\mathcal{S}_c$. Data at 30 mK and 3 mK are shown, as well as conductivities found by the extrapolation of $T^{1/2}\to 0$. (Thomas *et al.* 1983)

transition as opposed to a discontinuous jump in σ to zero at a value σ_{min} (σ_{min} is estimated to be 20 $(\Omega\ \mathrm{cm})^{-1}$ for Si:P). The second feature is that the data fits the form of equation (9) with $\nu=0{\cdot}51\pm0{\cdot}05$ and $\sigma_0=20\sigma_{min}$ in agreement with earlier work of Rosenbaum *et al.* (1981a).

Experiments on other systems also show continuous transitions below the estimated values of σ_{min} (see § 5.2). It is now generally accepted (Mott 1984) that although σ_{min} is a useful parameter for describing the conductivity (particularly the temperature-dependent conductivity of insulating samples, as well as the prefactor in equation (9) for metallic samples) it does not represent a finite minimum metallic conductivity as originally proposed. Mott (1984) now finds theoretical justification for $\sigma_{min}\to 0$ for samples in zero magnetic field but proposes that in non-zero fields σ_{min} is finite. There is controversy over this latter claim as is discussed in the next section.

5.2. *Magnetic field tuning*

The application of a strong magnetic field, H, introduces a new length scale into the system, the cyclotron radius, $L_H\equiv(\hbar c/eH)^{1/2}$. Recently, Mott (1984) has proposed that the expression for σ_{min}, equation (5), be modified to

$$\sigma_{min} = 0{\cdot}03e^2/\hbar L' \qquad (12)$$

where $L' = L_H$ for $L_H > a$, or $L' = a$ for $L_H < a$. This model, if true, would explain the absence of any observed σ_{min} in the stress-induced transition in Si:P which occurred in zero field ($\sigma_{min} \sim 0$).

Strong magnetic fields have induced M–NM transitions in certain systems via two different methods: reducing disorder so that insulators are driven into the metallic state, and shrinking donor wave functions so that metals become insulators. von Molnar and co-workers (1983) have investigated the first method in the system $Gd_{3-x}v_xS_4$, where v represents a randomly distributed vacancy. The samples change from ferromagnetic metals ($x=0$) to antiferromagnetic insulators ($x=1/3$) as x increases. The disorder in the system results from the random potentials of the vacancies, as well as the magnetic interactions of the conduction electrons with the Gd^{3+} ions. Magnetic fields reduce the disorder and insulating samples with x just below 1/3 can be driven metallic for moderate field strengths. In Figure 5 we see the data of von Molnar *et al.* for a sample with $x=0{\cdot}321$, corresponding to $n=2{\cdot}5\times10^{20}$ cm^{-3}. σ_{min} is estimated at 50 $(\Omega \text{ cm})^{-1}$ using equation (5). The transition is clearly continuous below this value of σ_{min}.

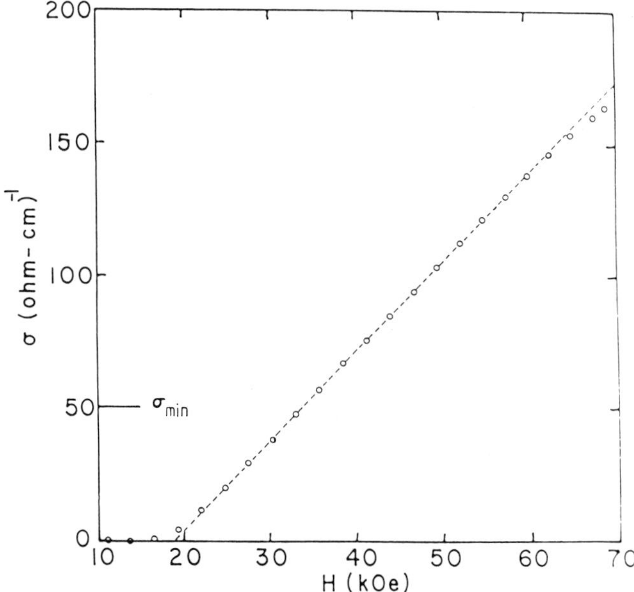

Figure 5. Conductivity versus magnetic field for a sample of $Gd_{3-x}v_xS_4$ with $x=0{\cdot}321$ at $T=300$ mK. The magnetically induced transition shows finite conductivities below σ_{min}. (Von Molnar *et al.* 1983)

However, Mott's new theory suggests that $\sigma_{\min}$ should be calculated from equation (12). For this system $L_H \gg a$, reducing $\sigma_{\min}$ calculated in this fashion below the limits of detectability.

Very recently, members of this same group (Washburn *et al.* 1985) have extended the study of this system down to 6 mK and again find that the transition is continuous. In these latest experiments rounding of the transition due to inhomogeneities does not occur until a value of $\sigma(0)$ well below the $\sigma_{\min}$ calculated from equation (12).

The second method of field-induced transition is caused by wave function shrinkage which drives metals into insulators, as demonstrated by Robert *et al.* (1980) in n-InSb. The magnetic field alters the donor wave function, producing Bohr radii $a_\perp$ and $a_\parallel$ for directions perpendicular and parallel to $\vec{H}$, respectively. The Mott condition for the transition (equation (1)) now becomes

$$n_c^{1/3}(a_\perp^2 a_\parallel)^{1/3} \simeq 0{\cdot}26 \tag{13}$$

Increasing the field strength shrinks the perpendicular orbit $a_\perp$ until a metallic sample is driven insulating. This technique has been used successfully on several systems (primarily III–V compounds) but not yet conclusively with silicon or germanium since their relatively small Bohr radii necessitate very high magnetic fields to drive the transition. Since this type of field-induced transition occurs only for $L_H < a$, $\sigma_{\min}$ is given by $0{\cdot}03e^2/\hbar a$.

In Figure 6 we see the data of Mansfield *et al.* (1985) for an n-InSb sample

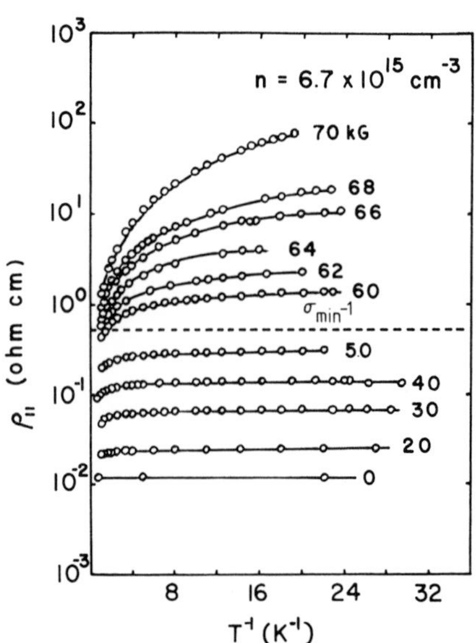

Figure 6. Resistivity parallel to the magnetic field plotted against $1/T$ for a sample of n-InSb. The transition is induced by the strong field. (Mansfield *et al.* 1985)

with carrier concentration of $6{\cdot}7 \times 10^{15}$ elec. cm^{-3}. They present the data in the standard way with the resistivity parallel to the field, $\rho_{\parallel}$, plotted against T^{-1}. The bottom few curves represent metallic behaviour, since decreasing temperature leads to no change in $\rho_{\parallel}$ (that is, $\sigma_{\parallel}(0)$ is finite). The top curves exhibit insulating behaviour since $\rho_{\parallel}$ continues to increase as T decreases. These authors find that the transition occurs at a critical field strength of 6·4 T, although this value is difficult to extract from this type of plot. To find the critical field more precisely they plot $\sigma_{\parallel}$ at constant field strength against $T^{1/2}$. At very low temperatures the conductivity has a temperature correction given by equation (11), making it straightforward to find $\sigma_{\parallel}(0)$ by extrapolating the curve to zero temperature.

Comparing equations (1) and (13) we can write an expression analogous to equation (9),

$$\sigma(0) = \sigma_0 (a_{\perp}^2 a_{\parallel}/(a_{\mathrm{H}}^*)^3 - 1)^{\nu} \tag{14}$$

Mansfield and co-workers find their data on two n-InSb samples fit this form as shown in Figure 7. The solid curves are best fits to equation (14) with $\nu = 1{\cdot}2$. They also find finite values of $\sigma(0)$ below $\sigma_{\min}$ calculated from equation (12).

Recent experimental work by Long & Pepper (1984a) on a metallic n-InP sample with dopant concentration of 7×10^{16} elec. cm^{-3} also shows the $T^{1/2}$ dependence of σ. However, when the transition to the insulating state is induced by magnetic fields of order 9·5T they find that $\sigma(0)$ drops discontinuously to

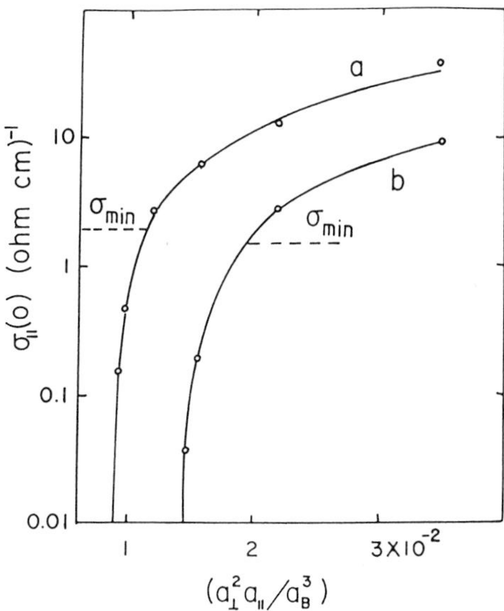

Figure 7. Conductivity parallel to the magnetic field is plotted against a normalized effective Bohr radius for two samples of n-InSb with different donor concentrations. In both cases, the magnetically induced transition is continuous below $\sigma_{\min}$. (Mansfield *et al.* 1985)

zero. Long & Pepper (1984b) have also studied Si:Sb samples with a range of dopant concentrations and find that some samples appear to be driven insulating at high field. They also find evidence for a magnetically produced non-zero value of σ_{min} for these samples, in agreement with their results on n-InP. The seemingly contradictory results of Mansfield *et al.* and Long & Pepper on the existence of σ_{min} may be due to an experimental problem: the difficulty of simultaneously achieving high magnetic fields and very low temperatures. Both groups had low temperatures limits of 30 to 50 mK, depending on field strength. Since temperature corrections to σ are important near the transition, data at lower temperatures may be required to determine whether or not magnetic fields can produce a finite value of σ_{min}.

6. Value of the critical exponent

Many systems have now been investigated which exhibit a continuous transition in the form of equation (9). The values obtained for the critical exponent, ν, tend to cluster near 1 or $\frac{1}{2}$. As previously mentioned, the non-interacting, scaling theory of localization predicts $\nu = 1$. A recent calculation by Castellani & Lee (1984) considering coulomb interactions with disorder, but neglecting localization, also yields $\nu = 1$. On the other hand, a scaling theory for interacting electrons by Grest & Lee (1983) predicts $\nu = 0{\cdot}6$. These theories suggest that the value of ν might be a good indication of the relative importance of disorder compared to coulomb interaction effects for a particular system. As an example, the Si:P system, which has been studied to within 0·1% of n_c by stress tuning, has a critical exponent of $0{\cdot}50 \pm 0{\cdot}08$ and interaction effects are known to be important for the temperature (Rosenbaum *et al.* 1981a, 1983) and magnetic field (Rosenbaum *et al.* 1981b) corrections to σ.

Systems such as the amorphous semiconductor–metal alloys and $Gd_{3-x}v_xS_4$ have many more randomly distributed scattering centres than charge carriers. These systems should be more influenced by localization effects than by coulomb interactions between the charge carriers. Indeed it is found that $\nu \sim 1$ for these cases, as is shown in Table 1. For the III–V semiconductors, samples are generally (unintentionally) compensated, which leads to a lower concentration of charge carriers than scattering centres. Again, $\nu \sim 1$ for these systems. The doped elemental semiconductors, however, are usually uncompensated, yielding one charge carrier for each scattering centre (charged impurity). For these cases interaction effects are large and the data yields $\nu \sim \frac{1}{2}$.

One method which can reduce the effects of coulomb interactions, while having little influence on disorder, is to reduce the concentration of charge carriers by the introduction of compensating impurities. Thomas *et al.* (1982a) measured the value of ν for Ge:Sb for a series of samples with varying values of compensation, K. Their results are shown in Figure 8 with the results of other systems shown for comparison. Increasing compensation does cause ν to increase

Table 1 (*a*) Systems with one scattering site per charge carrier. Critical exponent $\nu \sim \frac{1}{2}$.

System	Reference
Si:P	Rosenbaum *et al.* 1981a, 1983
Si:P (stress induced transition)	Paalanen *et al.* 1982
Si:As	Newman & Holcomb 1983a
Si:P,As	Newman & Holcomb 1983b
Si:Sb	Long & Pepper 1984b
Ge:Sb	Thomas *et al.* 1982a

(*b*) Systems with several scattering sites per charge carrier. Critical exponent $\nu \sim 1$.

System	Reference
Ge:Sb (compensated)	Thomas *et al.* 1982a
GaAs:n (compensated)	Morita *et al.* 1985
InSb:n (compensated)	Morita *et al.* 1985
InSb:n (field induced transition)	Mansfield *et al.* 1985
$Gd_{3-x}v_xS_4$ (field induced transition)	von Molnar *et al.* 1983
a-Si:Au	Nishida *et al.* 1985
a-Si:Nb	Bishop *et al.* 1985
a-Si:Al	Bishop *et al.* 1985
a-Ge:Au	Mochel 1984
a-Si:Mo	Yoshizumi 1984

After Thomas & Paalanen (1985).

from $\sim \frac{1}{2}$ toward the localization value of ~ 1. These results, plus others shown in Table 1, suggest that coulomb interactions play an important role in reducing the value of ν.

Newman & Holcomb (1983b) studied the effect on ν of increasing the disorder by using two impurities of different ionization energies. They investigated the double-donor system Si:P,As and found that the increased vertical disorder does not measurably increase ν compared to the single-donor systems Si:P and Si:As. It has very recently been suggested (Ruckenstein *et al.* 1984, Paalanen *et al.* 1984) that the coulomb interaction effects which apparently are responsible for the reduction of ν to $\sim \frac{1}{2}$ in Si:P may be analogous to the correlated electronic behaviour calculated by Brinkman & Rice (1970) for ordered systems near the Mott transition. Important features of Brinkman–Rice systems are the enhancement of the electron effective mass and the reduction of spin diffusion relative to charge diffusion, a well known characteristic of Si:P (see Milligan *et al.* 1985). Thomas & Paalanen (1984) have suggested that a Brinkman–Rice model extended to disordered systems should lead to an enhanced density of states. This could account for the observed anomalously large specific heat of metallic Si:P at low T (Thomas *et al.* 1981) as well as the low T magnetic susceptibility enhancement observed in Si:P (Andres *et al.* 1981; see Milligan *et al.* 1985). Clearly a complete theoretical investigation of disordered systems which includes both localization and coulomb interaction effects would

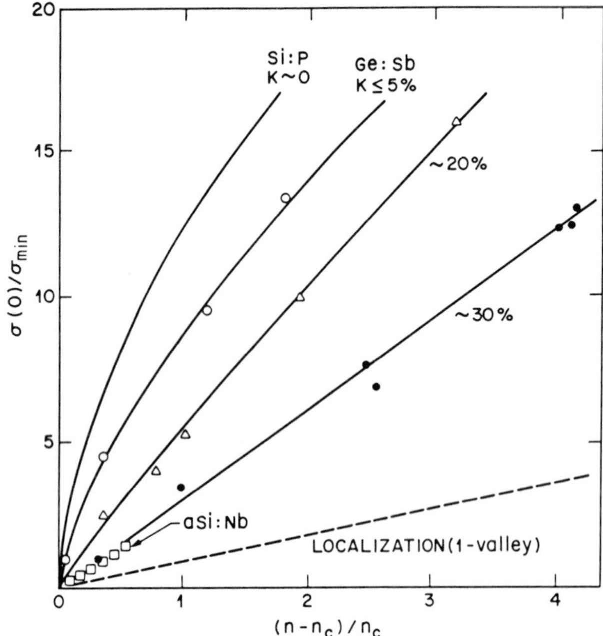

Figure 8. Normalized zero temperature conductivity is plotted against normalized donor concentration for several systems of different compensation, K. The critical exponent, ν, changes from $\frac{1}{2}$ to 1 as compensation increases. In the compensated Ge:Sb and the amorphous Si:Nb alloy there are several scattering sites per charge carrier. The dotted line shows the result calculated from scaling theory of localization assuming no interactions. (Thomas *et al.* 1982a)

be very useful in providing a better understanding of many aspects of the M–NM transition in semiconductors.

7. Conclusion

Over the past several years significant theoretical and experimental progress has been made in understanding the metal–nonmetal transition in doped semiconductors. Theoretical work on the scaling theory of localization and interaction effects in disordered systems has been particularly helpful in describing the approach to the transition and the temperature and field corrections to the conductivity. The recent advances in smoothly tuning through the transition, by either applying external stress or magnetic fields, as well as the increasing availability of He-3 dilution refrigerators for low temperature work, promise to bring the experimentalist still closer to this zero T, critical transition.

References

Abrahams, E., Anderson, P.W., Licciardello, D.C. and Ramakrishnan, T.V., 1979, *Phys. Rev. Lett.*, **42**, 673.

Altshuler, B.L. and Aronov, A.G., 1979a, *Zh. Eksp. Teor. Fiz.*, **77**, 2028 (*Sov. Phys. – JETP*, **50**, 968).

Altshuler, B.L. and Aronov, A.G., 1979b, *Solid St. Commun.*, **36**, 115.

Altshuler, B.L., Aronov, A.G. and Lee, P.A., 1980a, *Phys. Rev. Lett.*, **44**, 1288.

Altshuler, B.L., Khmelnitskii, D., Larkin, A.I. and Lee, P.A., 1980b, *Phys. Rev.*, B**22**, 5142.

Anderson, P.W., 1958, *Phys. Rev.*, **109**, 1492.

Andres, K., Bhatt, R.N., Goalwin, P., Rice, T.M. and Malstedt, R.E., 1981, *Phys. Rev.*, B**24**, 244.

Bhatt, R.N., 1982, *Phys. Rev.*, B**26**, 1082.

Bishop, D.G., Spencer, E.G. and Dynes, R.C., 1985, *Solid-St. Electron.*, **28**, 73.

Brinkman, W.F. and Rice, T.M., 1970, *Phys. Rev.*, B**2**, 4302.

Castellani, C. and Lee, P.A., 1985, *Solid-St. Electron.* (in press).

Cuevas, M. and Fritzsche, H., 1965a, *Phys. Rev.*, A**137**, 1847.

Cuevas, M. and Fritzsche, H., 1965b, *Phys. Rev.*, A**139**, 1628.

Edwards, P.P. and Sienko, M.J., 1978, *Phys. Rev.*, B**17**, 2575.

Fritzsche, H., 1962a, *Phys. Rev.*, **125**, 1552.

Fritzsche, H., 1962b, *Phys. Rev.*, **125**, 1560.

Fritzsche, H., 1978, in *The Metal–Nonmetal Transition in Disordered Systems*, edited by L.R. Friedman and D.P. Tunstall (Edinburgh: Scottish Universities Summer School in Physics), p. 193.

Gorkov, L.P., Larkin, A.I. and Khmelnitskii, D., 1979, *JETP Lett.*, **30**, 228.

Grest, G.S. and Lee, P.A., 1983, *Phys. Rev. Lett.*, **50**, 693.

Ioffe, A.F. and Regel, A.R., 1960, *Prog. Semiconductors*, **4**, 237.

Lee, P.A. and Ramakrishnan, T.V., 1982, *Phys. Rev.*, B**26**, 4009.

Long, A.P. and Pepper, M., 1984a, *J. Phys. C: Solid St. Phys.*, **17**, 3391.

Long, A.P. and Pepper, M., 1984b, *J. Phys. C: Solid St. Phys.*, **17**, L425.

Mansfield, R., Abdul Gader, M. and Fozooni, P., 1985, *Solid-St. Electron.*, **28**, 109.

Milligan, R.F., Rosenbaum, T.F., Bhatt, R.N. and Thomas, G.A., 1985, in *Electron–Electron Interactions in Disordered Systems*, edited by A.L. Efros and M. Pollak (Amsterdam: Elsevier Science Publishers B.V.), p. 231.

Mochel, J.M., 1984, private communication.

Morita, S., Mikoshiba, N., Koike, Y., Fukase, T., Ishida, S. and Kitagawa, M., 1985, *Solid-St. Electron.*, **28**, 113.

Mott, N.F., 1949, *Proc. Camb. Phil. Soc.*, **32**, 281.

Mott, N.F., 1956, *Proc. Phys. Soc. London*, A**62**, 416.

Mott, N.F., 1972, *Phil. Mag.*, **26**, 1015.

Mott, N.F., 1974, *Metal–Insulator Transitions* (London: Taylor & Francis).

Mott, N.F., 1984, *Phil. Mag.*, B**49**, L75.

Newman, P.F. and Holcomb, D.F., 1983a, *Phys. Rev.*, B**28**, 638.

Newman, P.F. and Holcomb, D.F., 1983b, *Phys. Rev. Lett.*, **51**, 2144.

Nishida, N., Furubayashi, T., Yamaguchi, M., Morigaki, K. and Ishimoto, H., 1985, *Solid-St. Electron.*, **28**, 81.

Paalanen, M.A., Rosenbaum, T.F., Thomas, G.A. and Bhatt, R.N., 1982, *Phys. Rev. Lett.*, **48**, 1284.

Paalanen, M.A., Ruckenstein, A.E. and Thomas, G.A., 1985, *Solid-St. Electron.*, **28**, 121.

Robert, J.L., Raymond, A., Aulombard, R.L. and Bousquet, C., 1980, *Phil. Mag.*, B**42**, 1003.

Rosenbaum, T.F., Andres, K., Thomas, G.A. and Lee, P.A., 1981a, *Phys. Rev. Lett.*, **46**, 568.

Rosenbaum, T.F., Milligan, R.F., Thomas, G.A., Lee, P.A., Ramakrishnan, T.V., Bhatt, R.N., DeConde, K., Hess, H. and Perry, T., 1981b, *Phys. Rev. Lett.*, **47**, 1758.

Rosenbaum, T.F., Milligan, R.F., Paalanen, M.A., Thomas, G.A., Bhatt, R.N. and Lin, W., 1983, *Phys. Rev.*, B**27**, 7509.

Ruckenstein, A.E., Paalanen, M.A. and Thomas, G.A., 1984, *Proceedings of the 17th International Conference of the Physics of Semiconductors, San Francisco*, edited by J.D. Chadi and P.J. Lansberg.

Thomas, G.A. and Paalanen, M.A., 1984, *Proceedings of the International Conference on Low Temperature Physics, Braunschweig.*

Thomas, G.A. and Paalanen, M.A., 1985, Proceedings of the International Conference on Localization, Interactions and Transport in Impure Metals, Braunschweig, edited by Y. Bruynseraede. In *Topics in Solid State Physics*, 1985 (Berlin: Springer).

Thomas, G.A., Ootuka, Y., Kobayashi, S. and Sasaki, W., 1981, *Phys. Rev.*, B**24**, 4886.

Thomas, G.A., Ootuka, Y., Katsumoto, S., Kobayashi, S. and Sasaki, W., 1982a, *Phys. Rev.*, B**25**, 4288.

Thomas, G.A., Kawabata, A., Ootuka, Y., Katsumoto, S., Kobayashi, S. and Sasaki, W., 1982b, *Phys. Rev.*, B**26**, 2113.

Thomas, G.A. and Paalanen, M.A., 1985, Proceedings of the International Conference on Localization, Interactions and Transport in Impure Metals, Braunschweig, ed. Y. Bruynseraede. In *Topics in Solid State Physics*, Springer-Verlag, Berlin, 1985.

Thomas, G.A., Paalanen, M.A., Rosenbaum, T.F., 1983, *Phys. Rev.*, B**27**, 3897.

von Molnar, S., Briggs, A., Floaquet, J. and Remenyi, G., 1983, *Phys. Rev. Lett.*, **51**, 706.

Washburn, S., Webb, R.A., von Molnar, S., Holtzberg, F., Flouquet, J. and Remenyi, G., 1985, *Phys. Rev.* B (in press).

Webman, I., Jortner, J. and Cohen, M.H., 1975, *Phys. Rev.*, B**8**, 2885.

Yoshizumi, S., 1985, *Solid-St. Electron.* (in press).

Ziman, J.M., 1964, *Principles of the Theory of Solids* (Cambridge: Cambridge University Press).

CHAPTER 8

quasi-one-dimensional organic conductors and conducting polymers

S.V. Subramanyam and Hemamalini Naik
Department of Physics, Indian Institute of Science, Bangalore 560012, India

1. Introduction

In the field of condensed-matter physics, the study of quasi-one-dimensional conductors has become very important and rewarding. By quasi-one-dimensional conductors, we mean materials which have a very large anisotropy in electrical conductivity—anisotropy of the order of 10^2–10^6. Such a large anisotropy is due to the peculiar stacking of the atoms and molecules in the crystalline phases of these materials. The electrons can easily move along certain directions, but crossing from one channel to another has to be by an activated or tunnelling process. The delocalization length (l) of the electron and the intermolecular distance (a) are related by $l_{\parallel} \gtrsim a_{\parallel}$ and $l_{\perp} \lesssim a_{\perp}$, where $\parallel$ and $\perp$ refer to directions parallel and perpendicular to the high-conducting axis.

Such systems have become exciting because of the tremendous physical possibilities—high-temperature superconductivity, novel magnetic ordering and new conduction processes through charge density waves, solitons, polarons; chemical possibilities—synthesis of new conducting systems with various chemical features; and technological possibilities based on molecular engineering for energy and entropy storage, transmission and dissipation. The subject has attracted such wide attention in the last ten years that more than 2000 papers have been published in this area, ten international conferences have been held, about twenty reviews and edited volumes have come out and one journal, *Synthetic Metals*, is devoted largely to this area (for example, Subramanyam 1981, Jerome & Schulz 1982, Hodina 1984).

Quasi-one-dimensional conductors are broadly classified into (a) linear metal-chain conductors, (b) organic charge transfer complexes, (c) conducting polymers and (d) inclusion compounds. We shall mainly be concerned with the organic charge transfer complexes and conducting polymers.

2. Physics of one-dimensional conductors

2.1. *Little's model of an excitonic superconductor*

The subject of one-dimensional conductors was given a big boost by the suggestion of Little (1964) that high temperature superconductivity was not improbable in such a system. Little considered a polyene chain to which is attached a suitable polarizable substituent. An electron moving in the chain forms certain excitations in the side group which mediates into an attractive electron–electron interaction. In such a case, calculations similar to those of the Bardeen–Cooper–Schreiffer theory give

$$kT_{\text{sup}} \sim 1{\cdot}14\hbar\omega_{\text{exciton}} \exp\left(-\frac{1}{N(E_{\text{F}})V}\right)$$

where $\hbar\omega_{\text{exciton}}$ is the excitation energy for excitons, V is the effective binding energy of the Cooper pairs for the electrons and $N(E_{\text{F}})$ is the electronic density of states at the Fermi energy. With typical values of the parameters, Little obtained superconducting transition temperatures of about 2000 K. However no system conforming to Little's requirements have been synthesised so far (Davis *et al.* 1976) and most of the one-dimensional systems show inherent lattice instabilities at low temperature.

2.2. *Peierls transition*

Peierls (1955) pointed out that a one-dimensional metal is unstable against a lattice distortion at low temperatures. The lattice will distort periodically and this results in the opening of a gap in the electronic spectrum. The distorted lattice will show either semiconducting or insulating properties depending on the size of the gap. The Peierls transition is essentially a manifestation of the Kohn anomaly in one dimension.

2.3. *Fröhlich conduction and charge density wave*

Peierls lattice distortion will result in redistribution of the charge density. The phase of the charge density will be related to the lattice distortion either in a commensurate way or in an incommensurate way. An incommensurate charge density wave (CDW) has the same energy as it propagates. The phase of the wave can be given any arbitrary value with respect to the lattice distortion. A superconducting state thus results. Fröhlich's calculations on this mechanism were at 0 K (Fröhlich 1954) and these have been modified by Kuper for finite temperatures (Kuper 1955). The charge density wave may be pinned to the lattice or may be free. Correspondingly a high-conductivity state may or may not be observed. Charge density waves have been observed in a number of organic metals such as TTF–TCNQ (Kagoshima *et al.* 1980) and metal trichalcogenides (Ong & Monceau 1977). The role of defects in pinning these waves and depinning of the CDW has also been investigated (Fukuyama & Lee 1978).

2.4. Solitons

In a linear metallic chain, the electronic excitations are inherently coupled to the lattice distortions. It is possible to have non-uniform lattice distortions—domains of Peierls distortions with localized interfaces of high energy, or non-linear excitation energy. In these cases we have domains with original commensurate distortion separated by domain walls. These are the kink solitons. In conjugated polymers, a soliton may become a kink in the bond alternation pattern. At one end the odd-numbered bonds may be single, and at the other end they may be double. Then to the left and right of the kink, bond alterations are out of phase with respect to each other. These solitons can behave like charged, mobile, trapped particles. Semiconductor–metal transitions in polymers observed at suitable dopant concentrations can be attributed to the soliton mechanism (Tomkiewicz *et al.* 1981, Su *et al.* 1980, Mele & Rice 1981).

3. Organic metals

3.1. Introduction

Electrical properties of TTF–TCNQ (tetrathiofulvalene–tetracyanoquinodimethane), studied by Coleman *et al.* (1973), showed giant, apparently divergent peaks at about 60 K in a small percentage of crystals. It was conjectured that these conductivity peaks were the percursors of superconductivity, but some instability prevented the superconducting state from stabilizing. This opened up a floodgate of work on organic metals. A number of such organic charge transfer complexes have been synthesized and studied (Khidekel & Zhilyaeva 1981, Andre *et al.* 1976 and Lyubovskaya 1983). Some of the donors and acceptors used in organic conductors are listed in Table 1.

The standard method for the preparation of these has been given by Melby *et al.* (1962). Good quality crystals can be grown by an electrochemical growth technique (Keller *et al.* 1980, Wheland & Gillson 1976, Bechgaard *et al.* 1981). Here the donor and acceptor solutions are allowed to interact by a gradual passage through filters. A steady current (current density of a few $\mu A\ cm^{-2}$) between the two enables the deposition of the conducting complex on the platinum electrode. Of all the thermodynamically favourable compounds which

Table 1. Typical organic quasi-one-dimensional conductors. (Kampar 1982, Khidekel and Zhlyaeva 1981, Lyubovskaya 1983).

DONORS

2 2′5 5′ Tetrathiofulvalene

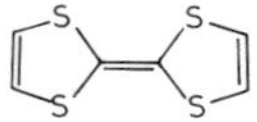

Tetraselenafulvalene

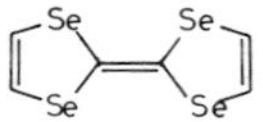

dimethyl TTF (cis)

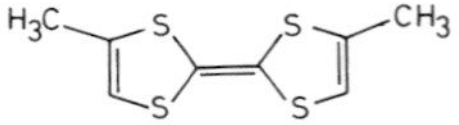

Tetramethyl TTF

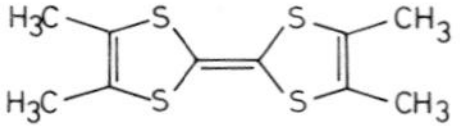

NNN′N′ Tetramethyl phenylenediamine (TMPD)

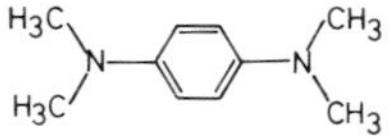

5,6,11,12 Tetrathiotetracene (TTT)

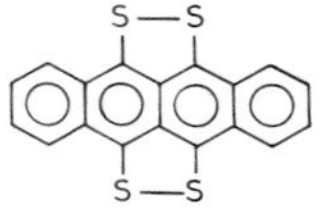

5,6,11,12 Tetraselenatetracene

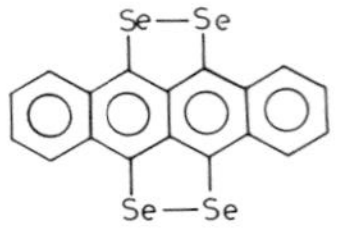

3 3′4 4′ Tetramethyl 2 2′5 5′ Tetraselena fulvalene

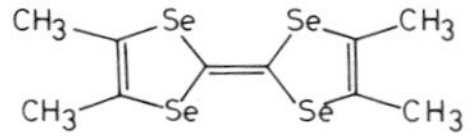

Hexamethylene Tetraselena fulvalene

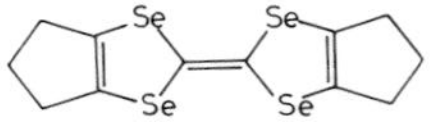

Nickel Phthalocyanine

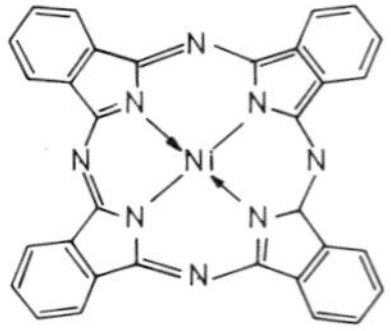

Perylene

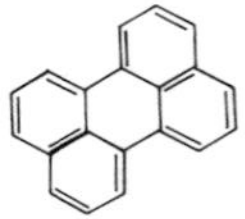

N-methyl acridinium

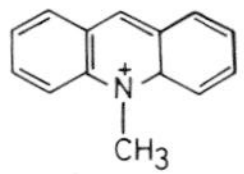

N-methyl phenazinium

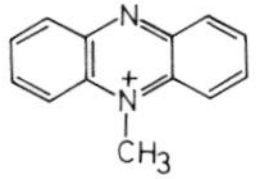

Pyrene

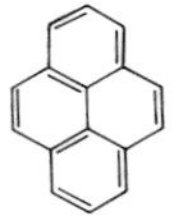

ACCEPTORS

7788 Tetracyanoquinodimethane

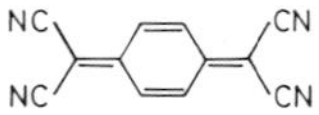

Tetracyanoethylene

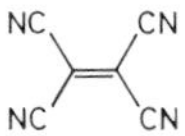

Tetracyanonaphthoquinodimethane (TNAP)

could crystallize from the solution, the metallic ones grow at a rapid rate. The basic feature of most of these TTF–TCNQ complexes is a reasonably high conductivity at room temperature, a conductivity peak at about 60–100 K and an insulating state at lower temperatures. However there exist a few such as HMTSF–TCNQ in which the metallic state is retained down to very low temperatures. Also, there are systems based on TMTSF and TTF which are superconducting at low temperatures. These will be discussed in detail later.

A characteristic feature of the conducting charge-transfer complex is the formation of stacks containing alternately donor (D) and acceptor (A) molecules. The distances between these molecules are usually less than the van der Waals' distance, so that a molecular orbital overlap can easily form. Such a stack structure gives rise to anisotropy in electrical conductivity. Quite a large number of charge-transfer complexes have been studied (Table 1).

The donor and acceptor charges can sometimes be completely separated even in the ground state. Such limiting cases of the conducting charge-transfer complexes are the radical ion salts. A characteristic feature of the radical ion salts is the formation of stacks of only donors or only acceptors. In contrast to the charge-transfer complex, the stacks are formed in this case from molecules belonging to the same species; the distances between these molecules is again less than the van der Waals' distance. Strong interactions arising within the stack are responsible for high conductivity. It is interesting that the conductivity of the radical ion salts is usually high compared to that of the charge-transfer complex, in which there is a mixed stack. (Only TTF–TCNQ, although forming a charge-transfer complex, has a segregated chain structure). Most empirical descriptions of 'how to design an organic metal' list the necessity of a segregated stack arrangement.

3.2. Guidelines for the formation of an organic metal

Many attempts have been made to establish the conditions under which an organic metal can be synthesized but these have resulted only in guidelines. (Garito & Heeger 1974, Perlstein 1977, Lyubovskaya 1983). These are as follows. (1) Their molecules must be planar in the neutral and ionic states. A slight deviation (2–4°) from planarity may facilitate close packing. (2) The molecules must be symmetrical. Asymmetry may lead to a random potential which may transform the system to a dielectric state. (3) The complex should form segregated stacks. This is possible if the donor–acceptor overlap is weak. Qualitatively, the donor–acceptor overlap can be made particularly weak by having the donor and acceptor of different size, shape and symmetry. (4) There must be unpaired electrons in the system. (5) The charge density should be located near the opposite ends of the molecular unit so that the e–e interaction is weak and the molecular unit becomes highly polarizable; otherwise the system will undergo transition to the Mott insulating state. (6) Partial charge transfer must take place.

Conductivity of the complex is quite low for ionic as well as neutral complex formation (Hatfield 1978). Moderately strong electron donors and acceptors satisfy this criterion. The difference between the oxidation–reduction potentials of the electron donor and acceptor must be minimum. (Torrance 1979.) (7) The distances between the molecules in the stack should be uniform and minimum, facilitating delocalization of the electron. The interstack distance should be small enough to stabilize the structure at low temperatures while large enough to reduce electron tunnelling probabilities.

3.3. Some general features of organic metals

There have been extensive experimental measurements on a large number of organic charge-transfer complexes and radical-ion salts (Jerome & Schulz 1982, Kampar 1982, Khidekel & Zhilyaeva 1981, Starodub & Krivoshei 1982, Haddon *et al.* 1984, Hatfield 1978, Shchegolev 1972, Andre *et al.* 1976). The absorption spectra of these complexes reveal a fundamental difference from the spectra of individual components and those of other molecular crystals. A strong absorption band characteristic of the intermolecular charge transfer appears. A typical Drude-like edge is observed in the 1 eV range for light polarized parallel to the conducting axis for organic metals.

The structures of the charge-transfer complexes are generally made up of chains of donor and acceptor molecules. In the single-stack donors, like TTF-halides, the TTF molecules are in stacks. The halide counter-ions are in the channels provided by the TTF stacks. In TMTTF and TMTSF salts, the planar molecules stack one above the other; there will be two donor molecules per unit cell. In single-stack acceptors like TCNQ salts, the intermolecular distances are not always uniform. In mixed stacks, the donor and acceptor molecules are interleaved within a single stack. For example, in TMPD–TCNQ there is a –donor–acceptor–donor–acceptor–structure. Sometimes the donor–acceptor molecules can also form double or segregated stacks. Then the degree of charge transfer becomes an important factor in deciding its conductivity.

The electrical conductivity of the charge-transfer complex is due to the donor stack or the acceptor stack or to both donor and acceptor stacks. One or both of the components may possess an unpaired electron, which may be formed by the charge transfer process. For example in the TTF–TCNQ stack, the TCNQ stack is the conducting one. In TSF–TCNQ both are conducting stacks. The charge transferred from TTF to TCNQ can be found by X-ray, Raman and photoemission spectroscopy. It is about 9×10^{-20} C.

Many of these charge-transfer complexes undergo the Peierls transition to semiconducting ground states. Coulomb localization, because of charge repulsion, may lead to a Mott–Hubbard semiconducting ground state.

Charge-transfer complexes so formed can have conductivities in the semiconducting range, intermediate range or metallic range. In the semiconducting charge-transfer complexes, for example, K–TCNQ ($\rho \sim 1\text{–}10^{10}\ \Omega$ cm), the conductivity is activated. The band gap results from the coulomb localization of a Mott–Hubbard transition. The localized spins on the adjacent molecules couple antiferromagnetically. Therefore the paramagnetic susceptibility decreases with a reduction in temperature. In alkali metal–TCNQ salts, there can be a transition to a dielectric phase resulting from a Peierls–spin density wave transition. In the intermediate conductors like MPht–TCNQ, Qn–TCNQ ($\rho \sim 1\text{–}10^{-2}\ \Omega$ cm), the conductivity is more like that of a semimetal, MPht = N-methyl-phthalazinium; Qn = Quinolium (Table 1). With the decrease in temperature, σ increases, reaches a maximum and then decreases (Graja *et al.* 1983). Probably there is an initial increase in the mobility of the charge carrier. The band filling and incomplete charge transfer reduces the coulomb localization. In NMP–TCNQ for example, the back charge-transfer makes about 10% of molecules neutral. The charge carriers are low and the spin–spin interaction also is low. In optical studies, the presence of neutral and charged molecules show up as additional mixed-valence transition bands. Organic metals have resistivity less than $10^{-2}\ \Omega$ cm and the resistivity decreases with a decrease in temperature. They either remain metallic down to very low temperatures, like HMTSF–TCNQ, or undergo transitions at which the conductivity is maximum and later become semiconducting like TTF–TCNQ. A certain amount of interchain coupling, as in HMTSF–TCNQ and $(TMTSF)_2ClO_4$, stabilizes the metallic state down to very low temperatures. The reduction in such interchain coupling makes the system more one-dimensional and susceptible for the Peierls transition as in TTF–TCNQ. The magnetic properties of the organic metals are complicated. There is no single model to explain the behaviour. As we saw earlier, in optical reflection studies these show a characteristic metallic plasma edge when the incident light is polarized parallel to the conducting axis. This reveals the anisotropic electrical conductivity of the material.

3.4. Some electrical conductivity studies

Measurements of electrical conductivity of one-dimensional conductors have been carried out extensively. D.c. and a.c. measurements provide valuable information on the nature and mobility of the charge carriers, the intra-chain and interchain effects, the mechanisms of conduction, including CDW and soliton, and the contributions from defects, impurities and so on. The experimental measurements have been obtained over a wide range of temperature, pressure, electric field and impurity concentration, using two-probe, four-probe and Montgomery techniques, the contactless method, electron microscopy and other methods.

Effect of pressure on conductivity

The effect of pressure on the electrical conductivity of organic conductors has been known for a long time. The conductivity rapidly increases with pressure

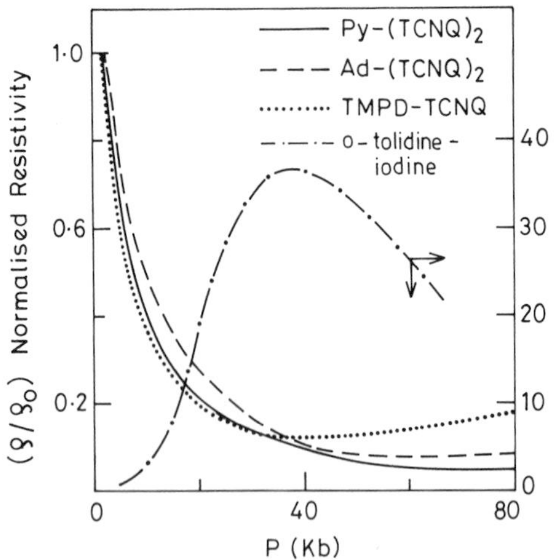

Figure 1. Effect of pressure on the electrical conductivity of an organic semiconductor.

until it reaches saturation; the conductivity at high pressures is nearly independent of pressure (Samara *et al.* 1962, Andrieux *et al.* 1979, Bandyopadhyay *et al.* 1980, 1981) (Figure 1). With increase in pressure, there is an increase in the molecular orbital overlap within the stack and consequently an increase in the charge transfer. With a further increase in pressure, the steric hindrance between the two molecules dominates and will prevent further decrease of the distance. The variation of the resistivity with pressure can be calculated in the tight-binding approximation. It is easily shown that

$$\rho/\rho_0 = \exp(Ap + Bp + Cp^2)$$

where A, B, C are constants related to the elastic coefficients of the system. The experimental data agree well with the above calculations (Bandyopadhyay *et al.* 1980).

But often, high pressures can cause coulomb repulsion to dominate, resulting in a lowering of the conductivity (Hemamalini Naik & Subramanyam, unpublished). It can also result in new phases of the complex—transition to metallic states (Kobayashi *et al.* 1980), transition to a superconducting state (Jerome *et al.* 1980) and transition to a polymeric phase (Nalini *et al.* 1976). Nalini has observed transitions to a high conducting polymeric phase in two complexes: benzidine–iodine and benzidine–DDQ (Figure 2). The interchain coupling becomes increasingly important at higher pressures and the system gradually loses its one-dimensionality. The three-dimensional ordering will also suppress the Peierls transition (Friend *et al.* 1978, Cooper *et al.* 1976). HMTSF–TCNQ therefore remains metallic down to very low temperatures at pressures exceeding 4 kbar, and 9 kbar suppresses the Peierls transition in $(TMTSF)_2PF_6$.

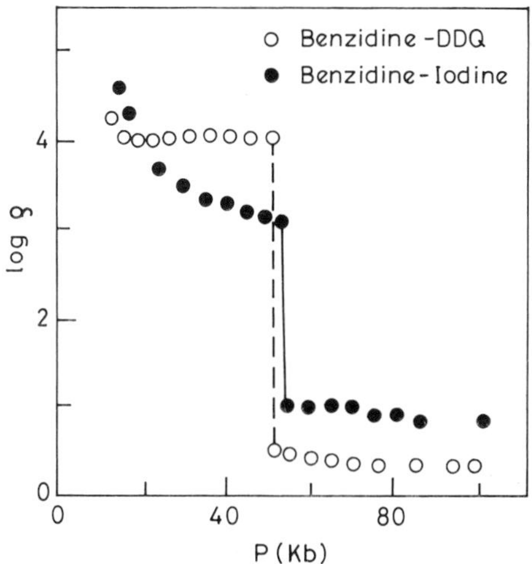

Figure 2. Pressure-induced semiconductor–metal transition in an organic conductor.

Superconductivity in organic conductors

An exciting development in the area of quasi-one-dimensional conductors has been the observation of superconductivity. While Little's model envisaged possibilities of high-temperature superconductivity, the Peierls transition suppressed even the metallic behaviour at low temperatures. The first observation of superconductivity came in the conducting polymer $(SN)_x$ with a superconducting transition at 0·3 K (Labes *et al.* 1979) (see §6.3). The superconductivity arises out of weak three-dimensional coupling and is easily explained.

Superconductivity has been observed in $(TMTSF)_2X$ compounds where $X = PF_6$, AsF_6, SbF_6, TaF_6, ClO_4, ReO_4, FSO_3 and in BEDT–TTF–X complexes; BEDT ≡ bis(ethylenedithiolo). In the $(TMTSF)_2X$ compounds, the TMTSF chains are metallic. In general, these undergo a transition from a high-conducting metallic phase to a semiconducting state around 10 to 180 K at normal pressures. This transition is not of Peierls type, as there is no lattice modulation observed. Instead, an antiferromagnetic ordering of the spins is observed as the temperature is reduced (Andrieux *et al.* 1981, Scott *et al.* 1981). This results in a spin density wave. If the pressure applied on the sample is increased, the metal–nonmetal transition is suppressed and the compound becomes type II superconducting at 0·9–1·3 K. Typical data for the metal–nonmetal transition and the superconducting transition are given in Table 2 and a typical phase diagram representing the metallic phase, the spin density wave (SDW) phase and the superconducting phase are given in Figure 3 (Jerome & Schulz 1982). In a small region around p_c, the SDW state and the superconducting state are found to coexist. The critical fields are anisotropic,

Table 2. Characteristics of organic superconducting salts.

Salt	$\sigma/(\Omega\,\text{cm})^{-1}$	$T_{M\text{-}NM}$/K	$T_{sup.}$/K	P/kbar
$(TMTSF)_2\ ClO_4$	400	—	1·4	—
$(TMTSF)_2\ ReO_4$	400	180	1·3	9·5
$(TMTSF)_2\ FSO_3$	1600	86	3·5	6·5
$(TMTSF)_2\ PF_6$	540	12–15	1·4	6·5–8·5
$(TMTSF)_2\ AsF_6$	430	12–15	1·4	9·5
$(TMTSF)_2\ SbF_6$	500	17	0·38	10·5
$(TMTSF)_2\ TaF_6$	—	—	1·38	<11
BEDT–TTF–ReO_4	200	—	2	4

(Lyubovskaya 1983)

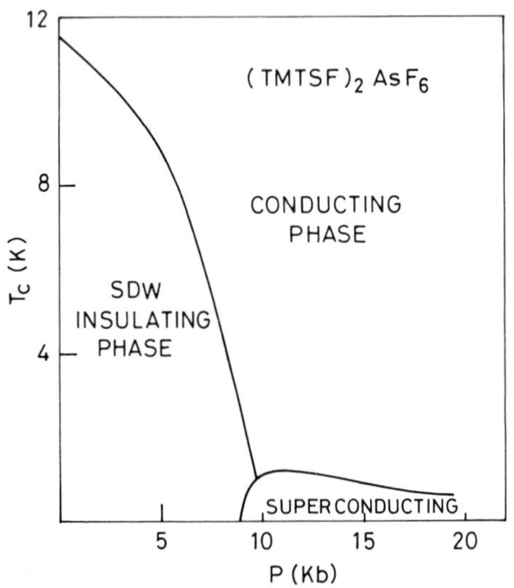

Figure 3. Superconducting phase of $(TMTSF)_2AsF_6$. (Jerome & Schulz 1982.)

with $H_c \sim 0{\cdot}0001$–$0{\cdot}1$ T along diverse directions. The complex $(TMTSF)_2ClO_4$ becomes superconducting at about 1·4 K at normal pressure. The spin density wave transition is not observed in all cases. For example, in $(TMTSF)_2FSO_3$, the FSO_3^- anion has a permanent dipole moment. The metal–nonmetal transition in this is associated with a charge density wave. The role of the anion in determining the superconducting properties of these complexes is not fully understood (Maaroufi *et al.* 1983).

BEDT–TTF–X is another class of organic conductors which exhibits superconductivity. Here the BEDT, TTF stacks have a dimeric structure with S–S distance of 3·3–3·45 Å (compared to the van der Waals' radius of the sulphur atom of 3·6 Å). The complex of BEDT–TTF with an electron acceptor like ReO_4 becomes superconducting at about 2 K and 4 kbar pressure. The structure of this salt is very similar to that of $(TMTSF)_2X$ although there is a difference in the anion arrangement. Instead of the sulphur, the selenium analogue of the complex BEDT–TTSF–X is also possibly superconducting.

The superconductivity in $(TMTSF)_2X$ and BEDT–TTF–X compounds is sensitive to impurities and disorder. A concentration of even 0·01% impurity can destroy the superconductivity. Fast cooled/quenched samples can introduce a sufficient amount of disorder to prevent the superconducting state from developing. (Tomic *et al.* 1982, Azevedo *et al.* 1983.) In view of these factors, there is a certain amount of divergence in the experimental data reported by several authors. But the formation of the superconducting state in these organic metals is a matter of much interest as it opens up an entirely new class of materials which can become superconducting.

Non-ohmic electrical conduction

Many organic conductors and other quasi-one-dimensional conductors exhibit non-ohmic conduction at fields exceeding a threshold value (Subramanyam 1981). While such a non-ohmic dependence is quite common in semiconductors, amorphous systems and multicomponent systems, its observation in one-dimensional conductors has certain characteristic features. These were studied earlier in transition metal trichalcogenides (Zettl *et al.* 1982). Here, below the metal–nonmetal transition, the conductivity becomes frequency dependent (Gruner *et al.* 1980) and large fluctuations are observed (Fleming & Grimes 1979). These are attributed to various mechanisms, such as the depinning of the charge density wave. Analogous to a classical situation, a rigid (but finite) incommensurate charge density wave can be depinned from impurity sites at suitable driving fields.

Similarly non-ohmic conduction has been observed for a number of organic conductors and conducting polymers: TTF–TCNQ (Gunning & Heeger 1978), $Qn(TCNQ)_2$ (Mihaly *et al.* 1979), $(TMTSF)_2PF_6$ (Chaikin *et al.* 1980), $(TMTSF)_2ClO_4$ (Murata *et al.* 1982), $(TTT)_2I_3$ (Venedek *et al.* 1982), $(CH)_x$ (Epstein *et al.* 1980, Mortensen *et al.* 1980). The model of the depinning of the charge density waves proposed by Bardeen (1979, 1980) explains some features of the non-ohmic conduction. Alternately, models based on space charge limited current (Farges *et al.* 1972), phonon-assisted hopping through random barriers, impurity pinning (Lee & Rice 1979) and sliding CDW (Frohlich 1954) are used. A typical plot of such non-ohmic behaviour is illustrated in Figure 4 (Hemamalini Naik & Subramanyam 1983). The non-ohmic part fits an equation of the type

$$I(E)=A(1-\exp(-BE))\exp(CE+DE^2)$$

reasonably well.

Electrical switching

A fast reversible/irreversible transition between a resistive state and a conducting state has been one of the fascinating problems in condensed matter physics (Ovshinsky 1968). This has attracted enormous theoretical and experimental attention. Such transitions are widely known in amorphous semiconductors and various other disordered systems. The onset of possible switching was indicated in the high fluctuation conductivity of transition metal trichalcogenides (Fleming & Grimes 1979). It had the origin in the depinning of the charge density wave. In organic conductors, the first observation of electrical switching came from Potember & Poehler (1979, Potember *et al.* 1982) in Cu–TCNQ. At a field of about 10^4 V cm^{-1}, the Cu–TCNQ sample switches from a high-impedance state to a low-impedance state. The mechanism of such electrical switching has been attributed to the formation of neutral TCNQ and a complex phase by the field-induced redox reaction

$$(Cu^+TCNQ^-)_n \rightleftharpoons Cu_x^0 + (TCNQ^0)_x + (Cu^+TCNQ^-)_{n-x}$$

This mechanism is supported by Raman spectroscopy also (Kamitsos *et al.* 1982). Similar switching has been observed with Cu, Ag and TCNE, TCNQ, $TCNQF_4$, TNAP.

Recently Bandyopadhyaya & Subramanyam (1979), Bandyopadhaya *et al.* (1980, 1981) and Hemamalini Naik & Subramanyam (1982, 1983) have observed electrical switching in several organic complexes. In all these cases, the material shows pronounced non-ohmic electrical conduction, and beyond a threshold field it undergoes switching (Figure 4). The ON state resistivity is only a few ohms and the switching field is about 10^3 V cm^{-1} for TM Bine–TCNQ (TMBine= tetramethylbenzidine) and *o*-tolidine–iodine and 10^2 V cm^{-1} for TMPD–TCNQ.

The switching is observed only under certain suitable initial conditions of temperature and pressure. The ON state is quite stable in all the samples and does not revert to the original OFF state as the driving current is removed. The OFF state can be re-established by the application of an a.c. pulse for TM Bine–TCNQ or by heating for TMPD–TCNQ. The ON state is metallic, while the OFF state is activated. The switching is not due to electrode effects and thermal effects but is electronic in origin. The phenomenon of electrical switching in these materials is not fully understood.

Effect of impurities and disorder

The electrical properties of organic conductors are quite sensitive to small percentages of impurities and to lattice disorder. The Peierls transition, characteristic of a one-dimensional metal, can be suppressed by weak irradiation disorder (Forro *et al.* 1982). The effect of impurities on the Peierls transition has been studied by Chatterjee (1981), Bulaevskii & Sadovskii (1974) and Sen & Varma (1974). Assuming that the impurities introduce a random potential in the lattice, the free energy of the distorted lattice is written in a modified form. This free energy can be minimized with respect to lattice distortion to obtain the

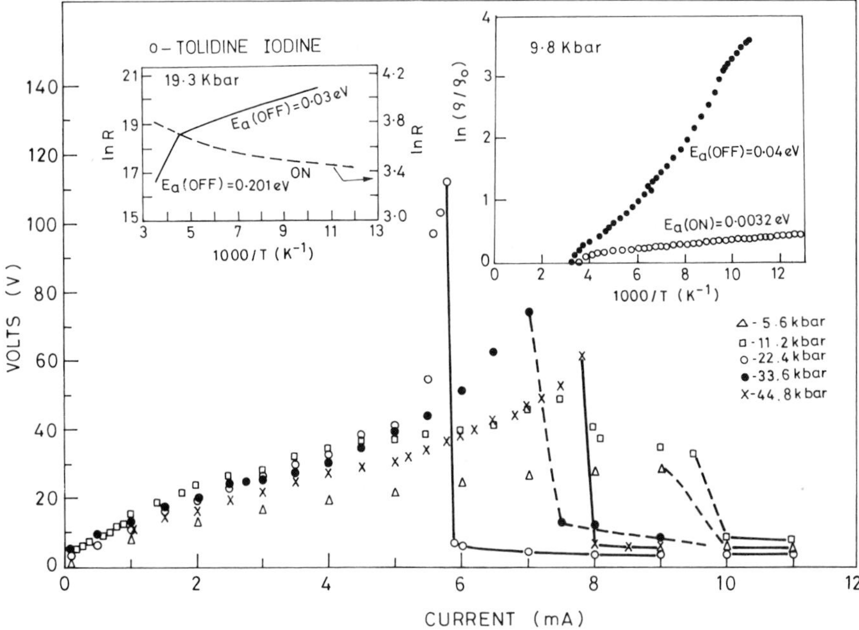

Figure 4. Non-ohmic conduction and switching in *o*-tolidine–iodine.

Peierls transition temperature. It has been shown by Chatterjee that the Peierls transition is suppressed if the impurity concentration exceeds a critical value.

Also the metallic chain gets fragmented by the irradiation and conductivity will then be due to a phonon-assisted hopping process (Zuppiroli *et al.* 1980, Przybylski & Graja 1981). A change in conductivity by an order of 10^6 can even be observed. Mott & Twose (1961) have pointed out that even traces of impurity can localize the electrons in an ideal one-dimensional system. The localization leads to an insulating state at 0 K. The electrical properties of one-dimensional conductors with a known introduction of dopant/disorder has therefore been a matter of extreme importance. Often, during crystal growth, traces of solvent can be trapped in the crystal and these affect the electrical properties drastically (Bandyopadhyay & Subramanyam 1981). Deliberate introduction of disorder by X-ray, ion-implanation, neutron radiation and impurity addition has been done (Gunning *et al.* 1979). Many of the non-linear effects and switching observed in organic conductors can be suppressed by randomly introduced disorder. Even fast thermal cycling of the sample can introduce microcracks which can mar the conductivity behaviour (Murata *et al.* 1982).

Several theoretical calculations are available on the role of defects/disorder in the propagation of charge density waves (Tsuzuki & Sasaki 1980) and electron localization (Teranishi & Kubo 1979). At higher fields the electrons have to tunnel across the localized states and the tunnelling probability depends on the strength and distribution of impurities. Based on Mott's tunnelling calculation, it has been shown that a cross over from three-dimensional coupling to one-dimensional impurity tunnelling is also possible at low temperatures (Chatterjee 1981). The possibility of the depinning of a charge density wave by an external a.c. field has also been studied (Chatterjee 1982).

3.5. Applications

The applications of organic conductors are numerous. The low energy gaps make them suitable for energy absorption and conversion, as in solar cells, reprography, etc. They can also be used as catalysts. The switching finds potential applications in Q-switches for laser pulses in electronics. The sensitive dependence of electrical properties on temperature can be used for thermal control. Photo-oxidation of donors in halocarbon solvent can be used for optical information processing. Superconducting mechanisms in these offer newer types of tunnel junctions. The possibilities of chemical tailoring of these materials opens up a new vista of molecular electronics, wherein novel designs of molecular structures can be produced for a wide variety of microelectronic, optoelectronic and chemical applications.

4. Metal chain compounds

It is possible to synthesize a large number of inorganic metal chain compounds with high anisotropy in electrical conductivity. For example, in $K_2Pt(CN)_4Br_{0.3}3H_2O$, the distance between Pt–Pt atoms is 2·89 Å compared to 2·77 Å in Pt metal. The overlap of d_{z^2} orbitals along the chain makes it almost metallic in the chain direction. Perpendicular to the chain direction, the Pt atoms are well insulated by the four cyano groups which surround each Pt atom in a plane and the space between the chains is occupied by the bromine which partially oxidizes the Pt chain. KCP–Br shows a broad metal–insulator transition at about 200 K and remains an insulator at lower temperatures. Similar square planar complexes can be formed in a wide variety of systems with Pt, Ir, Rh and Ni chains (Krogman 1969, Miller 1982).

Mercury chains can be formed in many cationic polymercury complexes and strong electron acceptors can be introduced into these. $Hg_{2.86}AsF_6$ and $Hg_{2.91}SbF_6$ are examples of these. The mercury atoms form orthogonal parallel strands in two dimensions with short equivalent interchain Hg–Hg distance of 2·64 Å and an interchain spacing of 3·085 Å. The Hg–Hg distance depends on the relative ratio of XF_6 molecules to Hg atoms and is therefore incommensurate with respect to XF_6 lattice. The electrical conductivity is large in the Hg–Hg plane and is small in the perpendicular direction. At low temperatures, below 4 K, apparently the complexes show very high conductivity. These anomalous results have been explained by Datars *et al.* (1978), who showed that Hg extrudes out of the crystal below 200 K. The extruded mercury forms a sheath of filamentary superconductor. Similar filamentary superconductors are also formed in In, Ga and Sn complexes (Bogomolov *et al.* 1980, 1977).

The metal trichalcogenides (MX_3) where M = Nb or Ta and X = S or Se are also quasi one-dimensional conductors (Rouxel 1983). The clear observation of charge density waves is a striking phenomenon in these. $NbSe_3$ and $TaSe_3$ exhibit superconductivity at low temperatures and pressure effect on the superconducting transition has also been studied (Yamaya *et al.* 1983).

5. Inclusion compounds

Novel one-dimensional chains can also be formed by inserting molecules/ions in the channel cavities of some host structures (MacNicol *et al.* 1978). The host structures normally used for this purpose are clathrates, urea, thiourea, cyclodextrin, coumarin, amylose and zeolite. The remarkable property of these materials is the formation of a cage-like structure with reasonably large cavities. For example, in cyclodextrin, the channels are of about 5 Å diameter. Complexes of copper(II)-montmorillonite with cyclodextrin (Kijima *et al.* 1984) and metal iodides in cyclodextrins (Domb *et al.* 1979) have been studied.

Oza (1980) prepared polyiodide chains in host lattices of α-cyclodextrin, amylose and coumarin. The room temperature resistivities of these were high ($\sim 10^4\ \Omega$ cm). But at high pressures, of about 29 kbar, $coumarin_4$–KI–I_2 showed an abrupt jump in conductivity by several orders of magnitude (Figure 5). This has been explained by the formation of a linear iodine chain in the lattice. At higher pressures, the coulombic repulsions of the coumarin break the iodine chain.

6. Conducting polymers

6.1. Introduction

Little's model of a high-temperature superconductor was based on a polyene chain in which a delocalized electron moves freely through the chain. Attractive interaction between the electron and the excited state of the side group can result in superconductivity (Little 1964). As we saw earlier, the realization of such a model has many fundamental difficulties. There have been enormous efforts to make polymeric materials in which the conductivity is of the desired value—semiconducting/metallic. The approach is basically two-fold: (i) introducing a conducting filler into an insulating polymer, (ii) synthesizing a conjugated polymer chain and suitably doping it.

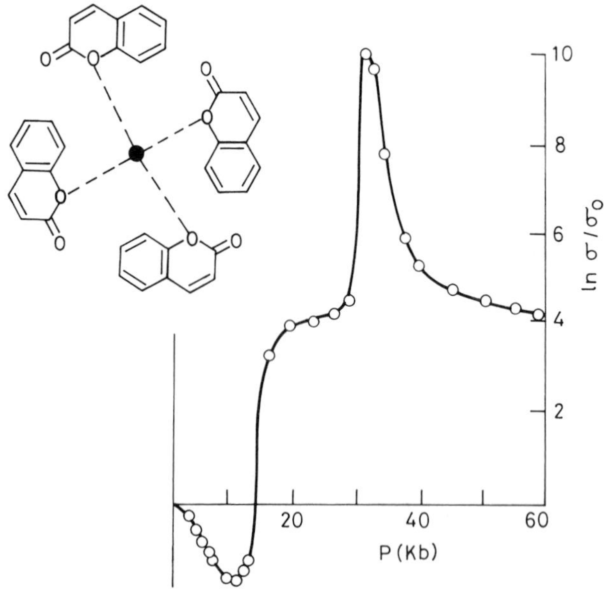

Figure 5. Variation of conductivity with pressure for a polyiodide chain in coumarin–KI–I_2 (I–I distance ≃ 3·15 Å).

6.2. *Composite polymers*

Composite polymers are generally formed by dispersing a metallic filler—a metal (Cu, Al, Ni, Ag, etc.), graphite or carbon black—in a dielectric. The conductivity of such a composite undergoes a sharp nonmetal–metal transition at a threshold concentration of the filler material (Figure 6). This threshold concentration depends on the type of distribution of the filler. To a first approximation, the distribution may be matrix type, statistical type or structured type.

The electrical conductivity of such composites can be calculated by simple methods (Odelvskii 1951) in the effective-medium model. For metal–nonmetal interfaces, the conduction is either by direct contact or by electron tunnelling. The mathematical modelling of these composites is based on the fluctuation tunnelling model (Sheng 1980) and the percolation model (Kirkpatrick 1973). In PVC–carbon composite for example, the fluctuation tunnelling model satisfactorily explains the conductivity–temperature relation (Sichel 1981). Electrons tunnel between clusters of closest contact. Tunnelling is promoted by the thermal energy fluctuations. The temperature dependence of the resistivity of the composites is

$$\rho = \rho_0 \exp[T_1(T + T_0)]$$

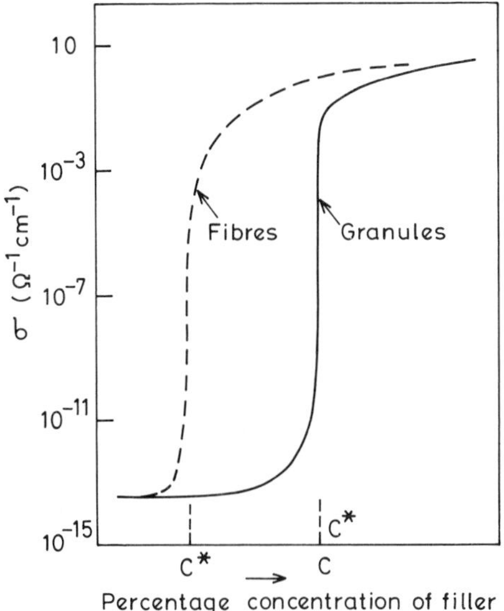

Figure 6. Conductivity of a filled polymer as a function of filler composition. The filler can be granular/fibrous.

where kT_1 is the electron tunnelling energy, T_0 is the temperature below which ρ is independent of T. In the percolation model, directed movement of the filler in a non-ordered polymer is considered (Kirkpatrick 1973). One obtains

$$\sigma \sim \sigma_0(c-c^*)^\beta \qquad \text{when } c>c^*$$

$$\sigma \sim \sigma_1(c^*-c)^\alpha \qquad \text{when } c<c^*$$

$$\sigma \sim \sigma_0 x^\delta \qquad \text{when } c=c^*$$

where c^*, β, $\delta=\beta/(\alpha+\beta)$ are constants of the model. A typical variation of σ with c is shown in Figure 6. At low concentrations ($c<c^*$), the system consists of insulated unconnected fillers. The conductivity is due to the thermally activated tunnelling of the charge carriers between the metallic particles. $\sigma \sim \exp(-A/T^{1/2})$ (Abeles *et al.* 1975). Near the percolation threshold, $c=c^*$, conducting chains begin to be formed. At higher concentrations, $c>c^*$, the metal forms a continuous phase ($c^* \simeq 0{\cdot}40$–$0{\cdot}5$ by volume).

If the filled material is dispersed in fibrous form, instead of chunky fragments, then the composite will be conducting at very much lower concentrations of filler. For example, even 5% fibrous filler with fibre diameter 0·01 mm, length 1 mm, will make the composite conducting (Figure 6). Numerous experimental data and statistical calculations are available for understanding the conductivity of such metalloplastics (Shevchenko & Ponomarenko 1983). It would be exciting if such fillers could be formed *in situ* in the dielectric matrix during the synthesis process. We shall largely be concerned with the synthetic polymers in which the conductivity is enhanced by doping.

6.3. *Synthetic conducting polymers: polysulphur nitride* $(SN)_x$

Polysulphur nitride is the first example of a conducting polymer. Although it was first synthesized by Burt in 1910 its electrical properties were measured only fifty years later. $(SN)_x$ crystallizes into a highly disordered form consisting of bundles of fibres. The structure of $(SN)_x$ as obtained by neutron diffraction is given in Figure 7. The two inequivalent structures alternate along the c-axis

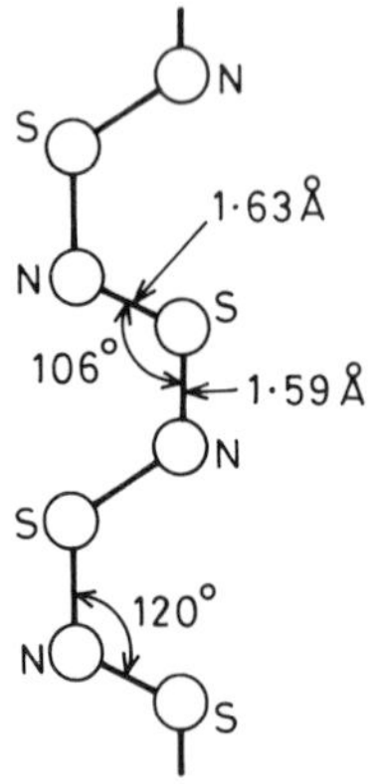

Figure 7. Structure of $(SN)_x$ along the chain axis.

whereas chains of the same type are adjacent along the a-axis. The weak interchain coupling allows intercalation of molecules into the fibre. The conductivity increases on intercalation by halogens. In particular, bromine is approximately five times more effective than other halogens (Kwak *et al.* 1979). The resistivity of $(SN)_x$ monotonically decreases with temperature and it becomes superconducting at 0·3 K (Greene *et al.* 1975). Band-structure calculations reveal the presence of three-dimensional interchain coupling. The superconducting properties are therefore of conventional type, but anisotropic. Brominated $(SN)_x$ also shows superconductivity. Excellent reviews of the properties of $(SN)_x$ are available (Labes *et al.* 1979).

6.4. *Synthetic conducting polymers—polyacetylene and its derivatives*

Basics

It was demonstrated during 1971–75 by Shirakawa and co-workers (1971, 1973, 1981) that conjugated films of diacetylene can be polymerized on a catalytic

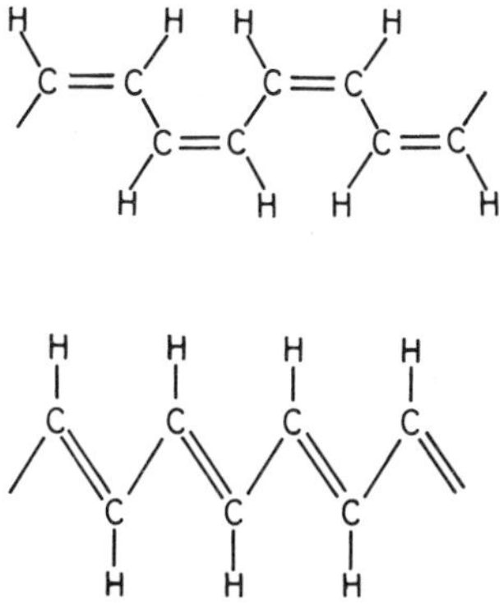

Cis $(CH)_X$

Trans $(CH)_X$

Figure 8. *Cis* and *trans* polyacetylene.

substrate. Polyacetylene is a non-pendant group polymer with an unsaturated backbone. The backbone consists of alternately singly and doubly bonded carbon atoms. The two possible structures, *cis* and *trans* types of $(CH)_x$, are shown in Figure 8. Conversion of the *cis* form into the more stable *trans* form takes place either during synthesis or during subsequent heat treatment. The films have a metallic lustre with conductivity between 10^{-9} $(\Omega\,cm)^{-1}$ (for the *cis* form) and 10^{-5} $(\Omega\,cm)^{-1}$ (for the *trans* form). On doping with powerful electron donors or acceptors, the conductivity increases by twelve orders of magnitude (from 10^{-9} $(\Omega\,cm)^{-1}$ to 10^{3} $(\Omega\,cm)^{-1}$). If the films are oriented, σ increases to 10^{5} $(\Omega\,cm)^{-1}$. These studies have aroused enormous interest and a number of other conducting polymers such as polyphenylene, polypyrrole, polymeric kapton, polyphenylenesulphide, polyheptadiyne, and so on, have been obtained. We shall consider these polymers in a subsequent section.

As in organic metals, the conjugated polymer systems must satisfy certain requirements to become semiconducting or metallic.

(a) The polymers must have high molecular weights and continuous conjugated chains.

(b) It must be possible to obtain the polymers in the form of homogeneous films or crystals.

(c) It must be possible to distribute uniformly strong electron donors and acceptors.

(d) The chemical structure must facilitate electrical conduction.

(e) The nature and properties of the material must be stable and withstand ambient changes.

Polymerization methods

There are a number of methods of obtaining high molecular weight conjugated polymers with donor/acceptor dopants (Matmishiyah & Kobryanski 1983, Aldissi 1984)—ionic polymerization, radical polymerization, pyrolysis, high pressure shear quenching (Zharov 1984) mechanical rolling (Vaneso *et al.* 1982), vapour deposition polymerization (Snow 1981), electrochemical polarization (Przyluski *et al.* 1982) and quenching from the liquid state (Rault *et al.* 1980) are some of the methods normally adopted for the preparation of conducting polymers. The most widely studied polymer, polyacetylene, is obtained by the polymerization of the monomer via the formation of a Π-complex involving an ionic or coordinative mechanism (Aldissi 1984). There are some difficulties in the ionic polymerization of acetylene to obtain high molecular weight compounds because of isomerisation reactions. The coordinative polymerization has been widely used particularly using the Ziegler–Natta polymerization. Apart from Ziegler–Natta catalysts, transition metal derivatives, lanthanide derivatives and metathesis catalysts are also employed. Co-polymerization is also done with organometallic polymers and anionic polymers. The choice of the particular catalyst determines the *cis* isomer content of polyacetylene. The molecular weight, molecular dispersity and the mechanical properties are also influenced by the catalytic activity. The morphology of the polymer also is affected by the catalyst. While the polyacetylene films are generally fibrillar, the size (diameter and length) of the fibril varies with the nature and concentration of the catalyst. The diameter of the fibril can vary from a few ångströms to a few microns (when the structure tends to be globular) and the polymer chain axis is the fibre axis (Chien *et al.* 1982).

The stability of polyacetylene is one of the major problems. The backbone of the polymer and its functional groups react with oxygen and/or moisture, making it age with time. There have been several attempts at stabilizing polyacetylene. These were by chemical doping, ion implantation (Zuppiroli & Friend 1978), plastification by using a protecting polymer such as polypyrrole or using anti-oxidizing agents (like benzoquinone) (Aldissi 1984). This is one of the major technological problems in the applications of polyacetylene.

Conductivity of polyacetylene

As remarked earlier, $(CH)_x$ consists of weakly coupled C–H units in its chain, with alternating single and double bonds. It can occur in two forms—*cis*$(CH)_x$ and *trans* $(CH)_x$. Both appear as silvery, flexible films. The *cis* form gradually converts into the more stable *trans* form. The ratio of *cis/trans* forms can be suitably controlled during synthesis. Room temperature conductivities of the crystalline forms of $(CH)_x$ vary from 10^{-5} $(\Omega\,\mathrm{cm})^{-1}$ for the *trans* material to 10^{-9} $(\Omega\,\mathrm{cm})^{-1}$ for the *cis* isomer. Polyacetylene can be doped with a wide variety of semiconductors and metals, such as I, AsF_5, IBr, SbF_6, ClO_4, Na, and so on, by chemical/electrochemical methods (Wegner 1977). The conductivity of the doped polyacetylene is approximately 10–10^3 $(\Omega\,\mathrm{cm})^{-1}$. The dopants can be electron donors or electron acceptors. It is thought that the dopant causes a conversion into the *trans* form and generates free-radical defects. These radical defects, while interacting with the dopant, form charged dopants and charged defects.

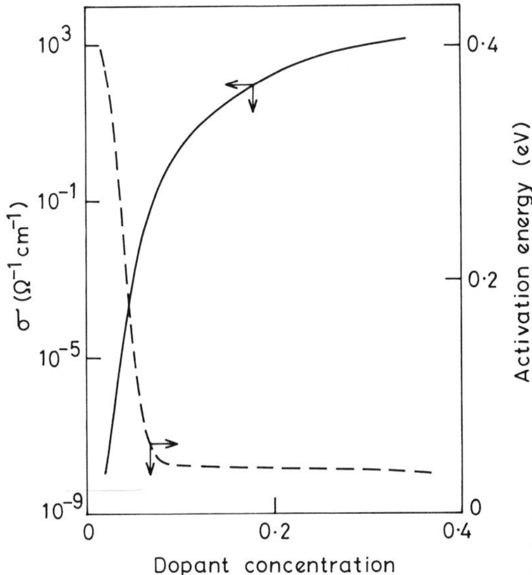

Figure 9. Conductivity and activation energy of $(CH)_x$ as a function of dopant concentration.

As shown in the Figure 9, the conductivity of polyacetylene increases rapidly with the amount of dopant and then it saturates. The insulator–metal transition occurs at about 4 mole % of the dopant. Here the conductivity also changes from an activated hopping process to an unactivated process. This is seen in the same figure. There have been extensive work on the absorption edge (MacDiarmid & Heeger 1979), Hall effect, thermopower and mobility measurements. The experiments indicate that the charge carriers are predominantly holes, while electron states also play an important part.

The conductivity of doped polyacetylene has been extensively studied as a function of pressure, temperature, frequency and electric field. On applying external pressure, the conductivity increases by an order of magnitude up to about 30 kbar and then gradually decreases (Ferraris *et al.* 1980). Initially the fibrils come closer and at higher pressures either the dopant may seggregate or the chains may get cross-linked or broken. The conductivity decreases with temperature roughly as a function of $\exp(T^{-1/4})$ (Audemaert *et al.* 1981). This behaviour is attributed by Sheng (1980) to the fluctuation-induced tunnelling model between regions of conducting islands. The conductivity is largely frequency independent up to 500 MHz around room temperature, while it becomes frequency dependent at low temperatures (Epstein *et al.* 1980, 1981). The non-ohmic nature of electrical conduction in doped $(CH)_x$ is different in the regions of low and high dopant concentration. For low concentrations of dopant, the field dependence can be explained by the above metallic island model. At higher dopant concentrations, there are some non-linear effects, perhaps attributed to the interfibre interactions. The intercrystalline contact resistances give a hopping type temperature dependence. When this contribution is reduced, the doped $(CH)_x$ behaves more like a metal (Meixiang *et al.* 1983).

Soliton and polaron mechanisms of conduction

Many electrical properties of $(CH)_x$ and doped $(CH)_x$ cannot be explained by electron–hole conduction mechanisms. Also the unique one-dimensionality of $(CH)_x$ enables novel types of electrical conduction. The ground state of the *trans* $(CH)_x$ is doubly degenerate. It can exist in one of the two forms depicted in Figure 10. By symmetry arguments, the single and double bonds could be interchanged

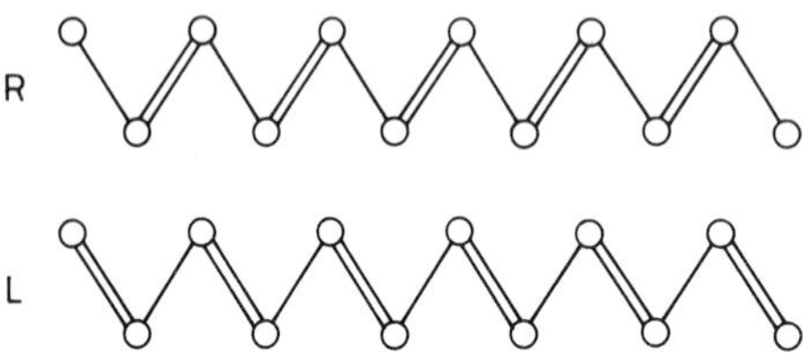

Figure 10. Symmetric, degenerate forms R, L of $(CH)_x$.

without altering the energy of the system. Such a double degenerate ground state system can be topologically distributed. The excited states of such a system will tend to be non-linear. The bond alternation at the kink or domain wall of the different topological states can appear as solitons and result in unexpected properties (Su *et al.* 1979, 1980, Su & Schrieffer 1980).

During the conversion of *cis* to *trans* form of $(CH)_x$, the formation of L or R forms of chain is random. In fact, some segments of $(CH)_x$ chain can be of L type and some of R type. At the boundary of L and R configurations a free radical is formed. Such a kink gives rise to a characteristic spin signal. In the energy-band diagram, the kink state will lie midway between the conduction and valence bands. It will be neutral, with spin 1/2, and will show paramagnetism. On addition of a dopant like an electron acceptor, the electron at the kink site is removed and the soliton will become positively charged. Then the spin will be zero and will be non-magnetic. An electron donor will result in a zero spin, negatively charged state. These are represented in the Figure 11.

The movement of the soliton is from band alternations at adjacent sites. The 'mass' of such a moving soliton can also be estimated (Su *et al.* 1979, 1980). On the addition of a dopant during the *cis–trans* conversion process, there are two possibilities—formation of a soliton (neutral/charged); formation of an electron–hole. That which requires the lower energy will be preferentially formed. Several calculations show that the formation of a soliton is energetically more favourable (Su *et al.* 1979, 1980, Takayama *et al.* 1980). The formation of such solitons has

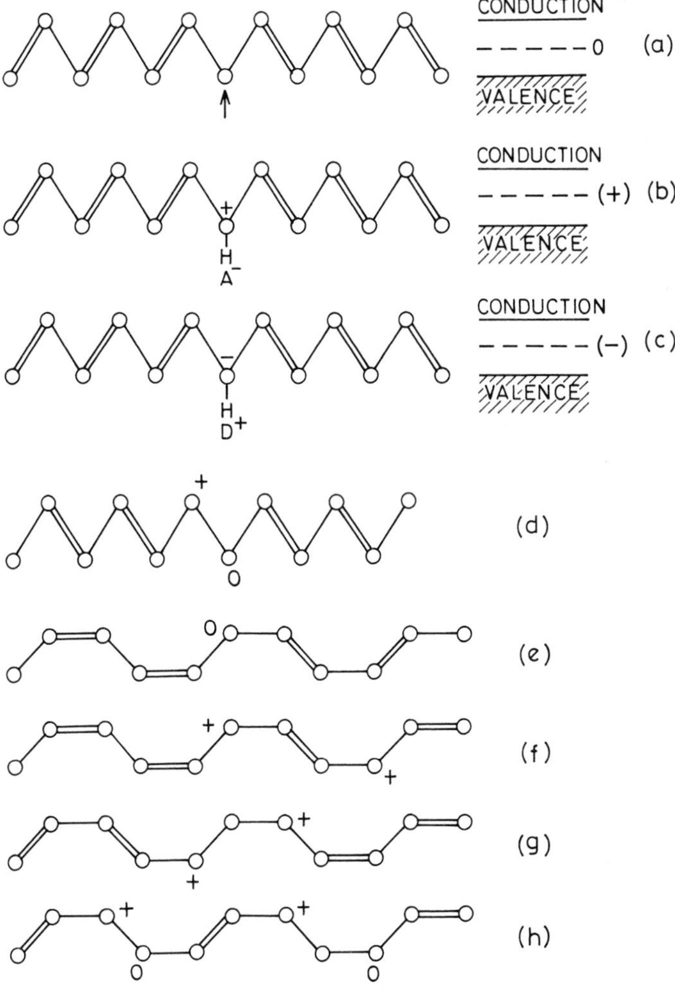

Figure 11. Solitons, polarons in $(CH)_x$. Also the energy levels of solitons are shown. (*a*) One kink, neutral soliton, (*b*) one kink, positive soliton, (*c*) one kink, negative soliton, (*d*) polaron = neutral + charged soliton, (*e*) $(CH)_x$ *cis* neutral soliton, (*f*) two charged solitons, (*g*) bipolaron/bisoliton, (*h*) two polarons.

also been experimentally demonstrated by absorption spectroscopy (Heeger *et al.* 1984).

In *trans* $(CH)_x$, an incident photon of suitable energy can create a soliton pair. The soliton pair so created can move in an applied electric field and give rise to photoconductivity. This has a characteristically different behaviour from photoconductivity in semiconductors. Heeger *et al.* (1984), Conwell (1984) and Bishop *et al.* (1984) have made a detailed study of such photoexcitation of solitons.

The topological excitations in a linear conjugated chain can also take up more complicated forms (Figure 11). There can be combinations of neutral, positively charged and negatively charged solitons. If these are able to interact with each other, then the dynamics of the system become complicated. For example a charged and a neutral soliton can combine. Then we have a polaron. Two polarons can combine to form a bipolaron in which the neutral parts will annihilate and the charged parts form the bisoliton. Such polaron, bipolaron formation is more easily achieved in *cis* $(CH)_x$. The propagation of polarons and bipolarons is under detailed study now (Scott *et al.* 1984, Chance *et al.* 1984).

6.5. Other conducting polymers

There are a large number of other polymers which have conductivities in the semiconducting range (Mort 1980, Duke & Gibson 1984). Some typical values of conductivity for these are given in Table 3.

Some of the semiconducting polymers are obtained by direct pyrolysis. The pyrolysis temperature and atmosphere contribute to the final conductivity. These polymers can also be prepared with a metal dopant. Molecular doped polymers of the type listed in Table 3 have generally higher conductivity. Polyvinyl

Table 3. Typical conducting polymers.

Polymer	Dopant Dopant	σ $(\Omega\ \mathrm{cm})^{-1}$ at 25°C	Reference
Polyparaphylene	AsF_5	500	Ivory *et al.* 1979
Polythienylene	ClO_4	10^3	Hotta *et al.* 1983b
Poly-*p*-phenylenesulphide	AsF_5	1	Chance *et al.* 1981
Poly-2,5-diylpyrrole	BF_4	10^2	Kanazawa *et al.* 1980
Polythiophene	AsF_5	10	Kobayashi *et al.* 1984
Polymer kapton	AsF_5	10^2	Brim *et al.* 1980
Polypyrrole	ClO_4	40	Hotta *et al.* 1983a
Polyvinyl chloride	SO_3	10^{-3}	Soga *et al.* 1983
Poly-*m*-pyrrolylene	BF_4	10^2	Diaz & Castillo 1980
Poly-*m*-pyrrolylene	ClO_4	10^2	Tourillon & Garnier 1982
Poly-metallophthalocyanine	—	10^{-1}	Inabe *et al.* 1984

pyridine, polyvinyl imidazole and ionene polymers can be complexed with TCNQ salts to give conductivity as high as 10^{-2} $(\Omega\,\mathrm{cm})^{-1}$. Polyphthalocyanines with metal containing groups (Si, Ge, Ni, etc.) have conductivities in the range 10^{-1}–10^{-6} $(\Omega\,\mathrm{cm})^{-1}$ (Inabe *et al.* 1984). There have been conductivity, Hall effect and e.s.r. measurements on many of the listed polymers at various dopant levels. In general, the conductivity increases with acceptor dopant concentration in all of them as indicated in Figure 9.

6.6. Applications

The achievement of polymers with tailored electrical properties makes them suitable for a variety of applications. The charge-transfer complexes based on polyvinylpyridine and iodine or poly *n*-alkyl-pyridinium polyiodide have been used as cathodes in batteries for heart pacemakers and electronic watches. The synthetic polymeric conductors have also been used as rectifiers (Chiang *et al.* 1978), Schottky barrier junctions (Etemad *et al.* 1982, Tsukamoto *et al.* 1982) solar cells (Weinberger *et al.* 1982, Shirakawa *et al.* 1981) and batteries (Nigrey *et al.* 1981) as well as current carriers to replace metallic wires. The instability of these polymers has prevented their large-scale application, but this is, nonetheless, a strongly pursued area of technology.

Acknowledgements

We thank our colleagues A.V. Nalini, A.K.T. Oza, A.K. Bandyopadhyay and S. Chatterjee for their significant contributions in this area. We also thank Professor E.S. Raja Gopal, Professor Kumar and other colleagues for helpful discussions and Professor C.N.R. Rao for his encouragement. Financial support received from DST and ISRO–IISc Space Technology Cell is gratefully acknowledged.

References

Abeles, B., Sheng, P., Coutts, M.D. and Arie, Y., 1975, *Adv. Phys.*, **24**, 407.
Aldissi, M., 1984, *Synthetic Metals*, **9**, 131.
Andre, J.J., Bieber, A. and Gautier, F., 1976, *Ann. Phys., Paris*, **10**, 145.
Andrieux, A., Duroure, C., Jerome, D. and Bechgaard, K., 1979, *Journal de Physique Letters*, **40**, L381.
Andrieux, A., Jerome, D. and Bechgaard, K., 1981, *J. Physique Lett.*, **42**, L87.
Audemaert, M., Gusman, G. and Deltom, R., 1981, *Phys. Rev.*, **24**, 7380.
Azevedo, L.J., Williams, J.M. and Crompton, S.J., 1983, *Phys. Rev.*, B **28**, 6600.
Bandyopadhyay, A.K., 1981, *PhD Thesis*, Indian Institute of Science.
Bandyopadhyay, A.K. and Subramanyam, S.V., 1979, *Nucl. Phys. and Solid St. Phys. (India)*, **22C**, 251.
Bandyopadhyay, A.K. and Subramanyam, S.V., 1980, *Nucl. Phys. and Solid St. Phys. (India)*, **23C**, 218.
Bandyopadhyay, A.K. and Subramanyam, S.V., 1981, *Pramana*, **16**, 151.
Bandyopadhyay, A.K., Chatterjee, S., Subramanyam, S.V. and Bulka, B.J., 1980, *J. Phys. C: Solid St. Phys.*, **13**, L803.
Bandyopadhyay, A.K., Chatterjee, S. and Subramanyam, S.V., 1981, *Chemica Scripta*, **17**, 47.
Bardeen, J., 1979, *Phys. Rev. Lett.*, **42**, 1498.
Bardeen, J., 1980, *Phys. Rev. Lett.*, **45**, 1978.

Bechgaard, K., Cameiro, K., Olsen, M., Rasmussen, F.B. and Jacobsen, K.S., 1981, *Phys. Rev. Lett.*, **46**, 852.
Bishop, A.R., Campbell, D.K., Lomdahe, P.S., Horovitz, B. and Phillpot, S.R., 1984, *Synthetic Metals*, **9**, 223.
Bogomolov, V.N., Kishin, N.A. and Kumazerov, Yu.A., 1977, *JETP Letters*, **26**, 71.
Bogomolov, V.N., Kolla, E.V., Kumazerov, Yu.A., Kuneva, N.M. and Projodin, V.N., 1980, *Solid State Communications*, **35**, 363.
Brim, H.B., Tomkiewicz, Y., Aviram, A., Broers, A. and Sumnae, B., 1980, *Solid St. Commun.*, **35**, 135.
Bulaevskii, L.N. and Sadovskii, M.V., 1974, *Sov. Phys.—Solid St.*, **16**, 743.
Burt, F.P., 1910, *J. Chem. Soc. Transactions*, **97**, 1171.
Chaikin, P.M., Gruner, G., Engler, E.M. and Greene, R.L., 1980, *Phys. Rev. Lett.*, **45**, 1874.
Chance, R.R., Shacklette, L.W., Eckhart, H., Sowa, J.M., Elsenbaumer, R.L., Ivory, D.M., Miller, G.G. and Baughman, R.H., 1981, in *Conductive Polymers*, edited by R.B. Seymour (New York: Plenum).
Chance, R.R., Bredas, J.L. and Silbey, R., 1984, *Phys. Rev.* B, **29**, 4491.
Chatterjee, S., 1981, *PhD Thesis*, Indian Institute of Science.
Chatterjee, S., 1982, *Solid St. Commun.*, **41**, 541.
Chiang, C.K., Gau, S.C., Fincher, C.R., Park, Y.W., MacDiarmid, A.G. and Heeger, A.J., 1978, *Appl. Phys. Lett.*, **33**, 18.
Chien, J.C.W., Yamashita, Y., Hirsch, J.A., Fan, J.L., Schen, M.A. and Karasz, F.E., 1982, *Nature*, **299**, 608.
Coleman, L.B., Cohen, M.J., Sandman, D.J., Yamagishi, F.G., Garito, A.F. and Heeger, A.J., 1973, *Solid St. Commun.*, **12**, 1125.
Conwell, E.M., 1984, *Synthetic Metals*, **9**, 195.
Cooper, J.R., Weger, M., Jerome, D., Lefur, D., Bechgaard, K., Bloch, A.N. and Cowan, D.C., 1976, *Solid St. Commun.*, **19**, 749.
Datars, W.R., Van Schyndel, A., Lass, J.S., Chartier, D. and Gillespie, R.J., 1978, *Phys. Rev. Lett.*, **40**, 1184.
Davis, D.W., Gutfreund, H. and Little, W.A., 1976, *Phys. Rev.* B, **13**, 4766.
Diaz, A.F. and Castillo, J.I., 1980, *J. Chem. Soc.—Chem. Commun.*, 854.
Domb, E.R., Sellmeyer, D.J. and Sturgeon, G.D., 1979, *J. Phys. Chem. Solids*, **40**, 739.
Duke, C.B. and Gibson, H.W., 1984, in *Encyclopedia of Semiconductor Technology*, edited by M. Grayson (Chichester and New York: John Wiley & Sons).
Epstein, A.J., Gibson, H.W., Chaikin, P.M., Clark, W.G. and Gruner, G., 1980, *Phys. Rev. Lett.*, **45**, 1730.
Epstein, A.J., Rommelmann, H., Abkowitz, M. and Gibson, H.W., 1981, *Phys. Rev. Lett.*, **47**, 1549.
Etemad, S., Heeger, A.J. and MacDiarmid, A.G., 1982, *Annual Rev. Phys. Chem.*, **33**, 443.
Farges, J.P., Brau, A. and Gutman, F., 1972, *J. Phys. Chem. Solids*, **33**, 1723.
Ferraris, J.P., Webb, A.W., Weber, D.C., Fox, W.B., Carpenter, E.R. and Brandt, P., 1980, *Solid St. Commun.*, **35**, 15.
Fleming, R.F. and Grimes, C.C., 1979, *Phys. Rev. Lett.*, **42**, 1423.
Forro, L., Janossy, A., Zuppiroli, L. and Bechgaard, K., 1982, *J. Physique*, **43**, 977.
Friend, R.H., Jerome, D., Fabre, J.M., Giral, L. and Bechgaard, K., 1978, *J. Phys., C: Solid St. Phys.*, **11**, 803.
Frohlich, H., 1954, *Proc. R. Soc.* A, **233**, 296.
Fukuyama, H. and Lee, D.A., 1978, *Phys. Ref.* B, **17**, 535, 542.
Garito, A.F. and Heeger, A.J., 1974, *Accounts of Chemical Research*, **7**, 232.
Graja, A., Przybylski, M., Sekretarczyk, G., Pukacki, W., Rajchel, A., Willis, M.R., Wallwork, S.C., Mihaly, G. and Mihaly, L., 1983, *J. Physique*, **44**, C3, 1365.
Greene, R.L., Street, G.B. and Suter, L.J., 1975, *Phys. Rev. Lett.*, **34**, 577.
Gruner, G., Tippie, L.C., Sanny, J., Clark, W.G. and Ong, N.P., 1980, *Phys. Rev. Lett.*, **45**, 935.

Gunning, W.J. and Heeger, A.J., 1978, *Solid St. Commun.*, **27**, 843.
Gunning, W.J., Chiang, C.K., Heeger, A.J. and Epstein, A.J., 1979, *Phys. Rev. Lett.*, **96**, 145.
Haddon, R.C., Kaplan, M.L. and Wull, F., 1984, in *Encyclopedia of Semiconductor Technology*, edited by M. Grayson (Chichester and New York: John Wiley & Sons).
Hatfield, W.E., 1978, *Molecular Metals* (New York: Plenum).
Heeger, A.J., Blanchet, G., Chung, T.C. and Fincher, C.R., 1984, *Synthetic Metals*, **9**, 173.
Hemamalini Naik and Subramanyam, S.V., 1982, *National Symposium on Instrumentation* (Bangalore: Instrument Society of India).
Hememalini Naik and Subramanyam, S.V., 1983, *Proc. Symposium on Amorphous Materials, Bombay* (Bombay: Department of Atomic Energy), p. 269.
Hodina, A.J., 1984, *Synthetic Metals*, **9**, B1, B97.
Hotta, S., Hosaka, T. and Shimotsuma, W., 1983a, *Synthetic Metals*, **6**, 319.
Hotta, S., Hosaka, T. and Shimotsuma, W., 1983b, *Synthetic Metals*, **6**, 69.
Inabe, T., Lomax, J.R., Lyding, J.W., Kannewurf, C.R. and Marks, T.J., 1984, *Synthetic Metals*, **9**, 303.
Ito, T., Shirakawa, H. and Ikeda, S., 1975, *J. Polymer Sci., Polymer Chemistry Edition*, **13**, 1943.
Ivory, D.M., Miller, G.G., Sowa, J.M., Shacklette, L.W., Chance, R.R. and Baughman, R.R., 1979, *J. Chem. Phys.*, **71**, 1506.
Jerome, D., Mazand, A., Ribane, T.M. and Bechgaard, K., 1980, *J. Phys. (Paris) Lett.*, **41**, L95.
Jerome, D. and Schulz, H.J. 1982, *Adv. Phys.*, **31**, 299.
Kagoshima, S., Ishiguro, T., Engler, E.M., Schultz, T.P. and Tomkiewich, Y., 1980, *Solid St. Commun.*, **34**, 151.
Kamitsos, L.I., Tzinis, C.H. and Risen, W.M., 1982, *Solid St. Commun.*, **42**, 561.
Kampar, V.E., 1982, *Russian Chem. Rev.*, **51**, 185.
Kanazawa, K.K., Diaz, A.F., Gill, W.D., Grant, P.M., Street, G.B., Gardini, G.P. and Kwak, J.F., 1980, *Synthetic Metals*, **1**, 329.
Keller, H.J., Nothe, D., Pritzkow, H., Wehe, D., Werner, M., Koch, P. and Schweitzer, D., 1980, *Molec. Cryst. Liquid Cryst.*, **62**, 181.
Khidekel, M.L. and Zhilyaeva, E.I., 1981, *Synthetic Metals*, **4**, 1.
Kijima, T., Tanaka, J. and Goto, M., 1984, *Nature*, **310**, 45.
Kirkpatrick, S., 1973, *Rev. Mod. Phys.*, **45**, 574.
Kobayashi, H., Kobayashi, A., Asuamik, K. and Minomura, S., 1980, *Solid St. Commun.*, **35**, 293.
Kobayashi, M., Chen., J., Chung, T.C., Moraes, F., Heeger, A.J. and Wudl, F., 1984, *Synthetic Metals*, **9**, 77.
Krogman, K., 1969, *Angewandte Chemie International Edition*, **8**, 35.
Kuper, C.G., 1955, *Proc. R. Soc.* A, **227**, 214.
Kwak, J.F., Greene, R.L. and Fuller, W.W., 1979, *Phys. Rev.* B, **20**, 2685.
Labes, M.M., Love, P. and Nichols, L.F., 1979, *Chem. Rev.*, **79**, 1.
Lee, P.A. and Rice, T.M., 1979, *Phys. Rev.* B, **19**, 3970.
Little, W.A., 1964, *Phys. Rev.* A, **134**, 1416.
Lyubovskaya, R.N., 1983, *Russian Chem. Rev.*, **52**, 736.
Maaroufi, A., Coulon, C., Flandrois, S., Delhaes, P., Mortensen, K. and Bechgaard, K., 1983, *Solid St. Commun.*, **48**, 555.
MacDiarmid, A.J. and Heeger, A.J., 1979, in *Molecular Metqls*, edited by W.E. Hatfield (New York: Plenum).
MacNicol, D.D., McKendrick, J.J. and Wilson, D.R., 1978, *Chem. Soc. Rev.*, **7**, 65.
Matmishiyah, A.A. and Kobryanskii, V.M., 1983, *Russian Chem. Rev.*, **52**, 751.
Melby, L.R., Harder, R.J., Hertler, W.R., Maheler, W., Benson, R.E. and Mochel, W.E., 1962, *J. Am. Chem. Soc.*, **84**, 3374.
Mele, E.J. and Rice, M.J., 1981, *Phys. Rev.* B, **23**, 5397.

Meixiang, W., Ping, W., Yong, C., Renynan, Q., Fosong, W., Xiogian, Z. and Zhi, G., 1983, *Solid St. Commun.*, **47**, 759.

Mihaly, G., Janossy, A., Kurti, A., Ferro, J. and Gruner, G., 1979, *Phys. Stat. Solidi* B, **94**, 287.

Miller, J.S. (ed.), 1982, *Extended Linear Chain Compounds*, Volumes 1 & 2 (New York: Plenum).

Mort, J., 1980, *Science*, **208**, 819.

Mortensen, K., Thewak, M.L.W., Tomkiewich, Y., Clarke, T.C. and Street, G.B., 1980, *Phys. Rev. Lett.*, **45**, 490.

Mott, N.F. and Twose, 1961, *Adv. Phys.*, **10**, 107.

Murata, K., Ukachi, T., Anzai, H., Saito, G., Kajimura, K. and Ishiguro, T., 1982, *J. Phys. Soc. Japan*, **51**, 1817.

Nalini, A.V., Oza, A.K.T. and Subramanyam, S.V., 1976, *Nucl. Phys. and Solid St. Phys.* (*India*), **19C**, 41.

Nigrey, P.J., MacInnes, D., Nairns, D.P., MacDiarmid, A.G. and Heeger, A.J., 1981, in *Conductive Polymers*, edited by R.B. Seymour (New York: Plenum).

Odelvskii, V.I., 1951, *Sov. Phys.—Tech. Phys.*, **21**, 667.

Ong, N.P. and Monceau, P., 1977, *Phys. Rev.* B, **16**, 3443.

Ovshinsky, S.R., 1968, *Phys. Rev. Lett.*, **21**, 1450.

Oza, A.K.T., 1980, *PhD Thesis*, Indian Institute of Science.

Peierls, R.E., 1955, *Quantum theory of solids* (Oxford: Oxford University Press).

Perlstein, J.H., 1977, *Angewandte Chemie International Edition*, **16**, 519.

Potember, R.S. and Poehler, T.O., 1979, *Appl. Phys. Lett.*, **34**, 405.

Potember, R.S., Poehler, T.O., Rappa, A., Cowan, D.O. and Bloch, A.N., 1982, *Synthetic Metals*, **4**, 371.

Przybylski, M. and Graja, A., 1981, *Physica*, **104B**, 278.

Przyluski, J., Zagorska, M., Conder, K. and Pron, A., 1982, *Polymer*, **23**, 1872.

Rault, J., Sotton, M., Rabourdin, C. and Robelin, E., 1980, *J. Physique*, **41**, 1459.

Rouxel, J., 1983, *VII International Conference on Solid Compounds of Transition Element, Proceedings*, Grenoble, France, PL13, 1.

Samara, G.A. and Drickamer, H.G., 1962, *Journal of Chemical Physics*, **37**, 474.

Samara, G.A., Durouse, C., Jerome, D. and Bechgaard, K., 1979, *J. Physique* (*Paris*) *Lett.*, **40**, L381.

Scott, J.C., Pedersen, H.J. and Bechgaard, K., 1981, *Phys. Rev.* B, **24**, 475.

Scott, J.C., Bredas, J.L., Yakushi, K., Pfluger, P. and Street, G.B., 1984, *Synthetic Metals*, **9**, 165.

Sen, P.N. and Varma, C.M., 1974, *Solid St. Commun.*, **15**, 1905.

Shchegolev, I.F., 1972, *Phys. Stat. Solidi*, a, **12**, 9.

Sheng, P., 1980, *Phys. Rev.*, **21**, 2180.

Shevchenko, V.G. and Ponomarenko, A.T., 1983, *Russian Chem. Rev.*, **52**, 757.

Sichel, E.K., 1981, *Appl. Phys. Commun.*, **1**, 83.

Shirakawa, H. and Ideka, S., 1971, *Polymer J.*, **2**, 231.

Shirakawa, H., Ito, T. and Ikeda, S., 1973, *Polymer J.*, **4**, 460.

Shirakawa, H., Ikeda, S., Aizawa, M., Yoshitake, J. and Suzuki, S., 1981, *Synthetic Metals*, **4**, 43.

Snow, A.W., 1981, *Nature*, **292**, 40.

Soga, K., Nakamura, M., Kobayashi, Y. and Ikeda, S., 1983, *Synthetic Metals*, **6**, 275.

Starodub, V.A. and Krivoshei, I.V., 1982, *Russian Chem. Rev.*, **51**, 439.

Su, W.P. and Schrieffer, J.R., 1980, *Proc. National Academy of Sciences, USA*, **77**, 5626.

Su, W.P., Schrieffer, J.R. and Heeger, A.J., 1979, *Phys. Rev. Lett.*, **42**, 1698.

Su, W.P., Schrieffer, J.R. and Heeger, A.J., 1980, *Phys. Rev.* B, **22**, 2099.

Subramanyam, S.V., 1981, in *Preparation and Characterization of Materials*, edited by R. Honig and C.N.R. Rao (London: Academic Press).

Takayama, H., Lin-Liu, Y.R. and Macik, 1980, *Phys. Rev.* B, **21**, 2388.
Teranishi, N. and Kubo, R., 1979, *J. Phys. Soc. Japan*, **47**, 720.
Tomic, S., Jerome, D., Menod, P. and Bechgaard, K., 1982, *J. Phys. (Paris) Lett.*, **43**, L839.
Tomkiewicz, Y., Schultz, T.D., Borm, H.B., Taranko, A.R., Clarke, T.C. and Street, G.B., 1981, *Phys. Rev.* B, **24**, 4348.
Torrance, J.B., 1979, *Accounts of Chemical Research*, **19**, 79.
Tourillon, G. and Garnier, F., 1982, *J. Electroanalytical Chem.*, **135**, 173.
Tsukamoto, J., Ohigashi, H., Matsumura, K. and Takahashi, A., 1982, *Synthetic Metals*, **4**, 177.
Tsuzuki, T. and Sasaki, K., 1980, *Solid St. Commun.*, **33**, 1063.
Vaneso, G., Egyed, O., Pekker, S. and Janossy, A., 1982, *Polymer*, **23**, 14.
Venedek, I.B., Ermolenko, A.N., Esipov, V.V., Rodionav, A.G. and Serebryakova, E.A., 1982, *JETP Lett.*, **35**, 117.
Wegner, G., 1977, *Pure and Appl. Chem.*, **49**, 443.
Weinberger, B.R., Akhtar, M. and Gan, S.C., 1982, *Synthetic Metals*, **4**, 187.
Wheland, R.C. and Gillson, J.L., 1976, *J. Am. Chem. Soc.*, **98**, 3916.
Yamaya, K. and Oomi, G., 1983, *Journal of Physical Society of Japan*, **52**, 1886.
Zettl, A., Jackson, C.M., Janossy, A., Gruner, G., Jacobsen, A. and Thompson, A.H., 1982, *Solid St. Commun.*, **43**, 345.
Zharov, A.A., 1984, *Russian Chem. Rev.*, **53**, 140.
Zuppiroli, L. and Friend, R.H., 1978, *Phil. Mag.* B, **37**, 321.
Zuppiroli, L., Bouffard, S., Bechgaard, K., Hilti, B. and Mayer, C.W., 1980, *Phys. Rev.* B, **22**, 6035.

CHAPTER 9

the Mott transition for binary compounds, including a case study of the pyrite system $Ni(S_{1-x}Se_x)_2$

John A. Wilson
H.H. Wills Physics Laboratory, University of Bristol, Bristol, BS8 1TL, UK

Foreword

Examination is made of how elusive a Mott, correlation-driven, metal–nonmetal transition is to find within the realm of solid binary compounds. With regard to transition metal compounds there is a very wide variety of effects, such as crystal structure, spin structure, coordination geometry, principal quantum number, valence and electronegativity difference, which affect the proximity of an individual compound to the boundary separating Mott insulating and metallic regimes. Only the systems $(V/Cr)_2O_3$ and $Ni(S/Se)_2$ come close to satisfying what is being sought. The work on the sesqui-oxide system has been frequently reviewed, but I am not aware of any comprehensive discussion of the pyrite system based on NiS_2. NiS_2 is a particularly good material for the present work because it can be forced across the metal–nonmetal boundary under readily accessible pressures, without recourse to doping. As demonstrated for the mixed systems based on NiS_2, the additional effects of disorder add serious complications to an event one is seeking to strip of all complication.

It is good to be able to review the work on NiS_2 on the occasion of Professor Mott's 80th birthday, since our initial discovery of this system was made in 1970, at the time when I was working in Cambridge as a young researcher much influenced by the activities and tutelage of Professor Mott.

A great deal of theoretical progress and data accumulation has occurred since the famous conference at San Francisco in 1968 (*Rev. Mod. Phys.*, **40**), and, indeed, since my own efforts in this field in 1969–72 prior to moving on to Bell Laboratories and CDW work. It would be nice to report now that the matter was fully settled, with many examples described. However, the topic is a major one not lightly disposed of, as the new band-structural discussions highlight. Experimentally, the situation is so sensitive, especially for binaries, that new examples remain lacking. It is hoped that the present review of the situation in NiS_2, the best proposition to date, will encourage somebody to attempt new, good quality, high-pressure work, with really well characterized samples, to yield that definitive data on the Mott transition so long sought. Much further exploration is also required of whether a modern 'one-electron' spin-polarized band structure can come close to describing the ground-state properties of a material like NiS_2, even if not of a hard Mott insulator like CoO. Perhaps the answer will be in the book celebrating Professor Mott's 90th birthday.

1. Introduction to the materials and phenomenology

Rock-salt structured d^7CoO is the classic Mott insulator (Mott 1974). It is a material that is a nonmetal despite its odd number of electrons per primitive unit cell (for which the basis is just one molecule). Although there are empty spin and space orbitals degenerate with those occupied (which in a free-electron situation would build into a part-filled band of Bloch states), exchange and correlation interactions in the prevailing narrow-band, poor-overlap situation produce localized, nonmetallic electrons. Such electrons display a paramagnetic moment. For cases like CoO, where there is more than one electron per site, they interact as in a free atom to establish *S*, *L* and *J* quantum numbers. In a crystal the actual set of coupled states arising reflects the local crystal structure and chemistry. The low-intensity, narrow-line, optical spectra derived from these ground and excited states are termed ligand field spectra (see Lever 1968, Jorgensen 1971). They characterize transition metal salts, such as blue $d^9CuSO_4 \cdot 5H_2O$. Figure 1 shows the ligand field spectra of CoO (Pratt & Coelbo 1959) and $CoCl_2$ (Ferguson *et al.* 1953).

For the f electrons in the rare-earth elements and their compounds, an extremely highly correlated localized condition is the norm (Jorgensen 1977, Hulliger 1979), the spatially semicore character of the 4f electrons precluding significant overlap of these wavefunctions between metal sites or hybydrization with the coordinating ligand states. The semicore character arises because the 4f states are not restrained from entering the inner reaches of the atom by any orthogonal shell of 3f electrons.

The situation, it follows, is somewhat less extreme for the 5f actinides, and among the 5f elements prior to americium the f electrons clearly contribute to the bonding forces and to the electrical conduction (see Johanssen *et al.* 1981). Thus the crystal structure of α-U is complex, its carriers are heavy and it shows charge density wave formation (Lander 1982). Nonetheless in the vast majority of actinide compounds the 5f electron states remain localized, for example, f^3UI_3, f^2UO_2 and USe_2, f^1UCl_5 (Erdös & Robinson 1983). Here, the higher the valence forced upon the cation, the smaller is the residual f^n core. For some divalent and many trivalent uranium compounds (such as metallic US) the situation becomes more complex (Damien & de Novion 1981, Schoenes *et al.* 1984) because 6d electrons can be present simultaneously with the f electrons at the Fermi level. (Note: ThS is a non-magnetic d^2 metal with only 6d electrons in the Fermi sea and so is comparable to the $5d^1$ metals LaS, LuS.) The unusual behaviour one

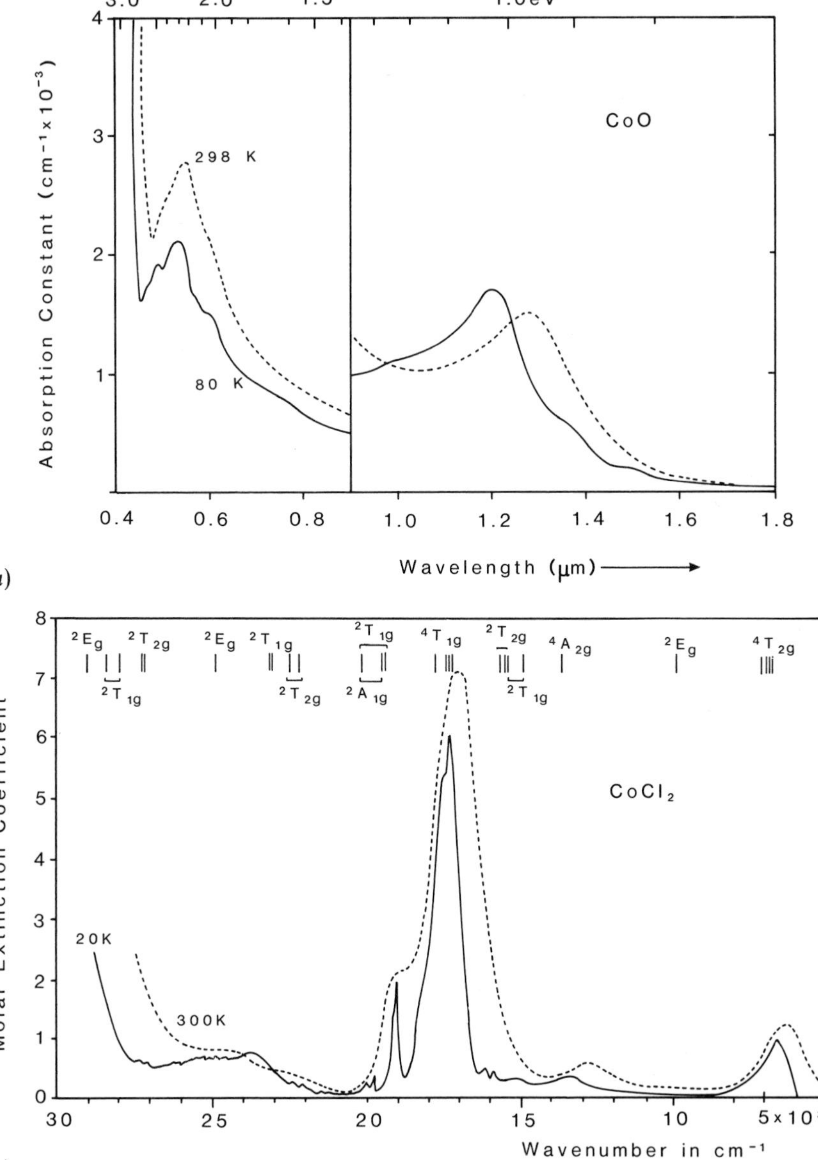

Figure 1. Ligand field spectra of high-spin Co^{2+} compounds: (*a*) CoO (Pratt & Coelho 1959)—note scale change; (*b*) $CoCl_2$ (Ferguson *et al.* 1963).

actually observes with US and PuS arises, as for SmS under pressure (Figure 2: Wilson 1977), not from a simple Mott transition within the f-band, but from the effects of configuration crossover in the f- and d-electron counts (see, for example, Wilson 1977a,b).

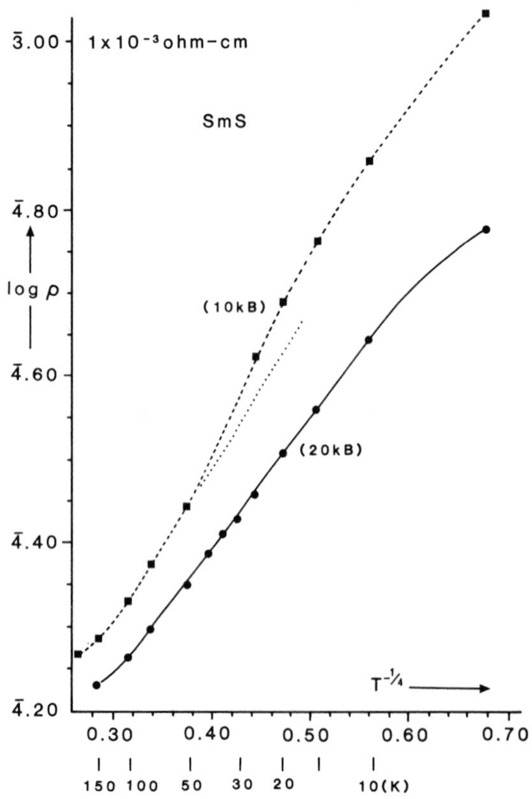

Figure 2. Resistivity in SmS under pressure in interconfiguration fluctuation regime, showing increase in log ρ approx. as $T^{-1/4}$ at low temperatures. (Wilson 1977)

Configuration crossover in transition-metal compounds between sub-bands of the d-band manifold (for example, t_{2g} and e_g with octahedral coordination for high-spin/low-spin effects) likewise can strongly affect the lattice parameter of a compound (see Figure 3: Wilson 1972) and greatly alter the conduction characteristics. For example $4d^5$ low-spin α-$RuCl_3(t_{2g}^5)$ is actually smaller than high-spin $FeCl_3(t_{2g}^3e_g^2)$; and low-spin $IrO_2(t_{2g}^5)$ is metallic when high-spin $MnTe(t_{2g}^3e_g^2)$ is an insulator. The metallic character of CoS_2, structurally closely related to CoO, is due in part to its low-spin configuration $(t_{2g}^6e_g^1)$ against the high-spin $(t_{2g}^5e_g^2)$ of CoO.

With octahedral coordination as in these compounds, the three t_{2g} orbitals are non-bonding, being directed between M–X bonds, whilst the two e_g states are elevated to higher energy from bonding/antibonding interaction with the ligand valence states. The e_g states point in the M–X bond directions.

Adoption of the high-spin, half-filled shell configuration $t_{2g}^3e_g^2$ is, as elaborated upon previously (Wilson 1972), the key to Mott-insulating character, holding even for the tellurides MnTe and $MnTe_2$. Despite considerable p–d

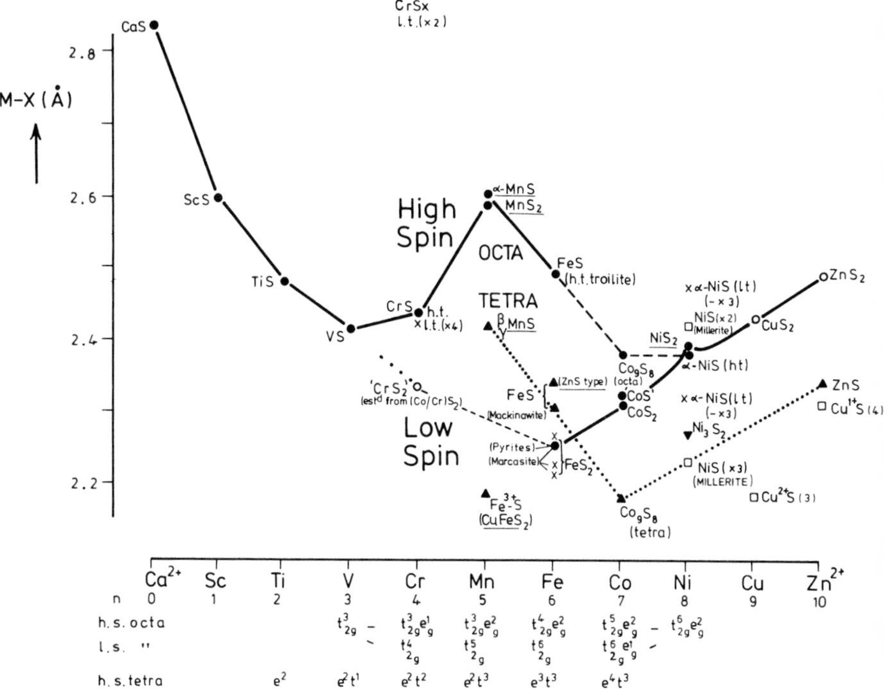

Figure 3. Effect of t_{2g} vs. e_g occupation on M–X bond length for high- and low-spin chalcogenides having octahedral and tetrahedral coordination (Wilson 1972). Note local maximum around NiS_2.

hybridization being incurred in these tellurides through the proximity of ligand 5p and cation 3d energies, the resulting 'd' bandwidths still remain insufficient to overcome the large exchange and correlation energies operative with the high-spin d^5 Hund's-rule ground state $^6A_{1g}$.

In compounds where, with increasing bandwidth (W) and diminishing exchange and correlation (loosely monitored by an overall Hubbard U), delocalization is first encountered (for example, in d^2CrO_2, d^3CrP, high-spin d^6FeS or low-spin d^7CoS_2), those metals are magnetic. Non-magnetic, superconducting ground states are by contrast found in these metallic transition metal compounds with 'early' and uncontracted d states such as d^1ScS or TiN, or, subsequently, when there is heavy d-band overlap with the main p-valence band, as in $CoTe_2$, PdTe, CuS, CuS_2 and CuV_2S_4.

Of the common transition metal compounds, about half are Mott insulators (like $NiSO_4$ and $NiCl_2$) and half are metallic (such as NiSe and NiAs). Typically, transfer from 'salt-like' to 'alloy-like' behaviour occurs in the vicinity of the oxides and nitrides. The situation is illustrated in Figure 4. The construction of this

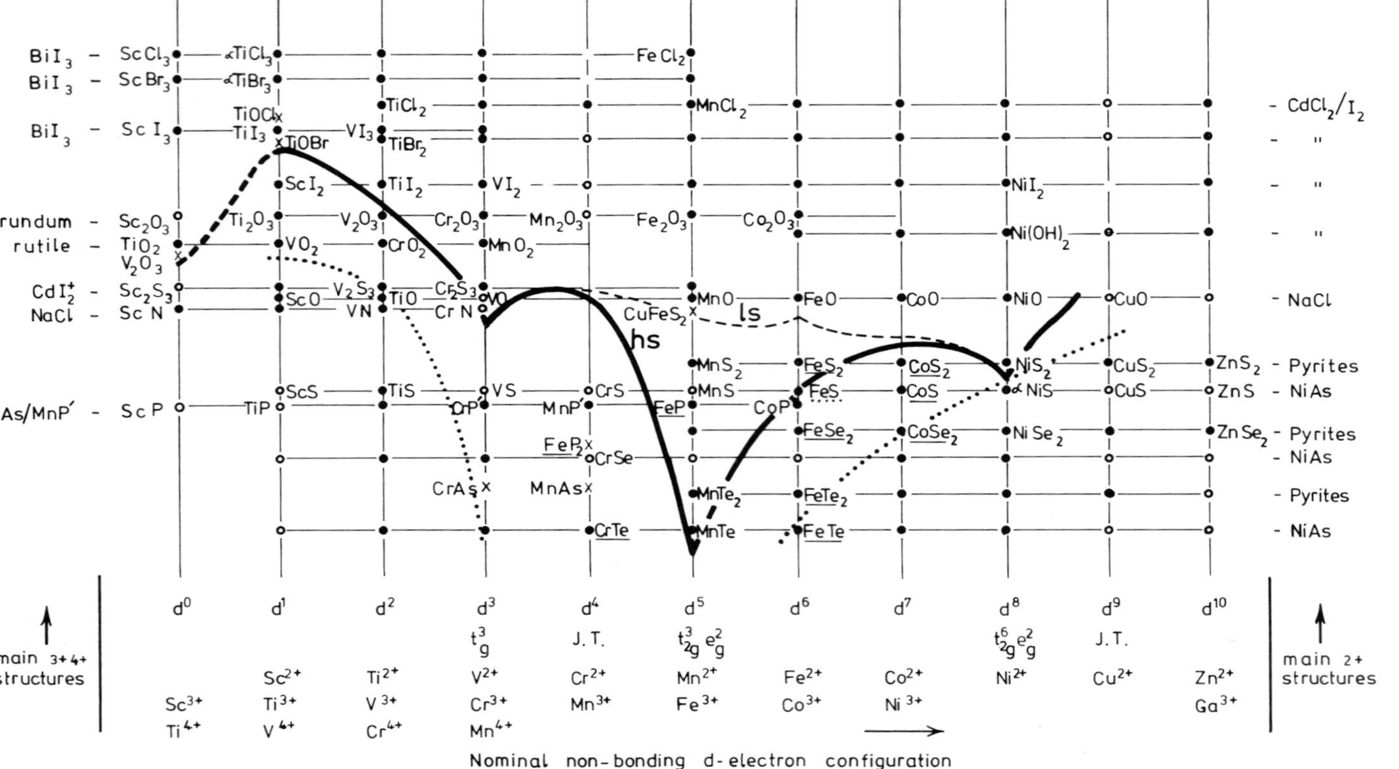

Figure 4. Binary 3d compounds arrayed by anion electronegativity and structure against cation 'd-band' electron number. The full curve separates Mott insulators from metals (or broad-band semiconductors). The broken line relates to low-spin materials. The dotted curves in the metallic regime separate the magnetic metals like CoS_2 and CrO_2 from the momentless metals like $NiSe_2$ or TiS. For extended discussion of this

diagram has been discussed at length in Wilson 1972, 1973. One should take note of the Mott-insulating character of $3d^2TiI_2$, d^3CrN and MnO_2, d^6FeO, d^7CoO and d^8NiS_2, in contrast with the metallic character of $5d^2ThI_2$, d^3CrP, high-spin d^6FeS, low-spin d^7CoS_2, and d^8NiS and $NiSe_2$. What has been particularly frustrating in examining this potentially very interesting delocalization boundary is that the change in electronic character entails a change in bonding which in almost all cases leads to a change in structure type. Thus, while Mott-insulating CrN and FeO are NaCl-type, metallic CrAs and FeS are NiAs-structured. A further 'deceit' is found between MnO_2 and MnS_2: the valence changes from $4+(d^3)$ in rutile-structured MnO_2 to $2+(d^5)$ in pyrite-structured MnS_2, so causing the latter to remain Mott-insulating. Moreover $ReO_2(5d^3)$, though metallic, has its rutile-related structure severely distorted by M–M bonding.

Even if a structure were preserved, there is no guarantee mixed crystals could be produced for all compositions to cross the boundary continuously. There occurs a significant volume collapse into the metallic state, with the dielectric properties of the latter significantly changed. This can lead to state-immiscibility, as, for example, happens within the tungsten bronzes Na_xMO_2 (Delmas *et al.* 1980). Mixed crystals never are an ideal medium in which to examine electronically fragile phenomena; the disorder even when introduced on the anion site has a strongly perturbing effect.

Only two pseudo-binary systems to date come close to traversing the Mott transition at all satisfactorily. These are the much-investigated systems $d^2/d^3(V/Cr)_2O_3$ and $d^8Ni(S/Se)_2$ (for references to early work see Wilson 1972, 1973; for later work on V_2O_3 see Honig in this volume (pp. 261–285), and for NiS_2 see §2 of the present paper). The latter pyrite system, though less often cited than the former corundum system, is already better suited to the role of prototype for the reason just given above—namely, that it involves anion not cation substitution. Moreover, because the cations are in a face-centred cubic array, there is no adjustment in some structural free parameter for the cation, such as arises with the rhombohedral structure of the sesqui-oxide system. Still more important is that pure NiS_2 is on the insulating side of delocalization, while V_2O_3, although needing only about 4% Cr to become Mott-insulating, is itself metallic. Thus, only with unsubstituted NiS_2 is it possible under pressure to traverse the Mott transition from insulator to metal ($p_1(300\ K) \sim 30$ kbar, Wilson & Pitt 1971). We shall return to discuss the properties of the $Ni(S/Se)_2$ system at length in §2. Here it remains to note that one can readily render NiS_2 metallic by d^7Co and d^9Cu doping, additional to Se substitution or the application of pressure.

Now there do exist other nickel compounds even more covalent than $NiSe_2$ which are not metallic, and one must take care not to confuse such compounds with Mott insulators. An example is divalent NiP_2. The divalency arises because the phosphorus is present in pseudo-chalcogen chain form $(-P-P-)^{2-}$, and the semiconducting character of NiP_2 arises because the $Ni^{2+}(d^8)$ sites possess square-planar coordination. This 4+2 coordination is typical of many d^8

compounds, bringing semiconductivity to PdO, PdS, PdS_2 (Collins *et al.* 1979) as well as to NiP_2. By contrast with '2+'d^8NiP_2, $NiAs_2$ is '4+'d^6 and one of a large set of t_{2g}^6 low-spin, octahedrally coordinated semiconductors which includes '4+'PtS_2, '3+'CoP_3 and '2+'FeS_2. Such materials have been discussed at length in Wilson 1972, 1984, Wilson & Yoffe 1969 and Hulliger 1968, where with the application of a little crystal chemistry it may be seen how these d-band electron counts arise from the rather strange looking stoichiometries.

A d^8 electron count arises for NiS_2 because the pyrite structure contains the pseudo-halogen pairs $(S–S)^{2-}$. Similarly pairs procure the d^0 semiconductivity of ZrS_3 (as $Zr(S_2)S$) to match 4+ZrO_2, ZrS_2, etc. Directly of relevance to our present topic is that semiconductivity thence extends to d^1NbS_3, as to NbO_2, under pair formation in the cation sublattice. There opens a bonding/anti-bonding gap within the d-band (Wilson 1979). Figures 5(*a*) and 5(*b*) show the band structures (Caruthers *et al.* 1973) for semiconducting, monoclinic, cation-paired VO_2, and for metallic, Curie–Weiss paramagnetic, unpaired VO_2—the state reached above the unpairing transition at 340 K. A corresponding unpairing occurs within Ti_2O_3 around 430 K (see Castellani *et al.* 1979, Paquet & Leroux-Hugon 1980). Unpairing is not encountered in $4d^1NbO_2$ until 1100 K (Pynn *et al.* 1978). No such effects are found for the M–M bonded chains in $4d^2MoO_2$ or $5d^3ReO_2$, which despite the bonding are metallic. Recall from Figure 4 that the simply structured rutiles $3d^2CrO_2$ (Chamberland 1977) and $3d^3MnO_2$ (Ohama & Hamaguchi 1971) are ferromagnetic metal and Mott insulator respectively.

The fact that paired VO_2 or Ti_2O_3 cannot be doped as classical semiconductors (Rice *et al.* 1970, Chandrashekar *et al.* 1974) re-emphasizes that for narrow-band materials the electronic and structural disorder inherent in doping is to be avoided if pressure might achieve the desired effect. All disorder simply enhances the intertwining of Mott localization with Anderson disorder-induced localization (Mott 1972, Ritala & Kurkijarvi 1982). This intertwining is much in evidence for those materials like shear-structure Ti_4O_7 with mixed valence problems on two or more sites. Then correlation as detected by ρ, χ and C_V is clearly enhanced (Penson *et al.* 1979, Hodeau & Marezio 1979).

In this section it now remains to comment briefly on the extreme rarity of s- and p-band metallic compounds outside the real alloy category. SnAs provides one such example of an open s/p band, superconducting metal. It has the same rock-salt structure as 'lone-pair', semiconductive SnTe. Car *et al.* (1978) have given the similar but more complex 'lone-pair' type band structure of layered SnSe. Also in contrast with SnAs, layered GeAs and isoelectronic GaSe emerge as semiconductors under M–M pair formation (see Robertson 1979), while chain-structured $GeAs_2$ is semiconductiving via the structural route $Ge^{4+}(\text{-As-})^{1-}As^{3-}$ (see Hulliger 1976, Fig. 80). TlSe disproportionates to $Tl^{1+}Tl^{3+}(Se^{2-})_2$. All these various ways of arriving at closed-band behaviour derive from the large spatial extent of the s and p wavefunctions and their dominant structure-determinating role. Only for 2p states (not orthogonal to any 1p) is there any sign at normal

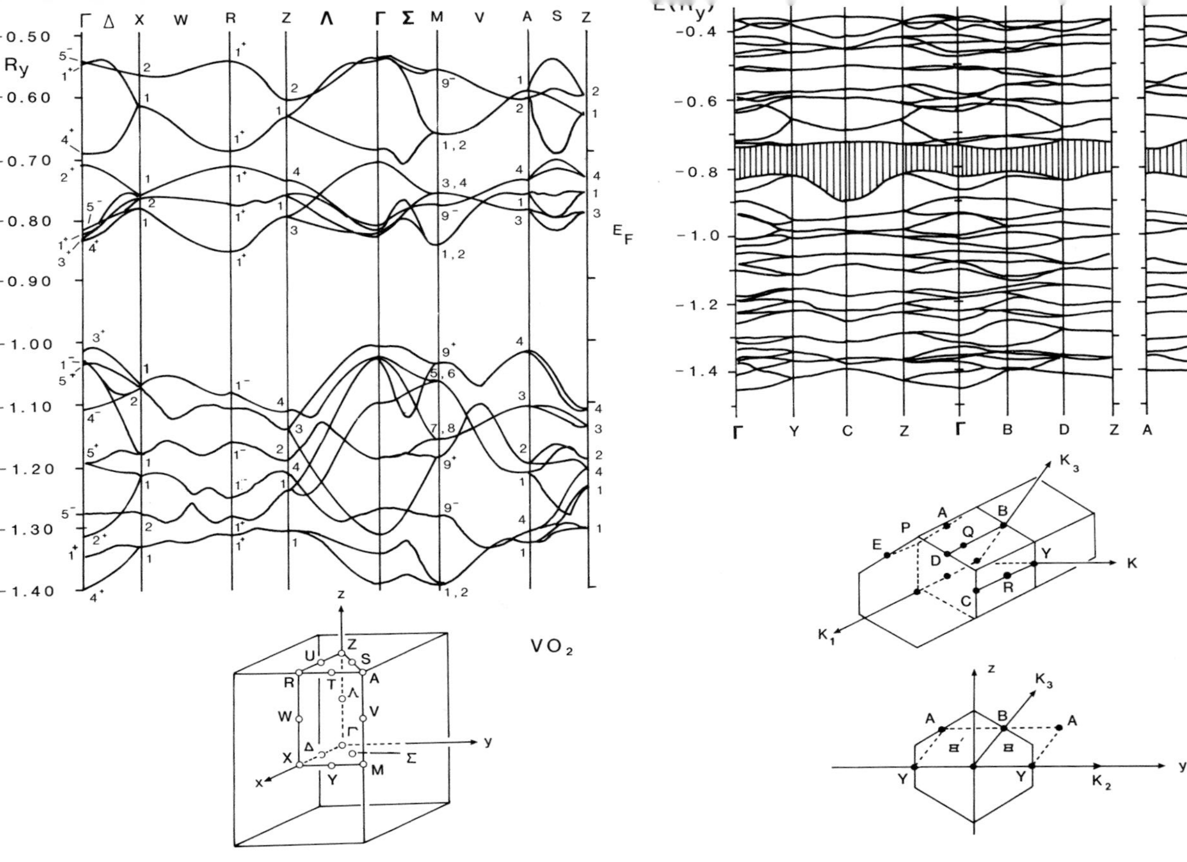

Figure 5. Band structure of d^1VO_2 in (*a*) tetragonal metallic phase and (*b*) V–V paired monoclinic semiconducting phase (Caruthers *et al.* 1973). The two zones are in parallel orientation.

densities of achieving passably stable paramagnetic entities, such as NO_2, N_2O_3 or NO, or $B_{12}C_{11}$. Particularly to be noted, therefore, are antiferromagnetic, insulating, pyrite- and marcasite-structured NaO_2 (Brüesch *et al.* 1976, Zielinski & Parlinski 1984).

Though s and p elements of themselves accordingly figure only little in our present topic (for example, expanded caesium above the critical point, Hensel 1984) they do offer impressive latitude to the topic of Mott insulation when extended to ternaries like the perovskites and spinels, much explored by Goodenough (1963, 1968, 1971a,b, 1984) and to be presented elsewhere in this volume. The interesting properties of $LaNiO_3$ as against $LaCuO_3$ (Goodenough *et al.* 1973), the contrast between $SrRuO_3$ (Pauli paramagnetic metal) and $CaRuO_3$ (Curie–Weiss metal) (Cox *et al.* 1983), the properties of $CsFeS_2$ (Nishi *et al.* 1983), of $BaVS_3$ (Sayetat *et al.* 1982) and of $LiTi_2O_4$ (Watanabe *et al.* 1984) are all cases in point.

2. The electronic properties of NiS_2 and derivatives

2.1. Some basics

The crystal structure of NiS_2 is the pyrites structure. This can be thought of as rock-salt with the simple anions replaced by the axial complex $(S–S)^{2-}$. These units acquire orientations parallel to all four body diagonals, so that the $z=4$ cell remains primitive and the simple cubic space group $Pa3(T_h^6)$ shows no tetrads. Diagrams of the structure are to be found in Wilson & Yoffe 1969, Figure 7, and Wilson & Pitt 1971a, Figure 9. The coordination octahedra show a single M–X bond length (2·40 Å for NiS_2), but each octahedron is slightly compressed down one or other trigonal axis (X–M–X = 86·5° for NiS_2; Furuseth *et al.* 1969). The octahedra are linked together both by direct corner-sharing and through the X–X links. The cations form a face-centred cubic sublattice, with a minimum M–M spacing of 4·02 Å in NiS_2. The S–S pairing at 2·06 Å is tighter in NiS_2 than in yellow $d^{10}ZnS_2$, and hence the antibonding states from the S–S pair are ejected from the valence band to well above the Fermi level in d^8NiS_2, as also in semiconducting low-spin $t_{2g}^6FeS_2$.

NiS_2 is directly flanked on the one hand by e_g^1 ferromagnetic CoS_2 (see Inoue *et al.* 1980), a metal of direct relevance to narrow-band magnetism (see Takahashi & Tano 1982), and on the other by e_g^3 superconducting CuS_2, which for good measure shows a CDW/PSD (charge density wave/periodic structural distortion) below 160 K (Krill *et al.* 1976c, Vanderschaeve & Escaig 1976a,b).

These pyrite disulphides have been discussed at length in relation to their optical spectra in Wilson 1973, and more recent data from Suga *et al.* (1983a,b) continue to support the band scheme of the 1973 paper which is reproduced in Figure 6. Fairly recent band calculations by Bullett (1982) are on the limit of what presently is trustworthy in detail but certainly come close to this general picture,

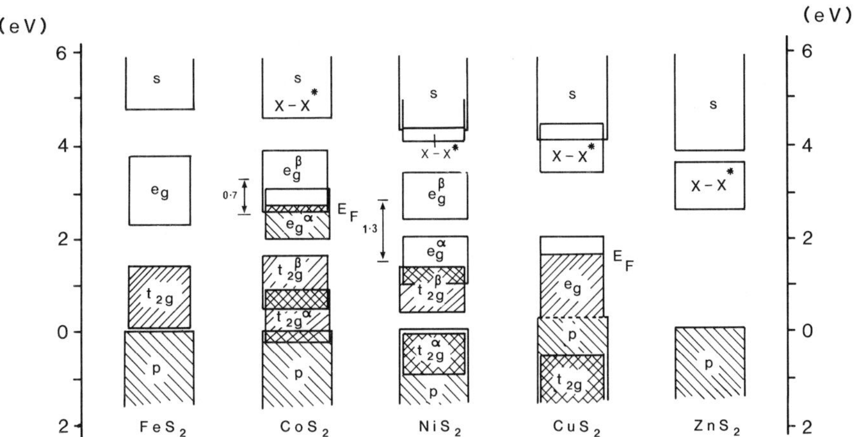

Figure 6. Disposition of energy bands for the 3d pyrite disulphides. The bands for CoS_2 and NiS_2 show α, β site-spin polarized condition. This spin splitting plays a dominant role in establishing a Mott–Hubbard gap within the e_g band. (For the derivation of this figure from optical data, etc., see Wilson 1973.) The band (X–X)* is the antibonding p-band from the chalcogen pair.

though the calculations were not geared to show the exchange splitting in CoS_2 or the Mott–Hubbard gap of NiS_2.

Between the sulphides and selenides, the main chalcogen p-based valence band rises toward the d-band by $\sim 1/3$ eV (cf. MnSe vs. MnS) producing a very marked percentage reduction in the separation of the valence band and the Fermi level (see Wilson, 1972, Figure 6). The augmented p–d hybridization is sufficient to cause $NiSe_2$ to be a metal with rather simple Pauli paramagnetism (Inoue *et al.* 1980). The p–d mixing in d^8NiSe_2 is clearly more marked than in d^7CoSe_2, where despite losing long-range ordering the metal does retain a Curie–Weiss susceptibility, and indeed moment saturation at high temperatures ($T > 400$ K). Again, I have given an early discussion of this situation in Wilson (1972, 1973). How the situation has evolved theoretically since then may be found in Moriya (1983), and Takahashi & Tano (1982).

As was first explored by Wilson & Pitt (1971), it is possible to close the Mott–Hubbard gap in NiS_2 by the application of pressure (~ 40 kbar at 300 K) or by substituting Se for S (about 30% at 180 K), which leads to the rudimentary phase diagram for the system shown in Figure 7. More recent work will be discussed in detail below, but note that the general form of the diagram (as for the equivalent diagram for V_2O_3:Cr (McWhan *et al.* 1971) reflects the following physics. (i) Between the AFI and PI phases, T_N rises with P as Anderson's dynamic super-exchange formula $T_N \propto t^2/U$ (U diminishing and t, the transfer integral $\propto W$, increasing). (ii) Between the M and PI phases the Mott line rises more steeply because of a very small entropy decrease into the highly correlated metal. The sign of the line's gradient follows from the volume contraction into the

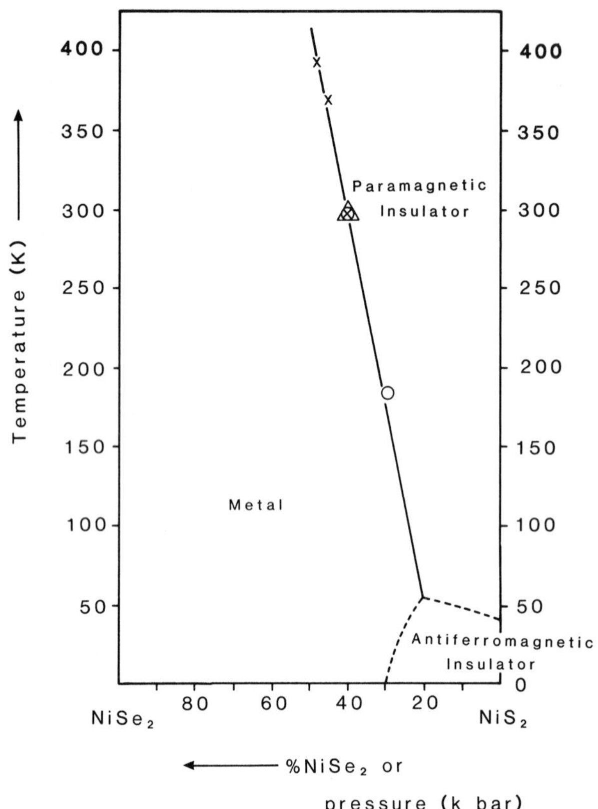

Figure 7. Phase diagram for the Ni(S/Se)$_2$ system (Wilson & Pitt 1971a). For a recent refinement see Figure 17. There is an approximate equivalence between Se doping and pressure following 1% Se ≡ 1 kbar.

metallically screened phase (see Figure 17, Wilson 1972). (iii) The third line is that between the M phase and an antiferromagnetic phase. The latter is definitely metallic before its disappearance as we shall see below. This AFM phase brings further interest to the system as compared with $(V/Cr)_2O_3$. The complete decay of the AFM phase in the nickel pyrite system has not yet been thoroughly tracked, but T_N clearly is falling from its peak value, reached close to the Mott line intersection as expected (see Mori & Takahashi 1983; note in the neutron work by Panissod *et al.* (1979) that despite what is claimed it is the AFI phase which is being observed).

It is now possible to calculate the phase diagram of Figure 7, for example, from a functional integral treatment of transverse and longitudinal spin fluctuations (Moriya 1979) in the half-filled band condition as for $e_g^2NiS_2$. Calculable also are the variations in χ with T at various t/U, as in Wilson (1972) Figure 15a, tracking the dissolution of the local moment with increasing t/U.

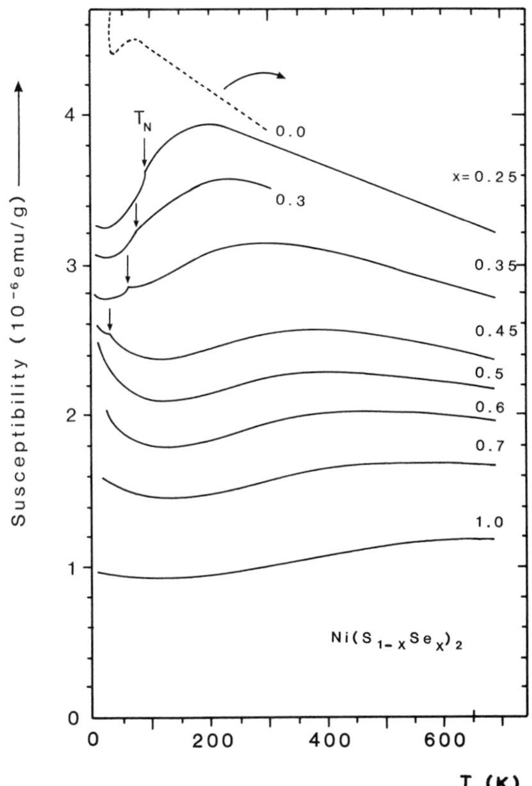

Figure 8. Magnetic susceptibility for the system $Ni(S/Se)_2$ (from Ogawa 1979, Ogawa *et al.* 1974). Note the maximum in χ occurs well above the Néel point for the compositions $0{\cdot}25 < x < 0{\cdot}45$ (which are all metals at low temperature).

Ideally at this point one would like to measure the magnetic susceptibility of NiS_2 to high pressures, but as yet one must be satisfied with measurements through the mixed system $Ni(S/Se)_2$. Figure 8 shows such susceptibility data from Ogawa 1979.

These curves show clearly that, as for pure NiS_2, the Néel point is not the point at which χ is a maximum. (Note: T_N is further identifiable from the high-pressure dilatometry reported by Mori & Takahashi 1983.) The maximum in χ is highly sample dependent, as first noted by Krill *et al.* 1976a. This latter matter now deserves some attention, since it provides the main detraction from NiS_2 being the ideal system.

2.2. Disorder and nonstoichiometry

CoS_2 is a ferromagnetic metal, but isoelectronic $(Fe_{0{\cdot}5}Ni_{0{\cdot}5})S_2$ is neither ferromagnetic nor metallic (Adachi *et al.* 1976a). Its room temperature resistivity is around 1 Ω cm and rises further on cooling, while the magnetic susceptibility, though of appropriate effective moment, presents a large *negative* Curie–Weiss constant much as for NiS_2 itself. In the same way it is not to be expected that $(Co_{0·5}Cu_{0·5})S_2$ will behave as NiS_2, either magnetically or electrically. With these compounds the narrow-band situation prevailing is too fragile for simple rigid-band behaviour to hold.

The transport properties are even more sensitive with regard to this matter of disorder than are the magnetic properties. Illustration of this is provided by the system $(Fe/Co)S_2$, where, although a linear rise is found in the saturation magnetization from e_g^0 towards e_g^1, the resistivity still is observed to *rise* strongly toward low temperatures (Ogawa *et al.* 1975, Figure 10). There is no doubt that here the transport problems were being multiplied by the use of polycrystalline samples, but the tendency remains for crystalline samples.

In keeping with the above, Krill *et al.* (1976) found that, working within the Mott-insulating regime, powdered samples of NiS_2 gave a resistivity below 40 K which follows a $T^{-1/4}$ variation, as for variable-range hopping in a disordered material. It is now known that such behaviour is of somewhat more general origin and occurrence (Ritala & Kurkijarvi 1982, Fleishman & Anderson 1980), common to correlated, disturbed bands, as for example in the 1T-TaS_2 CDW system, whether pure (Uchida & Tanaka 1984), irradiated (Mutka 1983) or substituted (DiSalvo *et al.* 1976). A further case was given earlier in Figure 2 for metallic SmS in its interconfiguration fluctuation (ICF) regime.

For NiS_2 the effect of Co and Cu substitution potentially is of prime interest when seeking to move away from the half-filled band position and gain delocalization. Figure 9 shows 300 K resistivity–pressure data for the system $(Ni/Cu)S_2$ up to 10% substitution (Krill *et al.* 1976a). For these samples it is seen from Figures 10 and 11 that below the Néel point the resistivity can display a metal-like increase with temperature (see also Ogawa *et al.* 1974, Figure 10). Such transport data from polycrystalline compacts is significantly different, however, from that obtained in the single-crystal work reported by the MIT group and given below (Mabatah *et al.* 1980, Co-doping; Kwizera *et al.* 1981, Cu-doping). The latter group have made a very substantial effort to reach a detailed theoretical fit to their data, simultaneously coping with their complex set of resistivity and Seebeck curves from both of the above cation-doped systems and from the mixed anion system $Ni(S/Se)_2$ (Kwizera *et al.* 1980). Prior to examining that analysis, however, it is necessary to be more aware of the situation for 'pure' NiS_2. Once again from the work of Krill *et al.* (1976a,b), we seem to be presented with a point of departure less simple than might be desired.

Figure 12 summarizes the findings of Krill *et al.* (1976) derived from density and lattice parameter measurements plus a chemical analysis of their samples. It appears (for this batch of samples at least) that the compound displays a complex defect structure with vacancies present on each sublattice, so that stoichiometric NiS_2 has 3 to 4% Ni and S site vacancies. The various samples show strong and systematic change in lattice parameter and in their resistivity behaviour under

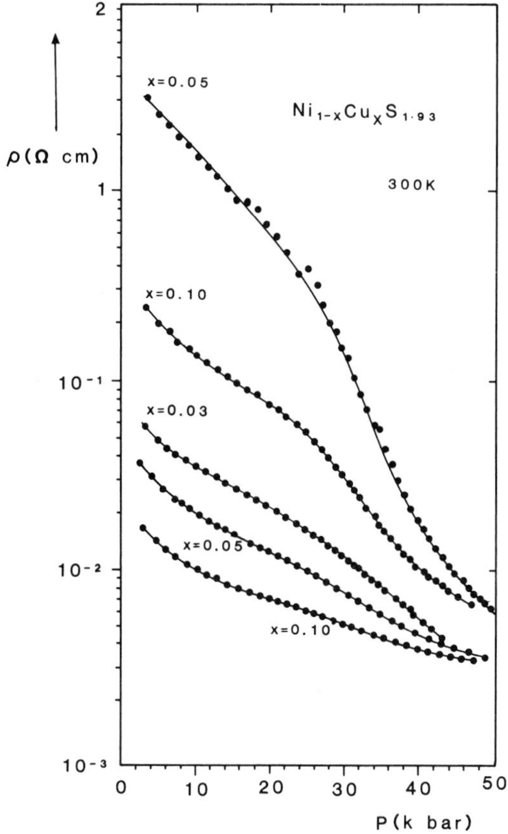

Figure 9. Log resistivity vs. pressure at 300 K for copper-substituted 'NiS_2'. (Krill *et al.* 1976a)

pressure (see Figures 13 and 14). The resistivity is seen to be highest at low sulphur content (around $x = 1{\cdot}9$), just where the lattice parameter is greatest and the cation sublattice is intact. (This behaviour might indicate some tendency towards a Mott-insulating rock-salt (c.f. α-MnS).) With increase in the sulphur content above $x = 1{\cdot}9$, one sees the rapid development of nickel vacancies. The latter inevitably lead to the appearance of p-type carriers, which are evident in the Seebeck measurements of Kwizera *et al.* 1980. The carriers thus introduced are sufficiently free at 300 K to reduce very significantly the height of the d^8 lattice parameter peak, though not to a value as low as that yielded by extrapolation between d^7CoS_2 and d^9CuS_2 or from their fully *metallic* Ni-substituted derivatives (see Ogawa *et al.* 1974, Figures 1 and 16, Wilson & Pitt 1971a, Figure 1). The effect of these defects on the pressure delocalization event (see Figure 14) clearly bears direct comparison with the effect of Cu or Co doping shown in

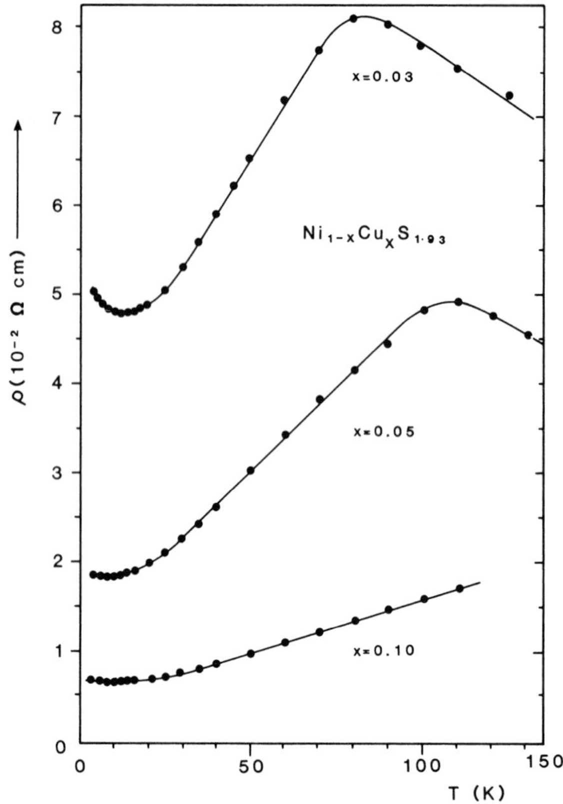

Figure 10. Low-temperature resistivity behaviour for copper-substituted 'NiS_2'. (Krill *et al.* 1976a)

Figure 9. p-Type carriers dominate the Seebeck coefficient below 300 K for up to 2% Co doping, 10% Cu doping and 40% Se substitution (Mabatah *et al.* 1980, Kwizera *et al.* 1981).

It is rather odd that only the paper by Krill *et al.* (1976a) draws attention to the problem of stoichiometry and vacancies in these materials. Single-crystal data from Bouchard *et al.* (1973) and from Kwizera *et al.* (1980) on the $Ni(S/Se)_2$ system both include curves for NiS_2 where the resistivity reaches 10^2 Ω cm at 4·2 K. The work of Krill *et al.* would in fact indicate such material to be $NiS_{1\cdot91}$. Clearly this problem has to be readdressed.

Before finally turning to the Mott transition in the system, one ought now to be made fully aware of the low-temperature magnetic conditions which prevail.

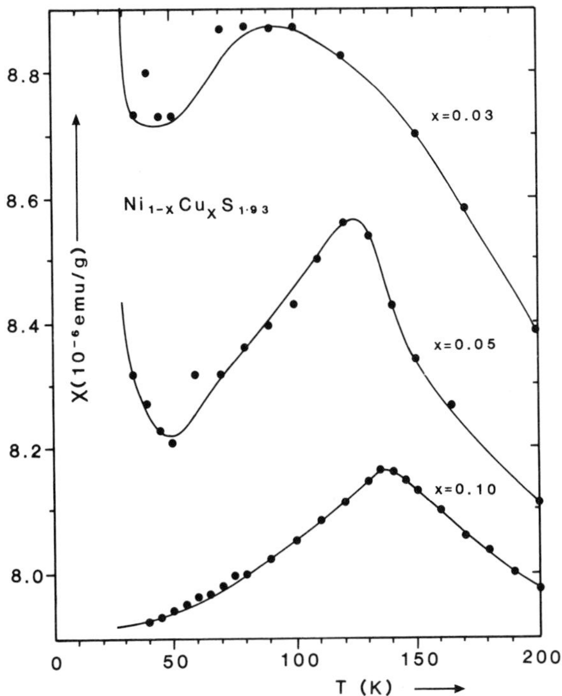

Figure 11. Low-temperature magnetic susceptibility for copper-substituted 'NiS_2'. (Krill *et al.* 1976a)

2.3. *Magnetism and magnetic disorder in NiS_2 derived systems*

'NiS_2'

The magnetic character of NiS_2 is considerably more complex than that of NiO, and is strongly reflective of the critical position held by NiS_2. Figure 8 shows the susceptibility up to high temperatures. The proximity of NiS_2 to delocalization leads to a low absolute magnitude for the susceptibility (730×10^{-6} cgs units/mole), a very large Curie–Weiss $\theta\sim -1000$ K, and an effective moment at 300 K that is considerably greater than the free-ion value for Ni^{2+} (namely, $2{\cdot}83\ \mu_B$); behaviour as shown in Wilson (1972) Figure 15a. Furuseth *et al.* (1969) found $\mu_{C\text{-}W}=3{\cdot}15\ \mu_B$ around 300 K, falling to the free-ion value only above 400 K. Bither *et al.* (1968) noted that the moment size is very sample dependent; a feature characteristic too of the long-range magnetic order in NiS_2.

The neutron diffraction work of Hastings & Corliss (1970) gave the first indication of the complex magnetic order adopted by NiS_2. Their results suggested to them an ordering of the 'first kind' at 40 K (T_{N1}), followed at 30 K (T_{N2}) by an additional ordering of the 'second kind' over the face-centred cubic Ni

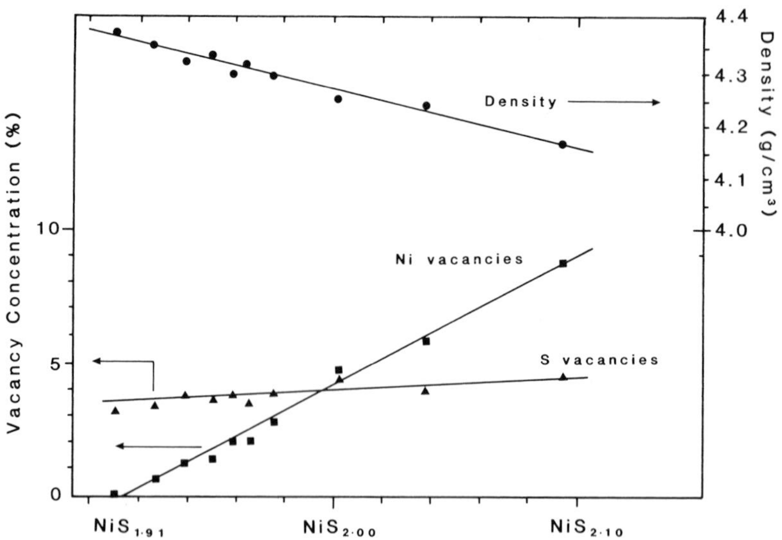

Figure 12. Analysis of defect structure for 'NiS_2' as function of stoichiometry. When the nickel sublattice is complete at $NiS_{1 \cdot 91}$ the sulphur sublattice is 4% defective according to this analysis (and for this means of sample preparation). (Krill *et al.* 1976a)

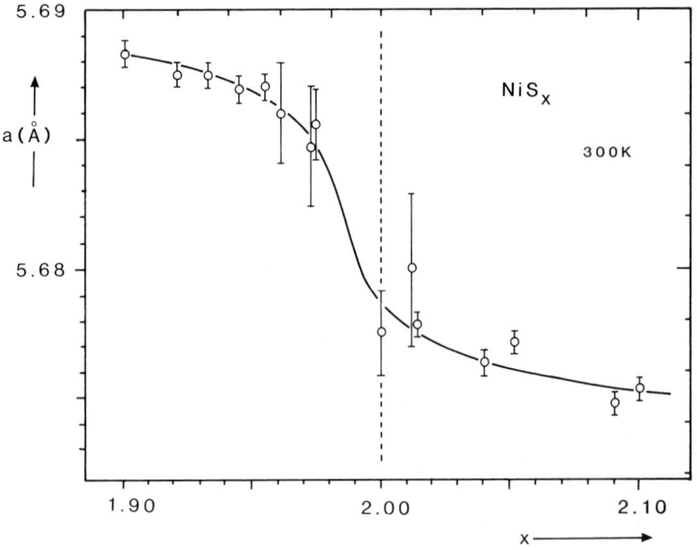

Figure 13. Lattice parameter vs. composition for the samples of Figure 12, showing lattice collapse and metallization around $x = 2{\cdot}00$, where *both* sublattices are about 4% defective. (Krill *et al.* 1976a)

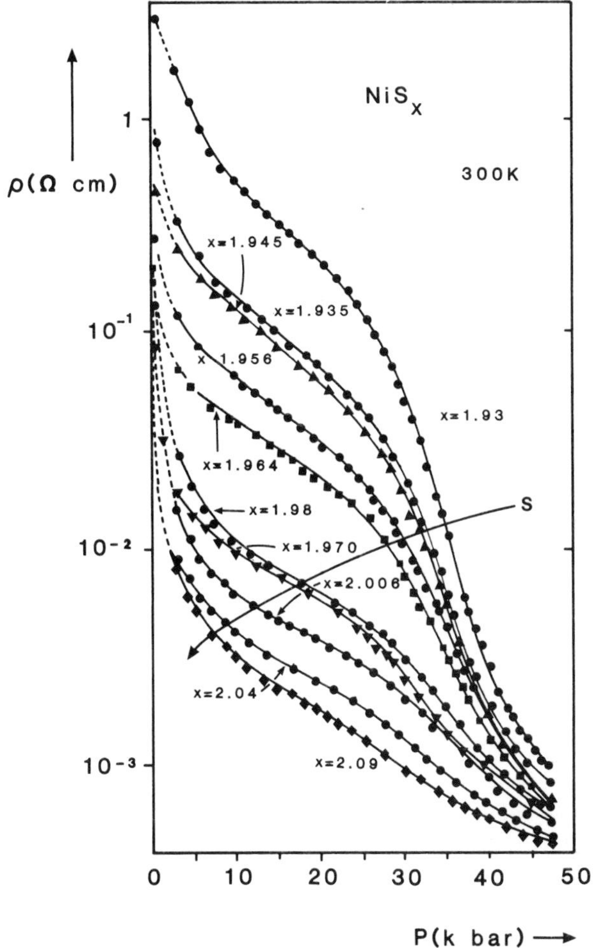

Figure 14. Effect of stoichiometry on pressure behaviour of resistivity for 'NiS_2' at 300 K. (Krill *et al.* 1976a)

lattice with a $2a_0$ cell. At 4·2 K the deduced moments were $\langle \mu_1 \rangle = 1{\cdot}0\ \mu_B$ and $\langle \mu_2 \rangle = 0{\cdot}6\ \mu_B$, combining if presumed independent to yield 1·17 μ_B. That the two orders are in fact not fully independent is seen in Figure 15, from the work of Panissod *et al.* (1979). Miyadai *et al.* (1978) have confirmed that the nature of the M1 order is essentially preserved below T_{N2}, and that the rise in M1 magnetic scattering is due to a rapid improvement in the level of M1 order attained with the development of the supplementary M2 component. Mössbauer work using ^{57}Fe detects a small spin 'cant' angle, which increases below T_{N2} (being about 20° from the triads $\langle 111 \rangle$ below T_{N2}; Nishihara *et al.* 1975a,b).

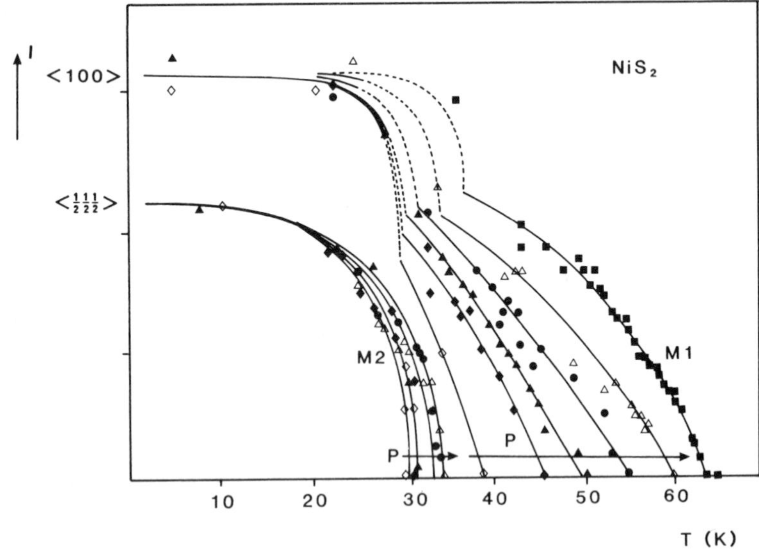

Figure 15. Neutron diffraction intensities for magnetic scattering from NiS_2 antiferromagnetic orderings as a function of pressure up to the metal–insulator transition. M1 order measured through $\langle 100 \rangle$ intensity; M2 through $\langle \frac{1}{2}, \frac{1}{2}, \frac{1}{2} \rangle$ indexed for simple a_0 cell. (from Panissod *et al.* 1979)

Resolving the details of these non-collinear magnetic structures by neutron work alone is rendered impossible by the complex magnetic domain structure (see the sequence of papers by Kikuchi *et al.* 1978a,b; Miyadai *et al.* 1975). In their 1978 paper, these workers proposed models of a four-sublattice M1 $MnTe_2$-type structure, and a still more complex M2 magnetic structure with eight sublattices and two Ni site types (the latter seemingly detected in ^{61}Ni Mössbauer work by Czjzek *et al.* (1974, 1976)). Three degenerate modes combine for an M1 structure but four modes are needed for an M2 structure, with the result that an M1 + M2 combination must be orthorhombic. Indeed such a loss of symmetry has been detected in NiS_2 below T_{N2} by strain gauge dilatometry (Nagata *et al.* 1976). In good material the transition is a first-order one involving a minute volume collapse of 0·02%. The transition is associated with a very appreciable latent heat (see Krill *et al.* 1976a, Figure 5) which Kikuchi (1979) has shown to be largely magnetic in origin. Kikuchi worked from the Clapeyron equation for the change in T_{N2} with magnetic field. Complementary to this, Mori & Watanabe (1978) worked with the change of T_{N2} with pressure. By either means, ΔS, at around 1 J mol^{-1} deg^{-1}, is revealed to be a very sizeable fraction of the total magnetic entropy.

The latter actually is found via calorimetric work (Ogawa 1976) to amount to only $R \ln 2$ (as for $s = \frac{1}{2}$) rather than to $R \ln 3$ (as for $s = 1$). Why this value is so suppressed (as similarly is the neutron-detected moment) remains one of the mysteries of this system. It may well relate to the fact that there are four Ni atoms per cell and hence four e_g bands. It is to be noted that the same feature of a very low ($\sim 0{\cdot}4\ R \ln 2$) magnetic entropy recurs with metallic $e_g^1CoS_2$, although now the ferromagnetically ordered moment is close to 1 μ_B as expected. It is worth noting too that Hastings and Corliss actually used $s = \frac{1}{2}$ Brillouin functions to describe their neutron data from NiS_2. Similar problems occur for the t_{2g}^2 system V_2O_3.

A further complexity presented by NiS_2 is that the coexistence of M2 with M1 order is unable to constitute the ground state (versus M1) for a system governed simply by mean field behaviour and a Heisenberg-type two-spin interaction between an ion and its nn, nnn ((next)-nearest neighbour), etc. As is well known, a Hamiltonian of the Heisenberg form will ensue from second-order perturbation theory and the Hubbard Hamiltonian, a procedure that yields also the Anderson super-exchange formula for $T_N \propto t^2/U$ (see Brandow 1977, Chao & Berggren 1977, Cyrot 1977, Yoffa & Adler 1979, and references therein). Going beyond two-spin interactions into third- and fourth-order perturbation theory, in order to incorporate significant three-spin and four-spin closed path interactions, naturally must complicate matters enormously. That, however, is the direction in which the results for NiS_2 point. The theory for four-spin ring interactions has been developed in relation to NiS_2 using the Hamiltonian $R[(\mathbf{S}_1 \cdot \mathbf{S}_2)(\mathbf{S}_3 \cdot \mathbf{S}_4) + (\mathbf{S}_4 \cdot \mathbf{S}_1)(\mathbf{S}_2 \cdot \mathbf{S}_3) - (\mathbf{S}_3 \cdot \mathbf{S}_1)(\mathbf{S}_2 \cdot \mathbf{S}_4)]$ by Yoshimori & Fukuda (1979) and Yosida & Inagaki (1981). The analysis proceeds by decomposing the $2a_0$ fcc lattice into the four component, two-atom ferromagnetic sublattices and defining spin-vector addition and subtraction vectors $\mathbf{S}_i$ and $\mathbf{T}_i$ for each of these. Above T_{N2}, all spin 'twins' within a ferromagnetic sublattice are internally parallel (each sublattice with its spins at some small common angle to the various (4) high-symmetry triads). Below T_{N2}, under the action of the four-spin interaction, the cell is enlarged to $2a_0$ as the sublattice spin twins internally diverge and strongly augment the cubic symmetry breaking, giving finite spin-difference vectors $\mathbf{T}_i$. Since the sum and difference vectors are mutually perpendicular, we have the diffraction effects of M1- and M2-type scattering coexisting. The final result of such four-spin interaction and **T** vector formation below T_{N2} is to produce a small resultant ferromagnetic moment along a magnetostrictive unique axis forming parallel to the original cube edge (001).

The size of the weak ferromagnetic moment detected both by magnetization measurements (Adachi *et al.* 1972, Kikuchi 1979) and neutron diffraction (Kikuchi *et al.* 1978b) is of approximately $0{\cdot}02\,\mu_B$ per Ni, giving a low-temperature saturation magnetization along $(001) \sim 1{\cdot}0$ emu/g. This ferromagnetism is associated again with domain formation, and slow annealing in a 1 T field is required to pole the sample. In such samples it is possible then to detect first-order thermal hysteresis at T_{N2} (Kikuchi 1979, Figure 8). The theory of a four-spin origin for this ferromagnetism is strongly supported by the fact that Kikuchi's results follow the relation $M_f(T) \propto \mu(\mathrm{M1}) \times [\mu(\mathrm{M2})]^2$ predicted by Yoshimori & Fukuda (1979).

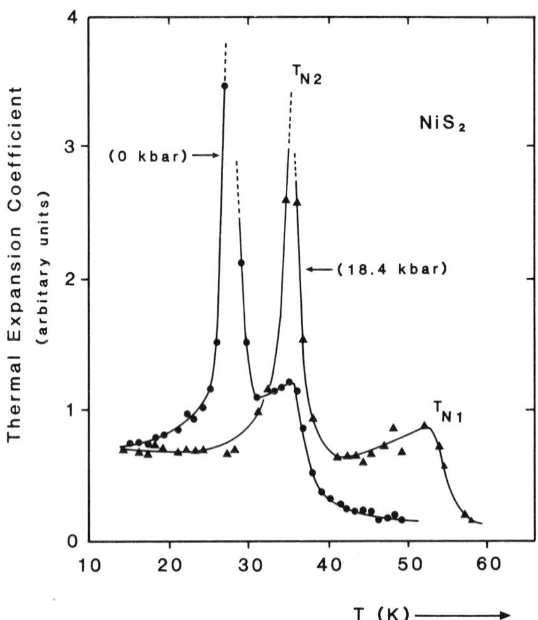

Figure 16. Thermal differential of dilatometry signal from NiS_2 at two pressures in the AFI regime, showing rapid rise in T_{N_1} and sensitivity at T_{N2}. (from Mori & Watanabe 1978)

As a consequence of the above, T_{N2} is particularly strongly marked in all magnetic measurements such as the susceptibility (see Figure 8), as also in calorimetry and dilatometry. The latter is a particularly convenient means for tracking T_N under pressure, or in substitution experiments. Figure 16 (from Mori & Watanabe 1978) complements the neutron data of Figure 15, showing the rise in both T_{N1} and T_{N2} under pressure. Figure 17 (from Mori & Takahashi 1983) gives the low-temperature part of the phase diagram as obtained by such means. The rise in T_N on approaching delocalization is expected from the Anderson formula. The subsequent fall in T_N beyond delocalization, requiring pressures in excess of 40 kbar, unfortunately has not been studied for pure NiS_2 but only in substitution experiments (with, for example, Se, As, Co, Cu) as will be discussed shortly.

What should finally be observed here is the finding of the defect study by Krill *et al.* (1976a) that defects (Ni vacancies, see Figure 12) rapidly suppress the size of the weak ferromagnetic moment, while leaving $T_{N2}(\equiv T_C)$ more or less unchanged. This makes the effect of a defective Ni sublattice much more akin to Co substitution than to slow metallization through Se substitution (Miyadai *et al.* 1983a—see below). In the case of T_{N1}, Ni vacancies secure a very rapid rise (as occurs following all routes to metallization). This accounts for the great variability in the literature for T_{N1}, 38 K seeming to mark the optimal lower limit.

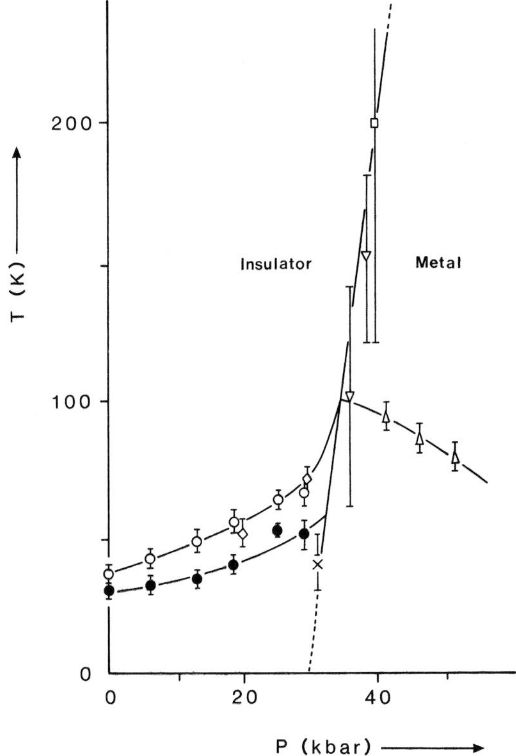

Figure 17. Low-temperature phase diagram for NiS_2 under pressure and approximate $p=0$ equivalent for $Ni(S/Se)_2$ using 1·2 kbar ≡ 1% Se, as obtained by strain gauge dilatometry (from Mori & Takahashi 1983). Compare Figure 7.

Under substitution that strongly perturbs the cation sublattice

We have just noted that the ferromagnetic moment and the M2 order are very easily lost by perturbation of the cation sublattice. As little as 4 or 5% of either Ni vacancies (Krill *et al.* 1976a) or iron, cobalt or copper substitution (Miyadai *et al.* 1983b, Krill *et al.* 1976a, Ogawa *et al.* 1974) suffices for this. The effect is to be seen in Figure 11 (which should be viewed in conjunction with Figures 10 and 9, and 14). While M2 order is being suppressed M1 order flourishes, with T_{N1} rising by up to a hundred degrees. Betsuyaku *et al.* (1974), by neutron diffraction on $Ni_{0\cdot 9}Cu_{0\cdot 1}S_2$, find an average moment little different from that in NiS_2, but with a T_{N1} of 85 K. Note that this, as with the results in Figure 8, places T_{N1} well below the peak in χ. Ten percent Cu (or Co) doping is effectively the limit to the low-temperature semiconductive regime, characterized by anomalously large lattice parameters (see Figure 3; also Figure 13).

Mössbauer ^{61}Ni hyperfine measurements (Krill *et al.* 1976a) quickly reveal, for a cation-perturbed lattice in particular, that a simple uniform medium description is quite inadequate. For samples containing 5% and 10% Cu, T_N would appear to lie in the vicinity of 130 K, but the spectrum contains signals corresponding to two sites, α and β. By a concentration analysis it was convincingly shown that β sites, which give the higher hyperfine signal, possess at least one nearest neighbour Cu atom. (Be careful here not to confuse site types α and β with the two types of site appearing in M2 ordering.)

A very similar result was found for samples with nickel vacancies (Krill *et al.* 1976a, Figures 20, 15 and 16). For the latter samples the absolute magnitude of χ remains very sensitive above T_{N1} (see Gautier *et al.* 1973a,b, Figure 1), accounting for the sizeable discrepancy between the values given in Ogawa *et al.* (1974) and in Krill *et al.* (1976a) (compare present Figures 8 and 11). It might be noted here that the remarkable properties of nickel monosulphide are also very strongly modified by its readily altered defect content (Coey *et al.* 1976).

Metallization of NiS_2 can also be secured by anion substitution using As (Adachi & Kimura 1977) (which has the apparent advantage over Se substitution that the lattice parameter is virtually invariant). However As substitution introduces d^7Ni^{3+} sites and the results become all but equivalent to Co doping, showing drastic effects on M2 order. T_{N1} remains detectable to about 17% substitution, where each Ni has on average one nearest neighbour As or two nearest neighbour Ni^{3+} sites. In the study of narrow-band magnetism, pure Ni(AsS) makes a very interesting comparison with isoelectronic ferromagnetic CoS_2. Perhaps heavy fermion behaviour is what below 25 K finally drives χ up strongly from its Pauli paramagnetic value of higher temperatures. The structure in fact is slighly distorted at 300 K (to T^4–$P2_13$).

Matching the NiAsS work, Adachi *et al.* (1976b) have revealed further lack of strict parallel between CoS_2 and the 50/50 mixed system $Fe_{1/2}Ni_{1/2}S_2$. Though of the appropriate moment the latter material now has a negative θ_p, and moreover a resistivity which by 4·2 K has climbed to 10^3 Ω cm.

Yet a further study by Adachi *et al.* (1976a) examined the 'percolation' problems encountered both for magnetic and transport properties within the e_g^0/e_g^1 system $Co(As/S)_2$.

Well worth mastering at this point is Figure 18, reproduced with minor modification from Adachi & Kimura (1977).

$Ni(S_{1-x}Se_x)_2$

Let us now examine the situation in the system that is of prime interest for the Mott transition, $Ni(S/Se)_2$.

As seen from Figures 7 and 17, the 'triple' point in the phase diagram occurs at about 25% Se substitution. Up to this Se content χ has fallen only by about 3% from its value in NiS_2 (Miyadai *et al.* 1983b, Figure 2), but beyond here with metallization χ decreases rapidly, as is evident in Figure 8 of the present paper (from Ogawa 1979).

By $x \sim 0{\cdot}15$, well before delocalization, M2 ordering and the accompanying

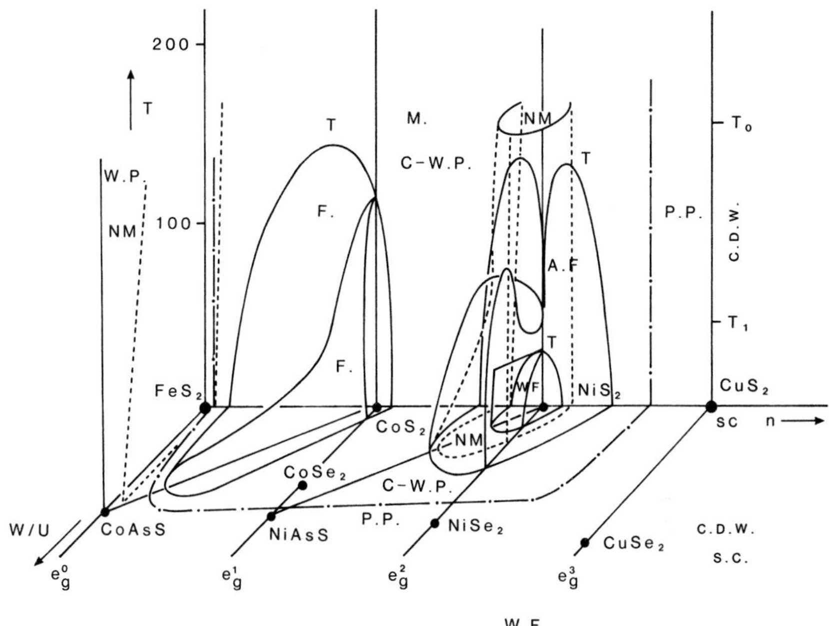

Figure 18. Phase diagram for 3d pyrite materials indicating magnetic and electronic characteristics (from Adachi & Kimura 1977, with minor adjustments and additions). M, metallic; NM, non-metallic; F, ferromagnetic; AF, antiferromagnetic; CWP, Curie–Weiss paramagnetic; PP, Pauli paramagnetic; WP, weak paramagnet; WF, weak ferromagnetism; SC, superconducting; CDW, charge density wave.

ferromagnetism are completely suppressed (Miyadai *et al.* 1983a,b). Through this range of x, T_{N1} rises, as it did under pressure (see Figure 17, taken from Mori & Takahashi, 1983), despite the above slow fall in χ with x (at $p=0$) and an accompanying fall in the ordered moment $\bar{\mu}_1$ (see Figure 19, taken from Miyadai *et al.* 1983b). By $x=0{\cdot}25$, $\bar{\mu}_1$ is only $\sim 0{\cdot}6\,\mu_B$. Figure 19 indicates that $\bar{\mu}_1$ continues to decrease smoothly beyond $x=25\%$, when one gains the antiferromagnetic metallic (AFM) state, not formed in the system based on V_2O_3.

Figure 8 gives Ogawa's susceptibility results from the AFM regime, which show that both T_{N1} (as $\bar{\mu}_1$) and Curie–Weiss behaviour are lost by around $x=0{\cdot}50$. This termination is evident also in Mössbauer and n.m.r. work (^{77}Se n.m.r., Gautier *et al.* 1975; ^{61}Ni Mössbauer, Czjzek *et al.* 1976; ^{57}Fe Mössbauer, Nishihara *et al.* 1977; ^{61}Ni n.m.r., Kitaoka *et al.* 1982). The latter experiments indicate no gross change in the geometry of the M1 ordering, as also does the neutron work.

The AFM regime is, as already apparent, attained also with Co and Cu doping $\gtrsim 10\%$ (magnetically oriented references to such work may readily be traced from sections D and E in Table 1). Special attention is drawn to the comprehensive set of results in Ogawa *et al.* (1974). The calorimetric data from

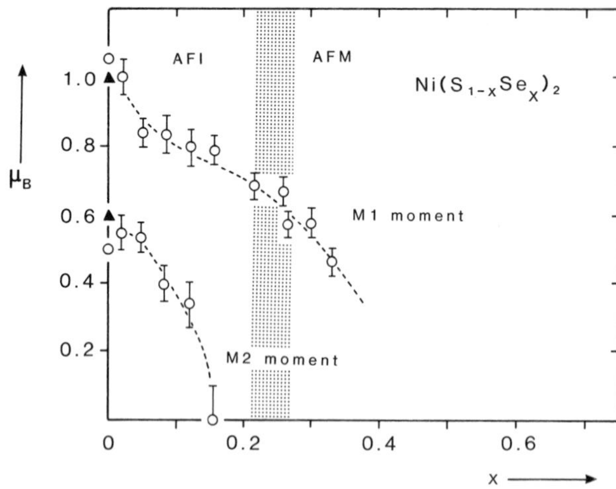

Figure 19. Neutron diffraction results for the system $Ni(S/Se)_2$ showing loss of $\bar{\mu}$ for M2 and M1 orderings. M1 order passes smoothly through the metal–insulator transition composition at ~25% Se at low temperatures. (from Miyadai *et al.* 1983b)

Ogawa (1976), shown in Figure 20, affirm that some form of magnetic order exists up to large x values in the mixed cation systems, with $T_0 > 100$ K. Evident also from the calorimetry (and confirmed by ^{57}Co n.m.r.; Yasuoka, unpublished) is that the cobalt in such a mixed system based on NiS_2 rather surprisingly carries no moment. Hence it behaves magnetically as a rather simple diluent, much as low-spin Fe^{2+} does in the Mössbauer work, though Nishihara *et al.* (1977) did point to evidence of cluster formation in their Mössbauer study on $Fe_{0\cdot5}Ni_{0\cdot5}S_2$. The chemical effect of the neighbourhood was noted earlier when discussing the mixed anion system $Ni(S/As)_2$.

$Ni(S/Se)_2$ clearly does not escape this problem, but its effects are minimized as far as possible (evident in that it now takes 15%, not 5% doping to destroy M2 ordering). Local environment effects were not apparent from the $\frac{1}{2}$% ^{57}Fe Mössbauer work of Nishihara *et al.* 1975a,b, 1979, which relies on picking up a supertransferred hyperfine field from the Ni atoms of the host. However, we have seen that direct ^{61}Ni Mössbauer work yields a signal that appears to be from two types of site, α and β. Krill *et al.* (1976b, Figure 4) analysed their results binomially in the manner introduced by Jaccarino and Walker for Nb–Fe to show that β sites, the magnetic sites, hold a moment by virtue of no more than one of their six nn chalcogens being a Se. All Ni sites with a count of two or more Se in their coordination unit are deemed non-magnetic. Employing this finding, it was then demonstrated that the falling average moment $\bar{\mu}_1$ detected by neutron diffraction derives from β sites upholding a constant moment of $1\cdot2\ \mu_B$ at all x. More recently, Kitaoka *et al.* (1982), using ^{61}Ni n.m.r., have indicated that the situation is probably somewhat more complicated, with only Ni sites possessing five or six Se nn atoms being completely non-magnetic. The neutron data are not sufficiently

Table 1. Papers on NiS_2-based systems arranged by date of submission and topics.

	Paper	Origin	t c/p	ρ, S, R_H	EM a_0, α C_V L	χ, M, N n.m.r., Mossb.	IR Raman Opt. X-ray	Pages
NiS_2								
	S 84	TJ	c				O	4
	SITSSST 83	TJ	c				O	9
	STSSSSY 83	TJ	c				OX	3
	B 82	E	t					12
	YI 81	TJ	t			M		10
	HHBH 80	E	c660				X	9
	KMI 80	SJ	c			M		3
	PKVM 79	F	c			N(hp)		4
	YF 79	J	t			M		2
	K 79	SJ	c			M(F)		7
	F 78	SJ	c	R_H(F)				8
	MKI 78	SJ	pc			N(F)		3
	KMIF 78	SJ	c			N(F)		6
	KMFIT 78	SJ	c			M(F), N		6
	NOW 78	TJ	p			Mossb.(Fe)		5
	MKI 77	SJ	c			χ, N(F)		2
	SUSM 77	TJ	c				R	6
	MW 77	O	p600			Mossb.		5
	NIM 76	SJ	c		α			2
	NM 76	SJ				M(F)		
	VE 76	E	c(hp)		EM			4
	VE 76	E	c(hp)		EM			8
	SSKG 76	TJ	c?				X	5
	WH 76	O	p600			Mossb.		2
	KLGRCFS 76	F	pc	ρ	C_V, Stoich.	χ, Mossb.		22
	NOW 75	TJ	p740			Mossb.		7
	CFSKGLR 74	F	p			Mossb.		3
	MTNIMH 75	SJ	c			N		7
	ASS 74	NJ	c			M(F)		7
	OYW 74	TJ	c?				X	1
	LJEF 74	U	c				X	3
	LW 74	E	p				IR	7
	GKLR 73	F	p(hp)		Stoich.	χ, M		4
	W 73	U	t					17
	MMTU 73	SJ	p			χ, M		2
	GKLR 73	F	pc(hp)	ρ(p)	Stoich.	χ		3
	KDAL 72	U	c	ρ			O	5
	ASYO 72	NJ	p			χ, M		1
	G 72	U	t					9
	W 72	E	t					55
	WP 71	E	cp(700)	ρ(P)				14
	G 71	U	t					13
	KN 71	E	p		a_0			11
	BK 70	E	p		a_0			20
	HC 70	U	p			N		2
	FKA 69	E	p		str.	χ		10
	AST 69	NJ	p, 650			χ, N, n.m.r.		8
	JCBBFG 68	U	c, p	ρ		χ		4
	B 68	U	p650	ρ	a_0	χ		8
	BBCDS 68	U	cp(hp)	ρ	a_0	χ	O	13
	H 68	E	p	ρ	review str.			147
$NiSe_2$ and $Ni(S/Se)_2$								
	MST 83	SJ	p			M(F)		5
	KYO 82	TJ	p			n.m.r.		2
	MSTMM 83	SJ	p700			M, N, χ		2
	TO 81	TJ	t					2
	WKO 82	TJ	p		α	χ		3
	LGSC 80	O	c				R	8
	MCA 80	TJ	t					4
	IYO 80	TJ	p		C_V	χ, n.m.r.		7
	KA 80	F	p				X	8
	KDA 80	U	c	ρ, S, R_H				8
	HHBH 80	E	pc				X	9
	NO 79	TJ	p	ρ		Mossb.		7
	O 79	TJ	p	ρ	γ	χ		4

Table 1—*contd.*

Paper	Origin	c/p	ρ, S, R_H	EM a_0, α C_V L	χ, M, N n.m.r., Mossb.	IR Raman Opt. X-ray	Pages
KMM 77	SJ	p700	ρ(p)				5
MW 77	O	p550			Mossb.		5
CFSKLPGR 76	F	p			n.m.r.		3
KPLGRCFSK 76	F	pc			χ, N, n.m.r., Mossb.		5
PK 75	F	p			N		2
GKLPRCFS 75	F	p600		a_0	χ, N, n.m.r., Mossb.		3
JBGJMW 73	U	c750	ρ(p)	a_0, DTA	χ		4
BGJ 73	U	c750	ρ(p)				8
WP 71	E	c, p700	ρ(p)				14
FKA 69	E	p		a_0, α	s		10
NOAT 77	TJ	p			Mossb.		4
AUTS 76	NJ	c(hp)	ρ		χ		2
CB 73	U	c660			e.p.r.		8
BDCBY 70	U	hp	ρ		χ		8
$(Ni/Co)S_2$							
MST 83	SJ	p700		M(F)			4
NO 80	TJ	p			Mossb.		7
SC 80	E	t					11
YMEAD 79	U	c	ρ, S, R_H				4
O 79	TJ	p			M		4
MYEDA 80	U	c	ρ, S, R_H			O	14
O 77	TJ	p		C_V			2
MYEDA 77	U	c	ρ, S				4
KO 76	TJ	p		α			2
O 76	TJ	p		C_V			8
MMY 74	SJ	p	ρ(p)				5
KLGRCFS 76	F	p	ρ(p)		χ, Mossb.		22
YMMO 74	SJ	p	ρ(p)				1
OWT 74	TJ	pc	ρ	a_0	χ, M		12
JCBBFG 68	U	c, p	ρ		χ		4
B 68	U	p, 650		a_0			8
$(Ni/Cu)S_2$							
MST 83	SJ	p700			M(F)		4
KMDA 81	U	c	ρ, S, R_H				6
MMY 74	SJ	p	ρ(p)				5
BHO 74	TJ	p			N		3
YMMO 74	SJ	p	ρ(p)				1
OWT 74	TJ	pc compacts	ρ		χ, M		12
$Ni(S/As)_2$							
AK 77	NJ	p	ρ		χ, M		7
ATK 76	NJ	p725	ρ		χ		5
Pressure experiments							
MT 83	SJ	p	(Se)	α			2
PKVM 79	F	c		α	N		4
MW 78	SJ	p		α			3
KMM 77	SJ	p700	ρ(Se)				5
WMM 76	SJ	p	ρ(Co)				3
KLGRCFS 76	F	pc	ρ(Stoich.) (Cu)				22
MMY 74	SJ	p	ρ(Co, Cu)				5
MMYO 74	SJ	p	ρ(Co)				4
EMM 73	SJ	c		a_0			2
MMY 73	SJ	p	ρ				3
YMMO 73	SJ	p	ρ Co Cu				1
GKLR 73	F	pc(hp)	ρ				3
WP 71	E	pc	ρ(Se)				14

Key:
J = Japan. NJ = Nagoya. SJ = Sapporo.
TJ = Tokyo. F = France.
E = Other Europe. U = USA.
O = Other. p = Powder or sinter.
c = Crystals. 650, etc. = growth temp.
hp = High pressure synthesis.
t = Non-experimental paper.
F = Ferromagnetic phase.
M = Magnetization.
N = Neutron work.
EM = Electron microscopy.
str = Structural.
Stoich. = Study of stoichiometry.

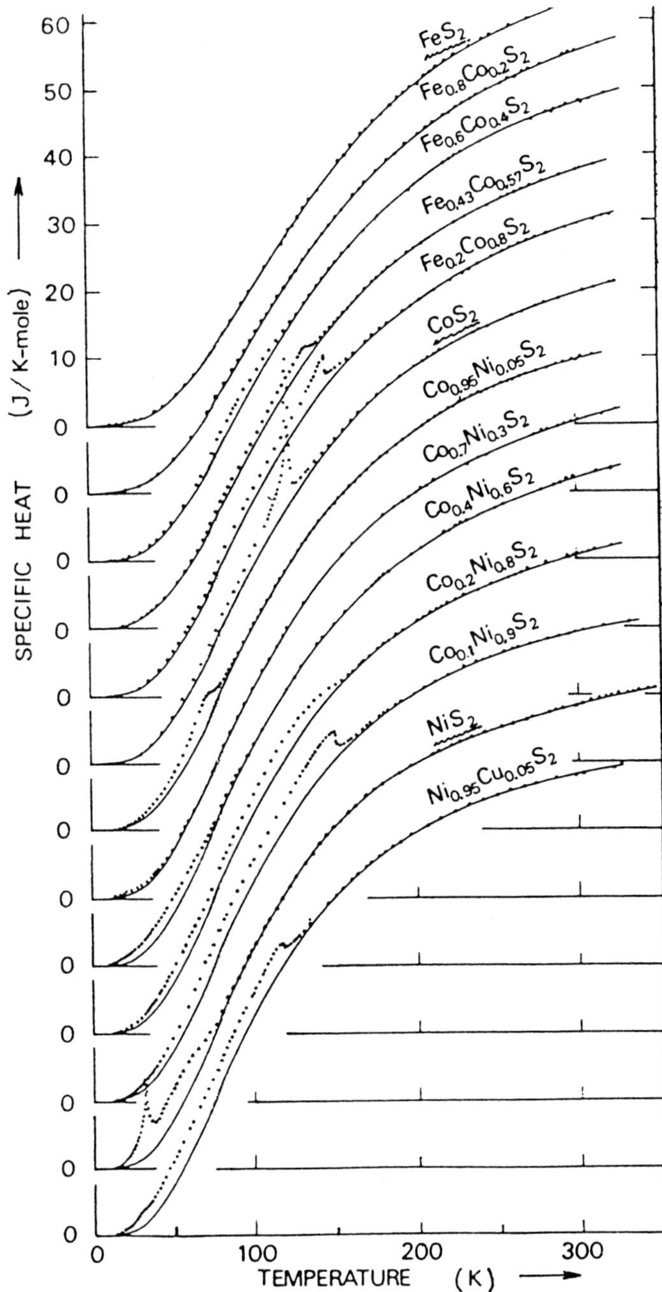

Figure 20. Calorimetric data from Ogawa (1976). Solid lines indicate calculated electronic and phonon contributed background. Their subtraction reveals magnetic contribution to data.

accurate to substantiate this. One possible problem here is structural inhomogeneity noted by Jarrett *et al.* (1973) for samples with x close to the M–NM boundary where there is abnormal X-ray line broadening.

2.4. Enhancement effects in the AFM phase

As seen in Figure 8, $NiSe_2$ is a Pauli paramagnet and unlike $CoSe_2$, with which comparison is made in Waki *et al.* (1982) and Inoue *et al.* (1979), it shows very little Stoner enhancement of χ (multiplicative factor $f_s = 2$ for $NiSe_2$ vs 8 for $CoSe_2$. If $NiSe_2$ does display any feature of interest it is the knee in the Knight shift and possible small peak in $1/T_1$ at 30 K (Inoue *et al.* 1979, Figure 4). This is reminiscent of a CDW. Both $CuSe_2$ (Krill *et al.* 1976c) and $CoSe_2$ (Panissod *et al.* 1976) are known to have some form of structural instability.

Returning to Figure 8 we see that χ rises strongly as sulphur is added to $NiSe_2$, being doubled by the 50/50 composition at which low-temperature ordering first emerges. For relevant theoretical discussion see Moriya (1979) and Chao & Berggren (1977).

More striking still than this is the observation made by Ogawa (1979) that the electronic specific heat also doubles over the above composition range. Compared with what might be expected from Chao and Berggren's work (1977) this is a very large enhancement, for which paramagnon scattering is to be held responsible.

Strong evidence for spin-fluctuation effects within the metallic range of x also comes from resistivity measurements. As we originally noted (Wilson & Pitt 1971) the resistivity of a sample with $x = 0{\cdot}4$ closely follows a T^2 variation from 10 K up to 250 K just below the metal–insulator transition.

Mori *et al.* (1975) likewise found T^2 behaviour for pure NiS_2 when just taken into the AFM range by the application of 36 kbar pressure. More recently, work by Kamada *et al.* (1977) and by Ogawa (1979) has shown that the coefficient of the non-linear low-temperature term in $\rho - \rho_0$ appears to diverge where $\bar{\mu}_1$ vanishes—near NiSSe at $p = 0$ (though the relevant temperature range diminishes). This is in line with the SCR theory presented by Ueda (1977) (see also Moriya 1979, §2.5). It is difficult from pressure measurements to confirm that the temperature index becomes 1·5, but there clearly is some departure from T^2.

Watanabe *et al.* (1976), using a combination of Co doping and pressure ($x = 0{\cdot}07$ and p from 20 to 80 kbar), again discovered a divergence in the 'T^2' scattering within the metallic regime, appearing for the given doping at 35 kbar. It obviously would be of much interest to know under what (considerably higher) pressure the AFM phase is eliminated for pure NiS_2. From Figure 17 the Se equivalence rate suggests $p_c \sim 80$ kbar.

2.5. Lattice effects at the metal–insulator transition

Bouchard (1968) first drew attention to the 0·45% excess in the lattice parameter of NiS_2 over the value to be calculated by extrapolating from Co and

Cu doped samples set well within their respective metallic regimes at $x>0{\cdot}25$. Note the base-line here issues only from compositions well beyond those at which the activation energy for conduction (at 300 K) has already gone to zero ($x\sim0{\cdot}1$; see Ogawa *et al.* 1974).

The above Δa_0 of about $\frac{1}{2}\%$ in a_0 (partially evident also in Figure 13) is an order of magnitude greater than for the magnetostrictive change occurring in pure NiS_2 around T_N (Nagata *et al.* 1976).

The above 'excess' of $\sim1{\cdot}5\%$ in the volume of NiS_2 at ambient conditions is the total volume needing to be lost to reach the metallic state starting from ambient NiS_2. Now the 300 K compressibility data from Endo *et al.* (1973), shown in Figure 21, if combined with our insulator–metal transition pressure of 35 kbar at 300 K (Wilson & Pitt 1971), will yield a volume jump there of only about $-0{\cdot}6\%$. This is just the discontinuity recorded in fact by Jarrett *et al.* (1973) for a 25% Se doped sample when cooled around 70 K (at $p=0$) through its $I\rightarrow M$ transition. As is to be seen from Figure 21 the threefold difference between the above two ΔV values arises from the fact that the compressibility of the metallic phase is only three quarters that of the insulating phase.

It has sometimes been questioned as to whether by 300 K we are not above the critical point terminating a first-order Mott line. However, the 300 K Mössbauer data from Nishihara & Ogawa (1979, Figure 9) still show a sharp

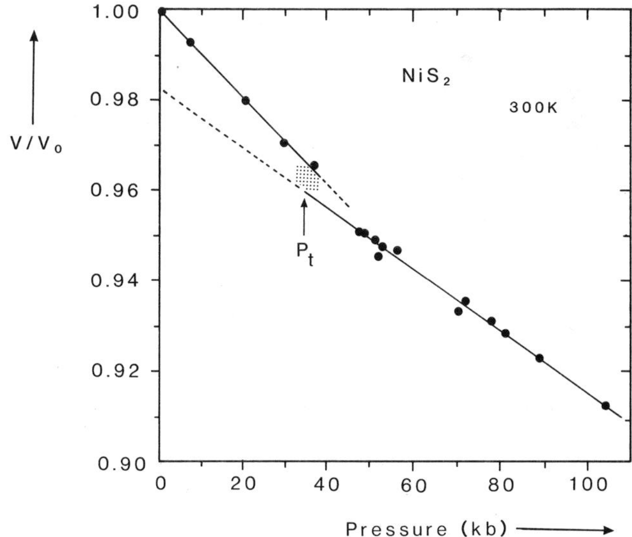

Figure 21. Room temperature compressibility data for NiS_2 from Endo *et al.* (1973), imdicating break around 35 kbar where resistivity data show rapid change (see Figure 14). Volume difference $\sim0{\cdot}6\%$ between phases at this pressure.

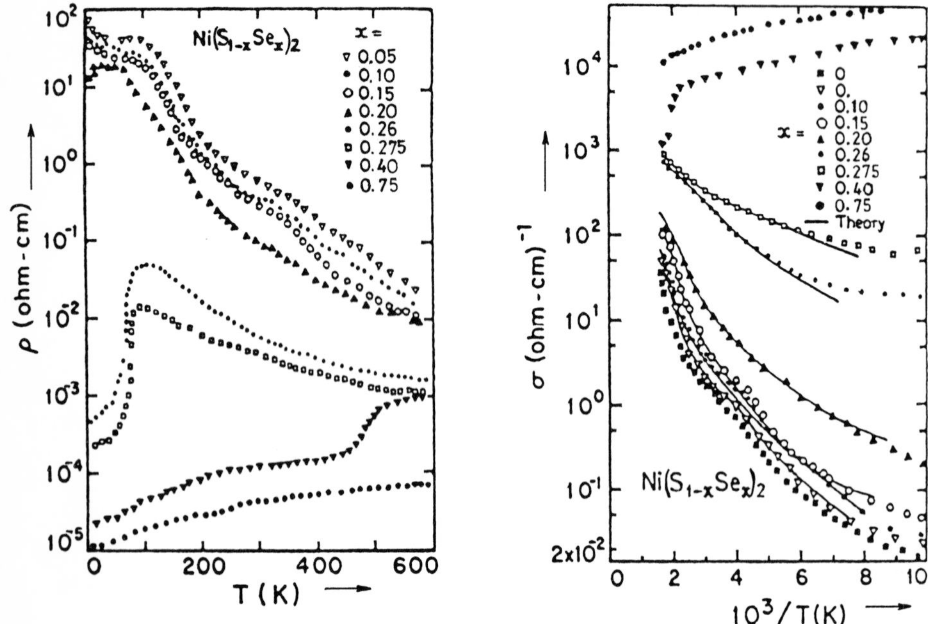

Figure 22. (*a*) Resistivity data from Kwizera *et al.* (1980) for $Ni(S/Se)_2$, showing the metal→insulator transition for samples with about 27% Se at around 100 K. (*b*) Same data in σ vs. $1/T$ plot (down to 100 K only, note). Solid curves are theoretical fits (see §2.6). Clearly data from samples with 23% and 34% Se should be most interesting.

break in quadrupole splitting at the 30% Se doping level appropriate to a room temperature transition. The change in quadrupole splitting was shown to be governed largely by the lattice change. Raman work is apparently not sensitive enough to pick up a discontinuity at the insulator–metal transition. Present spectra from $Ni(S/Se)_2$ show a lot of disorder broadening (Lemos *et al.* 1980). Also it ought to be noted that the X–X bond-breathing mode (A_{ig}) shows two-mode behaviour from S–S and S–Se pairs. A $27\frac{1}{2}$% Se sample was reported to give no spectral sharpening on cooling, and moreover to give no apparent change on passing the M—I transition temperature of (for a bulk sample) about 100 K. A large transition around this temperature *is* present in the dilatometry of Mori & Takahashi (1983), and it clearly would be very illuminating to follow the change to higher temperatures through samples of slightly greater Se concentration. From resistivity measurements, we reported T_M at 180 K for $x = 30\%$ (Wilson & Pitt 1971), and Kwizera *et al.* (1980) (unlike Bouchard *et al.* 1973) indicate some residual action up to 500 K in a 40% Se sample (see Figure 22). It now is time to look at transport measurements in detail.

2.6. Electrical transport measurements near the Mott line

Pure NiS_2: pressure results

The strong effects resulting from the disorder of mixed crystals (whether mixed cation or mixed anion) direct one to high-pressure work. Unfortunately, as we have seen, the pressures necessary to delocalize NiS_2 ($\sim$35 kbar at 300 K; $\sim$30 kbar at 0 K) are too high for the application of truly hydrostatic pressures. It is not clear, then, whether the breadth of transition obtained in our own resistivity work (Wilson & Pitt 1971a) or in that by Krill *et al.* (1976a) (see Figure 9) is to be ascribed to this cause. Perhaps at 300 K we are indeed above the critical point. An extensive set of measurements on NiS_2 using low temperatures in conjunction with high pressures is available from Mori *et al.* (1974) but, alas, only on compacted powder. The result is that with pure NiS_2 the transition has never been seen as sharply as in selenium-substituted material (Figure 22(*a*)). No high-pressure optical work on NiS_2 has been attempted.

The semiconductor-like edge apparent in NiS_2 under ambient conditions in the infrared at 0·4 eV decreases significantly upon warming, and also upon doping with Co (see Mabatah *et al.* 1980, Figure 14). Figure 6 from Wilson (1973) displays the assessed location and interpretation of this gap in the e_g band. This gap opened by many-body electron–electron interaction leads to a semi-conductive phase with significantly different transport properties to those of a standard d-band semiconductor like MoS_2 or t_{2g}^6 $FeS_2/Se_2/Te_2$. The latter materials are Van Vleck paramagnets, unlike Curie–Weiss paramagnetic NiS_2, and may be doped n-type or p-type with mobilities at low temperatures well in excess of 100 $cm^2\ V^{-1}\ s^{-1}$ (Echarri & Sanchez 1974). NiS_2 is, by contrast, always p-type even though the Ni:S content ratio is $>\frac{1}{2}$, and moreover the mobility of the carriers is so low that it has not yet led to a detectable Hall constant, whether in pure or Se-doped NiS_2 (Kwizera *et al.* 1980), or in Co-doped (Mabatah *et al.* 1980) or Cu-doped (Kwizera *et al.* 1981) material. This puts the value of μ_H at well below 1 $cm^2\ V^{-1}\ s^{-1}$ within the activated and small-polaron range.

Warming leads to a widening of the effective band e_g bandwidths in NiS_2, bringing a reduction of about 50% in the gap by 500 K, and dropping the Seebeck coefficient to a much lower (and negative) value. This change is not, however, to be confused with the splitting of the α and β bands going to zero. Indeed, as is to be seen in the phase diagram of Figure 7, increasing the temperature heads one away from the Mott transition for NiS_2, not towards it. Heating NiS_2 also leads around 400 K to the saturation point of amplitude fluctuations for the paramagnetic moment, as first noted by Furuseth *et al.* (1969) and Adachi *et al.* (1969), and explained by Moriya (1979). Furuseth *et al.* (1969) note that the expansion coefficient changes sharply at this point, and the resistivity and Seebeck coefficient appear to respond too, as we shall see.

It will be very interesting to find what happens to the optical gap at T_N. It may well change by no more than 0·1 eV, though as we have seen in §2.3, the ordered moment is oddly reduced in the antiferromagnetic phase. Normally the position of the 'spin-polarized' bands is little altered by the onset of long-range order, being largely determined by on-site interaction, as discussed in Wilson (1972, 1973) and illustrated by the case of $NaCrS_2$ (t_{2g}^3 Mott insulator). There the $p \rightarrow t_{2g}^{\beta}$ edge changes by only +0·02 eV at T_N (18 K) (Blazey & Rohrer 1969). I will delay further discussion of my construction of the site-spin polarized band structures of Figure 6 until §3, except to say that they were drawn up in the 1973 paper in close comparison with the optical results for the 3d pyrite sulphide family obtained by Bither *et al.* (1968), and that recent optical work for example, by Sato (1984), or UPS and XPS (ultraviolet/X-ray photoemission spectra) work, for example, by van der Heide *et al.* (1980), or 'simple' band-structural calculations (Bullett 1982) call for no significant modification to the scheme.

In transport measurements the activation energy detected at high temperatures in NiS_2 is around 0·3 eV for good quality material, making this somewhat less than for the optical gap, as if diminished by the p-type defect structure of NiS_2. The defect character of the compound prevents ρ rising to very high values at low temperatures, a limiting value of around 10^2 Ω cm being reached by 100 K (see Figure 22). At low temperatures there is evidence of $\exp(T^{-1/4})$ behaviour (Krill *et al.* 1976a) as in other electronically disturbed narrow-band systems (such as ICF behaviour, see Figure 2, or disordered CDWs).

Under pressure, the high-temperature activation energy in NiS_2 extrapolates to zero around 40 kbar at 300 K (Mori *et al.* 1974) as the α and β bands are overlapped. Perhaps supporting the idea that there is no first-order collapse into this metallic condition at 300 K is the fact that the above decrease in E_g appears remarkably linear with pressure (although remember these results are on powder compacts).

Doping with Co or Cu rapidly eliminates the temperature range for which a large activation energy is seen, the typical activation energy being 0·1 eV or less by 5% doping. The changes are somewhat more rapid for Cu than for Co doping. These small activation energies extrapolate to zero for pressures progressively smaller with x than the above 40 kbar. The MIT group have attempted a detailed modelling of the various small activation energies and the effect these have on their extensive resistivity and Seebeck coefficient data. This work will now be examined.

Se, Co and Cu doped results and the MIT analysis

The principal data upon which the MIT analysis is founded are the resistivity and Seebeck coefficient measurements (to 600 K) which are shown in Figures 22 to 27. The conductivity results for Co and Cu doping appear rather similar, as likewise do the Seebeck results for Co and Se doping (though at very different x). Note that: (*a*) only for Se doping are those discontinuities present which we associate with a classic Mott transition (see Figure 22(*a*), $x = 27\frac{1}{2}\%$,

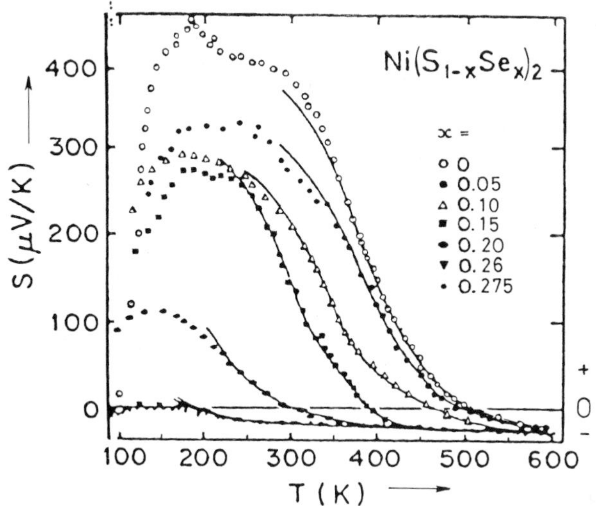

Figure 23. Seebeck coefficient for $Ni(S/Se)_2$ from Kwizera *et al.* (1980). Note lines of theoretical fit use same parameters as those in Figure 22, but do not cover the low-temperature regime.

$T \simeq 100$ K); (*b*) only for Cu doping does the Seebeck coefficient fail to convert through to small negative values at high *T*.

The solid curves in all these figures are generated with a uniform mode of modelling, with very limited adjustment between the two types of data. Although the fitting is not achieved over the entire temperature range, it is impressive. The model incorporates thermal activation of extrinsic and intrinsic carriers, an activated hopping mobility and the thermal closure under screening of the α/β separation (see below).

Points where the success in data-matching is not complete are, for example, (*a*) with $Ni_{0\cdot9}Co_{0\cdot1}S_2$ at very high or very low temperatures (see Figures 5 and 8 in Mabatah *et al.* 1980); (*b*) for pure NiS_2 at 'low' temperatures (see Figures 4 and 7 in Mabatah *et al.* 1980); (*c*) for $Ni(S/Se)_2$ in that most interesting range below 120 K (Figures 22 and 23). Clearly though, much of the theoretical fitting must be close to the correct functional form: this, I believe, disguises the fact that the interpretation offered is fundamentally in error.

The 'sway-back' with *T* that is apparent in the conductivity data for 5% Co doping is to be found also in the data of Ogawa *et al.* (1974) and Mori *et al.* (1974). As suggested by Mabatah *et al.* (1980, Figure 11) this doubtless comes from activated hopping, followed by carrier excitation, followed by termination of the latter and renewed hopping. What is in question is the identification of some of the empirical activation energies and in particular the largest. Mabatah *et al.* (1980) have tried to proceed with an extreme narrow-band approach (seemingly polaron induced) and a Hubbard *U* which is then effectively given by the α/β gap.

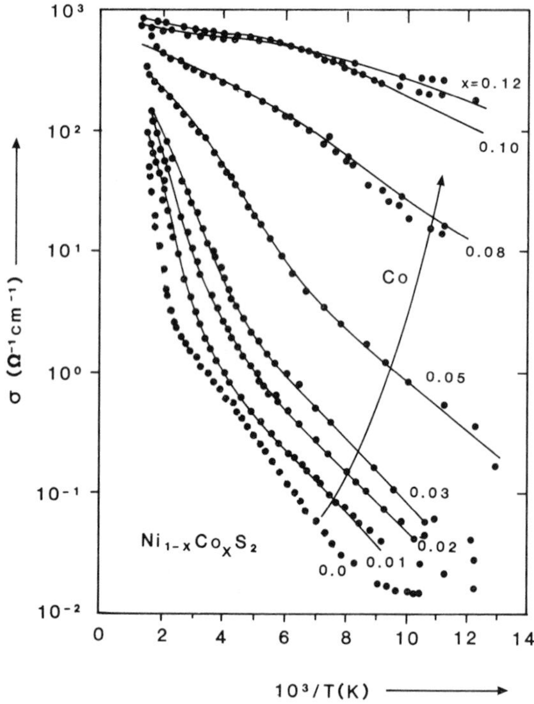

Figure 24. σ vs. $1/T$ plots for (Ni/Co)S_2 from Mabatah *et al.* (1980), including theoretical fits. There is no sharp insulator–metal transition.

U (plus E_g) is, it is claimed, brought to zero by combinations of doping and temperature increase. The course of analysis is dictated by a desire to see bandwidths held to extremely low values in Mott insulators (Adler 1978), and for carrier screening to carry U to *very* low values ($\ll 1$ eV) as discussed in great detail in Yoffa & Adler (1979). The term 'collapsed band' is used throughout to imply U is effectively brought to zero, not that the band gap between the α and β bands has been eliminated.

To the present author the above seem to be quite false premises, for is it not a finite U which continues to give to metallic CoS_2 its magnetic properties (as likewise for the AFM phase of $Ni(S/Se)_2$)? Also do not the optical spectra and even the photoemission results demonstrate bandwidths quite in keeping with Figure 6 and with the subsequent calculations of Bullett (1982)? Currently we are attempting to obtain dHvA (de Haas van Alphen) data from CoS_2 and $NiSe_2$, but one may already point to such data from metallic CuS_2 (Marcus & Bither 1970) or $NbSe_2$ or RuO_2 where the spectra are comparable. Indeed, even for a hard Mott insulator like NiO or CoO we do not believe that the appearance there of collective d^n set, 'ligand field' excitations is of necessity incompatible with a finite

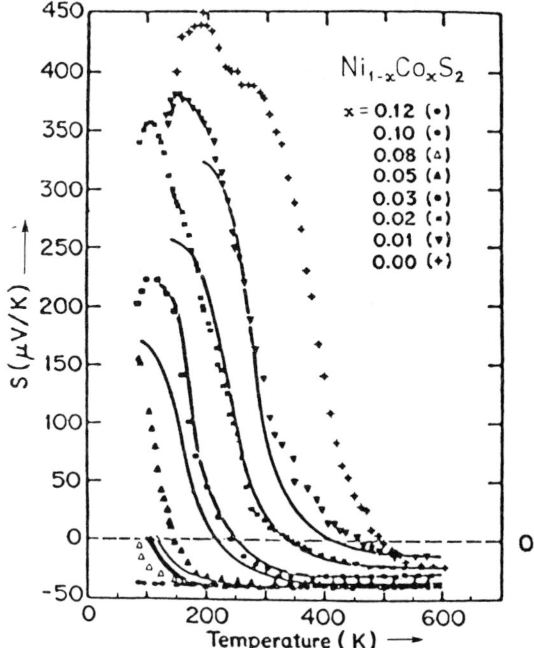

Figure 25. Seebeck data corresponding to Figure 24. Note coefficient negative and constant at high temperatures.

bandwidth, single-particle-like ground state of appropriately concocted form. I shall return to this in §3.

2.7. *Historical note on the $Ni(S/Se)_2$ system*

Hulliger's efforts culminating in the excellent review of 1968 set the stage for work to develop on $Ni(S/Se)_2$. The Oslo group with Kjekshus, the Osaka group under Adachi, and especially the du Pont group of Bither, Bouchard, Jarrett and co-workers immediately drove the 3d pyrite family to the fore. 1968 was the year too of my own PhD thesis, and the subsequent writing of the *Advance in Physics* article concerning the physical properties of the transition metal dichalcogenides as a whole.

It was in 1968 also that specifically metal–insulator work really accelerated, with, for example, Goodenough's paper on the perovskite systematics given at the Magnetism Conference, articles by Halperin & Rice and by Adler in *Solid State Physics*, and in particular the impact of the October San Francisco Conference (see *Review of Modern Physics*, **40**) with its papers by Mott, Adler and

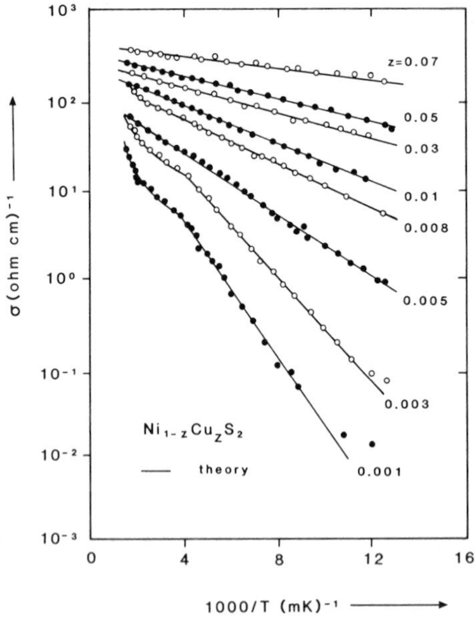

Figure 26. σ vs. $1/T$ plots for $(Ni/Cu)S_2$ from Kwizera *et al.* (1981). Note metallization occurs more rapidly than for Co doping.

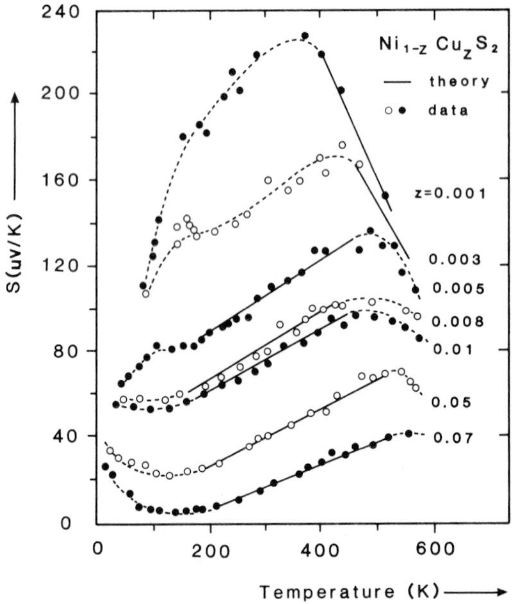

Figure 27. Seebeck data corresponding to Figure 26. Note coefficient positive at all temperatures.

Halperin & Rice. Without doubt the high point experimentally was provided by the work of McWhan, Rice and Remeika on the transition in the V_2O_3:Cr system published in *Physical Review Letters* in 1969.

Our subsequent attempt at reproducing the V_2O_3 pressure work for $Ni(S/Se)_2$ in 1970 (Wilson & Pitt 1971) proved it too to be a system of special note. After the synthesis of my 1972/3 papers, I dropped this topic to concentrate on CDWs. With Cyrot's theoretical work, and the French connections of Professor Mott and Professor Goodenough, from the CNRS conference at Aussois in 1971 to Autrans in 1976, the pyrite work was firmly taken up in France by Gautier, Krill, Panissod and others in parallel with related work on NiS, the vanadium oxides, the perovskites, and so on, at Grenoble and Bordeaux.

Meanwhile in Japan, under Adachi's and subsequently Ogawa's influence, the pyrite materials became steadily assimilated into the wide-ranging developments on magnetism by Moriya and collaborators (Moriya 1982, Moriya & Hasegawa 1980, Moriya & Takahashi 1984).

Following the unfortunate demise of the du Pont effort, and a comparable falling away of interest in the M–I topic at Bell Laboratories as elsewhere, and with CDW, ICF, ARPES, EXAFS, LDS and energy developments, NiS_2 work did not move forward again in the USA until adopted by the MIT group from 1977 to 1980. Today the pyrites have become almost uniquely the preserve of the various very proficient Japanese schools, as the present paper records.

A move is clearly now underway for band theorists to enter the metal–insulator field, bringing a revival of European input to the topic. Already we have first calculations on the pyrites from Bullett (1982) and Temmerman (unpublished). Briefly, then, we turn to see how the band-structure situation lies for narrow-band materials.

3. Energy band-structures for highly correlated materials

The Mott transition as such is an event of very rare occurrence. In contrast, Mott insulators constitute half of all binary compounds, and these are complemented by the numerous magnetic metals just to the other side of the localization/delocalization divide (see Figure 4). Hence it is a matter of much concern as to how the eigen states of these highly correlated materials are properly to be represented.

The question posed now is how much further is it possible to go beyond the local density approximation, $X\alpha$-type exchange and correlation, and the present 'simple' spin-polarized band-structure calculations, and yet reach meaningful and experimentally relevant results. Clearly, calculations on, say, ferromagnetic iron are already remarkably successful in yielding the correct ground-state properties (lattice parameter, magnetic moment or spin-split Fermi surface geometry as probed by dHvA techniques, and so on) (Hasegawa 1983, Gyorffy *et al.* 1984, Lonzarich 1984). Also there has been considerable success in accommodating the excitations of angular photoemission data (Gerhardt 1983). Whether, however, this can be done for ferromagnetic d^2CrO_2, or f^2d^2US, or d^3KCrS_2 or CrI_3 is a moot point. Even if the calculations can satisfactorily be brought to self-consistency, can a calculated Fermi surface to match experiment ever be obtained for metallic CrO_2, or even the moment for insulating MnO_2? With materials like US, as for SmS, it is well known that the act of f-electron photoemission abstracts the excited region from the band-structure milieu, under relaxation corrections greater than the relevant bandwidths. Multiplet structure results which is characteristic of the final-state electron count, not the initial (Campagna *et al.* 1976). Is then a band-structure calculation for, say, NiO, if not NiS_2, doomed to tell one little or nothing about the measurable d-state properties of the material? Initial attempts at producing quite sophisticated band structure do already in fact exist for NiO, from Mattheiss (1972), through Anderson *et al.* (1979) and Kunz (1981), to Terakura *et al.* (1984), but are these at all pertinent to a partially unquenched orbital magnetic moment, let alone to satisfactory description of the $p \rightarrow d$ charge-transfer spectra? The clear excitonic character of the latter type of spectra (see Figure 28) points to severe many-body reorganization in excitation. Indeed one has to look no further than silicon to find that adequate prescription of the transition energies and the detailed form of the spectrum (Hanke *et al.* 1984, Hybertson & Louie 1984) has a complexity withheld from our view ten years ago.

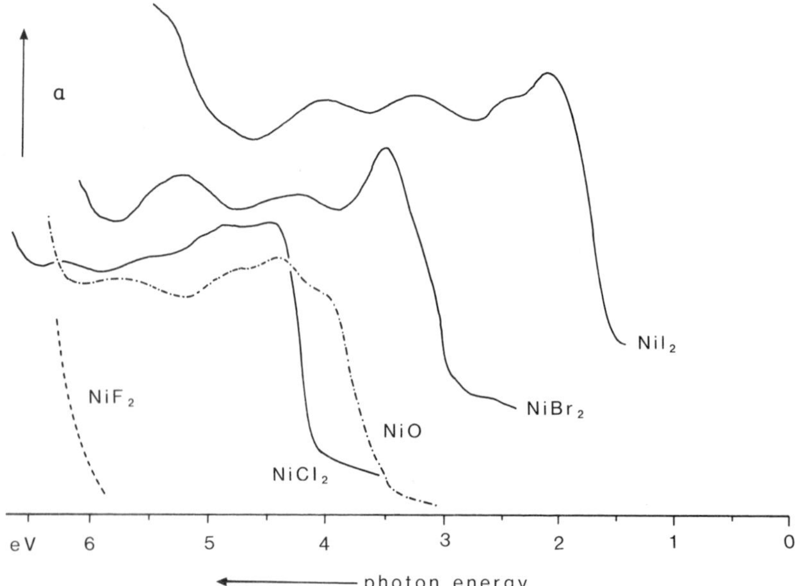

Figure 28. $p \rightarrow d$ charge transfer absorption spectra for various Ni^{2+} materials (Wilson 1970, Exeter IOP conf., unpublished). Excitonic components clearly substantial. Absorption coefficients $\sim 10^5\ cm^{-1}$.

It is my faith, based on extensive empiricism and expressed in Figure 6, that band-structure work properly 'concocted' will be able to reach meaningful and useful description up to the Mott transition of materials such as CoS_2 and CrO_2, even NiS_2 and MnO_2: but how much further? $CoPS_3$? CoI_2? CoO? The fact that $CoPS_3$ or CoI_2 show ligand field excitations (Figure 1) where the seven d electrons are excited as a coupled set does not necessarily preclude the description by the band structure of the $p \rightarrow d$ 'charge-transfer' transitions. For low-spin $t_{2g}^6 FeS_2$ the spectrum clearly still shows (as for PtS_2) $t_{2g} \rightarrow e_g$ band-to-band transitions of high intensity (Bither *et al.* 1968, Sato 1984): but will this be so with low-spin $t_{2g}^6 IrCl_3$?

One will know one is getting close to answering problems like this when one can show that FeS_2 is a low-spin semiconductor while FeS is a high-spin metal: or again that $CoPS_3$ and CoI_2 are high-spin insulators while CoS_2 is a low-spin metal. And when one finds $CoPS_3$, or CoI_2, or CoO or CoF_2, high-spin and insulating, will that insulating condition for all these various crystal structures really be expressed in the calculated 'band structure' (above as below some magnetic ordering point) by a gap at the Fermi level in the density of states? Does nature always actually provide routes, such as orbital superlattices (Spalek & Chao 1980), to secure this gapping, absent in more simply derived quasi-conventional band structures?

Whether this is reality or not for CoO (and my feeling is that it is not contrary to what has recently been anticipated by Terakura *et al.* (1984)), it is surely to be recognized that the insulating character of CoO (as indeed that of NiO and NiS_2) is the product of the advanced action of exchange and correlation, for which the appelation 'Mott insulator' is far more informative in regard to transport and magnetic characteristics than any extension to these of the label 'semiconductor'. Mott insulators are (largely) magnetic insulators and magnetic insulators are Mott insulators, by virtue of advanced exchange and correlation in narrow bands: bands which are made narrow by the energetic and spatial characteristics of the wavefunctions involved, and by the crystal geometry in which these are set and the possibilities for hybridization with states of greater delocalization.

Before closing, the above hinted proviso should be expressed regarding total identification of Mott insulators with magnetic insulators. It would be rather silly, surely, to exclude a material like $f^6 SmF_2$ because its J values happens to be zero; similarly $t_{2g}^6 IrCl_3$ is so akin to $t_{2g}^5 \alpha\text{-}RuCl_3$ (Rojas & Spinolo 1983) in its general character that it is clearly of the same dielectric type though non-magnetic. Alternatively, $\beta\text{-}RuCl_3$ and $\beta\text{-}RuBr_3$ (as $\beta\text{-}TiCl_3$) are non-magnetic through cation interaction, and once more are clearly not standard semi-conductors. $\beta\text{-}TiCl_3$ presents the same unique features in its 300 K optical spectrum that magnetic $\alpha\text{-}TiCl_3$ does and which would seem to monitor the charge fluctuations $d^1 + d^1 \rightarrow d^0 + d^2$ being promoted here optically prior to delocalization (Wilson 1972).

Undoubtedly ambiguities are going to occur, particularly away from the 3d series where M–M bonding readily arises as an alternative to antiferromagnetic order. Is PtI_2 with its Pt_6 clusters a Mott insulator like $PdCl_2$ or a semiconductor like PtO? Is either $ZrCl_2$ or ZrI_2, or both or neither, a dopable semiconductor like isoelectronic, isostructural 2H–MoS_2? A semantic problem and more indeed exists to be encountered, but it is not one to inject into the safe territory of the 3d-monoxides or disulphides.

The next ten years, it now appears, ought to see the answer to many of these questions.

Acknowledgements

The author would like to thank the Royal Society for their continued support of this research, together with related research on narrow-band phenomena in transition-metal and rare-earth compounds.

My thanks are also due to Professor Mott and to Dr A.D. Yoffe for introducing me respectively to the metal–insulator topic, and to those most illustrative of materials, the transition metal dichalcogenides.

References

Adachi, K. and Kimura, F., 1977, *J. Phys. Soc. Japan*, **43**, 838.
Adachi, K., Sato, K. and Takeda, M., 1968, *J. Appl. Phys.*, **39**, 900.
Adachi, K., Sato, K. and Takeda, M., 1969, *J. Phys. Soc. Japan*, **26**, 631.
Adachi, K., Sato, K., Yamauchi, K. and Ohashi, M., 1972, *J. Phys. Soc. Japan*, **32**, 573.
Adachi, K., Sato, K. and Sahashi, M., 1974, *J. Phys. Soc. Japan*, **37**, 1170.
Adachi, K., Ueno, T., Tohda, M. and Sawamoto, H., 1976a, *J. Phys. Soc. Japan*, **41**, 1069.
Adachi, K., Togawa, E. and Kimura, F., 1976b, *J. Physique Coll. 4*, **37**, C4–29.
Adler, D., 1968, *Solid St. Phys.*, **21**, 1.
Adler, D., 1969, *Rev. Mod. Phys.*, **40**, 714.
Adler, D., 1978, *Semiconductors & Insulators*, **3**, 367–92.
Anderson, O.K., Shriver, H.L., Nohl, H. and Johansson, B., 1979, *J. Pure Appl. Chem.*, **52**, 93.
Banda, E.J.K.B. and Wilson, J.A., 1985, *J. Phys. C: Solid St. Phys.* (in press).
Betsuyaku, H., Hamaguchi, Y. and Ogawa, S., 1974, *J. Phys. Soc. Japan*, **37**, 983.
Bither, T.A., Bouchard, R.J., Cloud, W.H., Donohue, P.C. and Siemons, W.J., 1968, *Inorg. Chem.*, **7**, 2208.
Bither, T.A., Donohue, P.C., Cloud, W.H., Bierstedt, P.E. and Young, H.S., 1970, *J. Solid St. Chem.*, **1**, 526.
Blazey and Rohrer, 1969, *Phys. Rev.*, **185**, 712.
Bouchard, R.J., 1968, *Mater. Res. Bull.*, **3**, 563.
Bouchard, R.J., Gillson, J.L. and Jarrett, H.S., 1973, *Mater. Res. Bull.*, **8**, 489.
Brandow, B.H., 1977, *Adv. Phys.*, **26**, 651–808.
Brostigen, G. and Kjekshus, A., 1970, *Acta Chem. Scand.*, **24**, 2993.
Brüesch, P., Bosch, M., Känzig, W., Ziegler, M. and Buhrer, W., 1976, *Phys. Stat. Solidi* b, **77**, 153.
Bullett, D.W., 1982, *J. Phys. C: Solid St. Phys.*, **15**, 6163.
Campagna, M., Wertheim, G.K. and Bucher, E., 1976, *Structure & Bonding*, **30**, 99.

Car, R., Ciucci, G. and Quartapelle, L., 1978, *Phys. Stat. Solidi* b, **86**, 471.
Caruthers, E., Kleinman, L. and Zhang, H.I., 1973, *Phys. Rev.* B, **7**, 3753, 3760.
Castellani, C., Feinberg, D. and Ranniger, J., 1979, *J. Phys. C: Solid St. Phys.*, **12**, 1541.
Chamberland, B.L., 1977, *Crit. Rev. Solid St. Sci.*, 1–30.
Chandler, R.N. and Bené, R.W., 1973, *Phys. Rev.* B, **8**, 4979.
Chandrashekar, G.V., van Zandt, L.L., Honig, J.M. and Jayaraman, A., 1974, *Phys. Rev.* B, **10**, 5063.
Chao, K.A. and Berggren, K.-F., 1977, *Phys. Rev.* B, **15**, 1656.
Coey, J.M.D., Roux-Buisson, H. and Brusetti, R., 1976, *J. Physique*, **37**, C4–1.
Collins, R., Kaner, R., Russo, P., Wold, A. and Avignant, D., 1979, *Inorg. Chem.*, **18**, 727.
Cox, P.A., Egdell, R.G., Goodenough, J.B., Hamnett, A. and Naish, C.C., 1983, *J. Phys. C: Solid St. Phys.*, **16**, 6221.
Cyrot, M., 1977, *Physica*, **91B**, 141–50.
Czjzek, G., Fink, J., Schmidt, H., Krill, G., Gautier, F., LaPierre, M.F. and Robert, C., 1974, *J. Physique*, **35**, C6–621.
Czjzek, G., Fink, J., Schmidt, H., Krill, G., Lapierre, M.F., Panissod, P., Gautier, F. and Robert, C., 1976, *J. Mag. Mag. Mater.*, **3**, 58.
Damien, D. and de Novion, C.H., 1981, *J. Nucl. Mater.*, **100**, 167.
Delmas, C., Fouassier, C. and Hagenmuller, P., 1980, *Physica*, **99B**, 81.
DiSalvo, F.J., Wilson, J.A. and Waszczak, J.V., 1976, *Phys. Rev. Lett.*, **36**, 885.
Echarri, A.L. and Sanchez, C., 1974, *Solid St. Commun.*, **15**, 827.
Endo, S., Mitsui, T. and Miyadai, T., 1973, *Phys. Lett.*, **46A**, 29.
Erdös, P. and Robinson, J.M., 1983, *The Physics of Actinide Compounds* in Physics of Liquids & Solids Series (New York: Plenum).
Ferguson, J., Wood, D.L. and Knox, K., 1963, *J. Chem. Phys.*, **39**, 881.
Fleishman, L. and Anderson, P.W., 1980, *Phys. Rev.* B, **21**, 2366.
Fukui, T., 1978, *J. Phys. Soc. Japan*, **45**, 1507.
Furuseth, S., Kjekshus, A. and Andresen, A.F., 1969, *Acta Chem. Scand.*, **23**, 2325.
Gautier, F., Krill, G., LaPierre, M.F. and Robert, C., 1973a, *Solid St. Commun.*, **11**, 1201.
Gautier, F., Krill, G., LaPierre, M.F. and Robert, C., 1973b, *J. Phys. C: Solid St. Phys.*, **6**, L320.
Gautier, F., Krill, G., LaPierre, M.F., Panissod, P., Robert, C., Czjzek, G., Fink, J. and Schmidt, H., 1975, *Phys. Lett.*, **53A**, 31.
Gerhardt, U., 1983, *Helv. Phys. Acta*, **56**, 105.
Goodenough, J.B., 1963, *Magnetism and the Chemical Bond* (New York: Interscience, Wiley).
Goodenough, J.B., 1968, *J. Appl. Phys.*, **39**, 403.
Goodenough, J.B., 1971a, *Prog. Solid St. Chem.*, **5**, Chap. 4.
Goodenough, J.B., 1971b, *J. Solid St. Chem.*, **3**, 26.
Goodenough, J.B., 1972, *J. Solid St. Chem.*, **5**, 144.
Goodenough, J.B., 1984, in *Physics and Chemistry of Electrons and Ions in Condensed Matter*, edited by J.V. Acrivos, N.F. Mott and A.D. Yoffe (Dordrecht: Reidel), pp. 1–44.
Goodenough, J.B., Mott, N.F., Demazeau, G. and Hagenmuller, P., 1973, *Mater. Res. Bull.*, **8**, 647.
Gyorffy, B.L., Kollar, J., Pindor, A.J., Staunton, J., Stocks, G.M. and Winter, H., 1984, *NATO ASI*, edited by P. Phariseau and W.M. Temmerman (New York: Plenum), Vol. 113, pp. 593–656.
Hanke, W., Gölzer, T. and Mattausch, H.J., 1984, *Solid St. Commun.*, **51**, 23.
Halperin, B.I. and Rice, T.M., 1968, *Solid St. Phys.*, **21**, 115.
Hasegawa, H., 1983, *J. Phys. F: Metal Phys.*, **13**, 1915.
Hastings, J.M. and Corliss, L.M., 1970, *IBM J. Res. & Dev.*, **14**, 227.

Hensel, F., 1984, in *Physics and Chemistry of Electrons and Ions in Condensed Matter*, edited by J.V. Acrivos, N.F. Mott and A.D. Yoffe (Dordrecht: Reidel), pp. 401–25.
Hodeau, J.L. and Marezio, M., 1979, *J. Solid St. Chem.*, **29**, 47.
Hulliger, F., 1968, *Structure & Bonding*, **4**, 83–229.
Hulliger, F., 1976, *Structural Chemistry of Layer-Type Phases*, Vol. 5 in 'Physics and Chemistry of Materials with Layered Structures', edited by F. Levy (Dordrecht: Reidel).
Hulliger, F., 1979 in *Handbook on Physics and Chemistry of Rare Earths*, edited by K.A. Gschneidner and L. Eyring (Amsterdam: North Holland), Chap. 33, pp. 154–235.
Hybertson, M.S. and Louie, S.G., 1984, *Solid St. Commun.*, **51**, 451.
Inoue, N., Yasuoka, H. and Ogawa, S., 1980, *J. Phys. Soc. Japan*, **48**, 850.
Jarrett, H.S., Cloud, W.H., Bouchard, R.J., Butler, S.R., Frederick, C.G. and Gillson, J.L., 1968, *Phys. Rev. Lett.*, **21**, 617.
Jarrett, H.S., Bouchard, R.J., Gillson, J.L., Jones, G.A., Marcus, S.M. and Weiher, J.F., 1973, *Mater. Res. Bull.*, **8**, 877.
Johansson, B., Skriver, H.L. and Andersen, O.K., 1981, in *Physics of Solids under High Pressure*, edited by J.S. Schilling and R.N. Shelton (North Holland: Amsterdam), pp. 245–62.
Jorgensen, C.K., 1971, *Modern Aspects of Ligand Field Theory* (Amsterdam: North Holland).
Jorgensen, C.K., 1976, in *Gmelin's Handbook of Inorganic Chemistry*, Vol. 39 (B1); *Rare Earths*, pp. 17–57.
Kamada, M., Mori, N. and Mitsui, T., 1977, *J. Phys. C: Solid St. Phys.*, **10**, L643.
Kasai, N. and Ogawa, S., 1976, *J. Phys. Soc. Japan*, **40**, 1789.
Kautz, R.L., Dresselhaus, M.S., Adler, D. and Linz, A., 1972, *Phys. Rev.* B, **6**, 2078.
Kikuchi, K., 1979, *J. Phys. Soc. Japan*, **47**, 484.
Kikuchi, K., Miyadai, T., Fukui, T., Ito, N. and Takizawa, K., 1978a, *J. Phys. Soc. Japan*, **44**, 410.
Kikuchi, K., Miyadai, T., Itoh, H. and Fukui, T., 1978b, *J. Phys. Soc. Japan*, **45**, 444.
Kikuchi, K., Miyadai, T. and Ito, N., 1980, *J. Mag. Mag. Mater.*, **15–18**, 485.
Kitaoka, Y., Yasuoka, H. and Ogawa, S., 1982, *J. Phys. Soc. Japan*, **51**, 2707.
Kjekshus, A. and Nicholson, D.G., 1971, *Acta Chem. Scand.*, **25**, 866.
Krill, G. and Amamou, A., 1980, *J. Phys. Chem. Solids*, **41**, 531.
Krill, G., LaPierre, M.F., Gautier, F., Robert, C., Czjzek, G., Fink, J. and Schmidt, H., 1976a, *J. Phys. C: Solid St. Phys.*, **9**, 761.
Krill, G., Panissod, P., LaPierre, M.F., Gautier, F., Robert, C., Czjzek, G., Fink, J., Schmidt, H. and Kuentzler, 1976b, *J. Physique*, **37**, C4–23.
Krill, G., Panissod, P., LaPierre, M.F., Gautier, F., Robert, C. and Nasr Eddine, M., 1976c, *J. Phys. C: Solid St. Phys.*, **9**, 1521.
Kunz, A.B., 1981, *J. Phys. C: Solid St. Phys.*, **14**, L455.
Kwizera, P., Dresselhaus, M.S. and Adler, D., 1980, *Phys. Rev.* B, **21**, 2328.
Kwizera, P., Mabatah, A.K., Dresselhaus, M.S. and Adler, D., 1981, *Phys. Rev.* B, **24**, 2972.
Lander, G.H., 1982, *J. Mag. Mag. Mater.*, **29**, 271.
Lever, A.B.P., 1968, *Inorganic Electronic Spectroscopy* (Amsterdam: Elsevier).
Lemos, V., Gualberto, G.M., Salzberg, J.B. and Cerdeira, F., 1980, *Phys. Stat. Solidi* b, **100**, 755.
Li, E.K., Johnson, K.H., Eastman, D.E. and Freeouf, J.L., 1974, *Phys. Rev. Lett.*, **32**, 470.
Lonzarich, G.G., 1984, *J. Mag. Mag. Mater.*, **45**, 43.
Lutz, H.D. and Willich, P., 1974, *Z. Anorg. Allg. Chem.*, **405**, 176.
Mabatah, A.K., Yoffa, E.J., Ecklund, P.C., Dresselhaus, M.S. and Adler, D., 1977, *Phys. Rev. Lett.*, **39**, 494.
Mabatah, A.K., Yoffa, E.J., Ecklund, P.C., Dresselhaus, M.S. and Adler, D., 1980, *Phys. Rev.* B, **21**, 1676.

McCann, V.H. and Ward, J.B., 1977, *J. Phys. Chem. Solids*, **38**, 991.
Marcus, S.M. and Bither, T.A., 1970, *Phys. Lett.*, **32A**, 363.
McWhan, D.B., Rice, T.M. and Remeika, J.P., 1969, *Phys. Rev. Lett.*, **23**, 1384.
Mattheiss, L.F., 1972, *Phys. Rev.* B, **5**, 290, 306.
Mazzafero, J., Ceva, H. and Alascio, B., 1980, *Phys. Rev.* B, **22**, 353.
McWhan, D.B., Remeika, J.P., Rice, T.M., Brinkman, W.F., Maita, J.P. and Menth, D., 1971, *Phys. Rev. Lett.*, **27**, 941.
Mitsui, T., Mori, N., Yomo, S. and Ogawa, S., 1974, *Solid St. Commun.*, **15**, 917.
Miyadai, T., 1979, *J. Mag. Mag. Mater.*, **10**, 29.
Miyadai, T., Miyahara, S., Takizawa, K. and Uchino, K., 1973, *Phys. Lett.*, **44A**, 529.
Miyadai, T., Takizawa, K., Nagata, H., Ito, H., Miyahara, S., Hirakawa, K., 1975, *J. Phys. Soc. Japan*, **38**, 115.
Miyadai, T., Kikuchi, K. and Ito, Y., 1977, *Physica*, **86–88A**, 901.
Miyadai, T., Kikuchi, K. and Ito, H., 1978, *Phys. Lett.*, **67A**, 61.
Miyadai, T., Sudo, S. and Takizawa, K., 1983a, *J. Phys. Soc. Japan*, **52**, 3308.
Miyadai, T., Sudo, S., Tazuke, Y., Mori, N. and Miyako, Y., 1983b, *J. Mag. Mag. Mater.*, **31–34**, 337.
Mori, N. and Takahashi, H., 1983, *J. Mag. Mag. Mater.*, **31–34**, 335.
Mori, N. and Watanabe, T., 1978, *Solid St. Commun.*, **27**, 567.
Mori, N., Mitsui, T. and Yomo, S., 1973, *Solid St. Commun.*, **13**, 1083.
Mori, N., Mitsui, T. and Yomo, S., 1975, *Proc. 4th Int. Conf. High Pressure, Kyoto, 1974*, p. 295.
Moriya, T., 1979, *J. Mag. Mag. Mater.*, **14**, 1–46.
Moriya, T., 1982, *J. Phys. Soc. Japan*, **51**, 2806–18.
Moriya, T., 1983, *J. Mag. Mag. Mater.*, **31–34**, 10–19.
Moriya, T. and Hasegawa, H., 1980, *J. Phys. Soc. Japan*, **48**, 1490.
Moriya, T. and Takahashi, Y., 1984, *Ann. Rev. Mater. Sci.*, **14**, 1–25.
Mott, N.F., 1969, *Rev. Mod. Phys.*, **40**, 677.
Mott, N.F., 1972, *Adv. Phys.*, **21**, 785.
Mott, N.F., 1974, *Metal–Insulator Transitions* (London: Taylor & Francis).
Mutka, H., 1983, *J. Physique*, **44**, C3–1713.
Nagata, H. and Miyadai, T., 1976, *Jap. J. Appl. Phys.*, **15**, 1507.
Nagata, H., Ito, H. and Miyadai, T., 1976, *J. Phys. Soc. Japan*, **41**, 2133.
Nishi, M., Ito, Y. and Ito, A., 1983, *J. Phys. Soc. Japan*, **52**, 3602.
Nishihara, Y. and Ogawa, S., 1979, *J. Chem. Phys.*, **71**, 3796.
Nishihara, Y. and Ogawa, S., 1980, *Phys. Rev.* B, **22**, 5453.
Nishihara, Y., Ogawa, S. and Waki, S., 1975a, *J. Phys. Soc. Japan*, **39**, 63.
Nishihara, Y., Ogawa, S. and Waki, S., 1975b, *J. Phys. C: Solid St. Phys.*, **11**, 1935.
Nishihara, Y., Ogawa, S., Adachi, K. and Tohda, M., 1977, *J. Phys. Soc. Japan*, **42**, 1180.
Ogawa, S., 1976, *J. Phys. Soc. Japan*, **41**, 462.
Ogawa, S., 1977, *Physica*, **86–88B**, 997.
Ogawa, S., 1979, *J. Appl. Phys.*, **50**, 2308.
Ogawa, S., Waki, S. and Teranishi, T., 1974, *Int. J. Magn.*, **5**, 349.
Ohama, N. and Hamaguchi, Y., 1971, *J. Phys. Soc. Japan*, **30**, 1311.
Ohsawa, A., Yamamoto, H. and Watanabe, H., 1974, *J. Phys. Soc. Japan*, **37**, 568.
Panissod, P., Krill, G., Lahrichi, M. and LaPierre, M.F., 1976, *Phys. Lett.*, **59A**, 221.
Panissod, P., Krill, G., Vettier, C. and Madar, R., 1979, *Solid St. Commun.*, **29**, 67.
Paquet, D. and Leroux-Hugon, P., 1980, *Phys. Rev.* B, **22**, 5284.
Penson, K.A., Ghatak, S. and Bennemann, K.H., 1979, *Phys. Rev.* B, **20**, 4665.
Plumier, R. and Krill, G., 1975, *J. Physique Lett.*, **36**, L249.
Pratt, G.W. and Coelho, R., 1959, *Phys. Rev.*, **116**, 281.
Pynn, R., Axe, J.D. and Raccah, P.M., 1978, *Phys. Rev.* B, **17**, 2196.
Rice, T.M., McWhan, D.B. and Brinkman, W.F., 1970, *Xth Int. Conf. Semiconductors*,

edited by S.P. Keller, J.C. Hensel and F. Stern (Cambridge, Mass.: U.S. Atomic Energy Commission), p. 293, see also McWhan *et al.*, 1972, *Phys. Rev.* B, **7**, 1920.
Ritala, R.K. and Kurkijarvi, J., 1982, *J. Phys. C: Solid St. Phys.*, **15**, 3101.
Robertson, J., 1979, *J. Phys. C: Solid St. Phys.*, **12**, 4777.
Rojas, S. and Spinolo, G., 1983, *Solid St. Commun.*, **48**, 349.
Sato, K., 1984, *J. Phys. Soc. Japan*, **53**, 1617.
Sayetat, F., Ghedira, M., Chenevas, J. and Marezio, M., 1982, *J. Phys. C: Solid St. Phys.*, **15**, 1627.
Schoenes, J., Frick, B., Vogt, O., 1984, *Phys. Rev.* B, **30**, 6578.
Spalek, J. and Chao, K.A., 1980, *J. Phys. C: Solid St. Phys.*, **13**, 5241.
Suga, S., Inoue, K., Taniguchi, M., Shin, S., Seki, M., Sato, K. and Teranishi, T., 1983a, *J. Phys. Soc. Japan*, **52**, 1848.
Suga, S., Taniguchi, M., Shin, S., Seki, M., Shibuya, S., Sato, K. and Yamaguchii, T., 1983b, *Physics*, **117–118B**, 353.
Sugiura, C., Suzuki, I., Kashiwakura, J., Goshi, Y., 1976, *J. Phys. Soc. Japan*, **40**, 1720.
Suzuki, T., Uchinokura, K., Sekine, I. and Matsuura, E., 1977, *Solid St. Commun.*, **23**, 847.
Takahashi, Y. and Tano, N., 1982, *J. Phys. Soc. Japan*, **51**, 1792.
Takano, H. and Okiji, A., 1981, *J. Phys. Soc. Japan*, **50**, 3835.
Terakura, K., Oguchi, T., Williams, A.R. and Kubler, J., 1984, *Phys. Rev.* B, **30**, 4734.
Uchida, S. and Tanaka, S., 1984, *J. Phys. Soc. Japan*, **53**, 667.
Ueda, K., 1977, *J. Phys. Soc. Japan*, **43**, 1497.
van der Heide, H., Hemmel, R., van Bruggen, C.F. and Haas, C., 1980, *J. Solid St. Chem.*, **33**, 17.
Vanderschaeve, G. and Escaig, B., 1976a, *Mater. Res. Bull.*, **11**, 483.
Vanderschaeve, G. and Escaig, B., 1976b, *J. Physique*, **37**, C4–105.
Waki, S., Kasai, N. and Ogawa, S., 1982, *Solid St. Commun.*, **41**, 835.
Ward, J.B. and Howard, D.G., 1976, *J. Appl. Phys.*, **47**, 388.
Watanabe, T., Mori, N. and Mitsui, T., 1976, *Solid St. Commun.*, **19**, 837.
Watanabe, M., Kaneda, K., Takeda, H. and Tsuda, N., 1984, *J. Phys. Soc. Japan*, **53**, 2437.
Wilson, J.A., 1972, *Adv. Phys.*, **21**, 143–98.
Wilson, J.A., 1973, in *Phase Transitions—1973*, edited by L.E. Cross (Oxford: Pergamon), pp. 101–15.
Wilson, J.A., 1977a, *Structure & Bonding*, **32**, 57–91.
Wilson, J.A., 1977b, *Valence Instabilities and Related Narrow Band Phenomena*, edited by R.D. Parks (New York: Plenum), p. 427.
Wilson, J.A., 1979, *Phys. Rev.* B, **19**, 6456.
Wilson, J.A., 1984, in *The Electronic Structure of Complex Systems*, edited by P. Phariseau and W.M. Temmerman (New York: Plenum), pp. 657–708.
Wilson, J.A. and Pitt, G.D., 1971, *Phil. Mag.*, **23**, 1297.
Wilson, J.A. and Yoffe, A.D., 1969, *Adv. Phys.*, **18**, 193–335.
Yoffa, E.J. and Adler, D., 1979, *Phys. Rev.* B, **20**, 4044–61.
Yoffa, E.J., Rodrigues, W.A. and Adler, D., 1979a, *Phys. Rev.* B, **19**, 1203.
Yoffa, E.J., Mabatah, A.K., Ecklund, P.C., Adler, D. and Dresselhaus, M.S., 1979b, *14th Int. Conf. Semiconductors* (Bristol: Institute of Physics), p. 473.
Yomo, S., Mori, N., Mitsui, T. and Ogawa, S., 1973, *J. Phys. Soc. Japan*, **35**, 1263.
Yoshimora, A. and Fukuda, H., 1979, *J. Phys. Soc. Japan*, **46**, 1663.
Yosida, K. and Inagaki, S., 1981, *J. Phys. Soc. Japan*, **50**, 3268.
Zielinski, P. and Parlinski, K., 1984, *J. Phys. C: Solid St. Phys.*, **17**, 3287.

CHAPTER 10

transitions in selected transition metal oxides

J.M. Honig
Purdue University, Department of Chemistry, West Lafayette, Indiana 47907, USA

1. Introductory comments

The study of metal–nonmetal transformations in transition metal oxides began roughly fifty years ago and has been steadily accelerating ever since. In this article, a cursory outline is provided for systems which exhibit such transformations; two interesting examples are discussed in some detail: V_2O_3 and Fe_3O_4. A full description of all oxides manifesting such transitions would require far more than the available space. Even in the discussion of vanadium sesquioxide and magnetite the author has had to be very selective. Apologies are extended to all workers whose research could not be described in this essay. Readers desiring more background in this general area are referred to several review articles and monographs (Goodenough 1967, Adler 1968, Mott 1974, Honig & Van Zandt 1975, Rao & Rao 1977, Rao 1984); most of these are by now rather dated but nevertheless provide alternative introductions and viewpoints to assist the reader.

Oxides provide a fertile field for the study of metal–nonmetal transitions because they tend to be rather ionic in character, so that the overlap of atomic wave functions is tenuous. Thus, relatively innocuous-looking changes, such as small alterations in temperature, pressure, or composition, suffice to bring about significant perturbations in the energy states of electrons close to the Fermi level, with enormous repercussions in a variety of physical properties.

It is worthwhile to review briefly several types of metal–nonmetal (M–NM) transitions in binary metal oxides which have been described in the literature; a few representative literature references are provided for background information.

(i) Metal–nonmetal transitions induced by pressure changes: As an example one may cite the conversion of NiO from the insulating to the metallic state by application of near hydrostatic pressures in the megabar range (Kawai & Mochizuki 1971). It is evident that for some materials close to the metal–nonmetal dividing line, pressures of this magnitude increase wavefunction overlaps between neighbours sufficiently to induce a change from localized to itinerant electron behaviour.

(ii) Metal–nonmetal transitions induced by composition changes: foremost in this category is vanadium monoxide which can exist over large composition ranges, VO_x, with $0{\cdot}78 \leqslant x \leqslant 1{\cdot}32$. At one time this compound was thought to exhibit a temperature-induced M–NM transition. It is now known that at ordinary temperatures VO_x is either metallic for $x<1$ or semiconducting for $x>1$ (Banus & Reed 1970). The change in characteristics has been rationalized by Goodenough (1972) in terms of changes of band structure in the vicinity of the Fermi level induced by alterations in the oxygen/metal ratio.

(iii) In $La_{1-x}Sr_xVO_3$, a metal–nonmetal transformation occurs in the range $x \approx 0{\cdot}225$ (Dougier & Hagenmuller 1975). This is generally interpreted as due to the creation of V^{4+} sites with increasing x, producing a fluctuating vanadium potential of sufficient magnitude to induce Anderson localization (Mott *et al.* 1975).

(iv) A very large metal–nonmetal transition is encountered in ferromagnetic, non-stoichiometric EuO specimens as a function of temperature near 50 K (Oliver *et al.* 1972, Penney *et al.* 1972). This phenomenon has been interpreted as arising from the disappearance of spin polarization band-splitting effects when the Curie point T_{Curie} of this material is approached from below. The lower 5d band edge moves rapidly upward in energy as the two spin-split portions of the band begin to coalesce near T_{Curie}. The upward sweep causes one of the oxygen vacancy trap levels to fall below the lowest 5d band edge; charge carriers thus cascade from itinerant to trap level states, resulting in a spectacular jump in resistivity of the material.

(v) Metal–nonmetal transitions can also be induced by transitions of a cation from a low-spin to a high-spin configuration. A representative example of this type is $LaCoO_3$, where such a transition is documented for Co on the basis of structural and transport data (Raccah & Goodenough 1967), as well as Mössbauer measurements (Bhide *et al.* 1972).

(vi) Two-dimensional effects are enountered in some special systems. For stoichiometric La_2NiO_4, which crystallizes in the K_2NiF_4 structure, planar conducting NiO layers in a perovskite configuration with La are sandwiched between insulating La–O rocksalt layers. When current flows along the basal planes, a metal–nonmetal transition is induced at 550 K which renders the material metallic at higher temperatures; no such transition exists when current flows along the orthogonal direction (Rao *et al.* 1984).

(vii) Transitions which involve charge-ordering phenomena, such as occur in Fe_3O_4, which are discussed below in some detail. Most of the Magnéli phases, V_nO_{2n-1} and Ti_nO_{2n-1} $(3 \leqslant n \leqslant 9)$ (Kachi *et al.* 1973), are also believed to fall in this category. The essential feature here is that a cation can exist in two distinct valence states randomly distributed among lattice sites above the transition; below the transformation an ordered arrangement of these states is encountered.

(viii) Lastly, there is a class of oxides in which a metal–nonmetal transition is induced by temperature variations; the transformation is accompanied by changes in structure and in a host of other physical properties. Ti_2O_3 (Van Zandt *et al.* 1968), VO_2 (Morin 1959, Ladd & Paul 1969) and V_2O_3 (McWhan *et al.* 1969, 1971, Kuwamoto *et al.* 1980), which is discussed below in greater detail, fall into this category. The theoretical interpretation of these transitions has been rather controversial; in general, two or more mechanisms must be invoked for an adequate description. Minimally, both electron correlation effects and lattice instabilities are involved in these changes of physical properties. For a listing of proposed mechanisms that deal with this type of metal–nonmetal transition see the review by Van Zandt & Honig (1974).

2. Transitions in the V_2O_3 alloy system

The V_2O_3 system has long been of interest because it manifests such a rich variety of metal–nonmetal transitions. The existence of a phase transformation seems to have been noticed first by Foëx (1946) in a study of the dilatation and electrical resistance of compacted specimens as a function of temperature; at 268 K he encountered sizeable discontinuities in both properties. The first comparable measurements on single crystals were published by Morin (1959); he observed a change in resistivity of nearly six orders of magnitude and a hysteresis of nearly twelve degrees. Feinleib and Paul (1967) later directed attention to a high temperature anomaly in the electrical properties of V_2O_3, discussed below in greater detail. A final impetus to the field was provided by McWhan and co-workers (1969, 1970, 1973) and by Gossard *et al.* (1971) who expanded on a series of unpublished observations by MacMillan (1960). Their work indicated that the transitions were enhanced by doping V_2O_3 lightly with Cr_2O_3 or Al_2O_3, while the incorporation of Ti_2O_3 or excess oxygen stabilized the metallic phase. These effects were expanded upon in later studies (Dumas *et al.* 1975, Dumas & Schlenker 1976, Kuwamoto *et al.* 1980, Shivashankar & Honig 1983). The great continuing interest in these types of phenomena is manifested by the roughly 500 papers that have been published since 1946 on V_2O_3 and its dilute alloys with other M_2O_3 oxides.

The subject is perhaps best elaborated in terms of the diagrams of Figure 1(*a*)–(*e*), where log ρ is shown schematically as a function of $1/T$; ρ is the electrical resistivity and T the absolute temperature. For the V_2O_3 (98·5%)–Cr_2O_3 (1·5%) alloy system of Figure 1(*a*), the system undergoes *three* transformations while the temperature is raised from the cryogenic region to 1000 K. At low temperatures the material is an antiferromagnetic insulator (AFI); a first-order transformation to a paramagnetic metallic (PM) phase then occurs in the 150 K range. At still higher temperature ($T > 250$ K) the system changes into a paramagnetic insulator (PI). On heating beyond 350 K the system gradually reverts back to a high-temperature metallic state (PM′). These various transitions may be dramatically altered by slight changes in alloy composition. For the V_2O_3 (97%)–

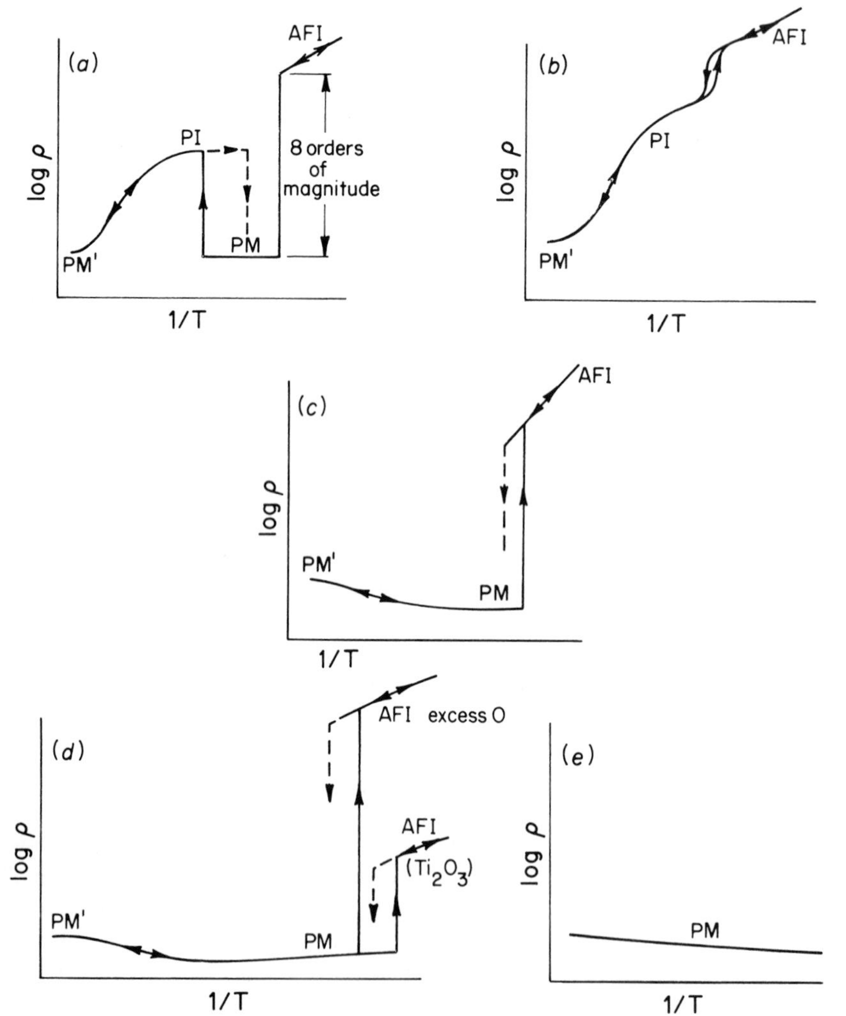

Figure 1. Schematic representation of electrical resistivity (log ρ) of V_2O_3 vs. inverse temperature ($1/T$) drawn roughly to scale. (*a*) $(V_{1-x}Cr_x)_2O_3$ or $(V_{1-x}Al_x)_2O_3$ with $0{\cdot}005<x<0{\cdot}018$; (*b*) same, with $0{\cdot}018<x<0{\cdot}1$; (*c*) V_2O_3; (*d*) $(V_{1-x}\square_x)_2O_3$ or $(V_{1-x}Ti_x)_2O_3$ with $0<x<0{\cdot}008$ for vacancies, $\square$, and $0<x<0{\cdot}05$ for Ti; (*e*) same as (*d*), with $x>0{\cdot}008$ or $x>0{\cdot}05$ respectively. Note hysteresis effects at first order transitions. AFI = antiferromagnetic insulator, PM = paramagnetic metal, PI = paramagnetic insulator, PM′ = high temperature metallic phase.

Cr_2O_3 (3%) system depicted in Figure 1(*b*), the metallic PM phase is lost altogether; on heating, the alloy proceeds directly from the AFI to the PI phase; on further heating one ultimately recovers the high-temperature PM′ phase. For the unalloyed V_2O_3 system sketched in Figure 1(*c*) one encounters only the low-temperature AFI–PM transformation and an anomalous increase in resistivity which gradually transforms the material from the PM to the PM′ phase. As shown in Figure 1(*d*), when Ti_2O_3 is incorporated or when the material is rendered non-stoichiometric the PM–AFI transition temperature is drastically reduced. Finally, with sufficient Ti_2O_3 content or with sufficient departure from ideal oxygen stoichiometry the material remains metallic at all temperatures, but undergoes an antiferromagnetic ordering process in the cryogenic temperature range, as shown in Figure 1(*e*). It should be noted that the effects of Cr_2O_3 or Al_2O_3 incorporation can be counteracted by use of Ti_2O_3, excess oxygen or pressure. Summarizing the above, one finds that as a function of temperature three, two, one or no metal–insulator transitions occur in the V_2O_3 system as a function of temperature, depending on compositional details.

No fewer than eight different transition mechanisms have been proposed to account for the above observations (Kuwamoto *et al.* 1980). Most of these models are no longer tenable in light of more recent experimental data. For a brief review of these theories the reader is referred elsewhere (Van Zandt & Honig 1974). One particular explanation that at one time had gained considerable popularity, was to ascribe the sharp first order phase changes in Figure 1 to the Mott transition. There are many reasons which mitigate against such a view (Honig & Van Zandt 1975, Kuwamoto *et al.* 1980); here we mention merely one experimental finding which does not fit the Mott model: in both the AFI–PM and PM–PI transitions one encounters very large hysteresis effects which may range up to 70 K. Inasmuch as the Mott transition depends solely on electron density effects (Mott 1974), the transition should occur at a unique temperature. The fact that enormous hysteresis effects are encountered indicates that the lattice is heavily involved in the first order transitions schematized in Figure 1.

In what follows a model will be presented by which a very large body of experimental data may be rationalized in an elementary manner. Experimental evidence will then be presented to show the internal consistency of such an approach. Of course, consistency arguments do not prove that the model is correct. However, the scheme provided below seems to have fewer difficulties in coping with the data than alternative models.

2.1. Experimental data pertaining to electrical properties

At the outset it is important to delineate the experimental data one is trying to explain: these are shown in Figure 2. Part (*a*) pertains to the effect of incorporating Cr_2O_3 into the V_2O_3 host matrix. Parts (*b*) and (*c*) deal with the effects of incorporating Ti_2O_3 or excess oxygen. Figure 1 was constructed from these sets of experimental data.

A key to the understanding of the effects described above is furnished by the band structure considerations. At a very elementary level one may proceed by the

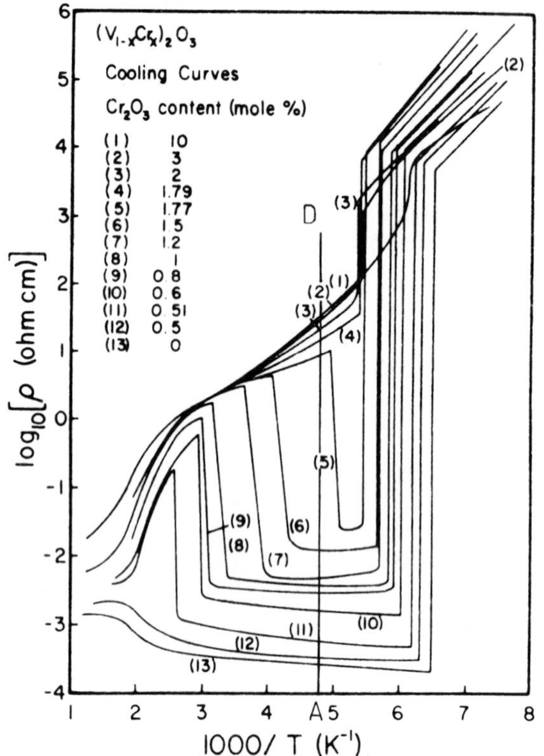

Figure 2. (*a*) Variation of resistivity with temperature in the $(V_{1-x}Cr_x)_2O_3$ system, $0 \leqslant x \leqslant 0{\cdot}1$ (after Kuwamoto *et al.* 1980).

method originally pioneered by Goodenough (1971). The fundamental scheme is depicted in Figure 3(*a*); on the left are shown the atomic 4s and 4p orbitals of vanadium as split in the rhombohedral crystal field of the corundum lattice. On the right are shown the corresponding anion levels for oxygen. The 4s and 4p cationic states interact strongly with the 2s and 2p anionic states to form rather broad hydridized bands indicated in the centre of the diagram. The exact nature of these crystal states is not of importance in the present discussion, since these particular levels are either completely filled or empty. What is of importance, however, are the crystal states of primarily 3d character which result from the interaction of the a_1 and e cationic states with the 2s and 2p states of oxygen. The development of the qualitative band-structure scheme is traced out in Figure 3(*b*). On the left-hand side are shown the e and a_1 levels derived from the ten 3d states of vanadium as split in the rhombohedral field of the corundum lattice. It must be recognized that the vanadium atoms are paired along the *c*-axis; this leads to formation of bonding and antibonding molecular states as indicated in Figure 3(*b*). Next, notice must be taken that the correct formula unit per unit cell is

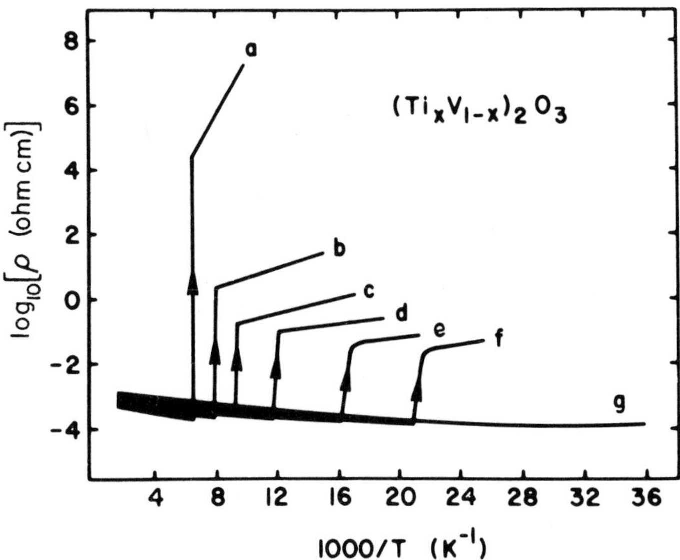

Figure 2. (*b*) Variation of resistivity with temperature in the $(V_{1-y}Ti_y)_2O_3$ system, $0 \leqslant y \leqslant 0{\cdot}05$ (after Shivashankar & Honig 1983).

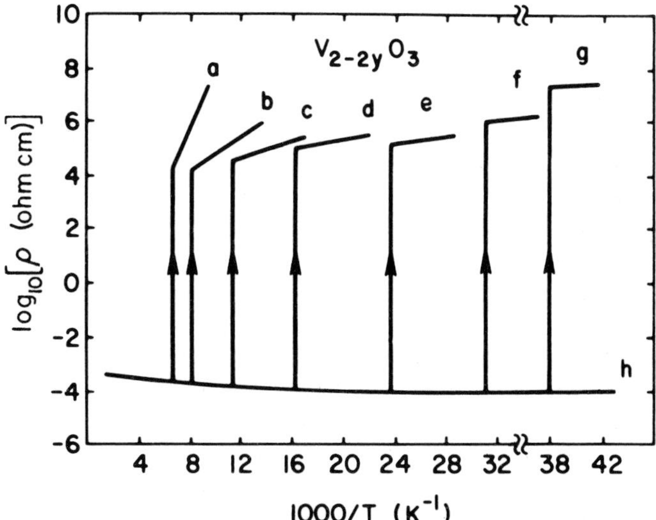

Figure 2. (*c*) Variation of resistivity with temperature for non-stoichiometry V_2O_3, $(V_{1-z}\square_z)_2O_3$, $0 \leqslant z \leqslant 0{\cdot}018$ (after Shivashankar & Honig 1983).

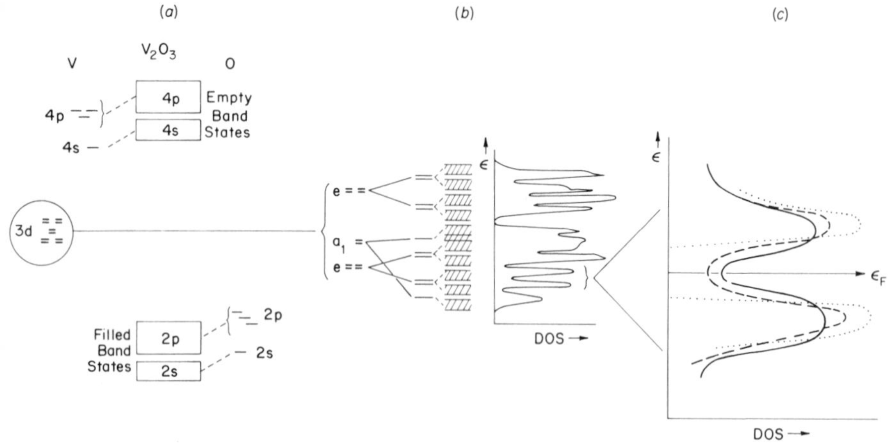

Figure 3. Energy levels for d electrons in V_2O_3. (*a*) Disposition of bands of primarily 4s and 2p, 2s character as obtained from atomic states. (*b*) Disposition of the 3d levels in the gap between the 4s and 2p bands. Note how the many sublevels give rise to alternating peaks and valleys. (*c*) Enlargement of the peaks near the Fermi level which falls in a minimum of the densities of states.

V_4O_6; on account of incipient overlap one obtains the narrow-band states shown in the centre of Figure 3(*b*). Finally, when account is taken of the overlap with anion states the density of states scheme shown on the right of Figure 3(*b*) is achieved. The indicated progression may not be correct in detail; however, the ultimate finding, namely, that the d-band structure is characterized by a sharp alternation of strong maxima and deep minima in the density-of-states (DOS), is qualitatively correct. All calculations concerning the band structure of V_2O_3 to date have shown such structural variations (Ashkenazi & Weger 1976, Castellani *et al.* 1978, Grodzicki *et al.* 1983); while differing in detail the qualitative features are reproduced in the above calculations. This should instil some confidence in the essential correctness of the picture presented in Figure 3(*b*).

The above results are based on the one-electron approximation. Electron correlation effects and electron–lattice interactions also play an important role in determining the band structure, and hence, the density of states of the V_2O_3 system. The general effect of these interactions is to narrow the bands, thereby accentuating the peaks and valleys in the density-of-states curve, Figure 3(*b*). For very strong interactions of this type one would expect to encounter quasi-isolated molecular states energy levels, resulting in an activated process for electron transport.

However, the available experimental evidence militates against electron localization in the d states. This matter has been discussed at length by Kuwamoto *et al.* (1980) for the PM, PI, and PM′ phases of the V_2O_3 and V_2O_3 alloy systems, on the basis of available electron transport data. The situation is less clear-cut for the AFI phase because of the paucity of Seebeck and optical data at low temperatures. However, the gradual decline in conductivity activation energy with increasing Ti or cation vacancy content shown in Figure 2(*b*) and (*c*), and the sudden reversion of the system to a metallic phase beyond a critical Ti or a critical V vacancy concentration makes it unlikely that electron correlation or electron–phonon interactions are strong enough to bring about electron localization.

Once it is accepted that charge carriers in the V_2O_3 system are delocalized, the explanation of all features in Figure 1 is fairly straightforward on the basis of Figure 3(*b*). The relevant portion near the Fermi level is shown on an enlarged scale in Figure 3(*c*).

As is shown, the Fermi level lies very close to the minimum of the second valley in the density-of-states curve. This feature immediately rationalizes the fact that the V_2O_3 system in the metallic state is a relatively poor conductor. For, the general considerations involving transport phenomena, as lucidly developed by A.H. Wilson (1954), show that for a complex band structure of the type shown above, the total conductivity is proportional to the density of states at the Fermi level. Since in the model of Figure 3(*b*) this quantity is small, the conductivity σ is expected to be low, as is consistent with a value of $\sigma = 10^3\ (\Omega\,\text{cm})^{-1}$ for undiluted V_2O_3. As shown later, the addition of Cr further depresses the density of states at the Fermi level, which immediately rationalizes the fact that the alloy containing 1·78 at. % Cr_2O_3 has the exceptionally high metallic resistivity of $\sigma^{-1} \equiv \rho = 0{\cdot}05\ \Omega\,\text{cm}$.

Moreover, whenever the Fermi level is situated in or near the minimum of the density of states, a band gap can be generated quite readily as depicted in Figure 4(*a*). The opening of such a gap is advantageous from the energy viewpoint: as is shown in block form in Figure 4(*b*), all electrons originally accommodated in the crosshatched region of the diagram before the phase transition are lowered in energy when the gap is created. This feature provides one driving force for the occurrence of a metal–nonmetal transition.

In the V_2O_3 system the opening up of the band gap may be achieved by three mechanisms: the first is the dilution of the cation sublattice by an essentially inert metal ion. Taking Al_2O_3 as an example, the substitution of Al for V introduces species into the cation sublattice whose 3d states are thermally inaccessible to the itinerant electrons in the host lattice. This is equivalent to the removal of 3d vanadium states; the effect is to narrow the 3d band states, thus accentuating the peaks and deepening the valleys in the density-of-states curves of Figure 3(*c*). With sufficient doping the gap finally opens: the V_2O_3 host has become a semiconductor. While the above arguments have been couched in terms of the Al–V_2O_3 system, similar considerations apply to the Cr–V_2O_3 system. In the latter case the d^3 configuration is exceptionally stable; energy levels of this type lie well below those thermally accessible to the V_2O_3 itinerant electrons. This qualitatively agrees with the astonishing similarity in electrical characteristics of Al-doped vs. Cr-doped V_2O_3 systems, as has been repeatedly demonstrated in the literature (Joshi & Honig 1982).

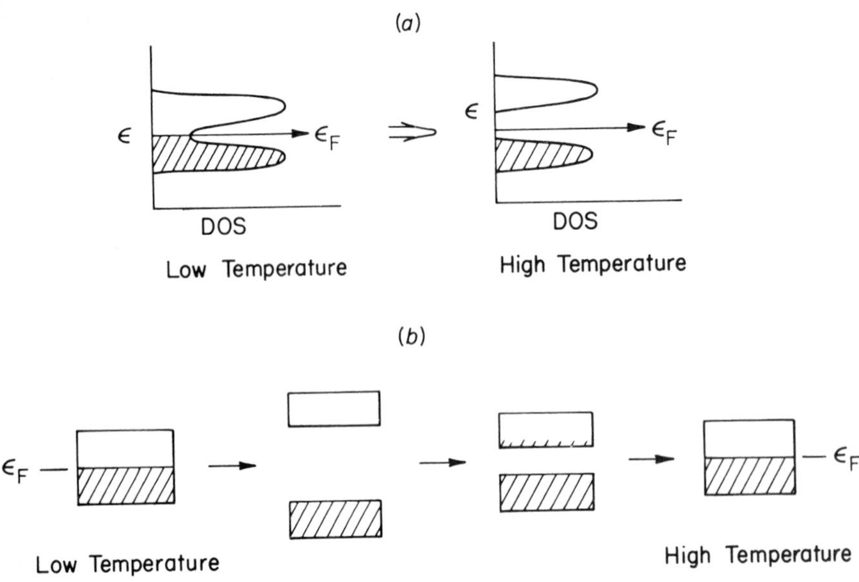

Figure 4. Schematic representation of the opening of a gap in the V_2O_3 alloy systems. (*a*) Density of states diagram, (*b*) block diagram for a partially filled band; ε_F is the Fermi level.

This situation is mirrored in the experimental data of Figure 2(*a*). If the temperature is kept constant (at 208 K, for example) and if successive amounts of Cr_2O_3 are added to the system, the resistivity steadily increases as shown by the straight line AD. At a critical doping level of $x_c = 0{\cdot}0178$, the system suddenly switches from the metallic to the semiconducting state. The incorporation of Cr also dilates the lattice, in that, at the same critical concentration, the pseudo-hexagonal lattice parameters change suddenly from $a_H = 4{\cdot}94$ Å, $c_H = 14{\cdot}01$ Å to $a_H = 5{\cdot}00$ Å, $c_H = 13{\cdot}93$ Å as x is increased past the x_c value. Thus, the crystal lattice has become distended, and a change in volume accompanies the opening up of the band gap and the lowering of the total energy for the itinerant electrons.

A second mechanism for opening up the band gap described above is through an increase in temperature. To assist in the descriptive process reference should be made to Figure 4(*b*). With a rise in temperature the phonon density of the material increases, thus potentially facilitating the distortion of the lattice needed to open up the band gap. Experimentally, it is known that at the critical temperature of the PM–PI phase transition of Figure 1, the unit cell volume increases by a remarkable 1·5%, essentially without changing the lattice symmetry (what 'essentially' means will be briefly discussed below). The resulting band gap of 0·3 eV, is sufficiently small so that even at room temperature some thermal promotion of charge carriers occurs, whereby holes are introduced in the valence band and electrons, in the conduction band. On increasing the temperature the thermal promotion is further facilitated. This has the effect of introducing a continually increasing density of carriers into the energetically elevated states of the conduction band at the expense of carriers in the valence band. Ultimately, the resulting energy increase balances out the original energy gain introduced by the lattice deformation. At this stage the crystal gradually reverts back to the starting state, which process is accompanied by a closure of the band gap with rising temperature, as schematized in Figure 4(*b*). Ultimately, at sufficiently high temperatures the band gap is completely closed and the alloy is back in a metallic state not too different from the PM configuration. The sequence shown in Figure 4(*b*) rationalizes the experimental observations that in the $(V_{1-x}Cr_x)_2O_3$ or $(V_{1-x}Al_x)_2O_3$ systems the resistivity drops gradually over a 400 K range so that the resistivity above 950 K is nearly the same as that of the corresponding PM phase near room temperature.

Obviously, independent evidence in favour of this model is necessary in order to render the description more credible. One such item is based on the systematic crystallographic studies carried out by Robinson (1975) as shown in Figure 5. Figure 5(*a*) shows the atomic positions in V_2O_3 and identifies the location of the cations M_1, M_2, M_3; Figure 5(*b*) shows how the M_1–M_2 and M_1–M_3 separation distances change with temperature in $(V_{1-x}Cr_x)_2O_3$ for $x=0$, 0·01. It is seen that the anomalous resistivity increase encountered between 400 K and 600 K for pure V_2O_3 (see Figure 2(*a*)) is mirrored by a corresponding

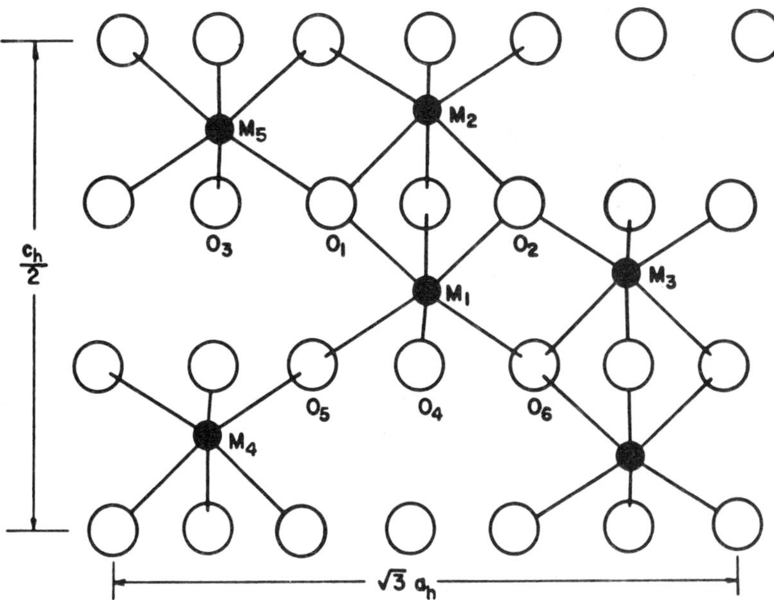

Figure 5. (*a*) Projection of atomic positions in V_2O_3. M_1–M_2 separation is parallel to the *c*-axis; M_1–M_3 separation is in the nearly orthogonal direction.

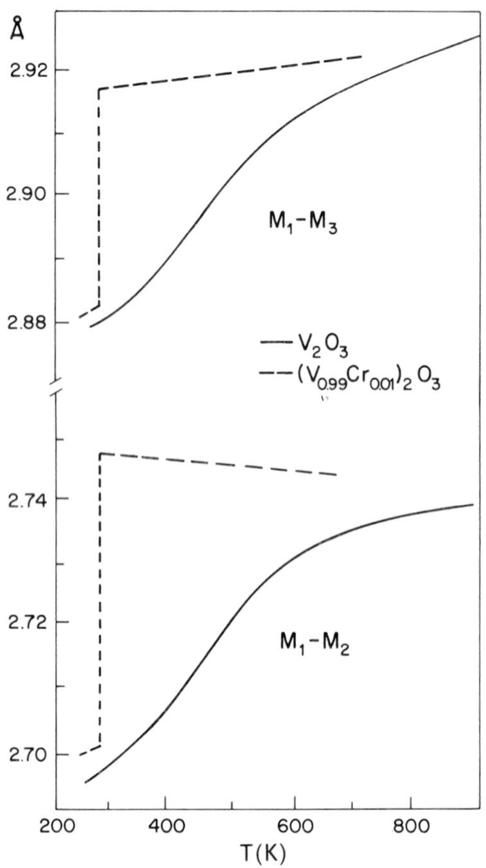

Figure 5. (*b*) M_1–M_2 and M_1–M_3 separation distance as a function of temperature in V_2O_3 (full curve) and $(V_{0\cdot985}Cr_{0\cdot015})_2O_3$ (broken curve) (after Robinson 1975).

anomalous increase in cation separation distances in V_2O_3 with temperature. The high temperature resistivity discontinuity shown in Figure 2(*a*) has as its counterpart the variation of M_1–M_2 and M_1–M_3 separations shown in Figure 5(*b*). At the PM–PI transition the atomic distances increase significantly; with a further rise in temperature these values remain either nearly constant or even decrease (!) so as to approach the values for pure V_2O_3 in the high temperature limit. This shows that the lattice deformation reverses its course as the temperature is increased and that the gap becomes smaller.

Additional evidence in favour of the model is derived from the X-ray photoelectron spectrum obtained by Hegde & Vasudevan (1979). At room temperature they report that the Fermi level clearly intersects the d band of the $(V_{1-x}Cr_x)_2O_3$ system. As the temperature is raised they encounter a gap of approximately 0·3 eV after the onset of the PM–PI transformation. With a further rise in temperature the gap narrows, and at the highest temperatures the Fermi level once again intersects the d band. These two experiments provide important clues in support of the model set forth above.

A final mechanism for opening up of the gap at the Fermi level is achieved through magnetic ordering. The exact scheme which is operative in the V_2O_3 system remains to be elucidated. However, it is sufficient to note that a change in lattice symmetry is encountered at the PM–AFI phase transformation. Whether this is due to magnetostrictive effects, or whether the lattice deformation drives the magnetic ordering process, is not important at this stage. The change in symmetry suffices to open the same gap as was generated by the other mechanisms. This rationalizes the observation that the gap of the AFI phase, about 0·3 eV, is the same as before.

Finally, it should be noted that a fourth mechanism exists for closing the gaps, namely, the application of hydrostatic pressure. It has been known for some time (McWhan *et al.* 1969) that V_2O_3 at a pressure $p \geqslant 26$ kbar remains metallic at all temperatures; larger pressures are needed to achieve the same state for the Cr- or Al-doped alloys.

A new feature must be invoked to explain the effects of Ti_2O_3 doping in the temperature and composition range where the AFI phase is stable. The replacement of V by Ti (which has one less 3d electron) results in a progressive diminution of the electron population within the d-band state system. In consequence, the Fermi level gradually moves toward lower energies, that is, toward the valence band edge. This rationalizes the steady decrease in slope of the straight line portions of the curves in Figure 2(*b*). Ultimately, the Fermi level is displaced below the valence band edge, at which point the PM–AFI transition disappears. A similar explanation holds for nonstoichiometric V_2O_3. With increasing excess oxygen, additional states in the 2p anion sublattice band become available. Electrons thus drop from the top of the 3d levels into the energetically lower lying 2p states. Once again, this causes a lowering of the Fermi level and produces the effects depicted in Figure 2(*c*).

While there is great similarity in the properties of the $(V_{1-x}Ti_x)_2O_3$ and the $(V_{1-x}\square_x)_2O_3$ systems, there is one striking difference: in the Ti–V_2O_3 alloy system the discontinuity in resistivity decreases with increasing Ti content, whereas in nonstoichiometric V_2O_3 the discontinuity first drops slightly and then increases markedly with increasing cation vacancy concentration. This difference in behaviour has been rationalized in a recent publication (Shivashankar & Honig 1983) on the basis of charge carrier mobility effects. For the Ti–V_2O_3 system it is assumed that the substitution of Ti for V does not seriously perturb the system, so that lattice scattering is thought to dominate the mobility. By contrast, the generation of cation vacancies does drastically perturb the charge carrier motion in non-stoichiometric V_2O_3. Now ionic impurity scattering is thought to dominate the scattering process; with increasing departures from stoichiometry this effect becomes increasingly noticeable, so that at the transition there is not only a change in charge carrier density but also a marked diminution in the mobility of charge carriers relative to the metallic phase. By use of quantitative theories a rather good fit has been obtained between the size of the resistivity discontinuity and the Ti content or the lattice vacancy concentration.

It is appropriate to conclude this section with a discussion of several experiments that further illuminate the role of the lattice in the PM–PI or the PM–AFI transitions of the V_2O_3 alloy systems.

Sinclair & Colella (1979) have studied the X-ray scattering in $(V_{1-x}Cr_x)_2O_3$ for $0{\cdot}005 < x < 0{\cdot}008$; they encountered three very weak Bragg satellites whose intensities were nearly four orders of magnitude below those of the corresponding main peaks. These satellites were present just above the PM–PI transition (that is, for the alloys in the PI insulating phase) but were absent from the spectrum of the metallic phase near room temperature. Although these subsidiary peaks were not always reproducible from sample to sample, wherever the extra peaks were observed they could be correlated with the PM—PI transition; moreover, their **Q** vectors along the c^* axis in reciprocal space were incommensurate with the reciprocal lattice vectors. These results constitute proof that slight structural perturbations do accompany the PM–PI transition. The fact that the extra peaks occur at incommensurate positions makes attractive a hypothesis by Overhauser (1979) that a charge density wave is present in the alloy system in the PI configuration. Clearly, additional studies are needed to place such an interpretation on a firmer footing, but the available data indicate that the lattice does change slightly in response to the PM–PI transition.

Further confirmatory evidence is supplied by Raman studies on $(V_{1-x}Cr_x)_2O_3$ with $x = 0$, 0·015, 0·02 and 0·03. Kuroda & Fan (1979) reported an upward displacement in the frequency of most Raman-active lines of pure V_2O_3 as the temperature was decreased past the PM–AFI transition. Honig (1984 unpublished) has estimated that the difference in entropy $S(\mathrm{AFI}) - S(\mathrm{PM}) \approx 0{\cdot}38$ e.u. on the basis of the reported Raman-active shifts; while the contributions from infrared active modes are unavailable, the above, incomplete analysis is, nevertheless, suggestive: the metallic phase of the lattice is 'softer' than the insulating phase. Essentially similar results were reported for the Cr-containing alloys (Tatsuyama & Fan 1980). Several of the Raman lines show a decrease in Stokes shifts as one passes from the PI to the PM phase. An almost compensating shift in the opposite direction occurred with decreasing temperature at the PM–AFI transition. Okamoto *et al.* (1983) extended the Raman studies from 300 to 600 K, to cover the region where the electrical anomaly occurs for pure V_2O_3, or in which the PI phase gradually reverts to the PM′ phase at elevated temperature for Cr-alloys with $x > 0{\cdot}005$. As the temperature was raised, there was first a softening of the A_1 modes, followed by a partial recovery; qualitatively, these findings agree with the theoretical prediction of such an effect by Zeiger (1975), who assumed that the a_1 and e portions of the band in Figure 2 shift extensively with rising temperature. In his calculations of the Hartree–Fock free energy, the electron correlation effects, the elastic energy, and the band energies were taken into consideration. In summary, several changes in the Stokes shift occur where the anomaly occurs for pure V_2O_3 or at the PM–AFI and PM–PI phase transitions; this clearly shows that the phonon structure of the lattice is altered. Also, judging from the Raman-active shifts, the lattice vibration entropy for the metallic phase is lower than that of the various insulating phases.

Additional evidence concerning lattice effects is provided by X-ray crystallography under pressure and by isothermal compression measurements. Finger & Hazen (1980) pointed out that V_2O_3 is unique among the sesquioxide corundum structures in that it is very anisotropic under pressure: the *a*-axis is almost three times more compressible than the *c*-axis. Also, in contrast to the other oxides, the atomic positions in V_2O_3 under pressure approach the ideal hexagonal close-packed arrangement.

Ultrasonic measurements confirm the above and provide considerable insight into the mechanical characteristics of the lattice. Among recent work may be cited the experiments of Sladek and collaborators (Nichols & Sladek 1981, Nichols *et al.* 1981, Yang *et al.* 1983). The anisotropy in isothermal compressibility as determined by ultrasonic techniques (Nichols & Sladek 1981) matches that cited by Finger & Hazen (1980). From the fact that certain shear-stiffness coefficients show a negative dependence on pressure it is deduced that V_2O_3 at high pressures (where these constants pass through zero) should exhibit a phase transition. Several of the elastic constants as well as the bulk modulus pass through a minimum in the range 500–600 K where the anomalous rise in electrical properties occurs for V_2O_3. A number of parameters decrease rather than increase with falling temperatures in the range 600–300 K, thereby presaging the PM–AFI transition near 150 K (Nichols *et al.* 1981). These results do not fit the standard models that invoke separate electronic and lattice contributions. What thereby seems well established is that the lattice is more than a passive partner in the course of the electrical anomaly and the phase transition exhibited by pure V_2O_3. Finally, in one of the most interesting measurements so far, Yang *et al.* (1983) determined the changes in C_{44} between 150 and 650 K for $(V_{1-x}Cr_x)_2O_3$. As shown in Figure 6, for $x=0$, this particular shear elastic constant varies only weakly with temperature for undoped V_2O_3; by contrast, C_{44} diminishes sharply with decreasing temperature for the alloys. In fact, for the specimen with $x=0{\cdot}03$, C_{44} extrapolates to zero precisely at the transition temperature for the PI–AFI transformation. The C_{44} curve for the alloy $x=0{\cdot}015$ runs parallel to the alloy with $x=0{\cdot}03$. Unfortunately, the ultrasonic signals of $(V_{0\cdot985}Cr_{0\cdot015})_2O_3$ are lost below 240 K, where the material transforms to the PM phase before undergoing the final transition to the AFI phase near 180 K. The authors interpret their findings in terms of continual change from rhombohedral to nearly hexagonal symmetry with decreasing temperature, before the alloys discontinuously transform to the monoclinic phase. The order parameters appropriate to such variations are also discussed in the article. Clearly, the first order PM–AFI transition of V_2O_3 is of quite a different character because it occurs without any premonitory effects in the elastic constant C_{44}. Nevertheless, several elastic constants show an abnormal increase

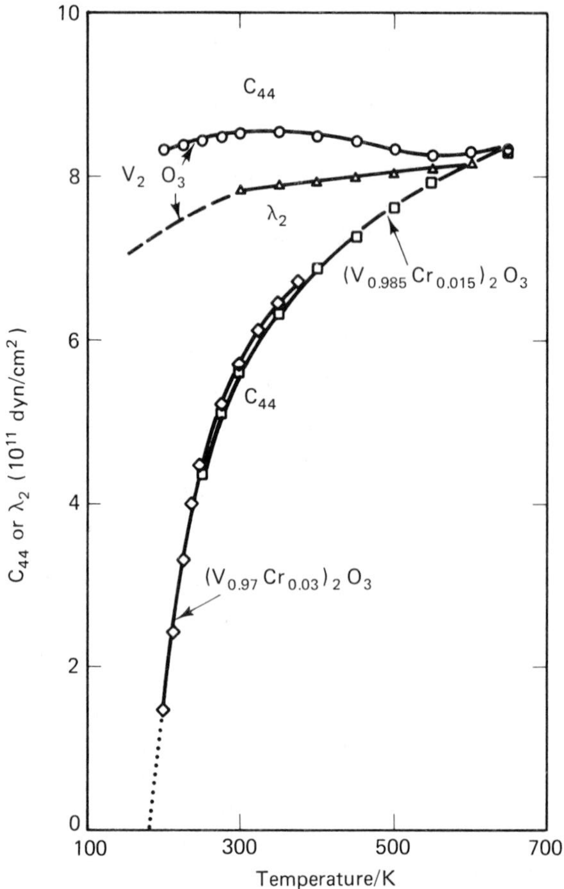

Figure 6. Variation of the shear parameter C_{44} with temperature of V_2O_3 and some of the Cr_2O_3 alloys. Note the softening of C_{44} for $(V_{0\cdot97}Cr_{0\cdot03})_2O_3$ (after Yang *et al.* 1983).

with diminishing temperature in the range 600–300 K, which points to anticipated instabilities at low temperatures.

In summary, the available data in the literature indicate that mechanical, electronic and magnetic interactions act jointly in bringing about the various transitions described above; it seems difficult to interpret the experimental findings by assigning one particular mechanism a pre-eminent role.

3. The Verwey transition in Fe_3O_4

The history of the Verwey transition in Fe_3O_4 dates back to 1926 when Parks & Kelley (1926) first detected an anomalous peak near 120 K in their measurements of the heat capacity of a natural crystal of magnetite; Millar (1929) reported a similar observation shortly thereafter. Weiss & Forrer (1929) noted an abrupt change in the magnetization of magnetite in the range 116–120 K. Okamura (1932) apparently was the earliest investigator to carry out several measurements, namely, heat capacity, thermal expansion, electrical resistivity, magnetization and X-ray diffraction, on a natural crystal of magnetite. These were in the nature of exploratory studies which indicated that all of the above properties exhibited anomalies in the range 105–115 K. Real interest in this system began to develop after Verwey and co-workers (Verwey 1939, Verwey & Haayman 1941, Verwey *et al.* 1947) initiated a series of systematic electrical measurements shich showed a near-discontinuity in the resistivity of ceramic specimens, close to the temperature where the various anomalies had been detected by the earlier workers. The electrical measurements included the effects of varying the Fe^{2+}/Fe^{3+} ratio in the octahedral sites. The associated phase transformation has continued to fascinate workers in the field during the ensuing decades. Since 1976 alone over 250 articles have been published in the literature, dealing with many types of studies that include electrical resistivity, Seebeck coefficient, magnetic after-effects, magnetization, optical reflectivity, thermal conductivity, ultrasonics, neutron-, X-ray and electron diffraction, Mössbauer effect and heat capacity; this listing is by no means exhaustive.

Despite the intensive activity in this area, the mechanism of the transition has remained elusive. Part of the problem arises from contradictory experimental information: for example, even such a fundamental property as the lattice symmetry has remained in doubt—the low temperature phase has been variously reported as being triclinic, monoclinic, rhombohedral or orthorhombic. The nature of the transition also is in dispute: there have been claims, based on heat capacity and Mössbauer studies, that Fe_3O_4 undergoes a multiplicity of closely spaced transitions, rather than the single, sharp transformation which has been claimed for the system by other workers. The resulting theoretical models are equally diverse: it is not known for sure whether the charge carriers are itinerant or localized, and while there is agreement that the Verwey transition involves charge ordering, there are many competing charge configurations which have been proposed to account for the observations.

The principal reason for the discrepancies seems to be the lack of adequate control over sample composition and homogeneity. A dramatic illustration of these effects is provided by measurements of the heat capacity anomaly that accompanies the transition. As detailed later, unannealed single-crystal specimens exhibit broad or bifurcated heat-capacity peaks. By contrast, homogenization of specimens leads to a dramatic sharpening of the heat-capacity anomaly. Similar results are encountered in electrical transition. As was already implicit in the earliest measurements (Verwey & Haayman 1941), non-homogeneous specimens undergoing the transformation near 120 K tend to show a broad electrical anomaly extending over several kelvins; properly treated samples exhibit very sharp transitions. Again, Mössbauer spectra of non-homogeneous specimens tend to be very complex, while those of properly annealed samples are much simpler.

These findings alert us to the fact that sample preparation techniques play a crucial role in the acquisition of reliable data. Much of the experimentation must necessarily be done on single-crystal specimens; thus, the experimentalist faces three problems: growth of single crystals of appropriate size, minimization of impurity content, and attainment of a specific composition which is uniform throughout the specimen. Generally, starting materials of high purity are available; one must thus guard against contamination during crystal growth. Vapour-transport techniques are useful for this purpose but the crystal size tends to be small. The other alternative is to use crucible-less techniques such as the skull melter (Harrison *et al.* 1984) for this purpose: 5–10 g single crystals can generally be obtained without difficulty.

The real source of trouble tends to be the homogenizing step; even single crystals that look perfect can vary widely in composition within the specimen. To eliminate this problem samples may be heated in CO/CO_2 buffer mixtures whose oxygen fugacity is measured by an appropriate electrolytic cell. Also, one must anneal at temperatures high enough for cation diffusion in the solid to occur on reasonable time-scales; higher temperatures also have the advantage of widening the stability range of the Fe_3O_4 phase field, thus permitting larger departures from the ideal 4/3 oxygen/iron ratio. On the other hand, one must be able to quench in the particular desired stoichiometry before any appreciable changes in composition or any exsolution processes set in; this sets an upper limit to the annealing temperature. In general, the best operating procedure involves use of thick specimens whose edges are later trimmed off; the core then has the desired uniformity of composition set by the oxygen fugacity at the temperature of the anneal. Obviously, the specimens should finally be checked for phase segregation, impurity content and structural integrity. Only samples subjected to these procedures can be assumed to yield reliable experimental information.

Unfortunately, in the majority of cases reported in the literature, measurements have been taken on samples that were not properly prepared. Even natural single crystals cannot reliably be used in experiments because they usually contain cationic impurities in amounts sufficient to affect the results. In short, much of the work cited in the literature—especially that which was carried out on unannealed samples—should be regarded with caution.

In view of the above problems it seems premature to attempt a unified description of the Verwey transition. Rather, two specific problems will be taken up that have been raised over several decades: the question whether charge carriers in Fe_3O_4 are in a band or localized regime, and the question whether the transition proceeds in one jump or in several stages. Both of these issues must be settled before the nature of the Verwey transition can be properly addressed.

The question of the itineracy of the carriers in the high temperature regime has divided investigators. For example, Camphausen *et al.* (1972), as well as Cullen & Callen (1973) have postulated that in the high temperature phase of

Fe_3O_4 electrons move more or less freely in band states. Strong electron–electron, or electron–phonon interactions would then be required to bring the calculated mobilities in line with those determined by experiment. Cullen & Callen (1973) proposed a model that included a variable intra-atomic coulomb repulsion between charge carriers. For certain values of this parameter a multiple-ordering scheme is obtained at the Verwey transition; as already mentioned, such an effect has been experimentally detected but only for samples that are inhomogeneous. Mott (1979, 1980) and Whall (1980) suggested that electrons are in localized states due to the random distribution of Fe^{2+} and Fe^{3+} ions on the octahedral interstices above the Verwey transition temperature T_V. Such localizations in random potentials was first proposed by Anderson (1956). An alternative model frequently invoked is the small-polaron regime; here, charge motion is impeded by self-trapping of the carriers via lattice deformations of their immediate surroundings. Several variants of this scheme have been considered: double exchange (Rosencwaig 1969), pair localization (Buchenau 1975), bipolaron transport (Chakraverty 1980) or the standard polaron model (Haubenreisser 1961, Camphausen 1972, Wu & Mason 1981). A combination of mechanisms has also been proposed (Lorenz & Ihle 1979, 1980; Ihle 1984).

One difficulty in the localized-electron model is the very low apparent conductivity activation energy, which approaches zero as the sample temperature is raised from 130 to 700 K; this effect has been noted by all investigators from the pioneering studies of Verwey (1939) to the most recent investigations (Kuipers & Brabers 1979, Aragón 1985 (unpublished)); a result, typical of roughly 30 reported measurements (Miles *et al.* 1957) is displayed in Figure 7. One escape mechanism from this dilemma is provided by going beyond the standard zero-order polaron theory (Yamashita & Kurosawa 1958, Holstein 1959, Reik & Heese 1967). In the elementary approximation the mobility u varies with temperature T as $T^{-3/2}\exp(-\varepsilon/kT)$, where ε is the effective activation energy. In the more sophisticated approach the mobility is proportional to the function

$$u \sim (\tau/T)\exp\{-\eta\tanh(\hbar\omega_0/4kT)\} \tag{1}$$

with

$$\tau^2 \equiv [\sinh(\hbar\omega_0/2kT)]/2(\hbar\omega_0/k)^2\eta \tag{2}$$

Here ω_0 is the relevant optical vibration frequency and η is related to the electron–phonon coupling constant (η may roughly be regarded as specifying the number of phonons in the lattice deformation accompanying the charge carrier motion). The above two-parameter expression provides an enormous flexibility to the temperature dependence of u: depending on the choice of η and ω_0 one obtains two limiting cases, namely, a nearly temperature-independent mobility, or a near-exponential dependence $u \sim T^{-3/2}\exp\{-\eta\hbar\omega_0/4kT\}$, with a range of

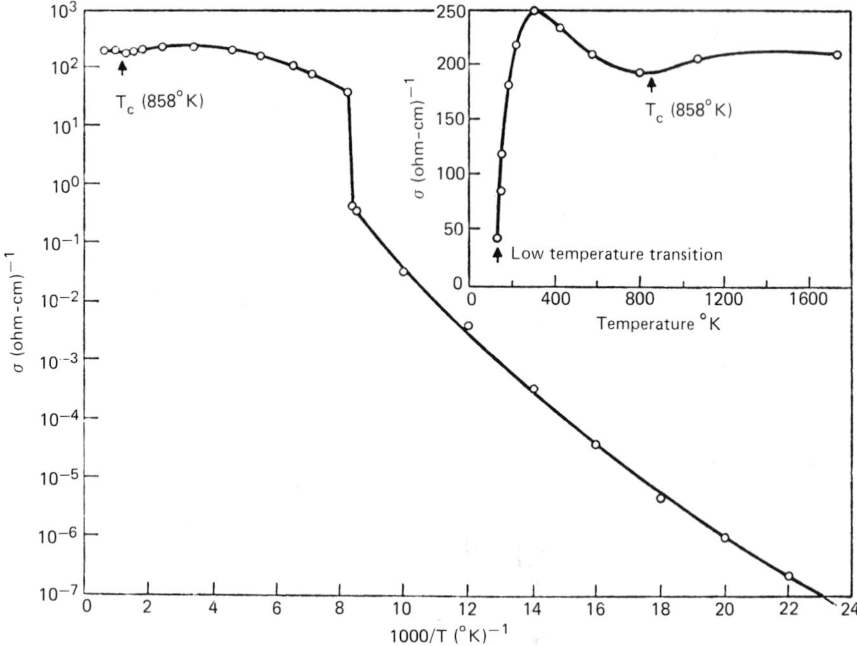

Figure 7. Conductivity of Fe_3O_4 as a function of temperature (after Miles *et al.* 1957).

intermediate behaviour. If one selects $1/\lambda = \omega_0/2\pi c$ in the range 1000–2000 cm^{-1} and η in the range 10–16, and under the assumption of a constant charge-carrier density, one can reproduce the temperature dependence of the conductivity of Figure 7 for $T = 130$–700 K. Thus, despite the low value of the effective conductivity activation energy ε in this temperature interval, the electron–phonon interaction parameters in equations (1) and (2) remain sizeable. Furthermore, the data below the Verwey transition may be roughly matched by the narrower choices of $1/\lambda = 300$–400 cm^{-1} and $\eta = 36$–40, so as to duplicate the observed slopes in the appropriate temperature range of Figure 7. If interpreted literally this would imply a noticeable softening of the lattice when magnetite is cooled below the Verwey transition temperature.

Since the two segments of Figure 7 may readily be accounted for on the basis of a well-accepted small-polaron theory, one should be cautious in interpreting departures from linearity of the data of Figure 7 as suggested by McKinnon *et al.* (1981), Shiozaki *et al.* (1981); for another viewpoint see Pai & Honig (1983). However, the above fit shows only that the small-polaron model is one of several candidate schemes for the interpretation of the transport properties in magnetite. Either this theory or the Hubbard electron-correlation model explains the constancy of the Seebeck coefficient above the Verwey transition (Kuipers & Brabers 1976, 1979; Whall 1980) exhibited in Figure 8. In the limit of constant charge-carrier density and with several additional assumptions, both the small-

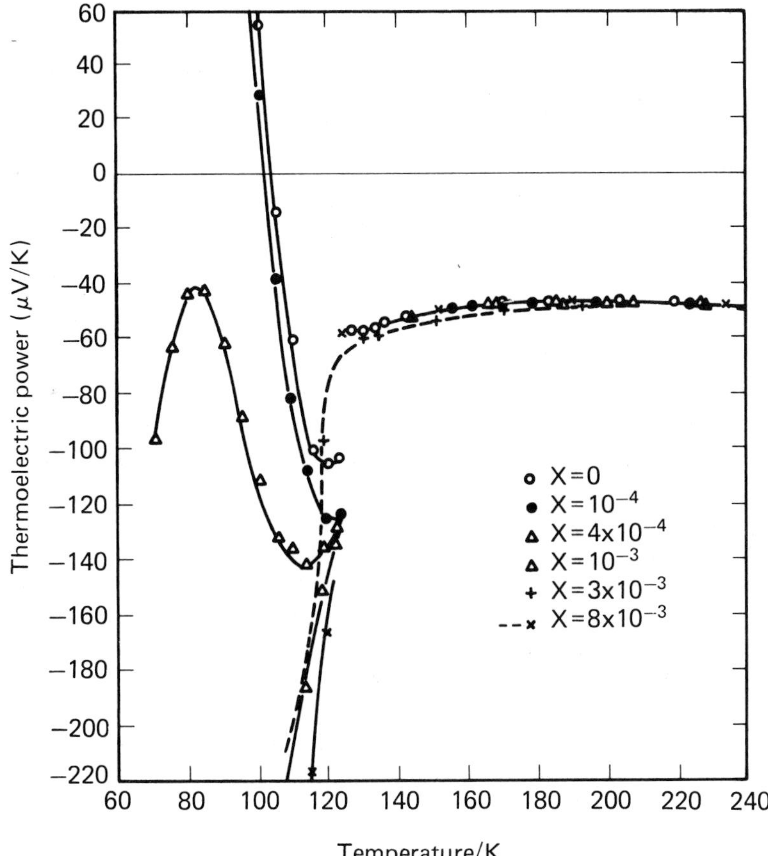

Figure 8. Seebeck coefficient versus temperature for undoped or lightly Ti-doped Fe_3O_4. The key designates the mole fraction of Ti in the specimens (after Kuipers & Brabers 1976, 1979).

polaron theory and electron-correlation effects (Chaikin & Beni 1976, Lewis 1976) lead to a modified Heikes formula for the Seebeck coefficient

$$\alpha = -(k/e)\ln[2(1-x)/x] \tag{3}$$

where x is the probability of occupancy by an electron of a transport level, and 2 is the spin degeneracy factor. For strictly stoichiometric Fe_3O_4, $|\alpha| \approx 60\ \mu V/deg$, rather close to the observed value: implicit in the above is the assumption that the charge carriers move only among the cations situated in the octahedral interstices.

The low-temperature regime depicted in Figure 8 is currently understood on the basis of a two-level scheme proposed by Kuipers & Brabers (1976, 1979) who were able to reproduce the observed temperature variations in detail.

In summary, there can be little doubt that charge carriers in Fe_3O_4 are localized rather than itinerant, but the exact cause of the localization remains to be established.

A second area of controversy centres on whether Fe_3O_4 undergoes a single or a multiplicity of Verwey transitions. Contradictory experimental evidence has been cited in favour of both alternatives. Specific heat measurements showing bifurcation effects and broad transitions have been reported by several groups of investigators (Westrum & Grønvold 1969, Evans & Westrum 1972, Rigo *et al.* 1978, 1980, 1983). Likewise, several Mössbauer investigations (Buckwald & Hirsch 1975, Galeczki *et al.* 1977) have led to the conclusion that the transformation occurs in stages. However, the most recent heat-capacity experiments on thoroughly annealed crystals (Matsui *et al.* 1977, Gmelin *et al.* 1983, Shepherd *et al.* 1984) have indicated that the transition in uniform specimens occurs in a single step over a temperature interval of not more than 0·4 K. Moreover, an increase in cation vacancy concentration produces a drop in the transition temperature (Gmelin *et al.* 1983) as shown in Figure 9. The entropy of the transition is near $R \ln 2$ per Fe_3O_4 formula unit (Shepherd *et al.* 1984), exactly half the value anticipated for a transformation from complete disorder to complete order of the Fe^{2+} and Fe^{3+} ions among the octahedral sites. This may be rationalized by postulating a chain formation such as $\ldots Fe^{2+}–Fe^{2+}–Fe^{3+}–Fe^{3+}–Fe^{2+}–Fe^{2+}–Fe^{3+}–Fe^{3+}\ldots$ along the various $\langle 110 \rangle$ directions.

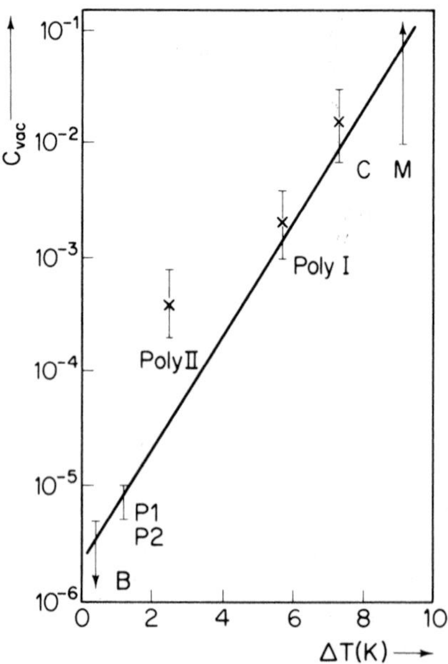

Figure 9. Decrease of Verwey transition temperature with increasing cation vacancy concentration in non-stoichiometric Fe_3O_4 (after Gmelin *et al.* 1983).

Such chain formation is consistent with structures proposed by Kita *et al.* (1983) on the basis of available X-ray and neutron diffraction studies; for an excellent review of the latter see Iizumi *et al.* (1982). The one-step transition is also in consonance with Mössbauer studies taken at closely spaced temperature intervals (Kündig & Hargrove 1969, Hargrove & Kündig 1970) in which no evidence was found for multistep transformations.

Finally, brief mention should be made of a very recent study in which it is claimed (Aragón *et al.* 1985) that beyond a critical cation vacancy concentration there is a sudden shift from a first-order to a second-order transition; representative results of several measurements are exhibited in Figure 10. These

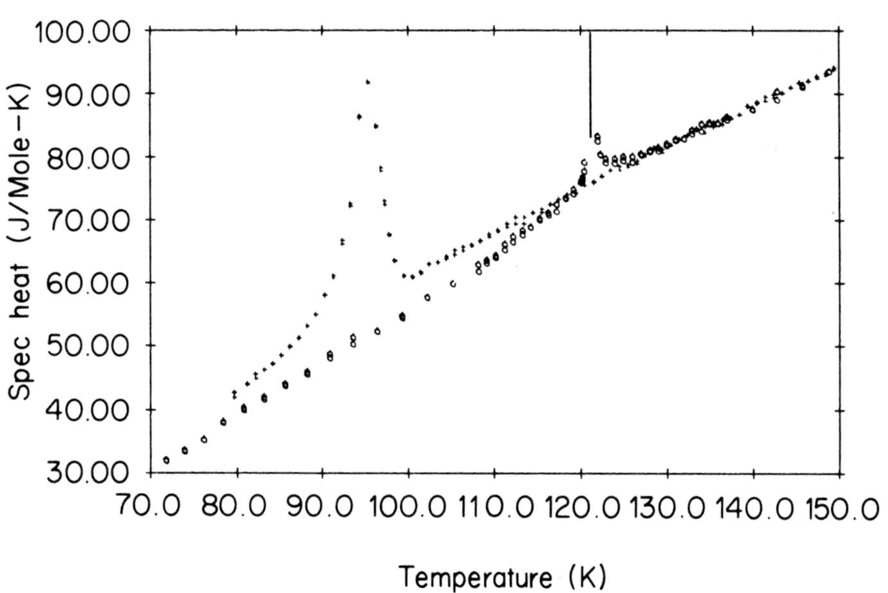

Figure 10. Specific heat anomalies in nearly stoichiometric and non-stoichiometric Fe_3O_4. Note first- and second-order transitions on right and left sides of the figure (after Aragón *et al.* 1985).

claims rest both on the type of heat-capacity anomaly that was encountered and on the measured cooling curves. Let the deviation from strict stoichiometry be represented by δ in $Fe_{3(1-\delta)}O_4$. Then for $\delta < \delta_{cr} \approx 0{\cdot}004$ one encounters the sharp spike shown on the right of Figure 10; also, a thermal arrest is found in the corresponding cooling curve. For $\delta > \delta_{cr}$ the heat capacity has the Λ-shape characteristic of second-order transitions shown on the left of Figure 10; no thermal arrests are encountered in the cooling curves. One should also note that the heat capacity anomalies are superposed on an apparently continuous background without any offset; this strongly suggests that no short-range ordering is present in well annealed specimens of Fe_3O_4. Further, whereas the two heat-capacity curves merge above the Verwey transition, the one for which $\delta > \delta_{cr}$ lies well above the other, indicating that the lattice of the former is softer. The results shown here are characteristic of five measurements taken below, and of four others above, the δ_{cr} value. Apparently, a new feature has been uncovered in these studies whose nature remains to be elucidated.

The above discussion should make it clear that a great deal still remains to be learned about the nature of the Verwey transition even though 45 years have passed since the pioneering studies by Verwey. However, at the very least, a clear understanding is now beginning to emerge concerning the strict precautions that must be taken to ensure composition uniformity; it is only with this type of sample that data can be collected which readily lend themselves to a proper theoretical interpretation.

Acknowledgements

The author is most indebted to Drs R. Aragón and J.P. Shepherd for many illuminating discussions concerning the Fe_3O_4 problem. Professor J. Appel was instrumental in providing a unified approach to the V_2O_3 alloy system; his penetrating insight into this problem is properly and gratefully acknowledged here. Finally, the author wishes to express his thanks to Professor A.W. Overhauser for his ready assistance in the theoretical interpretation of many of the experimental results described above.

References

Adler, D., 1968, *Rev. Mod. Phys.*, **40**, 714.
Anderson, P.W., 1956, *Phys. Rev.*, **102**, 1008.
Aragón, R., Shepherd, J.P., Koenitzer, J.W., Buttrey, D.J., Rasmussen, R.J. and Honig, J.M., 1985, *J. Appl. Phys.*, **57**, 3221.
Ashkenazi, J. and Weger, M., 1976, *J. Physique (Paris)*, **37**, C4–189.
Banus, M.D. and Reed, T.B., 1970, in *The Chemistry of Extended Defects in Non-metallic Solids*, edited by L. Eyring and M. O'Keeffe (Amsterdam: North Holland), p. 488.
Bhide, V.G., Rajoria, D.S., Rama Rao, G. and Rao, C.N.R., 1972, *Phys. Rev.* B, **6**, 1021.
Buchenau, U., 1975, *Phys. Stat. Solidi* b, **70**, 181.
Buckwald, R.A. and Hirsch, A.A., 1975, *Solid St. Commun.*, **17**, 621.
Camphausen, D.L., 1972, *Solid St. Commun.*, **11**, 99.
Camphausen, D.L., Coey, J.M.D. and Chakraverty, B.K., 1972, *Phys. Rev. Lett.*, **29**, 657.
Castellani, C., Natoli, C.R. and Ranninger, J., 1978, *Phys. Rev.* B, **18**, 4945, 4967, 5001.
Chaikin, P.M. and Beni, G., 1976, *Phys. Rev.* B, **13**, 647.
Chakraverty, B.K., 1980, *Phil. Mag.* B, **42**, 473.
Cullen, J.R. and Callen, E.R., 1970, *J. Appl. Phys.*, **41**, 879.
Cullen, J.R. and Callen, E.R., 1973, *Phys. Rev.* B, **7**, 397.
Dougier, P. and Hagenmuller, P., 1975, *J. Solid St. Chem.*, **15**, 158.
Dumas, J. and Schlenker, C., 1976, *J. Physique (Paris)*, **37**, C4–41.

Dumas, J., Schlenker, C., Tholence, J.L. and Tournier, R., 1975, *Solid St. Commun.*, **17**, 1215.
Evans, B.J. and Westrum, Jr., E.F., 1972, *Phys. Rev.* B, **5**, 3791.
Feinleib, J. and Paul, W., 1967, *Phys. Rev.*, **155**, 841.
Finger, L.W. and Hazen, R.M., 1980, *J. Appl. Phys.*, **51**, 5362.
Foëx, M., 1946, *C. R. Acad. Sci., Paris*, **223**, 1126.
Galeczki, R.A., Buckwald, R.A. and Hirsch, A.A., 1977, *Solid St. Commun.*, **23**, 201.
Gmelin, E., Lenge, N. and Kronmüller, H., 1983, *Phys. Stat. Solidi* a, **79**, 465.
Goodenough, J.B., 1967, *Mater. Res. Bull.*, **2**, 37, 49.
Goodenough, J.B., 1971, *Prog. Solid St. Chem.*, **5**, 145.
Goodenough, J.B., 1972, *Phys. Rev.* B, **5**, 2764.
Gossard, A.C., Menth, A., Warren Jr., W.W. and Remeika, J.P., 1971, *Phys. Rev.* B, **3**, 3993.
Grodzicki, M., Jepsen, O., Andersen, O. and Appel, J., 1983, personal communication.
Hargrove, R.S. and Kündig, W., 1970, *Solid St. Commun.*, **8**, 303.
Harrison, H.R., Aragón, R., Keem, J.E. and Honig, J.M., 1984, in *Inorganic Synthesis*, edited by S.L. Holt (New York: Wiley), Vol. 22, p. 43.
Haubenreisser, W., 1961, *Phys. Stat. Solidi*, **1**, 619.
Hegde, M.S. and Vasudevan, S., 1979, *Pramana*, **12**, 151.
Holstein, T., 1959, *Annals of Physics (New York)*, **8**, 343.
Honig, J.M. and Van Zandt, L.L., 1975, in *Annual Reviews of Materials Science*, edited by R.A. Huggins, R.N. Bube and R.W. Roberts (Palo Alto: Annual Reviews), Vol. 5, p. 225.
Ihle, D., 1984, *Phys. Stat. Solidi* b, **121**, 217.
Iizumi, M., Koetzle, T.F., Shirane, G., Chikazumi, S., Matsui, M. and Todo, S., 1982, *Acta Crystallogr.* B, **38**, 2121.
Joshi, G.M. and Honig, J.M., 1982, *Rev. Chim. Minérale*, **19**, 251.
Kachi, S., Kosuge, S. and Okinaka, H., 1973, *J. Solid St. Chem.*, **6**, 258.
Kawai, N. and Mochizuki, S., 1971, *Solid St. Commun.*, **9**, 1393.
Kita, E., Tokuyama, Y., Tasaki, A. and Siratori, K., 1983, *J. Magnetism and Magnetic Materials*, **31–34**, 787.
Kuipers, A.J.M. and Brabers, V.A.M., 1976, *Phys. Rev.* B, **14**, 1401.
Kuipers, A.J.M. and Brabers, V.A.M., 1979, *Phys. Rev.* B, **20**, 594.
Kündig, W. and Hargrove, R.S., 1969, *Solid St. Commun.*, **7**, 223.
Kuroda, N. and Fan, H.Y., 1977, *Phys. Rev.* B, **16**, 5003.
Kuwamoto, H., Honig, J.M. and Appel, J., 1980, *Phys. Rev.* B, **22**, 2626.
Ladd, L. and Paul, W., 1969, *Solid St. Commun.*, **7**, 425.
Lewis, A., 1976, *Phys. Rev.* B, **13**, 1855.
Lorenz, B. and Ihle, D., 1979, *Phys. Stat. Solidi* b, **96**, 659.
Lorenz, B. and Ihle, D., 1980, *Phys. Stat. Solidi* b, **98**, K63.
MacMillan, A.J., 1960, *Laboratory Report for Insulation Research* (Cambridge: Massachusetts Institute of Technology), p. 172.
Matsui, M., Tōdō, S. and Chikazumi, S., 1977, *J. Phys. Soc. Japan*, **42**, 1517.
McKinnon, W.R., Hurd, C.M. and Shiozaki, I., 1981, *J. Phys. C: Solid St. Phys.*, **14**, L877.
McWhan, D.B. and Remeika, J.P., 1970, *Phys. Rev.* B, **2**, 3734.
McWhan, D.B., Rice, T.M. and Remeika, J.P., 1969, *Phys. Rev. Lett.*, **23**, 1984.
McWhan, D.B., Menth, A. and Remeika, J.P., 1971, *J. Physique (Paris)*, **32**, C1–1079.
McWhan, D.B., Menth, A., Remeika, J.P., Brinkman, W.F. and Rice, T.M., 1973, *Phys. Rev.* B, **7**, 1920.
Miles, P.A., Westphal, W.B. and von Hippel, A., 1957, *Rev. Mod. Phys.*, **29**, 279.
Millar, R.W., 1929, *J. Am. Chem. Soc.*, **51**, 215.
Morin, F.J., 1959, *Phys. Rev. Lett.*, **3**, 34.
Mott, N.F., 1974, *Metal–Insulator Transitions* (London: Taylor & Francis).

Mott, N.F., Pepper, M., Pollitt, S., Wallis, R.H. and Adkins, C.J., 1975, *Proc. R. Soc.*, A**345**, 169.

Mott, N.F., 1979, *Festkörperprobleme*, **19**, 331.

Mott, N.F., 1980, *Phil. Mag.* B, **42**, 327.

Nichols, D.N. and Sladek, R.J., 1981, *Phys. Rev.* B, **24**, 3155.

Nichols, D.N., Sladek, R.J. and Harrison, H.R., 1981, *Phys. Rev.* B, **24**, 3025.

Okamoto, A., Fujita, Y. and Tatsuyama, C., 1983, *J. Phys. Soc. Japan*, **52**, 312.

Okamura, T., 1932, *Science Reports of the Tôhoku Imperial University*, **21**, 231.

Oliver, M.R., Dimmock, J.O., McWhorter, A.L. and Reed, T.B., 1972, *Phys. Rev.* B, **5**, 1078.

Overhauser, A.W., 1979, personal communication.

Pai, M. and Honig, J.M., 1983, *J. Phys. C: Solid St. Phys.*, **16**, L35.

Parks, G.S. and Kelley, K.K., 1926, *J. Phys. Chem.*, **30**, 47.

Penney, T., Shafer, M.W. and Torrance, J.B., 1972, *Phys. Rev.* B, **5**, 3669.

Raccah, P.M. and Goodenough, J.B., 1967, *Phys. Rev.*, **155**, 932.

Rao, C.N.R. and Rao, K.J., 1977, *Phase Transitions in Solids* (New York: McGraw-Hill).

Rao, C.N.R., 1984, *Accounts of Chemical Research*, **17**, 83.

Rao, C.N.R., Buttrey, D.J., Otsuka, N., Ganguly, P., Harrison, H.R., Sandberg, C.J. and Honig, J.M., 1984, *J. Solid St. Chem.*, **51**, 266.

Reik, H.G. and Heese, D., 1967, *Phys. Stat. Solidi*, **24**, 281.

Rigo, M.O. and Kleinclauss, J., 1980, *Phil. Mag.* B, **42**, 393.

Rigo, M.O., Kleinclauss, J. and Pointon, A.J., 1978, *Solid St. Commun.*, **28**, 1013.

Rigo, M.O., Mareche, J.F. and Brabers, V.A.M., 1983, *Phil. Mag.* B, **48**, 421.

Robinson, W.R., 1975, *Acta Crystallogr.* B, **31**, 1153.

Rosencwaig, A., 1969, *Can. J. Phys.*, **47**, 2309.

Shepherd, J.P., Koenitzer, J.W., Aragón, R., Sandberg, C.J. and Honig, J.M., 1984, *Phys. Rev.*, B, **31**, 1107.

Shiozaki, I., Hurd, C.M., McAlister, S.P., McKinnon, W.R. and Strobel, P., 1981, *J. Phys. C: Solid St. Phys.*, **14**, 4641.

Shivashankar, S.A. and Honig, J.M., 1983, *Phys. Rev.*, **28**, 5695.

Sinclair, F. and Colella, R., 1979, *Solid St. Commun.*, **31**, 359.

Tatsuyama, C. and Fan, H.Y., 1980, *Phys. Rev.* B, **21**, 2977.

Van Zandt, L.L., Honig, J.M. and Goodenough, J.B., 1968, *J. Appl. Phys.*, **39**, 594.

Van Zandt, L.L. and Honig, J.M., 1974, in *Annual Reviews of Materials Science*, edited by R.A. Huggins, R.H. Bube and R.W. Roberts (Palo Alto: Annual Reviews) Vol. 4, p. 19.

Verwey, E.J.W., 1939, *Nature*, **144**, 327.

Verwey, E.J.W. and Haayman, P.W., 1941, *Physica*, *8*, 979.

Verwey, E.J.W., Haayman, P.W. and Romeijn, F.C., 1947, *J. Chem. Phys.*, **15**, 174, 181.

Weiss, P. and Forrer, R., 1929, *Ann. Phys., Paris*, **21**, 279.

Westrum, Jr. E.F. and Grønvold, F., 1969, *J. Chem. Thermodynamics*, **1**, 543.

Whall, T.E., 1980, *Phil. Mag.* B, **42**, 423.

Wilson, A.H., 1954, *The Theory of Metals* (Cambridge: Cambridge University Press), Chap. 8.

Wu, C.C. and Mason, T.O., 1981, *J. Amer. Ceram. Soc.*, **64**, 520.

Yamashita, J. and Kurosawa, T., 1958, *J. Phys. Chem. Solids*, **5**, 34.

Yang, H., Sladek, R.J. and Harrison, H.R., 1983, *Solid St. Commun.*, **47**, 955.

Zeiger, H.J., 1975, *Phys. Rev.* B, **11**, 5123.

CHAPTER 11

chemical bond, crystal structure and the metal–nonmetal transition in oxide bronzes

J.P. Doumerc, M. Pouchard and P. Hagenmuller
Laboratoire de Chimie du Solide du CNRS, Université de Bordeaux, 1, 351, cours de la Libération, 33405 Talence Cedex, France

1. Introduction

It is generally accepted that the first reported oxide bronzes were sodium tungsten bronzes. They were obtained by Wöhler in 1823 by reducing in a hydrogen flow a mixture of sodium tungstate Na_2WO_4 and tungsten oxide WO_3. The reaction products had a metallic lustre and a gold-yellow colour which account for the term 'bronzes' used to qualify them later on. According to the ideas prevailing at that time, they were considered as stoichiometric compounds and only one century later Hägg (1935) proved the existence of a large compositional range ($0{\cdot}32 \leqslant x \leqslant 0{\cdot}93$) for Na_xWO_3. The high metallic-type electrical conductivity observed for the largest x-values and the possibility of varying their composition over a large range rapidly gave rise to many investigations concerning structural and physical properties. On the other hand, oxide bronzes of many transition elements other than tungsten have been prepared and studied more recently (see, for instance, Hagenmuller 1971).

For these historical reasons oxide bronzes have often been defined as compounds having the general formula M_xTO_n in which relatively electropositive metal atoms (M) are inserted into a more or less covalent network TO_n formed of oxygen and a transition element (T). A characteristic trend of the oxide bronzes which is of importance for the subject discussed here is the oxidation–reduction process between the metal M and the transition element T; it leads to the ionization of M and the partial reduction of T. Therefore, oxide bronzes can be considered as a particular class of *mixed valence compounds*. This property can be used as a more general definition, leading to the application of the name 'bronze' to any nonstoichiometric compound in which the formal oxidation state of a transition element can vary continuously or discontinuously as a function of chemical composition changes such as either cationic insertion or cationic or anionic substitution.

Therefore, to understand the physical properties of bronzes, it will first be necessary to consider whether the d-electrons of the transition element are localized or delocalized. This will depend on the interplay of many parameters, among which the spatial extension of the d-orbitals will of course play a major role: it is common to ascribe the semiconducting properties of most of the vanadium bronzes to the relatively short extension of the 3d orbitals, whereas the 5d orbitals of tungsten can overlap to a greater extent with 2p oxygen orbitals, explaining the metallic properties often shown by tungsten bronzes (Hagenmuller 1971). Molybdenum bronzes constitute an intermediate case. However, their highly anisotropic structures often determine their electronic properties and some of them can exhibit metal–nonmetal (M–NM) transitions with temperature variation.

In order to characterize the origin of M–NM transitions in bronze systems, we have to consider the various effects which can result from cationic or anionic substitutions or insertions.

(a) Crystal structure changes may lead to a modification of energy band diagrams, such as band splitting.
(b) Even when the lattice symmetry remains unchanged, a variation of unit cell parameters, and hence interatomic distances, may be observed. As the cation–cation distance reaches a critical value, a transition from delocalized to localized d-electrons may occur (Mott 1974, Goodenough 1971, 1984).
(c) The character of the chemical bonds will of course change with the nature of inserted or substituted atoms. For instance, increasing covalency will enhance the cation–anion–cation interactions and hence may lead to broadening of the corresponding energy bands.
(d) We have defined a bronze as a compound in which the oxidation state of a transition element can vary with the composition. Therefore, in the cases where electrons are delocalized and partially occupy a conduction band, a shift of the Fermi level may result from appropriate composition change.
(e) If substituted or inserted atoms are randomly distributed in their crystal sublattice, the resulting random potential may be large enough to give rise to Anderson localization.
(f) Starting from low insertion or substitution rates, the possibility of formation of an impurity band should also be considered.

As we shall see through the selected examples reported in this chapter, several of the above effects should generally be considered for each class of materials. This often leads to difficulties in explaining the electronic properties of a given bronze system, and may be the reason why several different and even conflicting theories have sometimes been proposed. We will concentrate on discussing their respective advantages and limits, in a field where Sir Nevill Mott's ideas have had a strong influence.

2. Tungsten bronzes

2.1. Crystal structure

Tungsten bronzes, M_xWO_3, can be described by considering the insertion of a metal M in a WO_3 sublattice. Two structural characteristics are always observed in tungsten bronzes: (i) tungsten atoms are always surrounded by six oxygen atoms forming a more or less regular octahedron, (ii) these octahedra are themselves corner-linked. An exception to this rule is the possible formation of shear planes along which octahedra share edges. These shear planes can be observed when the ratio O/W decreases below 3 and corresponds to annihilation of oxygen vacancies by cooperative atomic displacements (Wadsley 1963, Hyde *et al.* 1974).

The WO_6 octahedra can be arranged in different ways, depending on the nature and concentration of M. This leads to interstitial sites or tunnels of different sizes and with various coordination numbers. The structures to be considered may be easily deduced from one another by comparing the array of the octahedra and/or the distortions of the octahedra from the ideal symmetry.

The cubic phase of cation-deficient perovskite type

A cubic symmetry phase is observed for a large variety of oxide or oxyfluoride tungsten bronze systems, such as Na_xWO_3, Li_xWO_3, Ca_xWO_3, RE_xWO_3 (RE = rare earth element), $Na_xWO_{3-x}F_x$, $Na_xTa_yW_{1-y}O_3$ (Hagenmuller 1971, Doumerc 1978), and so on. It can be described as a ReO_3 network (Figure 1) where the 12-coordinate sites at the centre of the cube are increasingly occupied by the M^{n+} ions. When all these sites are filled up the ideal perovskite structure is obtained.

Two points have been discussed: (i) whether there is an ordering of M atoms into the A sites and (ii) the slight distortion from ideal cubic symmetry (sometimes not observed in earlier X-ray diffraction studies).

It seems that ordering in channels can occur only for special compositions such as, for example, $Na_{0{\cdot}75}WO_3$ (Atoji & Rundle 1960) or $Al_{0{\cdot}125}WO_3$ (Hagenmuller 1971).

Evidence for distortions in some 'cubic' bronzes was first given by observation of optical birefringence (Ingold & de Vries 1958), specific heat (Inaba & Naito 1975), neutron diffraction studies or local probe characterization such as n.m.r. of ^{23}Na nuclei. Using neutron diffraction, Wiseman & Dickens (1976) have shown for instance that the WO_6 octahedra are slightly tilted at room temperature in Na_xWO_3 and Li_xWO_3, whereas La_xWO_3 remains perfectly cubic. Similar distortions have been observed for pseudo-cubic hydrogen tungsten bronzes D_xWO_3 (Wiseman & Dickens 1973). N.m.r. studies on sodium nuclei allowed Bonera *et al.* (1971) and Tunstall (1975) to show that there were actually two non-equivalent sites for sodium in Na_xWO_3. In the X-ray diffraction study of

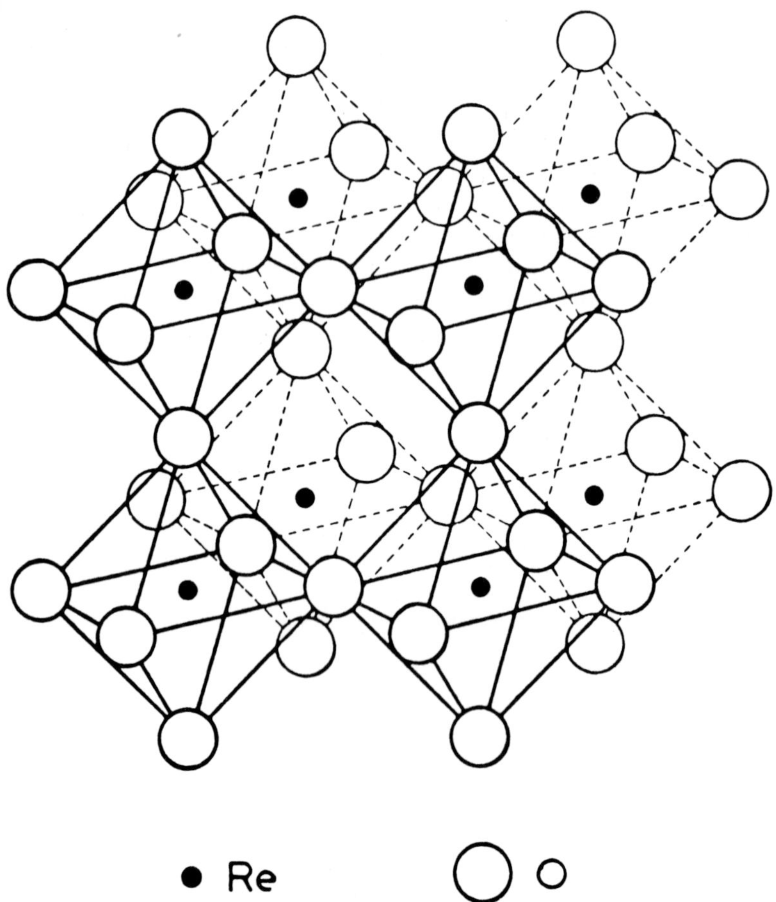

Figure 1. Crystal structure of ReO_3.

Clarke (1977) the existence of four distinct structural phases of pseudo-cubic Na_xWO_3 ($0{\cdot}6 < x < 1$) was suggested. The space group assignments and transition temperatures were confirmed by the Raman study of Flynn *et al.* (1977).

Structures related to WO_3 type phases

These are met with in most M_xWO_3 or $M_xWO_{3-x}F_x$ systems for a low insertion ratio, x (a few at.%).

When the temperature is increased, WO_3 exhibits successively five (at least!) allotropic forms: monoclinic (M_{II}) below 230 K (Diehl *et al.* 1978, Tanisaki 1960a), triclinic between 230 K and room temperature (Tanisaki 1960b, Diehl *et al.* 1978, Salje & Viswanathan 1975), again monoclinic (M_I) between room temperature and 600 K (Tanisaki 1960b, Loopstra & Rietveld 1969), orthorhomic between 600 K and ~1000 K and tetragonal above 1000 K (Kehl *et al.* 1952). The existence of a cubic phase at higher temperatures is not yet clearly established.

A detailed description of these phases is out of the scope of this chapter but it may be worthwhile to mention some general features of the orthorhombic and tetragonal phases, showing how they can be deduced from the ReO_3 structure by considering shifts of the tungsten atoms from the centre of the oxygen octahedra (Pouchard, private communication).

In the tetragonal form these shifts are parallel to one of the four-fold axes of the ideal cubic cell. Their direction is opposite for two neighbouring tungsten atoms as illustrated in Figure 2. This leads to a tetragonal unit cell with lattice

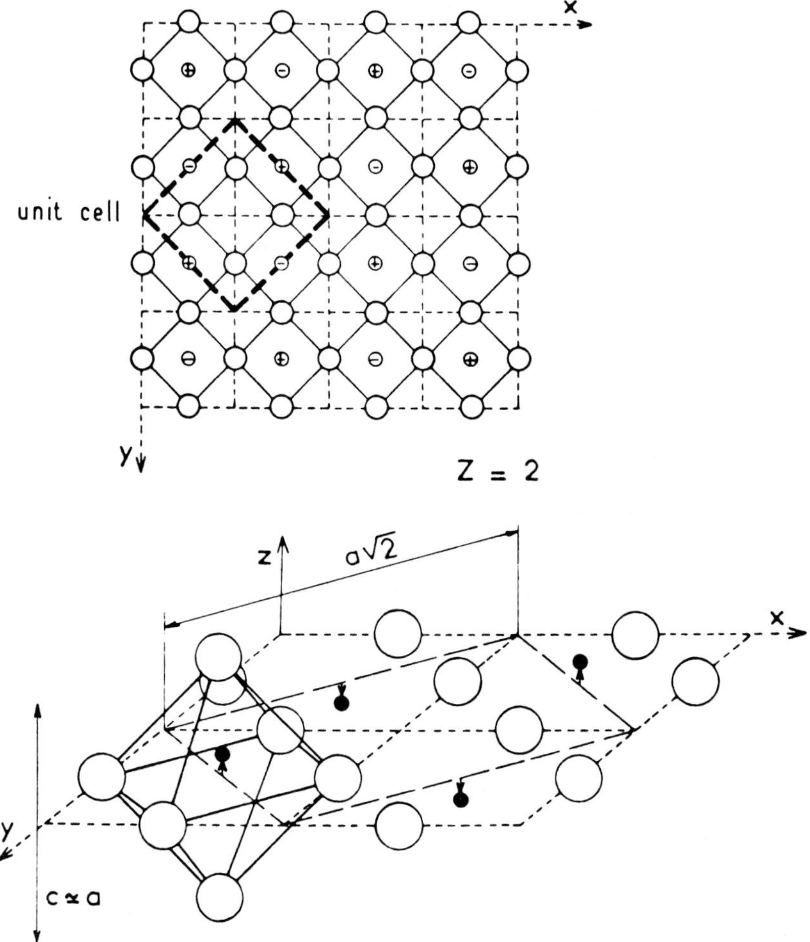

Figure 2. Relationship between the tetragonal form of WO_3 ($T > 1000$ K) and cubic ReO_3.

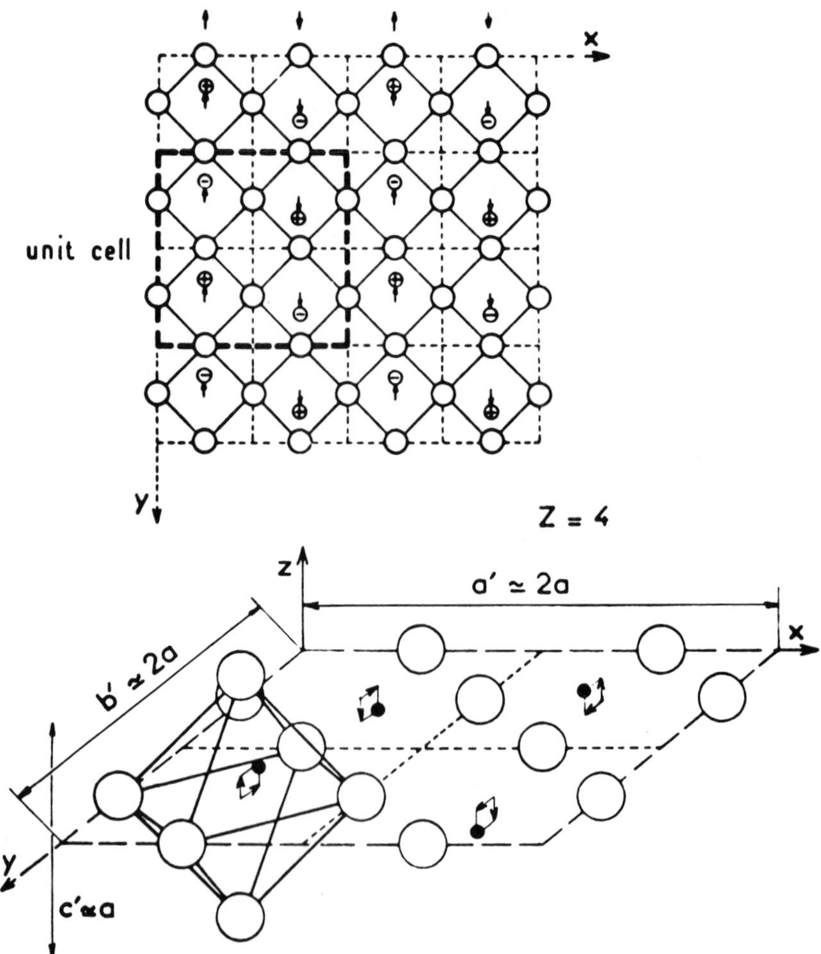

Figure 3. Relationship between the orthorhombic form of WO_3 ($600 < T < 1000$ K) and cubic ReO_3.

parameters $a \simeq a_c\sqrt{2}$ and $c \simeq a_c$ where a_c is the lattice parameter of the corresponding ideal cubic cell.

In the orthorhombic and monoclinic forms, tungsten atoms are displaced along two of the fourfold axes of the cubic cell as shown in Figure 3. This phenomenon implies a doubling of the lattice parameter *a*. It seems that in the monoclinic phase the octahedra are slightly tilted, giving rise also to doubling of *c* (Loopstra & Boldrini 1966).

The hexagonal potassium tungsten bronze (HKWB) phase

The structure of the hexagonal phase $K_{0 \cdot 28}WO_3$ was determined first by Magnéli (1953). As outlined above and illustrated in Figure 4, WO_6 octahedra are

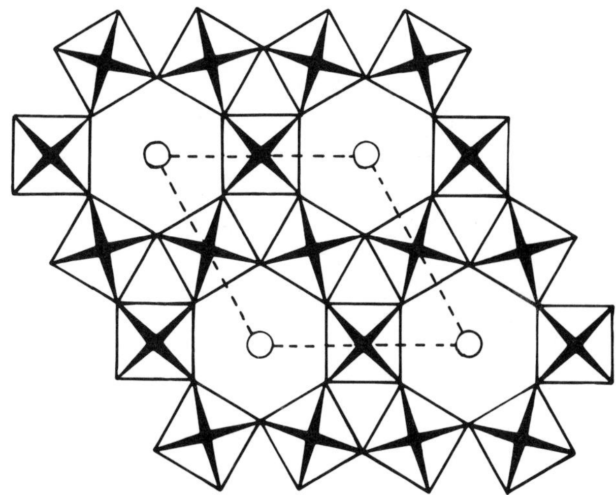

Figure 4. Projection on the basal plane of the structure of the hexagonal potassium tungsten bronze (HKWB) phase (Magnéli 1953).

corner-linked in all directions. The arrangement of the octahedra leads to channels of hexagonal and triangular cross section containing (12+6) and (6+3) coordinated sites respectively. Only relatively large ions such as potassium, rubidium, caesium, indium and thallium can occupy hexagonal tunnels, whereas lithium seems to be able to occupy triangular prismatic sites in K_xWO_3 (Banks & Goldstein 1968). When the hexagonal sites are filled up the formula MW_3O_9 is obtained. Either the substitution of oxygen by fluorine or that of tungsten by tantalum leads to an increase of the unit cell volume. However, in both cases it is mainly the c-parameter which increases, whereas a remains almost constant. This phenomenon has been attributed to a possible anionic ordering in the former case and in the latter one to a π(Ta–O) bond weaker than the π(W–O) bond, which is expected to have a stronger influence along c-axis, as the anisotropy of the structure suggests that π-bonding should be stronger in this latter direction (Doumerc *et al.* 1981).

Hussain (1978) has reinvestigated the M_xWO_3 bronzes with M = K, Rb, Cs and besides the hexagonal potassium tungsten bronze and tetragonal potassium tungsten bronze (TKWB) phases has found an intergrowth structure for $0{\cdot}06 < x < 0{\cdot}1$. The location of the alkali metal in the hexagonal potassium tungsten bronze structure has been discussed by Kihlborg & Hussain (1979).

A hexagonal form of WO_3 has been obtained by Gerand *et al.* (1979). These authors have also prepared hydrogen hexagonal bronzes (Figlarz & Gerand 1980) and lithium hexagonal bronzes (Gerand 1984) by chemical and electrochemical insertion.

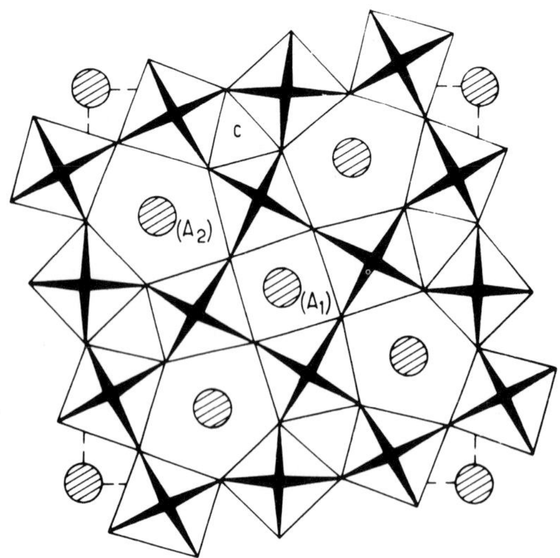

Figure 5. Projection on the (001) plane of the tetragonal potassium tungsten bronze (TKWB) phase (Magnéli 1949).

The tetragonal potassium tungsten bronze (TKWB) phase

The structure of the tetragonal potassium tungsten bronze phase as described by Magnéli (1949) is shown in Figure 5. There are three kinds of channels, all of them parallel to the *c*-axis, and of triangular, square and pentagonal cross sections, corresponding to sites labelled c, A_1 and A_2 respectively. Saturation of these sites leads to the formula $(A_1)_2(A_2)_4C_4W_{10}O_{30}$. Sodium and potassium can occupy the A_1 and A_2 sites, while lithium can enter the c sites (Takusagawa & Jacobson 1976). Superstructures have been identified for tin tungsten bronzes (Steadman 1972). Many niobates and tantalates exhibit this structure (Hagenmuller 1971).

2.2. Sodium tungsten bronzes

Phase diagram

The phase diagram of the system Na_xWO_3 has been studied by Ribnick *et al.* (1963) as a function of composition and temperature between room temperature and 1300 K. It is shown in Figure 6. The structures of the various phases observed in this system have been described above. For small *x*-values the gradual insertion of sodium into the WO_3 array leads to an evolution of structural types similar to that resulting from an increase of temperature: the M, O and T_I phases are isostructural with the corresponding WO_3 forms. The distortions progressively disappear as the sodium concentration increases until a cubic phase

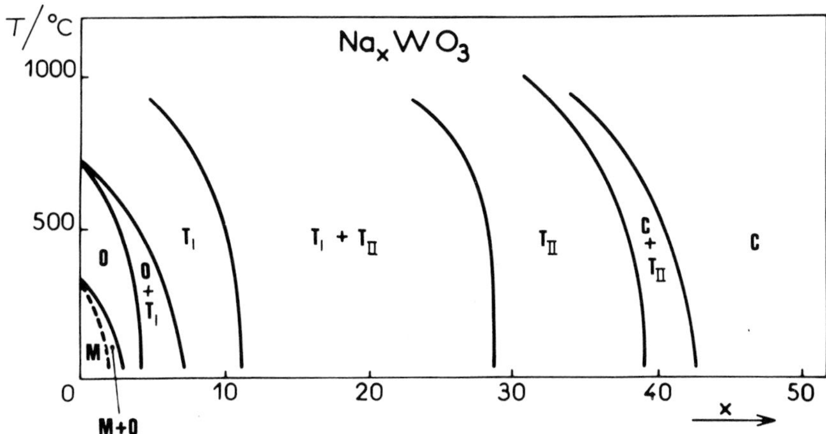

Figure 6. Phase diagram of the Na_xWO_3 system. (Adapted from Ribnick *et al.* 1963)

C is obtained. The evolution is not continuous: a phase T_{II} of tetragonal potassium tungsten bronze type appears before the cubic symmetry is reached, as well as more or less wide two-phase domains.

It is worthwhile to mention that it seems possible to obtain by quenching a metastable cubic phase at room temperature for x-values smaller than the equilibrium values given in Figure 6. McNeil and Conroy (1962), then Lightsey (1973) obtained crystals of cubic symmetry for x as small as 0·22, by diffusing sodium from crystals of higher Na-content.

Brown & Banks (1954) have shown that the lattice parameter of the cubic phase varies linearly with the insertion ratio x as $a(\text{Å}) = 3{\cdot}7845 + 0{\cdot}0820x$. This relationship seems, however, to break down as the M–NM transition is approached (Tunstall & Ramage 1980), that is, for $x \leqslant 0{\cdot}25$.

Physical properties

For high sodium contents ($x \geqslant 0{\cdot}25$) tungsten bronzes exhibit a metallic behaviour which has been characterized by intensive studies of their magnetic, electrical and optical properties (Hagenmuller 1971).

Measurements on bronzes with small insertion rate are less numerous, due to lack of crystals. However, some examples of semiconducting behaviour are known for $x < 0{\cdot}22$ in the Na_xWO_3 series and more generally for $nx \lesssim 0{\cdot}2$ in the case of other $M_x^{n+}WO_3$ bronzes.

Lightsey *et al.* (1976) have plotted as a function of composition the room temperature resistivity data obtained by different authors (McNeil & Conroy 1962, Muhlestein & Danielson 1967), including their own results (Figure 7). Using a band model which will be described in §2.4, Holcomb (1978) has determined, as shown in Figure 7, the products of the mean free path l and the

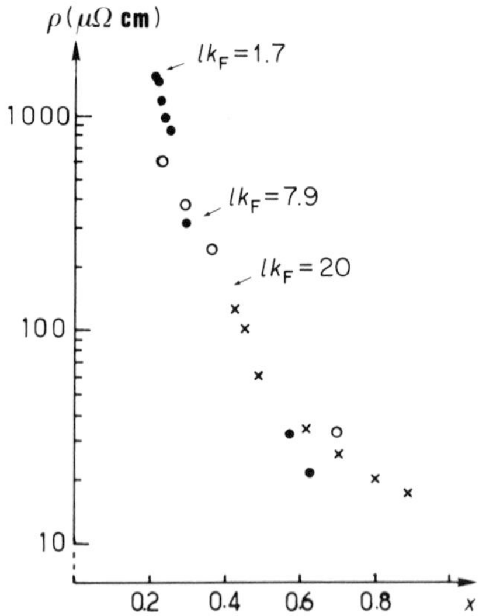

Figure 7. Room temperature resistivity of Na_xWO_3 variation vs. x. Some values of lk_F are given for the sake of comparison with the Ioffe–Regel criterion (Adapted from Holcomb 1978).

electron wavenumber k_F in order to test the experimental behaviour with respect to the Ioffe–Regel (1960) criterion, which predicts that metallic behaviour can be observed only if $lk_F > 1$. The Ioffe–Regel criterion led to the concept of a finite minimum metallic conductivity (σ_{min}) in disordered systems (Thouless 1978, Mott 1974, 1984, Mott & Kaveh 1983).

Hall effect measurements show that the number of carriers is roughly equal to the number of inserted alkali ions (Muhlestein & Danielson 1967, Lightsey *et al.* 1976). However, Muhlestein and Danielson found that the electronic properties of Na_xWO_3 cannot be explained quantitatively with a simple free-electron approximation and Lightsey *et al.* (1976) observed a Friedman anomaly for $x = 0{\cdot}22$ at 77 and 4 K.

Zumsteg (1976) measured heat capacity and magnetic susceptibility of cubic Na_xWO_3 for $0{\cdot}22 < x < 0{\cdot}60$. The observed Pauli paramagnetism is consistent with the metallic character (Sienko 1963). The results of Zumsteg are in agreement with the earlier heat capacity measurements for $x > 0{\cdot}50$ of Vest *et al.* (1958) and the magnetic susceptibility determination of Kupka & Sienko (1950) and Greiner *et al.* (1962). They show that the electronic density of states at the Fermi level $N(E_F)$ varies linearly with the Na concentration.

An interesting feature of metallic tungsten bronzes is the evolution of their bright colour with composition (Table 1). To explain it the shift of both plasma frequency and interband energy transitions have been considered as a function of composition (Dickens *et al.* 1968). Interference effects on the surface have also been invoked (Taylor 1969).

Table 1. Colour vs. composition for Na_xWO_3.

x	Crystal structure	Single crystal colour
0	monoclinic	yellow-green (translucent)
0·1	tetragonal (T_I)	black
0·2	two-phase	blue-black
0·3	tetragonal (TKWB)	blue
0·4	cubic	blue
0·5	cubic	violet
0·6	cubic	red-purple
0·7	cubic	red-orange
0·8	cubic	orange
0·9	cubic	yellow
1·0	cubic	orange (Chamberland 1969)

Optical properties have been studied by Lynch *et al.* (1973), Consadori & Stella (1970) and Camagni *et al.* (1977). Owen *et al.* (1978) made direct measurements of the optical dielectric constants which they interpreted on the basis of a rigid band model. Using polarized light, Atoji (1978) has investigated the birefringent twin-domain structure of bronze surfaces. He ascribed it to Na-deficient, epitaxial surface films.

Elastic constants have been measured, using Brillouin light scattering, by Benner *et al.* (1977, 1979) who found that the lattice constant of covalent W–O bonds increases with sodium concentration. McColm and Wilson (1978) have performed microhardness measurements and shown the dominant role played by the interactions of inserted cations with the WO_3 matrix in determining the mechanical properties of these bronzes.

N.m.r. measurements on ^{23}Na and ^{183}W nuclei as well as other investigations on Na_xWO_3 such as photoelectron spectroscopy will be mentioned in §2.4 which is devoted to the electronic structure of the materials.

2.3. Substituted sodium tungsten bronzes

The $Na_xWO_{3-x}F_x$ system

The phase diagram $NaF–WO_2–WO_3$ is shown in Figure 8, where we give also the evolution of the $W(O,F)_6$ octahedra as x increases (Doumerc & Pouchard 1970).

The M, O, T_I and C phases are analogous to the corresponding allotropic forms of WO_3 (§2.1). The T′ phase is isostructural with the tetragonal form of $BaTiO_3$; the tungsten atoms are shifted in a direction parallel to a fourfold axis of the idealized cubic cell. All these shifts keep the same direction, unlike in the T_I phase. The lattice constant of the C-cubic phase varies linearly with x and is given

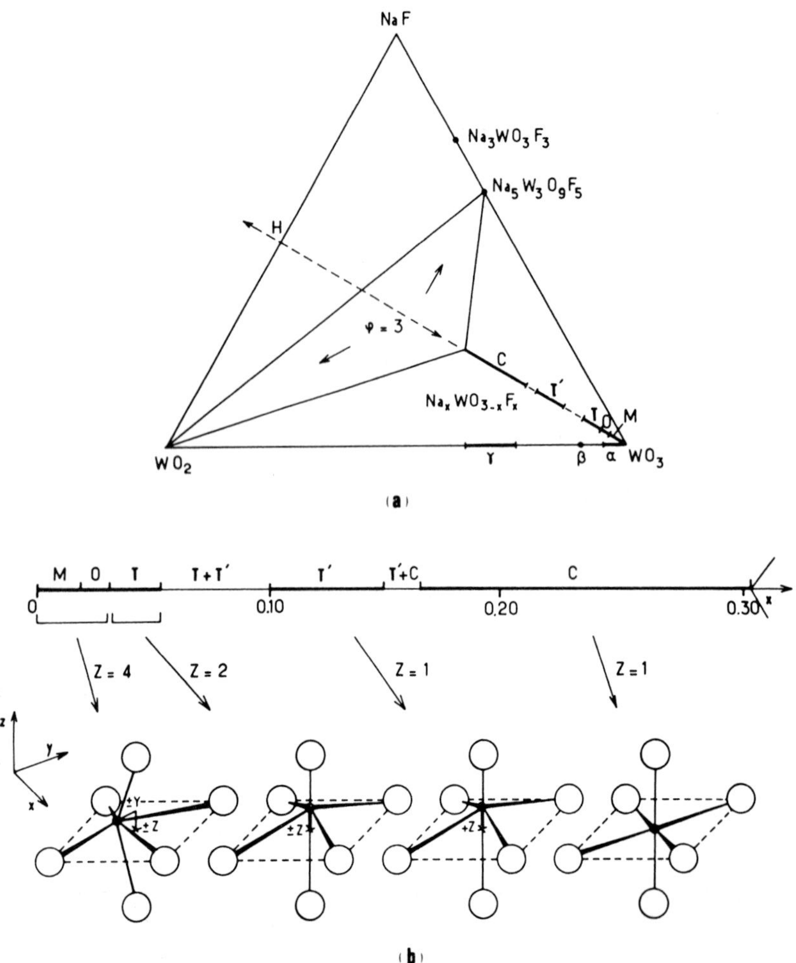

Figure 8. (*a*) Phase diagram of $Na_xWO_{3-x}F_x$ and (*b*) evolution of the octahedron distortion with x (Doumerc & Pouchard 1970).

by the equation

$$a_c\,(\text{Å}) = 0{\cdot}1022x + 3{\cdot}7938$$

The variation of the logarithm of the electrical conductivity σ with $1000/T$ and the variation of the thermoelectric power, S, with T are given in Figures 9 and 10 respectively, for polycrystalline samples of various compositions.

The magnetic susceptibility is nearly temperature independent, that is, of Pauli type. These results (Doumerc 1978) show that: (i) a metal–nonmetal transition seems to occur for a composition within the cubic phase domain, (ii)

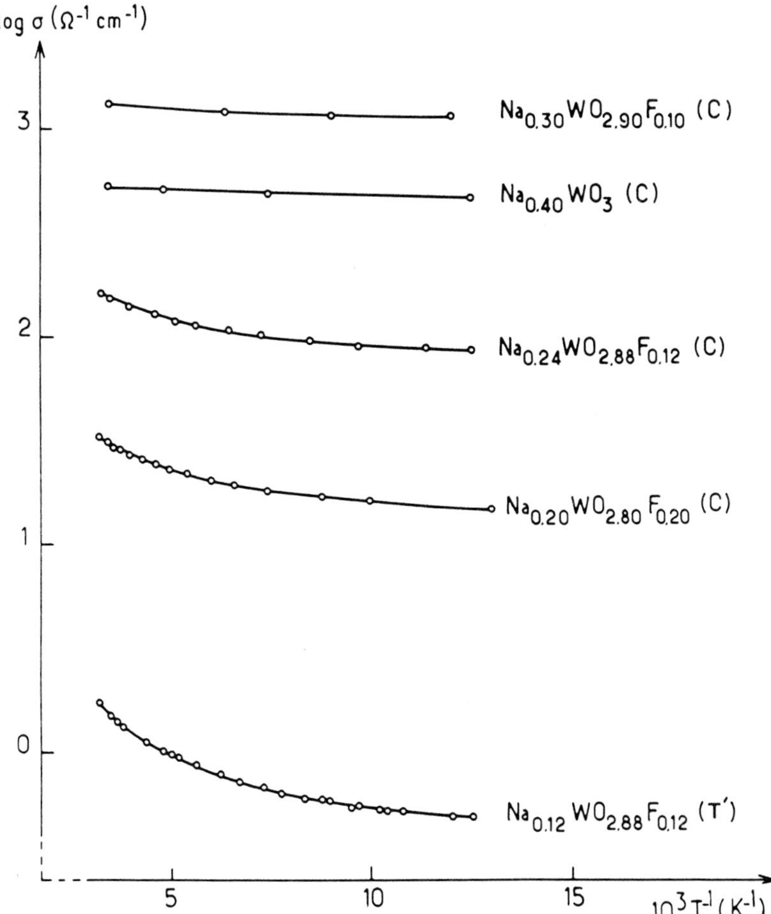

Figure 9. Variation of the logarithm of electrical conductivity vs. $1000/T$ for $Na_xWO_{3-y}F_y$.

there is no thermal activation of the number of carriers since $|S|$ increases with T (Mott & Davis 1979).

The system $Na_xTa_yW_{1-y}O_3$

Crystal growth by electrolytic reduction from Na_2WO_4–WO_3–Ta_2O_5 melts has been described by Weller *et al.* (1970), Doumerc *et al.* (1979) and more recently by Dubson (1984). The method is derived from that currently used for non-substituted bronzes (Wold *et al.* 1964). According to those works the quality and size of $Na_xTa_yW_{1-y}O_3$ single crystals obtained in this way was better for relatively high sodium content, that is, for $x \simeq 0{\cdot}6$, with y ranging between 0 and 0·25. Therefore low x-value crystals were prepared by deintercalation of sodium

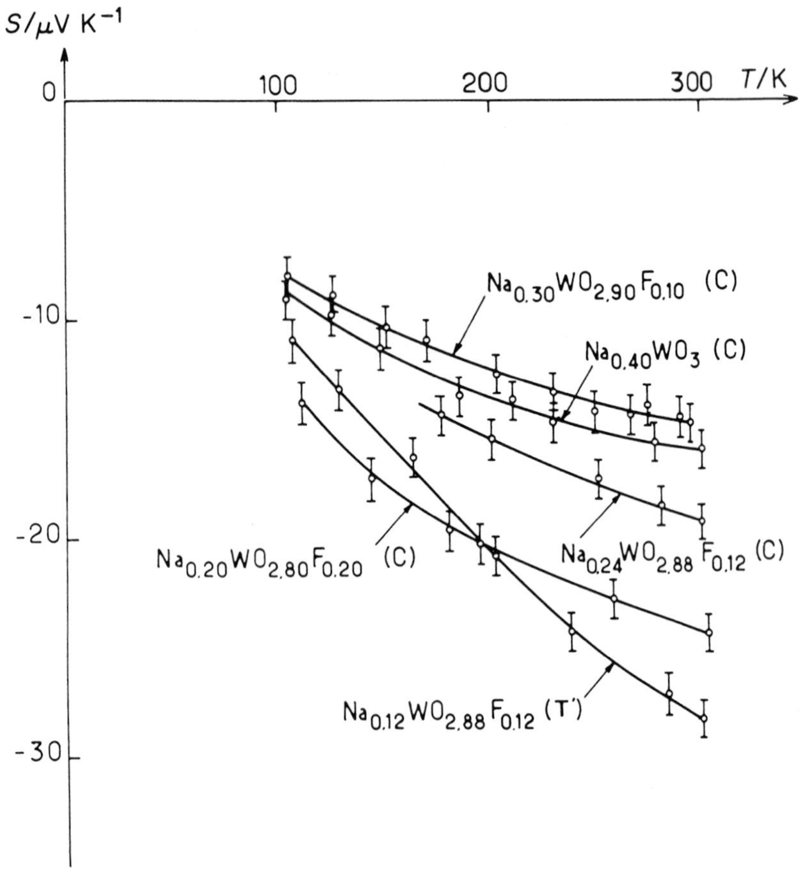

Figure 10. Thermoelectric power of $Na_xWO_{3-y}F_y$.

using a diffusion technique derived from that previously used for non-substituted bronzes by McNeil and Conroy (1962) and Lightsey (1973).

All samples studied by Doumerc *et al.* (1979) and Dubson and Holcomb (private communication) are cubic, at least in the limit of the routine X-ray diffraction analysis which cannot exclude very slight distortions similar to those found for Na_xWO_3 by neutron diffraction (Wiseman & Dickens 1976) as reported above (§2.1). The lattice constant increases with both sodium and tantalum concentrations.

The variation of log σ with $1000/T$ (Figure 11) shows that metallic behaviour is observed for single crystals corresponding to $x-y>0{\cdot}3$ (Doumerc *et al.* 1980). Figure 12 shows a plot of log σ vs. log T given by Dubson and Holcomb (private communication). This later work leads to a more precise determination of the critical composition corresponding to the M–NM transition at $(x-y)_c \simeq 0{\cdot}18$.

The thermoelectric power is negative and its absolute value increases with

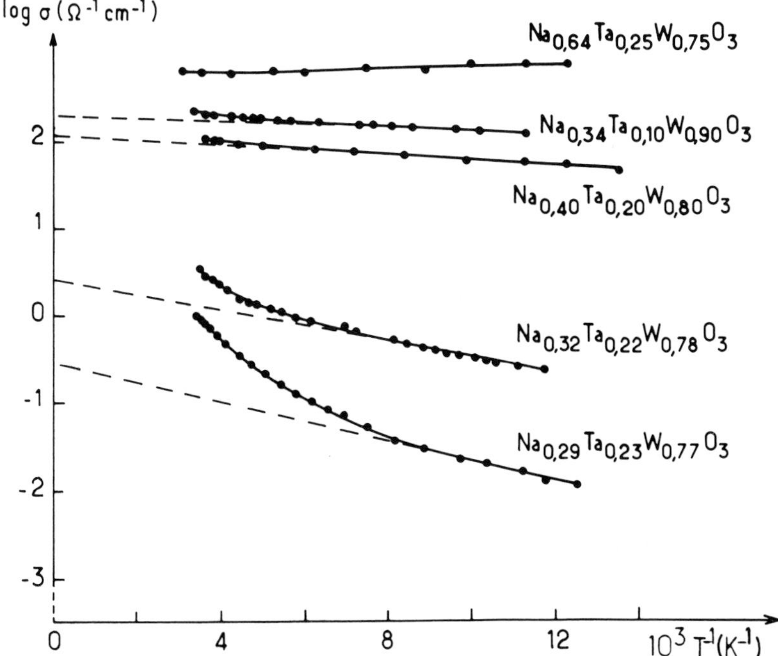

Figure 11. Variation of the logarithm of electrical conductivity vs. $1000/T$ for $Na_xTa_yW_{1-y}O_3$ (Dordor *et al.* 1983).

temperature (Figure 13), implying that there is apparently no activation of the number of carriers in the temperature range investigated and for the whole set of crystals including those for which σ increases with T.

The electronic and transport properties as well as the M–NM transitions in the systems described above will be discussed further in the following sections.

2.4. *Electronic structure of the tungsten bronzes*

One of the first electronic structures proposed for metallic cubic tungsten bronzes is that of Mackintosh (1963). Comparing the distance between sodium atoms in Na_xWO_3 ((3·7845 + 0·0820x) Å) (Straumanis 1949) and in sodium metal (3·72 Å), Mackintosh (1963) predicted a similar bandwidth for the 3s and 3p orbitals in both materials and assumed that electrons could partially occupy this band. Goodenough (1971) has given evidence against this model:

(i) from chemical point of view sodium is expected to reduce W^{6+} to a lower oxidation state;
(ii) crystals of ReO_3 are metallic and Pauli paramagnetic, thus showing that A atoms are not required and that the conduction band should have another origin;

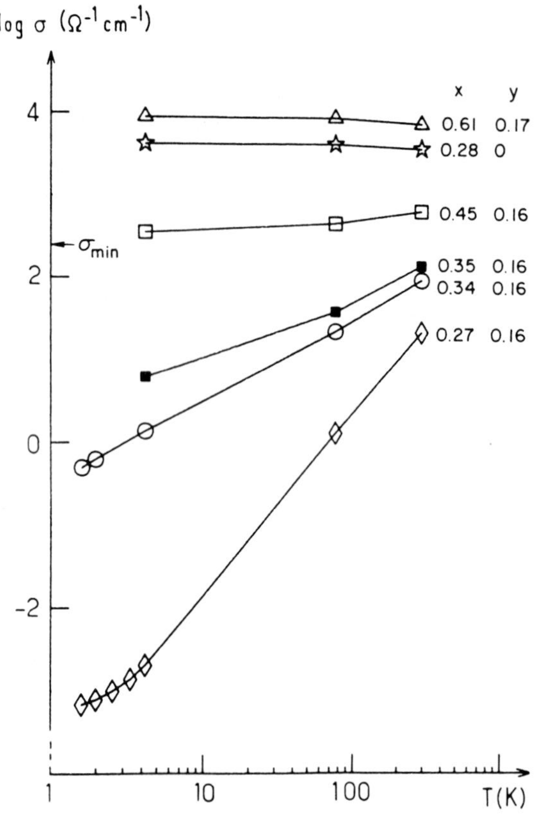

Figure 12. Electrical conductivity vs. temperature for $Na_xTa_yW_{1-y}O_3$ (Adapted from Dubson & Holcomb, private communication).

(iii) most of the n.m.r. studies have shown the absence or the existence of only a negligible Knight shift at the sodium nuclei (Jones *et al.* 1962, Narath & Wallace 1962, Tunstall 1975).

However, it is worth noting that more recently Tunstall & Ramage (1980) have observed a slight maximum in the Knight shift near $x \simeq 0{\cdot}3$ and a sharp decrease of the spin-lattice relaxation time T_1 for $x < 0{\cdot}3$. Although they confirm that the conduction band for large x-values consists mainly of tungsten 5d–oxygen 2p orbitals, they ascribe their results to the tendency of conduction electrons to move into sodium 3d orbitals as x decreases.

Goodenough's model

Metallic behaviour of cubic sodium bronzes Na_xWO_3 may be well understood using the schematic energy-band diagram of Goodenough (1971). This author has proposed for transition metal oxides energy diagrams built from

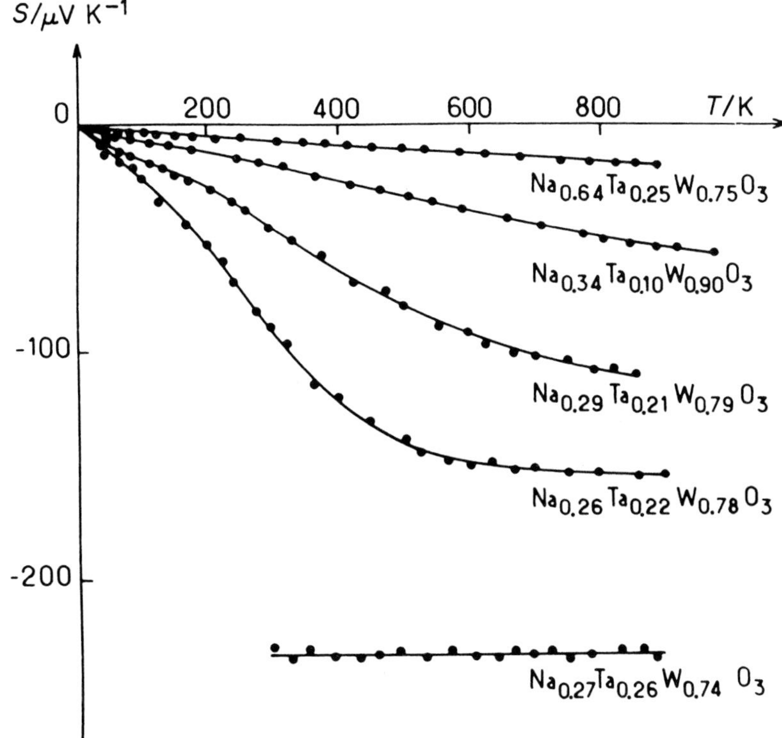

Figure 13. Thermoelectric power of $Na_xTa_yW_{1-y}O_3$.

atomic orbitals taking into account the iono-covalent character of the bonds and the symmetry of the crystal structure. One of the main advantages of such a model is that it allows comparisons of electronic properties of isostructural oxides when the composition is modified. A common feature of these energy-band diagrams is the existence of a large energy gap (5–10 eV) between 2p oxygen orbitals generally involved in σ or π bands and the ns–np cationic orbitals involved in σ^* antibonding orbitals.

Generally d-orbitals of either non-bonding, σ^* or π^* antibonding character lie in this energy gap. In the crystal these levels will or will not broaden into more or less wide energy bands, depending on the magnitude of direct cation–cation or indirect cation–anion–cation interactions.

The energy diagram proposed by Goodenough for ReO_3 is reproduced in Figure 14. In ReO_3 a rhenium atom is surrounded by six oxygen atoms, each of them being shared by two Re atoms (Figure 1). The splitting of d-orbitals of the transition elements can be deduced from classical ligand-field theories.

As pointed out by Goodenough, the shorter cation–cation distance (along the diagonal of a cube face in the ReO_3 structure) is too large ($\sim$5·5 Å) for

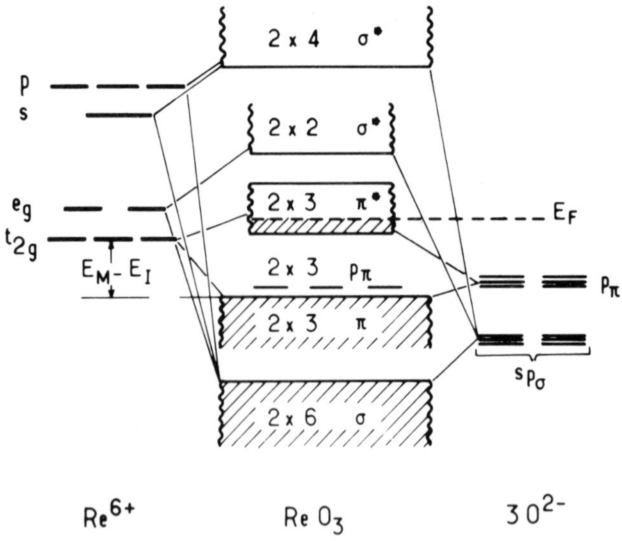

Figure 14. Energy-band diagram for ReO_3 (Adapted from Goodenough 1971).

allowing formation of a d-band in the cationic sublattice. However, the covalent bonding of cationic t_{2g} orbitals with oxygen p_π orbitals can give rise to π and π^* bands. In other words the conduction band for the ReO_3 and perovskite-type structures originates from the indirect overlapping of t_{2g} orbitals via oxygen 2p orbitals.

It may be worthwhile to report here the direct experimental confirmation of the model, given by Ferretti *et al.* (1965). They synthesized a Sr_2MgReO_6 compound which has an ordered perovskite structure in which the magnesium and rhenium atoms form an ordered array of two interpenetrating face-centred cubic lattices. This compound is a semiconductor, which implies a localization of the d-electrons of Re, thus excluding, as predicted, the possibility of a direct overlapping across a face diagonal of an elementary cube of the structure (Figure 1). On the basis of the energy diagram just described, it is theoretically possible to compare and to classify compounds with the perovskite structure using two parameters: one of them proportional to the overlap integral between cationic t_{2g} and anionic 2p orbitals and thus related to the formation and width of σ^* and π^* bands, the other being the number of d-electrons which are able to occupy these bands and thus are related to the Fermi-level position. (Electronic correlations should, of course, also be taken into account since they can, for instance, split into two Hubbard bands a half-filled mono-electronic band.)

In his extensive review of metallic oxides, Goodenough (1971) gave the basis of such a classification, which is outside the scope of this chapter. Let us, however, compare the oxides WO_3 and ReO_3. With 24 valence electrons per formula unit, all the bonding energy bands and energy levels consisting principally of anionic orbitals are completely occupied in the case of (idealized cubic) WO_3. Therefore, the Fermi level is situated in the energy gap between the π and π^* bands (Figure 14) and WO_3 is a semiconductor. As rhenium is just on the right hand side of tungsten in the periodic table, there is one more electron per formula unit and the π^* band is now partially occupied, which explains the metallic behaviour of ReO_3. The case of tungsten bronzes Na_xWO_3 is, one might say, intermediate between WO_3 and ReO_3: the ionization of sodium atoms transfers electrons to the WO_3 sublattice. They are delocalized in the π^* band, at least for x larger than a critical value x_c corresponding to the M–NM transition. The estimation of x_c is rather difficult and has been carried out using different and even conflicting theories, as we shall see in §2.5.

In his original paper, Goodenough (1965) actually proposed a more elaborate diagram for A_xWO_3, taking into account the A–O σ-bonds which can compete with the W–O π-bonds for the 2p oxygen orbitals, particularly for $x > 0{\cdot}75$, since for this composition all 2p oxygen orbitals are involved in bonds. This competition can lead to narrowing of the π^* conduction band. The magnitude of this effect depends on the acidity of the A cation. It has been invoked to explain that the resistivity tends to stop increasing with x in Na_xWO_3 for $x > 0{\cdot}75$, as clearly shown by the measurements of Ellerbeck *et al.* (1961). Hill & Egdell (1983) claim that they have observed experimentally, in their ultraviolet photoelectron spectra, a significant narrowing of the conduction band as x increases in Na_xWO_3. To support their model they also quote the increase with x of the effective mass m^* obtained from low-energy electron-energy-loss (LEELS) experiments. This behaviour of m^* was indeed observed in earlier determinations from magnetic susceptibility (Sienko & Truong 1961) and optical measurements (Owen *et al.* 1978).

An increase of m seems also to occur in $Na_xWO_{3-y}F_y$ with rising fluorine substitution rate (Doumerc 1978). The corresponding narrowing of the conduction band is easy to understand within the scope of Goodenough's model: as a result of the important role played by the anion in conduction band formation, it should be the consequence of the increasing number of W–F bonds, which are more ionic than the W–O bonds.

The rigid band model and the conduction band shape

For WO_3 indirect and direct optical band-gap values determined from photoelectrochemical measurements are equal to 2·7 and 3·5 eV respectively (Koffyberg *et al.* 1979). For metallic cubic bronzes Na_xWO_3 the energy difference between the top of the valence band and the bottom of the conduction band is much smaller (<2 eV) as shown by photoelectron spectroscopy measurements (see, for example, Hollinger *et al.* 1982) or band-structure calculations (Kopp *et al.* 1977, Bullett, 1983). Therefore, no rigid band behaviour can be expected between WO_3 and cubic Na_xWO_3. The existence of several phase transitions when x increases from 0 to 0·4 (Figure 6) was already 'a priori' a strong argument against the rigid band behaviour for $x<0{\cdot}4$, as there is no reason to expect the same energy-band diagram for the different observed structures (Wertheim & Chazalviel 1981).

Opinions are, however, much more divergent when x varies within the composition range of cubic Na_xWO_3 (that is, for $x<0{\cdot}4$). The progressive filling of a conduction band having a more or less constant shape is consistent with several experimental results: X-ray photoelectron spectra (Chazalviel *et al.* 1977, Wertheim & Chazalviel 1981), a photoemission investigation using synchrotron radiation (Hollinger *et al.* 1982), optical properties (Owen *et al.* 1978), specific heat, magnetic susceptibility (Zumsteg 1976), n.m.r. (Holcomb 1978), inelastic neutron scattering measurements (Kamitakahara *et al.* 1976) and electronic structure calculations (Bullet 1983).

On the other hand, Höchst *et al.* (1980, 1982) found no move of the Fermi level in the band structure when they determined the binding energies of oxygen and sodium core electrons as a function of x in Na_xWO_3. In their UPS spectra, Hill & Egdell (1983) found that the conduction band half-width remained roughly constant as x increases.

Although a combination of rigid-band and nearly-free-electron models was often used in pioneering works, there is now a general agreement to consider that it cannot account for some important features of the electronic properties of tungsten bronzes. For instance the magnetic susceptibility and the specific heat coefficient γ were found to vary linearly with x (Zumsteg 1976). N.m.r. studies of ^{183}W (Weinberger 1978, Holcomb 1978) and UPS studies (Hill & Egdell 1983) are also consistent with a linear x-dependence of the density of states at the Fermi level $N(E_F)$. This result cannot be explained with a spherical Fermi surface and a rigid parabolic density of states. The reason is that with $N(E)$ proportional to $E^{1/2}$ and assuming to that all sodium atoms are ionized, so that

$$\int_0^{E_F} N(E)\,\mathrm{d}E = x$$

$N(E_F)$ should be proportional to $x^{1/3}$ instead of being linear.

To account for a linear increase of $N(E_F)$ with x, at least three models have been worked out: percolation models, which will be referred to in the next section, and two models which will be examined now.

1. Rigid exponential band: The density of states obtained from band-structure calculations for ReO_3 by Mattheiss (1969) and more recently by Bullet (1983) suggests an exponential variation of $N(E)$ with E as pointed out by several authors (Silberglitt 1975, Tunstall 1975, Weinberger 1978, Zumsteg 1976, Holcomb 1978). They have shown that, with the assumptions that

$$N(E) = A\exp(E/E_0)$$

and that

$$x = \int_0^{E_F} N(E)\,dE$$

one immediately obtains

$$N(E_F) = x/E_0 + A$$

Calculations of Wolfram (1972) have shown that such a particular band shape could arise from the two dimensional character of the pd_π interactions between the $5dt_{2g}$ orbitals of tungsten and the 2p orbitals of oxygen. The two-dimensional character of the Fermi surface is consistent with the inelastic neutron scattering measurements of Kamitakahara *et al.* (1976).

The formation of a band tail, expected from the random potential due to randomly distributed Na-atoms, is an alternative (or simultaneous) explanation for the proposed variation of $N(E_F)$ (Mott 1977).

2. Non-rigid parabolic band: For $x > 0{\cdot}4$, Hill & Egdell (1983) fit their ultraviolet photoelectron spectra with a parabolic density of states. They account for the linear variation of $N(E_F)$ with x by a narrowing of the conduction band as the sodium content of the material increases. This narrowing cancels exactly the shift of the Fermi level with respect to the bottom of the conduction band, which otherwise would occur.

For $x < 0{\cdot}4$, Hill and Egdell found that the UV photoelectron spectroscopy band maximum is unambiguously below the Fermi level, which they ascribe to the formation of an impurity band. Hollinger *et al.* (to be published) have observed a similar phenomenon for $Na_{0{\cdot}1}WO_3$. To interpret their ^{23}Na n.m.r. study under high pressure, Tunstall & Ramage (1980) have also invoked the formation of an impurity band for low-x Na_xWO_3.

2.5. *The metal–nonmetal transition in tungsten bronzes*

The Na_xWO_3 system

The first question to ask may be: "What is the critical sodium concentration x_c below which Na_xWO_3 is no longer metallic?'. The value of x_c cannot be reached directly, since a two-phase domain exists between the metallic and the semiconducting phase, as may be seen from the equilibrium phase diagram (Figure 6). For a long time the experimental value of x_c agreed by most researchers was about 0·25. It was deduced from the fact that the conductivity, which increases with x, rises steeply at $x \simeq 0{\cdot}25$ (Shanks *et al.* 1963). However, the domain of the cubic metastable phase was extended down to $x = 0{\cdot}23$ by McNeil

Table 2. Critical compositions for the M–NM transition in tungsten bronzes and some related systems.

System	x_c	Number of d electrons per formula unit	Remarks and references
Na_xWO_3	0·25	0·25	(Shanks *et al.* 1963) Extrapolation of $\rho = f(x)$
	<0·22	<0·22	(Lightsey *et al.* 1976)
	0·17	0·17	(Webman *et al.* 1976) Percolation (effective medium theory)
	0·16	0·16	(Lightsey 1973) Site-percolation
	0·12	0·12	Mott criterion $n^{1/3}a_H = 0·25$ with $a_H = 3$ Å
Li_xWO_3	0·25	0·25	(Sienko & Truong 1961) Mott transition
Ba_xWO_3	<0·10	<0·20	(Conroy & Yokokawa 1965)
H_xWO_3 (amorphous)	0·10	0·10	(Wittwer *et al.* 1978) Optical measurements
WO_{3-x}	~0·10	~0·20	(Sahle & Nigren 1983)[a]
$WO_{3-x}F_x$	0·09	0·09	(Reynolds & Wold 1973)
$Na_xWO_{3-x}F_x$	0·20	~0·40	(Doumerc 1978)[a]
$Na_xTa_yW_{1-y}O_3$	$(x-y)_c = 0·30$	0·30	(Dordor *et al.* 1983)[a]
	0·18	0·18	(Dubson & Holcomb, private commun.) Scaling theory of localization
$La_{1-x}Sr_xVO_3$	0·22	0·22	(Dougier & Hagenmuller 1975)[a]
$NaNb_{1-x}W_xO_3$	0·44	0·44	(Miyamoto *et al.* 1983)[a]
$Na_{1-x}Sr_xNbO_3$	>0·50	>0·50	(Ellis *et al.* 1984)[a]

[a] Determined from the change of sign of the temperature coefficient of electrical resistivity.

& Conroy (1962) and later down to $x = 0·22$ by Lightsey *et al.* (1976) who attributed a metallic character to this composition. Therefore, it is now generally assumed that the transition occurs approximately for $x_c \simeq 0·20$ (Holcomb 1978). More precise values of x_c are actually yielded by the interpretation of the data in the light of various theories. Typical values are gathered in Table 2 for Na_xWO_3 as well as for several other tungsten bronzes and related systems.

The first attempts to interpret the M–NM transition in Na_xWO_3 and Li_xWO_3 were made about twenty years ago by Sienko & Truong (1961) and Mackintosh (1963). Both groups of authors considered the bronzes as highly doped semiconductors and described the transition as a Mott transition and hence used the well known Mott criterion (Mott 1961, 1974)

$$n_c^{1/3} a_H = 0·25 \quad (1)$$

Here n_c is the critical electronic concentration ($n_c = x_c/a^3$) and a_H the radius of a hydrogen-like orbit,

$$a_H = \hbar^2 \varepsilon_{st} / m^* e^2$$

where ε_{st} is the static dielectric constant. The value of the right hand side of equation (1) has been extensively discussed (see, for example, Berggren 1978). Taking $\varepsilon_{st}=6$, as can be deduced from the optical measurements of Sawada & Danielson (1959), and $m^*=m$ gives $a_H \simeq 3$ Å and $x_c \simeq 0{\cdot}12$. Mott (1977) has pointed out that for such a small value of a_H, which is not much larger than the interatomic distances, the hydrogen approximation is very rough and that relation (1) 'cannot have too much significance', but that 'the order of magnitude (of x_c) is correct'. Actually any value between about 0·1 and 0·22 can be accepted, since it belongs to the two-phase domain. Such a discontinuity at the transition was predicted by Mott (1974) for crystalline systems.

It might also be interesting to note that Edwards & Sienko (1978) have tested relation (1) for a wide variety of systems, including alkali–rare gas atom films as well as III–V or II–VI covalent semiconductors, covering a range of approximately 10^{10} in n_c and 600 Å in the characteristic Bohr radius. They found that (1) with the right hand side equal to $0{\cdot}26 \pm 0{\cdot}005$ is applicable over the entire range, provided that for compounds having low (or intermediate) dielectric constants (such as tungsten bronzes) a_H is replaced by an appropriate radius (a_H^*) associated with a 'realistic' wavefunction for the localized state.

In the model described up to now, the major role is played by the electronic correlations which lead to the localization of carriers. Once carriers are localized, their kinetic energy is lost, the entropy decreases and the insulating phase can accommodate a different structure where distortions and atomic rearrangements can help to minimize the free energy.

On the other hand, Goodenough (1965) suggested that the semiconducting behaviour of low-x sodium tungsten bronzes results form ferroelectric-type distortions which characterize the T_I phase. Such distortions are associated with a shift of the tungsten atoms towards one (or two) of the six oxygen atoms surrounding it. As a consequence, one (or two) of the W–O bonds is enhanced and d-electrons have a tendency to be localized in other orbitals rather than delocalized over the whole network.

A quite different approach to the problem is the use of percolation theories to predict the value of x_c as well as to explain the evolution of the physical properties as a function of the sodium content. Actually two types of model have to be distinguished:

(i) those in which percolation occurs between atomic sites,
(ii) those in which metallic domains of $NaWO_3$-composition of relatively large size are assumed to be embedded in a WO_3 matrix. The latter model was worked out, using an effective-medium theory, by Webman *et al.* (1976) who gave several arguments to support it, such as linear variation of magnetic susceptibility and specific heat with x. They also claim that $NaWO_3$ clusters of finite size (~ 100 Å wide) could be thermodynamically stable. Finally, they obtained a good fit for the x-dependence of electrical conductivity. However, some authors gave several arguments against this model. Most of the objections are gathered in a publication by Mott (1977).

The first objection comes from structural considerations: X-ray diffraction patterns are characteristic of a single cubic phase for $x > 0{\cdot}4$ with a lattice constant increasing linearly with x (Brown & Banks 1954). For $0{\cdot}28 < x < 0{\cdot}4$ a tetragonal potassium tungsten bronze phase with a different arrangement of octahedra (Figure 5) is obtained (Figure 6). This tetragonal potassium tungsten bronze type structure is neither that of WO_3 nor that of $NaWO_3$. Neutron diffraction studies did not give evidence of the existence of sodium clustering but rather of a random distribution of inserted atoms and of possible sodium ordering (Wiseman & Dickens 1976), in a homogeneous phase for particular x-values such as $x = 0{\cdot}75$ (Atoji & Rundle 1960). Other objections are raised by the n.m.r. measurements of Tunstall (1975) and Weinberger (1978), optical measurements (Owen *et al.* 1978) and photoelectron spectroscopy (Hollinger *et al.* 1982), studies which support a rigid band model at least for $x > 0{\cdot}4$ as we have already mentioned. Finally the linear x-dependence of magnetic susceptibility and specific heat can alternatively be explained by an exponential shape of the density of states (§2.4).

The treatment of the transport properties of Na_xWO_3 as a site-percolation problem with Na atoms randomly distributed was worked out by Fuchs (1965) and later by Lightsey (1973). In this model electrons are assumed to be trapped in the vicinity of Na^+ ions (actually they are, rather, located on tungsten atoms close to a Na^+ ion). For a simple cubic lattice the percolation threshold is reached for $x = 0{\cdot}33$ (Kirkpatrick 1971), which is a value much larger than the experimental one and corresponds to the metallic phase. However, from a fitting of experimental conductivity data with an exponential law (predicted by percolation theories) of the form:

$$\sigma \propto (x - x_c)^\beta$$

Lightsey obtained a value of $x_c = 0{\cdot}16$. He overcame the difficulty by considering that percolation paths are possible not only between first nearest neighbours but also between second and third nearest neighbours.

However, reviewing the problem, Mott (1977) pointed out that the tetragonal oxyfluoride bronze $WO_{2{\cdot}91}F_{0{\cdot}09}$ studied by Reynolds & Wold (1973) had a relatively high electrical conductivity ($\sigma(300\text{ K}) = 10^3\ (\Omega\text{ cm})^{-1}$).

He proposed two alternative explanations for the negative and weak temperature coefficient ($\partial\rho/\partial T$): this compound is either metallic and the increase of conductivity with temperature is due to an increase in $N(E_F)$ caused by phonon interactions as described by Brouwers & Brauwers (1975), or nonmetallic with a mobility edge just above E_F. In any case these remarks strongly suggest that the M–NM transition in Na_xWO_3 and other similar systems is rather of Anderson type and even that the two-phase domain between $x = 0{\cdot}1$ and $x = 0{\cdot}22$ could be between two metallic phases and would not be due to a discontinuous Mott transition as expected in ordered crystals.

Mott (1977) considers three possibilities for this Anderson-type transition:

(*a*) the random distribution of Na^+ ions gives rise to Anderson localization in the conduction band tail,
(*b*) an impurity band is formed, the states being localized for $x = x_c$,
(*c*) the impurity band is split into two Hubbard bands (Mott 1974) and localization occurs in a pseudo-gap between the two Hubbard bands (or the lower Hubbard band and the conduction band).

These three cases are illustrated in Figures 15(*a*), (*b*) and (*c*) respectively.

Most recent work supports the basic idea that the transition is of Anderson type, as suggested in each of the three cases proposed by Mott. There is however experimental and theoretical evidence in favour of each of these possibilities.

Since sodium atoms are fully ionized and electrons occupy a tungsten band,

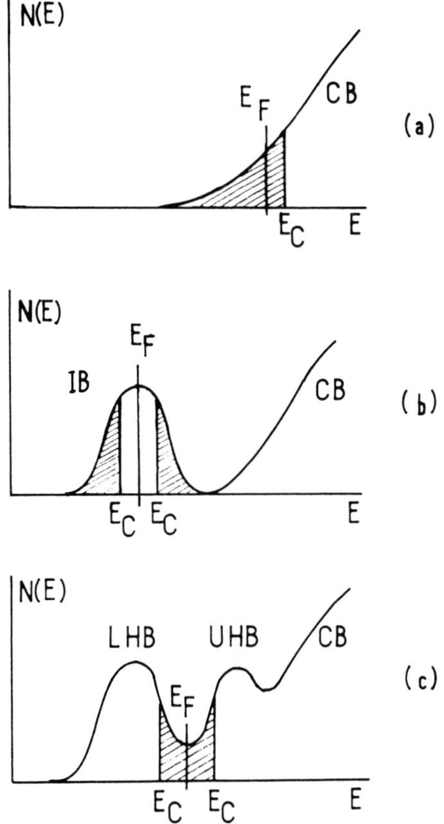

Figure 15. Three possibilities for Anderson localization in Na_xWO_3: (*a*) in a conduction band (CB) tail, (*b*) in an impurity band (IB) and (*c*) in a pseudo-gap between lower Hubbard band (LHB) and upper Hubbard band (UHB) or conduction band (adapted from Mott's model; see, for example, Mott & Davis 1979; Mott *et al.* 1975).

Mott (1977) considered Na^+ ions as donors 'with long range but not short range attractive force' and therefore that the formation of an impurity band is unlikely. Under such conditions (*a*) would be the correct description. All the experimental data available at that time were interpreted in the framework of this model, including the Hall effect measurements of Lightsey *et al.* (1976) who found that the Hall carrier density for $x = 0{\cdot}22$ at 77 K and 4 K was much below the free-electron value. Mott explained this Friedman anomaly (Friedman 1971) by the decrease of $N(E_F)$ associated with the broadening of the conduction band related to disorder. He also pointed out that this and other arguments, such as small values (relative to the free electron) of specific heat and magnetic susceptibility (Zumsteg 1976), are evidence against strong electronic correlations.

On the contrary, as we have outlined above, for Tunstall & Ramage (1980) and Hill & Egdell (1983) an impurity band is formed when x decreases towards x_c. For the former authors localization occurs in a narrow sodium 3d band (case (*b*)), whereas for the latter it occurs in a pseudo-gap (case (*c*)) where the density of states remains relatively high.

Substituted tungsten bronzes

As we have seen, the M–NM transition in Na_xWO_3 occurs for an x-value lying in a two-phase domain so that it is difficult to say whether it is a Mott transition of the kind observed in crystals where such a separation into two phases is expected, or an Anderson transition hidden by the two-phase domain. The latter hypothesis seems to be supported by the study of substituted tungsten bronzes.

In $Na_xWO_{3-y}F_y$, Doumerc (1978) observed a transition which has all the characteristics of an Anderson transition (see, for example, Mott *et al.* 1975) as can be seen from the variation of electrical conductivity and thermoelectric power with temperature (Figures 9 and 10). As the samples were polycrystalline, absolute values of conductivity should be considered with carefulness and the localization may originate from the disorder due to random distribution of Na^+ and F^-, within their respective sites in the $W(O,F)_3$ lattice, as well as from the disorder arising at grain boundaries. As already mentioned above, the substitution of fluorine for oxygen in $Na_xWO_{3-y}F_y$ was realized mainly to confirm the role played by the anion in the conduction mechanism. The fact that apparently an Anderson transition is directly observed in this system prompted the present authors to reinvestigate what would be the most appropriate compositional modifications in order to understand better the M–NM transition in tungsten bronze systems.

In the model described by Mott *et al.* (1975) for the Anderson transition in a conduction band tail, the disorder gives rise to localization of states with energies lower than a mobility edge E_c (Figure 15(*a*)). While the random substitution of fluorine for oxygen increases the disorder and therefore raises E_c, it also increases the number of d-electrons per tungsten atom, leading to an upwards shift of the

Fermi level according to Goodenough's band diagram. As both E_c and E_F are moving in the same direction it is pure chance that they can cross over, although this seems in fact to happen.

As Mott (private communication) suggested, it was worthwhile investigating a system in which a random substitution could lower the Fermi level rather than lift it. The substitution of tungsten by tantalum, leading to bronzes of formula $Na_xTa_yW_{1-y}O_3$, appeared more appropriate for two reasons:

(i) it modifies the W–O sublattice and, as the conduction band consists mainly of $5d_{t_{2g}}$ W orbitals, strong scattering of electrons by Ta^{5+} atoms can be expected and, ultimately, Anderson localization,
(ii) as tantalum is on the left hand side of tungsten in the periodic table, leading to an electronic configuration $5d^0$ for Ta^{5+} instead of $5d^1$ for W^{5+}, the Fermi level can be expected to lie at a lower energy.

A model has been set up on the basis of the following assumptions (Doumerc 1978).

(i) Eventual distortions not observable by the usual powder X-ray diffraction, of the type mentioned above, and the substitution of tantalum for tungsten have no large qualitative effect on the band structure (such as band splitting, and so on) but mainly a quantitative influence, leading, for instance to narrowing of the conduction band.
(ii) Sodium atoms and tantalum atoms are randomly distributed in their respective lattice, the random potential increasing with y for small y values.
(iii) If (i) is verified, Goodenough's band diagram (Figure 14), modified to account for disorder arising from (ii), can be used and the Fermi level position is determined by the number of d-electrons which is equal to $(x-y)$ per formula unit $Na_xTa_yW_{1-y}O_3$.

Mott (private communication) and Doumerc *et al.* (1980) described the M–NM transition in the following way: starting from the metallic side and keeping y more or less constant as $(x-y)$ decreases, the Fermi level is lowered and the Anderson transition occurs as E_F crosses E_c.

Experimental evidence supporting this model can be summarized as follows:

1. On both sides of the transition:
the transition occurs without (noticeable) structural change,
the magnetic susceptibility (Doumerc *et al.* 1979) is very weak and nearly temperature-independent (Pauli type),
the thermoelectric power is negative and its absolute value increases with T (Figure 13).

2. On the insulating side:
$\partial\sigma/\partial T$ decreases as T decreases (Figures 11 and 12) and Mott's law ($\sigma=\sigma_0 \exp(-B/T^{1/4})$) seems respected at low temperature (Dordor *et al.* 1983).

It is now necessary to emphasize that this model could need some (more or less important) corrections in the future to account for two recent experimental results. The first one is the observation of a very small density of states near the Fermi level, by photoelectron spectroscopy in $Na_xTa_{0.20}W_{0.80}O_3$ for $x<0.4$ (Hollinger *et al.* to be published). The second one is the observation of long relaxation times in n.m.r. of ^{183}W for samples near the M–NM transition (Dubson 1984).

Dordor *et al.* (1983) have proposed that the electrical conductivity in nonmetallic $Na_xTa_yW_{1-y}O_3$ is due to different hopping mechanisms. Variable range, Miller–Abrahams and Heikes hoppings are successively predominant as the temperature increases. It can be seen from Figure 13 that the temperature at which the Heikes mechanism becomes dominant (that is, when the thermoelectric power is nearly temperature-independent) increases with $(x-y)$. The authors have ascribed this behaviour to the fact that Heikes hopping is expected for $E_F<kT$ and that E_F increases with $x-y$.

Recently, Dubson and Holcomb have re-investigated the system $Na_xTa_yW_{1-y}O_3$. They extended the electrical measurements down to 1·6 K (Figure 12) and obtained good agreement with the results of Dordor *et al.* (1983). An interpretation of the transport properties using the scaling theory of

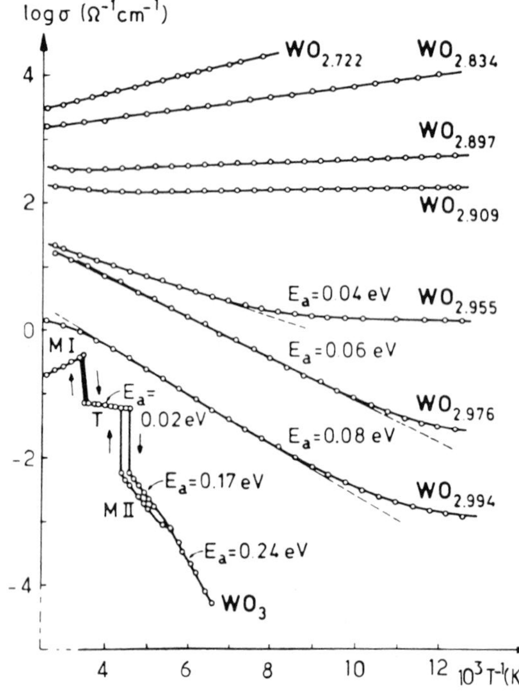

Figure 16. Conductivity variation vs. reciprocal temperature for WO_{3-x} (Adapted from Sahle & Nygren 1983).

localization (see, for example, Abrahams *et al.* 1979) leads to $(x-y)_c = 0{\cdot}18$ (Figure 12) (private communication).

On the other hand, Dordor *et al.* (1983) have proposed $(x-y)_c \simeq 0{\cdot}30$ from the change of sign of the temperature coefficient of electrical conductivity. It is interesting to note that the value of the conductivity for $x-y \simeq 0{\cdot}30$ is close to σ_{min}, the minimum metallic conductivity as defined by Mott (1974) and already mentioned in §2.2. To know whether a σ_{min} value exists or not in this system requires further investigations. For a general discussion of the problem, in particular of the negative temperature coefficient of resistivity in metals near the transition, see, for example, Mott & Kaveh (1983), Mott (1984) and references therein.

Table 2 shows that the value of x_c varies from system to system in spite of the rather large uncertainty of some determinations. This leads us to think that the M–NM transition is not directly driven by the electron concentration alone. Particularly for $WO_{3-x}F_x$ and amorphous H_xWO_3 it appears that $x_c(\simeq 0{\cdot}1)$ is relatively smaller than in other systems. In the latter system the Anderson transition was studied by optical measurements (Wittwer *et al.* 1978) and Mott's $T^{-1/4}$ law is obeyed over a wide range of temperature. This system is of obvious interest, as well as WO_{3-x}: in both cases the evolution of the $\log \sigma = f(1/T)$ curves with x suggests again an Anderson transition (see Figure 16 for WO_{3-x}). However, they will not be examined here in detail, since we think that the discussions given above can be applied to them without significant modifications.

3. Vanadium bronzes

3.1. *The system* $La_{1-x}Sr_xVO_3$

The phase diagram determined by Dougier & Hagenmuller (1975) is given in Figure 17. All phases exhibit structures derived from that of perovskite. Distortions are mainly due to tilting of octahedra along one or two of the four-fold axes of the perovskite unit cell. The lattice constants of M_I decrease as x increases (Dougier & Casalot 1970).

$SrVO_3$ is metallic, while in $LaVO_3$ the d electrons of vanadium are localized. Using Goodenough's model for the electronic structure (Figure 14), the π^* band is partially filled for both compounds. With an integer number of d electrons per vanadium atom, whether the π^* band is split into two Hubbard bands depends on the relative magnitude of the cation–anion transfer energy b_π (or overlap integral) compared to that of the intra-atomic repulsion U (Mott 1974). The dependence of b on structural distortions and on the covalency of the V–O bonds can explain the difference in behaviour observed for $LaVO_3$ and $SrVO_3$, considering the following effects (Goodenough 1971, 1984).

(i) The increase of covalency with the oxidation state, leads one to expect b_π to be larger for V^{4+} than V^{3+}.

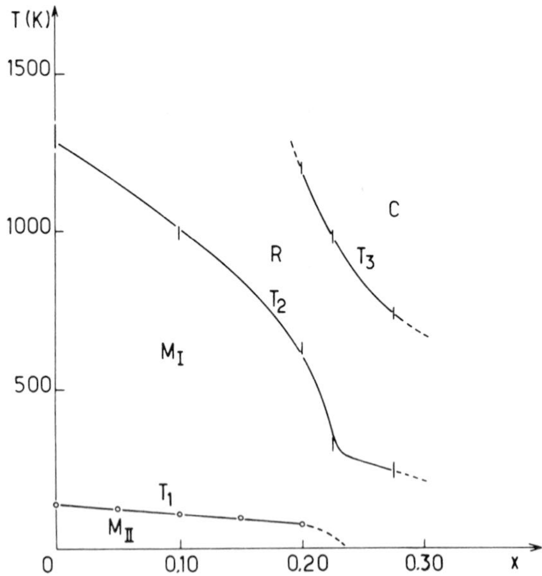

Figure 17. Phase diagram for $La_{1-x}Sr_xVO_3$ (Dougier & Hagenmuller 1975).

(ii) As already pointed out (Goodenough 1971), the A–O bonds compete with the V–O bonds for oxygen p_π orbitals. As Sr^{2+} is more basic than La^{3+}, V–O bonds are expected to be more covalent in $SrVO_3$ than in $LaVO_3$.
(iii) Distortions in $LaVO_3$ can lead eventually to a weaker overlapping of vanadium $3d_{t_{2g}}$ orbitals with oxygen 2p orbitals.

From these considerations, a Mott transition with, at low temperature, a possible two-phase domain between a metallic and a nonmetallic phase, would have been expected in $La_{1-x}Sr_xVO_3$. However, the phenomena are more complex since, in addition to the above effects, it is necessary to take into account:

(iv) The possibility of formation of an impurity band.
(v) A random distribution of La^{3+} and Sr^{2+} among the A sites of the perovskite structure, which can give rise to a random potential and hence to Anderson localization.

From the plot of log σ against $1/T$ given in Figure 18 (for polycrystalline samples) it can be shown that a M–NM transition occurs for $x_c \simeq 0{\cdot}225$. The thermoelectric power is positive for all compositions except $x = 0{\cdot}275$ (Figure 19).

Dougier & Hagenmuller (1975) and Sayer *et al.* (1975) have described the transition as an Anderson transition (Mott *et al.* 1975) and the evolution of transport properties as x increases can be explained as follows.

Sr^{2+} ions can be considered as carrying an effective negative charge with respect to the La^{3+} lattice and strontium atoms will act as traps with long range,

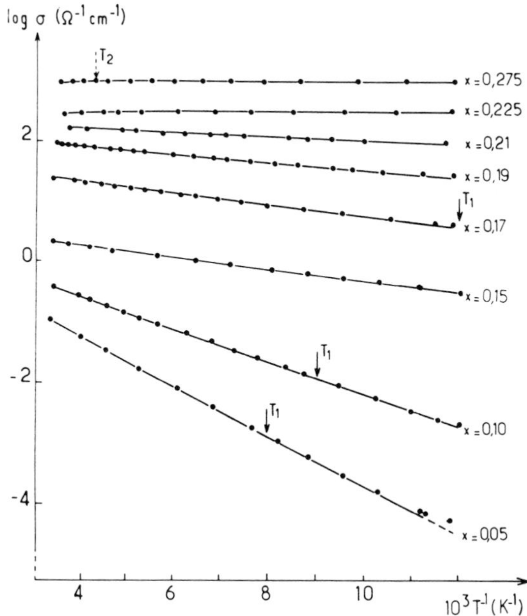

Figure 18. Variation of the electrical conductivity of $La_{1-x}Sr_xVO_3$ as a function of reciprocal temperature (Dougier & Hagenmuller 1975).

but not short range, attractive force (as proposed by Mott for Na^+ in WO_3), so that holes can be considered as V^{4+} ions stabilized in the vicinity of Sr^{2+} ions. These acceptor levels can broaden into an impurity band as x increases. The thermal activation of thermoelectric power for $x<0{\cdot}1$ corresponds to the excitation of these holes into the lower Hubbard π^* valence band. The larger activation energy of σ shows that, at least in this range of x-values, carriers in the π^* band are small polarons (Dougier & Hagenmuller 1975). As x increases, the upper Hubbard impurity band overlaps either the lower Hubbard impurity band or the valence band. Therefore the Fermi level lies either in a pseudo-gap or a band tail. Due to random distribution of La^{3+} and Sr^{2+}, states at the Fermi level are localized and conductivity results either from excitation to the mobility edge or from a hopping mechanism. In the temperature range investigated, the increase of the thermopower with T is in favour of a predominance of the latter mechanism (Mott & Davis 1979). At low temperature, variable range hopping is expected: Mott's law ($\sigma=\sigma_0 \exp(-B/T^{1/4})$) seems obeyed at low temperature for small x-values (Sayer *et al.* 1975).

The activation energy for electrical conductivity tends to zero as x tends to x_c. Figure 20 shows that its variation is proportional to $(x_c-x)^{1{\cdot}8}$ as predicted for an Anderson transition of that type (Mott 1974).

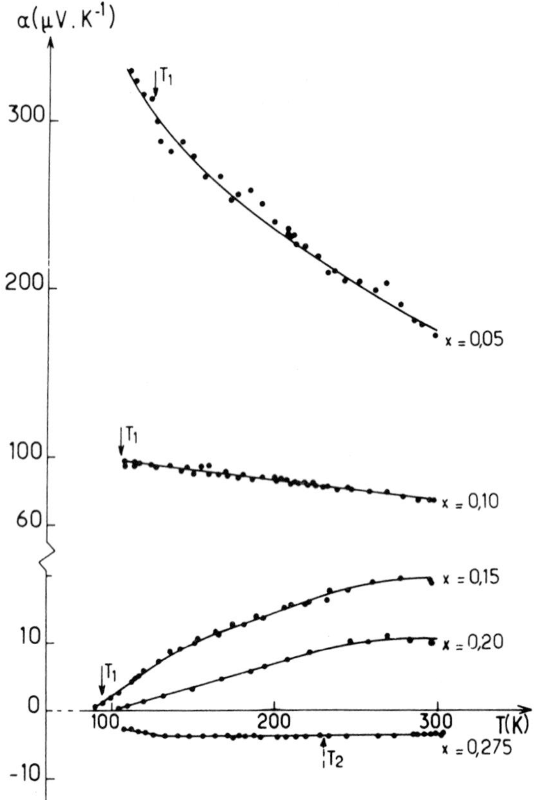

Figure 19. Thermoelectric power variation with temperature for $La_{1-x}Sr_xVO_3$ (Dougier & Hagenmuller 1975).

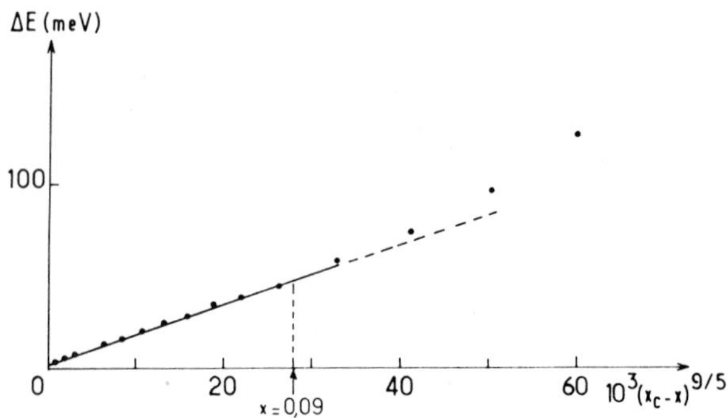

Figure 20. Variation of the activation energy of the electrical conductivity of $La_{1-x}Sr_xVO_3$ as a function of $(x_c - x)^{1 \cdot 8}$ (Dougier & Hagenmuller 1975).

3.2. β-copper vanadium bronzes

A review on vanadium bronzes has been given by Hagenmuller (1971) and the electronic structure discussed by Goodenough (1971). We will limit our discussion here to the case of β-$Cu_xV_2O_5$ bronzes in which an M–NM transition was clearly identified. Up to now we have considered oxide bronzes whose structures were almost or completely isotropic. On the contrary, structures of vanadium bronzes are more complex in general than those of tungsten bronzes. The structure of β-$Cu_xV_2O_5$ is highly anisotropic with a one-dimensional character. The monoclinic unit cell is given in Figure 21. There are three types of sites: two roughly octahedral sites, V_1 and V_2, and a triangular bipyramidal one, V_3. Both kinds of polyhedra form chains parallel to the y-axis of the monoclinic lattice by sharing common edges. These chains are linked by oxygen atoms in the other directions, giving rise to channels along the y-axis. The quasi-one-dimensional character was recognized from the anisotropy of the electrical conductivity (Ozerov 1959a,b,c, Wallis *et al.* 1977), of the optical reflectivity measurements (Kaplan & Zylberstein 1976) and of the e.s.r. measurements (Sperlich *et al.* 1975).

Most of the vanadium bronzes are semiconductors and their transport properties have been explained in terms of the localized-electron model (Goodenough 1971). However, studies on single crystals have permitted the observation of one-dimensional metallic behaviour for β-$Na_xV_2O_5$ above 200 K (Kobayashi 1979) and for β-$Cu_xV_2O_5$ (Villeneuve *et al.* 1976, Mori *et al.* 1981). According to Goodenough copper atoms are ionized to Cu^+ and electrons localized as V^{4+} at the V_1 sites. This author has pointed out that the V_1–V_1 distances are too large (> 3 Å) for the formation of a band from direct t_{2g} orbital overlapping through the edge common to two adjacent octahedra and that electronic delocalization can only occur in a π^* band arising from indirect

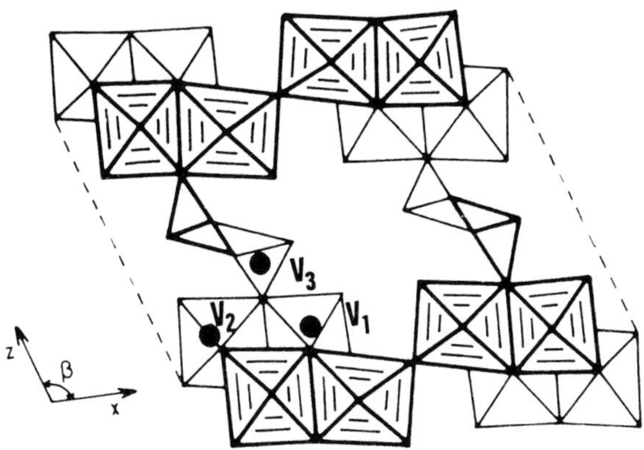

Figure 21. Idealized structure of the β-$Cu_xV_2O_5$ bronze.

overlapping between V_3 sites, through the oxygen 2p orbitals. In order to explain the metallic behaviour of $Cu_xV_2O_5$ for $x > 0{\cdot}55$, Villeneuve and co-workers have assumed that the Cu^+ ions are randomly distributed in the channels parallel to the y-axis (Figure 21) giving rise to broadening of the π^*–V_3 band. They have also considered that the V_1 states are overlapped by the band tail where Anderson localization is expected. For $x > 0{\cdot}55$ the metallic behaviour is explained by a Fermi level situated above the mobility edge. As the copper concentration decreases, an Anderson transition of the kind already described for other systems is observed.

For Mori *et al.* (1981) the bronzes $Cu_xV_2O_{5-\delta}$, with $x+2\delta<2/3$, exhibit a change from metallic to nonmetallic behaviour at a temperature which decreases as $x+2\delta$ increases. They have also shown that this temperature-dependent transition is not observed (at least above 1·5 K, the lowest temperature reached in their experiments) for $x+2\delta>2/3$, and emphasize that in this case the Fermi level will, anyway, lie in the π^*–V_3 band since V_1 sites are saturated for $x+2\delta=2/3$ (even without invoking the overlapping of both types of energy levels).

They claim that the low-temperature semiconducting behaviour is not due to a Peierls transition because of strong interactions between chains.

This last point distinguishes vanadium bronzes from blue molybdenum bronzes where a Peierls transition has been recognized and intensively studied.

4. Molybdenum bronzes

The molybdenum bronzes have been reviewed by Hagenmuller (1971). Recent developments concern the crystal growth methods, the preparation of new bronzes and the study of transport and optical properties.

Besides the usual electrolytic reduction (Wold *et al.* 1964) Greenblatt's group has developed a temperature gradient flux technique for crystal growth of alkali metal molybdenum bronzes (Ramanujachary *et al.* 1984). They have also obtained new lithium bronzes with a composition either similar to the already known sodium and potassium compounds ($K_{0{\cdot}30}MoO_3$, $K_{0{\cdot}33}MoO_3$, $K_{0{\cdot}9}Mo_6O_{17}$) or quite new, such as $Li_{0{\cdot}04}MoO_3$ (McCarroll & Greenblatt 1984).

Most of the molybdenum bronzes exhibit highly anisotropic electronic behaviour. A two-dimensional metallic-type conductivity has been recognized for $Li_{0{\cdot}9}Mo_6O_{17}$ by Greenblatt *et al.* (1984) and earlier for $K_{0{\cdot}9}Mo_6O_{17}$ by Buder *et al.* (1982).

The so-called blue bronze $K_{0{\cdot}30}MoO_3$ undergoes at 180 K an M–NM transition (Fogle & Perlstein 1972), whereas the red bronze $K_{0{\cdot}33}MoO_3$ is semiconducting at all temperatures (Bouchard *et al.* 1967). These two points will now be briefly examined.

4.1. The metal–nonmetal transition between the blue bronze $K_{0\cdot 30}MoO_3$ and the red bronze $K_{0\cdot 33}MoO_3$

The structure of $K_{0\cdot 30}MoO_3$ consists of groups of ten MoO_6 octahedra sharing common edges (Figure 22(*a*)). These clusters are linked by common oxygen atoms to form sheets and tunnels which are occupied by potassium atoms. In $K_{0\cdot 33}MoO_3$, sheets are also formed according to the same principle but with groups of only six MoO_6 octahedra (Figure 22(*b*)).

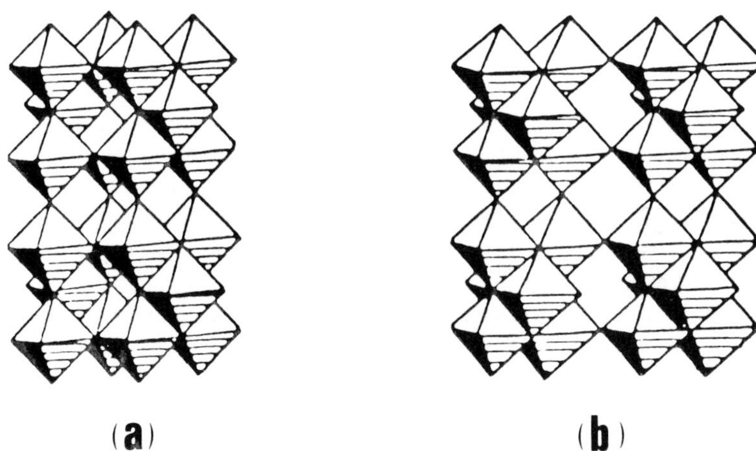

Figure 22. Idealized structures of the (*a*) red and (*b*) blue molybdenum bronzes.

Travaglini & Wachter (1983) have shown how the structures are related to each other. The structure of $K_{0\cdot 33}MoO_3$ formally results from the addition of two Mo atoms to the $K_3Mo_{10}O_{30}$ cluster of the blue bronze. As the coulomb repulsion destabilizes the resulting structure, the blue-bronze cluster tends to split into two subclusters, bringing the two additional Mo atoms as far apart from each other as possible. This produces two red-bronze clusters.

The blue bronze has one-dimensional metallic behaviour above 180 K and the red bronze is a one-dimensional semiconductor with a 0·5 eV energy gap. Travaglini and Wachter have proposed describing this qualitative change of electronic properties between the two compounds as a Mott transition. It is reported here as an example of the correlations between structural properties and localization of the d electrons, to illustrate the general statements made in the introduction.

The model of Travaglini and Wachter is based on the following considerations.

(i) In both compounds the Mo–Mo distances are too large to allow formation of a band of delocalized states by direct overlapping of the $4d_{t_{2g}}$ cationic orbitals. In

metallic $K_{0\cdot30}MoO_3$ the conduction band results from the overlapping of molybdenum $4d_{t_{2g}}$ and oxygen 2p orbitals (Travaglini *et al.* 1981).
(ii) Therefore, the transition from delocalized to localized states can be discussed in terms of a critical value for the overlap integral between 4d and 2p orbitals.

The quoted authors have carefully examined the distribution of Mo–O distances in both bronzes. The comparison shows that the Mo–O separation along the *b*-axis is larger for $K_{0\cdot33}MoO_3$ (1·92 and 2·02 Å) than for $K_{0\cdot30}MoO_3$ (1·88 and 1·96 Å). Therefore the critical value of the overlap integral is reached for an Mo–O distance lying between 1·96 and 2·02 Å. The dilatation of the red cluster along the *b*-axis can be explained by an increase of the repulsion due to the additional Mo atoms (Figure 22).

4.2. *Peierls transition in the blue bronze $K_{0\cdot30}MoO_3$*

The electronic properties of $K_{0\cdot30}MoO_3$ and other molybdenum bronzes have been intensively studied by the Grenoble 'Phase Transitions Group' (Schlenker *et al.* 1983).

Electrical conductivity measurements reveal an anisotropy of transport properties (Brusetti *et al.* 1982). The one-dimensional metallic behaviour has been confirmed by optical measurements using polarized light (Travaglini *et al.* 1981). Single crystals cleave parallel to the monoclinic *b*-axis and to the [102] direction. Travaglini *et al.* have measured the optical reflectivity for a light beam incident on the cleavage plane, polarized either parallel or perpendicular to the *b*-axis. For the former direction they observe at 300 K a metallic behaviour. The plasma frequency ω_p could not be directly observed on the reflectivity curves because of the screening by interband transitions. Using an appropriate method they obtained an unscreened ω_p value of 2·7 eV.

Assuming that each potassium atom contributes one electron to the conduction band leads to an effective mass equal to that of an electron. For light polarization perpendicular to the *b*-axis ω_p at 300 K is less than 0·03 eV, giving a carrier concentration four or five orders of magnitude less than in the metallic conduction direction. This value corresponds to an anisotropy much larger than that observed by conductivity measurements (Brusetti *et al.* 1982).

Using X-ray diffuse scattering, Pouget *et al.* (1983) have shown that the M–NM transition at 180 K is accompanied by a periodic lattice distortion and describe it as a Peierls type transition associated with an incommensurate charge density wave (CDW). A Peierls precursor was also observed at room temperature by Raman scattering (Travaglini *et al.* 1982). Neutron scattering studies have shown that the CDW wavevector smoothly approaches the commensurate value as the temperature is lowered and that a lock-in transition does not occur down to 6 K (Sato *et al.* 1983).

The transport properties of the CDW have been studied by Dumas *et al.*

(1984) and Dumas & Schlenker (1983). The incommensurate CDW may vanish as an electric field is applied, provided that its value is high enough to overcome the pinning of the CDW by impurities or crystal defects. The quoted authors have discussed the nonlinear, electrical noise and memory effects (Dumas *et al.* 1983). They have also extended their investigations to CDW behaviour in the two-dimensional metallic conductor $K_{0.9}Mo_6O_{17}$ (Dumas *et al.* 1984).

5. Conclusions

Oxide bronze type compounds offer a large variety of examples of M–NM transitions. Although the radial extension of the d orbitals increases from the first to the third transition element row and electronic delocalization is stronger for tungsten than for vanadium, metallic oxide bronzes can be found with vanadium as well as with molybdenum or tungsten.

When the composition changes discontinuously between phases of different crystal structures having a narrow non-stoichiometric range, the transition from metallic to nonmetallic state can clearly be explained by considering the variation of the overlap integrals of the atomic orbitals, giving rise to the Mott transition, for instance in molybdenum bronzes.

For systems in which large non-stoichiometric phase domains occur, as in tungsten bronzes, the interpretation of the M–NM transition seems more difficult. Recent experimental data obtained from the investigation of substituted bronzes support the idea that Anderson localization plays an important part in the origin of the M–NM transition and in the transport properties. However from the study and comparison of the numerous available data some fundamental problems remain unsolved:

(i) the precise determination of the critical composition and the existence of σ_{min},
(ii) the respective part played by electronic correlations and disorder, which are often simultaneously present in this kind of system.

As a metal is defined by $\sigma(T=0)\neq 0$, the first point requires electrical measurements at temperatures as low as possible, at least when it is possible to obtain samples with x near the critical transition value x_c.

The second problem requires further investigations in oxide bronze systems and particularly in new systems in which it would be possible to control the nature and value of the random potential, keeping the other parameters more or less constant.

Acknowledgements

The authors gratefully thank Sir Nevill Mott for having initiated the basic ideas of applicability of the concept of the Anderson transition to the systems described in this chapter. They acknowledge the helpful discussions, letters, comments and advice which he consented to give for many years.

We are also indebted to Professor D.F. Holcomb and Dr M. Dubson for communication of some recent results.

References

Abrahams, E., Anderson, P.W., Licciardello, D.C. and Ramakrishnan, T.V., 1979, *Phys. Rev. Lett.*, **42**, 673.

Atoji, M., 1978, *Solid St. Commun.*, **27**, 1227.

Atoji, M. and Rundle, R.E., 1960, *J. Chem. Phys.*, **32**, 627.

Banks, E. and Goldstein, A., 1968, *Inorg. Chem.*, **7**, 966.

Benner, R.E., Brody, E.M. and Shanks, H.R., 1977, *J. Solid St. Chem.*, **22**, 361.

Benner, R.E., Brody, E.M. and Shanks, H.R., 1979, *J. Solid St. Chem.*, **27**, 383.

Berggren, K.F., 1978, in *The Metal–Nonmetal Transition in Disordered Systems*, edited by L.R. Friedman and D.P. Tunstall (Edinburgh: Scottish Universities Summer School in Physics), p. 399.

Bonera, G., Borsa, F., Crippa, M.L. and Rigamonti, A., 1971, *Phys. Rev.* B, **4**, 52.

Bouchard, G.H., Perlstein, J.H. and Sienko, M.L., 1967, *Inorg. Chem.*, **6**, 1682.

Brouers, F. and Brauwers, M., 1975, *J. Physique Lett.*, **36**, L17.

Brown, B.W. and Banks, E., 1954, *J. Am. Chem. Soc.*, **76**, 963.

Brown, B.W. and Banks, E., 1963.

Brusetti, R., Chakraverty, B.K., Devenyi, J., Dumas, J., Marcus, J. and Schlenker, C., 1982, in *Recent Developments in Condensed Matter Physics*, edited by J.T. de Vreese, L.F. Lemmens, V.E. Van Doren and J. Van Royen, Vol. 2, p. 181 (New York: Plenum).

Buder, R., Devenyi, J., Dumas, J., Marcus, J., Mercier, J. and Schlenker, C., 1982, *J. Physique Lett.*, **43**, L59.

Bullett, D.W., 1983, *Solid St. Commun.*, **46**, 575.

Camagni, P., Manara, A., Campagnoli, G., Gustinetti, A. and Stella, A., 1977, *Phys. Rev.* B, **15**, 4623.

Chamberland, B.L., 1969, *Inorg. Chem.*, **8**, 1183.

Chazalviel, J.N., Campagna, M., Wertheim, G.K. and Shanks, H.R., 1977, *Phys. Rev.* B, **16**, 697.

Clarke, R., 1977, *Phys. Rev. Lett.*, **39**, 1550.

Conroy, L.E. and Yokokawa, T., 1965, *Inorg. Chem.*, **4**, 944.

Consadori, F. and Stella, A., 1970, *Lett. Nuovo Cim.*, **3**, 600.

Dickens, P.G., Quilliam, R.M.P. and Whittingham, M.S., 1968, *Mat. Res. Bull.*, **3**, 941.

Diehl, R., Brandt, G. and Salje, E., 1978, *Acta Crystallogr.* B, **34**, 1105.

Dordor, P., Doumerc, J.P. and Villeneuve, G., 1983, *Phil. Mag.*, **47**, 315.

Dougier, P. and Casalot, A., 1979, *J. Solid St. Chem.*, **2**, 396.

Dougier, P. and Hagenmuller, P., 1975, *J. Solid St. Chem.*, **15**, 158.

Doumerc, J.P., 1978, in *The Metal–Nonmetal Transition in Disordered Systems*, edited by L.R. Friedman and D.P. Tunstall (Edinburgh: Scottish Universities Summer School in Physics), p. 313.

Doumerc, J.P. and Pouchard, M., 1970, *C. R. Acad. Sci. Paris*, **270**, 547.

Doumerc, J.P., Marcus, J., Pouchard, M. and Hagenmuller, P., 1979, *Mat. Res. Bull.*, **14**, 201.

Doumerc, J.P., Dordor, P., Marquestaut, E., Pouchard, M. and Hagenmuller, P., 1980, *Phil. Mag.*, **42**, 487.

Doumerc, J.P., Kabbaj, F., Campet, G., Claverie, J. and Pouchard, M., 1981, *Solid St. Commun.*, **39**, 1045.

Dubson, M.A., 1984, *PhD Thesis*, Cornell.
Dumas, J. and Schlenker, C., 1983, *Solid St. Commun.*, **45**, 885.
Dumas, J., Schlenker, C., Marcus, J. and Buder, R., 1983, *Phys. Rev. Lett.*, **50**, 757.
Dumas, J., Escribe-Filippini, C., Marcus, J. and Schlenker, C., 1984, in *Physics and Chemistry of Electrons and Ions in Condensed Matter*, edited by J.V. Acrivos, N.F. Mott and A.D. Yoffe (Dordrecht: Reidel), p. 571.
Edwards, P.P. and Sienko, M.J., 1978, *Phys. Rev.* B, **17**, 2575.
Ellerbeck, L.D., Shanks, H.R., Slides, P.H. and Danielson, G.C., 1961, *J. Chem. Phys.*, **35**, 298.
Ellis, B., Doumerc, J.P., Pouchard, M. and Hagenmuller, P., 1984, *Solid St. Commun.*, **51**, 913.
Ferretti, A., Rogers, D.B. and Goodenough, J.B., 1965, *J. Phys. Chem. Solids*, **26**, 2007.
Figlarz, M. and Gerand, B., 1980, in *9th Symposium on Reactivity of Solids, Krakow*, (Krakow: Polish Academy of Sciences), p. 660.
Flynn, E.J., Solin, S.A. and Shanks, H.R., 1977, *Solid St. Commun.*, **25**, 743.
Fogle, W. and Perlstein, J.H., 1972, *Phys. Rev.* B, **6**, 1402.
Friedman, L., 1971, *J. Non. Cryst. Solids*, **6**, 329.
Fuchs, R., 1965, *J. Chem. Phys.*, **42**, 3781.
Gerand, B., 1984, *Thèse d'Etat*, University of Picardie.
Gerand, B., Nowogrocki, G., Guenot, J. and Figlarz, M., 1979, *J. Solid St. Chem.*, **29**, 429.
Goodenough, J.B., 1965, *Bull. Soc. Chim. France*, **4**, 1200.
Goodenough, J.B., 1971, *Prog. Solid St. Chem.*, edited by H. Reiss (Oxford: Pergamon), Vol. 5, p. 145.
Goodenough, J.B., 1971, *Progress in Solid State Chemistry*, edited by H. Reiss (Oxford: Pergamon), Vol. 5, p. 145.
Greenblatt, M., McCarroll, W.H., Neifeld, R., Croft, M. and Waszczak, J.V., 1984, *Solid St. Commun.* (in press).
Greiner, J.D., Shanks, H.R. and Wallace, D.C., 1962, *J. Chem. Phys.*, **36**, 772.
Hagenmuller, P., 1971, *Progress in Solid State Chemistry*, edited by H. Reiss (Oxford: Pergamon), Vol. 5, 71.
Hägg, G., 1935, *Nature*, **135**, 874; *Z. Phys. Chem.* B, **29**, 192.
Hill, M.D. and Egdell, R.G., 1983, *J. Phys. C: Solid St. Phys.*, **16**, 6205.
Höchst, H., Brigans, R.D., Shanks, H.R. and Steiner, P., 1980, *Solid St. Commun.*, **37**, 41.
Höchst, H., Brigans, R.D. and Shanks, H.R., 1982, *Phys. Rev.* B, **26**, 1702.
Holcomb, D.F., 1978, in *The Metal–Nonmetal Transition in Disordered Systems*, edited by L.R. Friedman and D.P. Tunstall (Edinburgh: Scottish Universities Summer School in Physics), p. 251.
Hollinger, G., Himpsel, F.J., Reihl, B., Pertosa, P., Doumerc, J.P., 1982a, *Solid St. Commun.*, **44**, 1221.
Hollinger, G., Himpsel, F.J., Martensson, N., Reihl, B., Doumerc, J.P., Akahane, T., 1982b, *Phys. Rev.* B, **27**, 6370.
Hussain, A., 1978, *Acta Chem. Scandinavica* A, **32**, 479.
Hyde, G.B., Bagshaw, A.N., Anderson, S., O'Keeffe, M., 1974, *Annual Review of Materials Science*, **43**.
Inaba, H. and Naito, K., 1975, *J. Solid St. Chem.*, **15**, 283.
Inaba, H. and Naito, K., 1976, *J. Solid St. Chem.*, **18**, 279.
Ingold, J.H. and de Vries, R.C., 1958, *Acta Metall.*, **6**, 736.
Ioffe, A.F. and Regel, A.R., 1960, *Prog. Semiconductors*, **4**, 237.
Jones, W.H. Jr., Gabarty, E.A. and Barnes, R.G., 1962, *J. Chem. Phys.*, **36**, 494.
Kamitakahara, W.A., Harmon, B.N., Taylor, J.G., Kopp, L., Shanks, H.R. and Rath, J., 1976, *Phys. Rev. Lett.*, **36**, 1393.
Kaplan, O. and Zylberstein, 1976, *J. Physique*, **37**, L123.
Kehl, W.L., Hay, R.G. and Wahl, D., 1952, *J. Appl. Phys.*, **23**, 212.
Kihlborg, L. and Hussain, A., 1979, *Mat. Res. Bull.*, **14**, 667.

Kirkpatrick, S., 1971, *Phys. Rev. Lett.*, **27**, 1722.
Kobayashi, H., 1979, *Bull. Chem. Soc. Japan*, **52**, 1315.
Koffyberg, F.P., Dwight, K. and Wold, A., 1979, *Solid St. Commun.*, **30**, 433.
Kopp, L., Harmon, B.N. and Liu, S.A., 1977, *Solid St. Commun.*, **22**, 677.
Kupka, F. and Sienko, M.J., 1950, *J. Chem. Phys.*, **18**, 1296.
Lightsey, P.A., 1973, *Phys. Rev.* B, **8**, 3586.
Lightsey, P.A., Lilienfeld, D.A. and Holcomb, D.F., 1976, *Phys. Rev.* B, **14**, 4730.
Loopstra, B.O. and Boldrini, P., 1966, *Acta Crystall.*, **21**, 158.
Loopstra, D.B. and Rietveld, H.M., 1969, *Acta Crystall.*, B, **25**, 1420.
Lynch, D.W., Rosei, R., Weaver, J.H. and Olson, C.G., 1973, *J. Solid St. Chem.*, **8**, 242.
Mackintosh, A.R., 1963, *J. Chem. Phys.*, **38**, 1991.
Magnéli, A., 1949, *Ark. Kem.*, **1**, 269.
Magnéli, A., 1953, *Acta Chem. Scandinavica*, **7**, 315.
Mattheiss, L.F., 1969, *Phys. Rev.*, **181**, 987.
McCarroll, W.H. and Greenblatt, M., 1984, *J. Solid St. Chem.*, **54**, 282.
McColm, I.J. and Wilson, S.J., 1978, *J. Solid St. Chem.*, **26**, 223.
McNeil, W. and Conroy, L.E., 1962, *J. Chem. Phys.*, **28**, 87.
Miyamoto, Y., Kume, S., Doumerc, J.P. and Hagenmuller, P., 1983, *Mat. Res. Bull.*, **18**, 1463.
Mori, T., Kobayashi, A., Sasaki, Y., Ohshima, K., Sazuki, M. and Kobayashi, H., 1981, *Solid St. Commun.*, **39**, 1311.
Mott, N.F., 1974, *Metal-Insulator Transitions* (London: Taylor & Francis).
Mott, N.F., 1984, in *Physics and Chemistry of Electrons and Ions in Condensed Matter*, edited by J.V. Acrivos, N.F. Mott and A.D. Yoffe (Dordrecht: Reidel), p. 287.
Mott, N.F. and Davis, E.A., 1979, *Electronic Processes in Non-Crystalline Solids*, 2nd edition (Oxford: Clarendon).
Mott, N.F. and Kaveh, M., 1983, *Phil. Mag.* B, **47**, L17.
Mott, N.F., Pepper, M., Pollitt, S., Wallis, R.H. and Adkins, C.J., 1975, *Proc. R. Soc.* A, **345**, 169.
Muhlestein, L.D. and Danielson, G.C., 1967, *Phys. Rev.*, **15**, 825.
Narath, A. and Wallace, D.C., 1962, *Phys. Rev.*, **127**, 724.
Owen, J.F. and Teegarden, K.J., Shanks, H.R., 1978, *Phys. Rev.* B, **18**, 3827.
Ozerov, R.P., 1959, *Kristallografiya*, **4**, 201; *Zurnal Neorganisceskoj Chimii*, **4**, 1047; *Russian J. Inorg. Chem.*, **4**, 476.
Pouget, J.P., Kagoshima, S., Schlenker, C. and Marcus, J., 1983, *J. Physique Lett.*, **44**, L113.
Ramanujachary, K.V., Greenblatt, M. and McCarroll, W.H., 1984, *J. Crystal Growth* (in press).
Reynolds, T.G. and Wold, A., 1973, *J. Solid St. Chem.*, **6**, 565.
Ribnick, A.S., Post, B. and Banks, E., 1963, *Non Stoichiometric Compounds*, Advances in Chemistry Series, Vol. 39, p. 246 (Washington: American Chemical Society).
Sahle, E. and Nygren, M., 1983, *J. Solid St. Chem.*, **48**, 154.
Salje, E. and Hoppmann, G., 1981, *Phil. Mag.* B, **43**, 105.
Salje, H. and Viswanathan, K., 1975, *Acta Crystall.* A, **31**, 356.
Sato, M., Fujishita, H., Hoshino, S., 1983, *J. Phys. C: Solid St. Phys.*, **16**, L877.
Sawada, S. and Danielson, G.C., 1959, *Phys. Rev.*, **113**, 1008.
Sayer, M., Chen, R., Fletcher, R. and Mansingh, A., 1975, *J. Phys. C: Solid St. Phys.*, **8**, 2059.
Schlenker, C., Filippini, C., Marcus, J., Dumas, J., Pouget, J.P. and Kagoshiuma, S., 1983, *J. Physique*, **44**, C3–1757.
Shanks, H.R., Slides, P.H. and Danielson, G.C., 1963, *Non Stoichiometric Compounds*, Advances in Chemistry Series, Vol. 39, p. 237.
Sienko, M.J. and Truong, T.B.N., 1961, *J. Am. Chem. Soc.*, **83**, 3939.

Sienko, M.J., 1963, *Non Stoichiometric Compounds*, Advances in Chemistry Series, Vol. 39, 224.
Silberglitt, R.S., 1975, *Bull. Am. Phys. Soc.*, **20**, 308.
Sperlich, G., Laze, W.D. and Bang, G., 1975, *Solid St. Commun.*, **16**, 489.
Steadman, R., 1972, *Mat. Res. Bull.*, **7**, 1143.
Straumanis, M.E., 1949, *J. Am. Chem. Soc.*, **71**, 679.
Takugasawa, F. and Jacobson, A., 1976, *J. Solid St. Chem.*, **18**, 163.
Tanisaki, S., 1960a, *J. Phys. Soc. Japan*, **15**, 573.
Tanisaki, S., 1960b, *J. Phys. Soc. Japan*, **15**, 566.
Taylor, G.H., 1969, *J. Solid St. Chem.*, **1**, 359.
Travaglini, G. and Wachter, P., 1983, *Solid St. Commun.*, **47**, 217.
Travaglini, G., Wachter, P., Marcus, J. and Schlenker, C., 1981, *Solid St. Commun.*, **37**, 599.
Travaglini, G., Mörke, I. and Wachter, P., 1982, *Solid St. Commun.*, **45**, 289.
Tunstall, D.P., 1975, *Phys. Rev.* B, **11**, 2821.
Tunstall, D.P. and Ramage, W., 1980, *J. Phys. C: Solid St. Phys.*, **13**, 725.
Vest, R.W., Griffel, M. and Smith, J.F., 1958, *J. Chem. Phys.*, **28**, 293.
Villeneuve, G., Kessler, H. and Chaminade, J.P., 1976, *J. Physique*, **37**, C4–79.
Wadsley, A.D., 1963, in *Nonstoichiometric Compounds*, edited by L. Mandelcorn (New York: Academic), p. 98.
Wallis, R.H., Sol, N. and Zylberstein, A., 1977, *Solid St. Commun.*, **23**, 539.
Webman, I., Jortner, J. and Cohen, M.H., 1976, *Phys. Rev.* B, **13**, 713.
Weinberger, B.R., 1978, *Phys. Rev.* B, **17**, 566.
Weller, P.F., Taylor, B.E. and Mohler, R.L., 1970, *Mat. Res. Bull.*, **5**, 465.
Wertheim, G.K. and Chazalviel, J.N., 1981, *Solid St. Commun.*, **40**, 931.
Wiseman, P.J. and Dickens, P.G., 1973, *J. Solid St. Chem.*, **6**, 374.
Wiseman, P.J. and Dickens, P.G., 1976, *J. Solid St. Chem.*, **17**, 91.
Wittwer, V., Schirmer, O.F. and Schlotter, P., 1978, *Solid St. Commun.*, **25**, 977.
Wöhler, F., 1823, *Ann. Chim. Phys.*, **43**, 29.
Wold, A., Kunnmann, W., Arnott, R.J. and Ferretti, A., 1964, *Inorg. Chem.*, **3**, 545.
Wolfram, T., 1972, *Phys. Rev. Lett.*, **29**, 1383.
Zumsteg, F.C., 1976, *Phys. Rev.* B, **14**, 1406.

CHAPTER 12

metal–nonmetal transitions in perovskite oxides†

C.N.R. Rao‡ and P. Ganguly
Solid State and Structural Chemistry Unit, Indian Institute of Science, Bangalore 560012, India

1. Introduction

Oxides of the general formula ABO_3 possessing the perovskite structure can be described by a three-dimensional network of corner-shared BO_6 octahedra similar to that in metallic ReO_3. Of interest to us in this article are oxides of the type $LnBO_3$ (Ln = La or rare-earth ion; B = trivalent first-row transition metal ion). Electronic and magnetic properties of these oxides can be understood in a simple way by extending the schematic energy diagram of ReO_3 (Figure 1) to the $(BO_3)^{n-}$ complex (Goodenough 1972, 1974). In such a picture, the e_g orbitals of the cation form σ-bonds with anion p_σ orbitals along the 180° B–O–B linkages while the cation t_{2g} orbitals form weaker π-bonds with anionic p_π orbitals. It is convenient to use the covalent B–O mixing parameter, $\lambda_{B\text{-}O}$ (σ or π), to define the cation–anion–cation transfer integral $b_{B\text{-}O\text{-}B}$ (proportional to $\lambda^2_{B\text{-}O}$). The width of the σ^* and π^* bands of e_g and t_{2g} parentage in Figure 1 would then depend on b (or λ) which would vary according to the following criteria:

(i) $\lambda_\sigma > \lambda_\pi$.
(ii) λ increases with the formal charge of B cation.
(iii) For the same formal charge, λ increases with the principal quantum number of the d orbitals.
(iv) Since the A cation competes with the B cation for charge transfer from the oxygen ion, the more basic the A cation, the larger is $\lambda_{B\text{-}O}$.
(v) A decrease in the B–O–B angle from 180° decreases λ.
(vi) For cations with the same principal quantum number for the d orbitals and the same formal charge, the radial extension of the d orbitals decreases with increasing atomic number, Z, of the B ion. Since the size of the B cation decreases with Z, one may expect a stronger B–O overlap with increasing Z; maximum

† Contribution No. 289 from the Solid State and Structural Chemistry Unit.
‡ To whom all correspondence should be addressed.

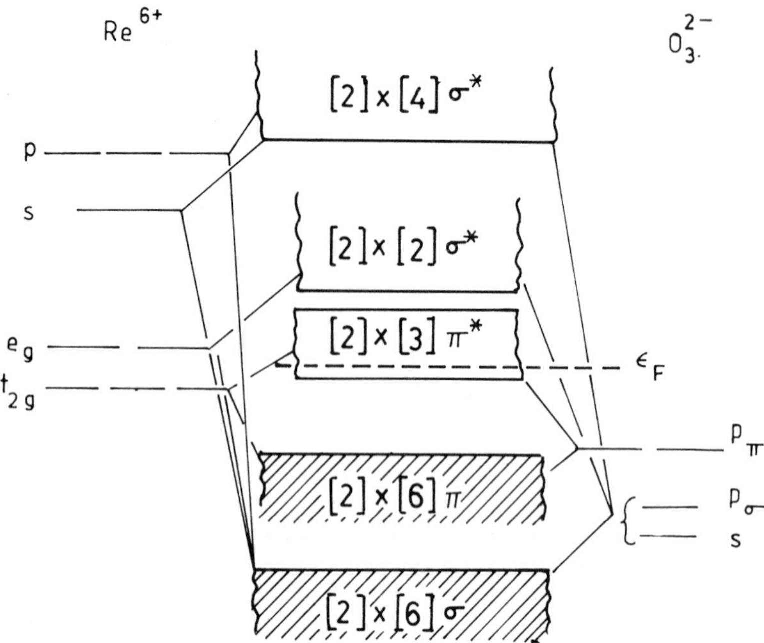

Figure 1. Energy-band scheme for ReO_3 (from Goodenough 1972).

B–O overlap may then be expected for the end-members of the transition elements in the same period.
(vii) For cations with the same formal charge containing d electrons with the same principal quantum number, intra-atomic exchange interaction decreases the radial extension as the total number of unpaired electrons on the cation increase (as S increases). The above generalizations have been employed to account for the properties of ABO_3 perovskites.

The variation in the room-temperature electrical resistivity, ρ, and the activation energy for electrical conduction, E_a, of the perovskite oxides (Ganguly *et al.* 1976) are shown in Figure 2. In accordance with (vi) and (vii) above, $LaTiO_3$ and $LaNiO_3$ (Ni ion in the low-spin state, $t_{2g}^6 e_g^1$) exhibit metallic (itinerant d electron) behaviour, the transition metal ions being $S=\frac{1}{2}$ ions; oxides with $S\geqslant 1$ for the B ions, such as those with V^{3+}, Cr^{3+}, Mn^{3+}, Fe^{3+}, are insulators exhibiting localized behaviour of the d electrons. The fairly strong covalent mixing in $LaCoO_3$ imposes a crystal field on the B cation which seems to be intermediate between that in $LaFeO_3$ (high-spin Fe^{3+}) and $LaNiO_3$ (low-spin Ni^{3+}) so that there is a spin-state equilibrium between high- and low-spin Co^{3+} ions (Heikes *et al.* 1964, Naiman *et al.* 1965, Jonker 1966). Electrical properties of $LaCoO_3$ are intimately connected to the spin-state population of the Co^{3+} ions and the material transforms from a semiconducting to a metallic state around 1200 K (Raccah & Goodenough 1965, 1967, Bhide *et al.* 1972). In a sense, the

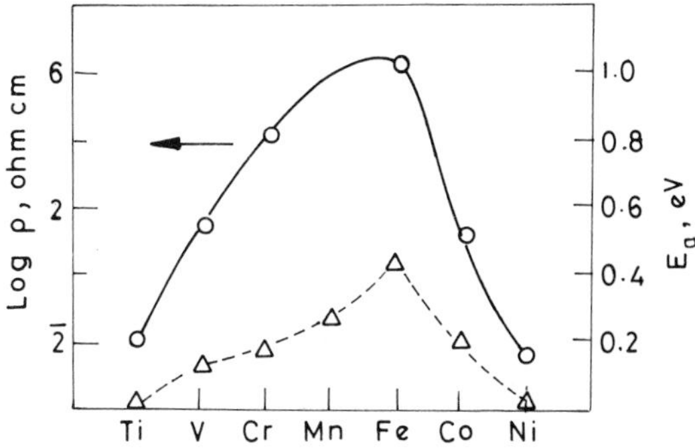

Figure 2. E_a and log ρ variation at 300 K for $LaBO_3$ compounds (from Ganguly *et al.* 1976).

behaviour of $LaCoO_3$ as a function of temperature may be looked upon as a composition-controlled metal–nonmetal (M–NM) transition, the composition of the various spin and valence states of the Co ions varying with temperature. We shall examine the behaviour of $LaCoO_3$ in some detail, a complete understanding of which seems to remain elusive.

In oxides such as $LaVO_3$ and $LaMnO_3$, replacement of La by Sr changes the oxidation states of the B ion from 3+ to 4+. Accordingly, in $La_{1-x}Sr_xBO_3$ (B = V, Mn, Co), the electrical conductivity approaches the metallic value when $x \geqslant 0{\cdot}25$ (Dougier & Casalot 1970, Jonker & van Santen 1950, 1953), the high conductivity being associated with the facile transfer of electrons from B^{3+} to B^{4+} ions. When $B = Mn^{3+}$, such an electron transfer taking place by the conservation of electron spin angular momentum leads to the stabilization of the ferromagnetic phase at low temperature according to Zener's double exchange mechanism (Zener 1951, Anderson & Hasegawa 1955). A wealth of information is available on $Ln_{1-x}Sr_xCoO_3$ at a microscopic level (Bhide *et al.* 1972, 1975, Rao & Bhide 1973); in this system, progressive increase in Sr^{2+} concentration results in the formation of an impurity band which is responsible for the itinerancy of the d electrons and the metallic character when $x \geqslant 0{\cdot}3$.

An interesting type of composition-controlled metal–nonmetal transition which we shall discuss is in oxides of the type $LaNi_{1-x}M_xO_3$ (M = Cr, Mn, Fe, Co). While $LaNiO_3$ in metallic, all the $LaMO_3$ perovskites are insulators (Ganguly *et al.* 1976). The $LaNi_{1-x}M_xO_3$ system is different from the $La_{1-x}Sr_xBO_3$ system since the one-electron energies of the M^{3+} levels are considerably different from those of the Ni^{3+} levels. In these composition-controlled oxide systems, the magnetic as well as the conducting species have strong d-electron character. We would therefore anticipate a correlation between magnetic and electrical properties in these oxides and would expect coulomb interaction to play a significant role in the metal–nonmetal transitions.

2. The metal–nonmetal transition in $LaCoO_3$

$LaCoO_3$ provides a striking example of a system in which the electrical and magnetic properties are directly correlated. Electron transport properties of $LaCoO_3$ are closely related to a complex temperature-dependent equilibrium of various spin and valence states of cobalt. The presence of predominantly diamagnetic Co^{III} (t_{2g}^6) ions at low temperatures renders this oxide a good insulator below 200 K. As the temperature is increased, the formation of paramagnetic spin states leads to a rapid increase in conductivity. The oxide eventually becomes metallic at high temperatures (>1200 K). $LaCoO_3$ is a unique example of a system in which a temperature-dependent compositional change of diamagnetic and paramagnetic cobalt species brings about a M–NM transition. Properties of $LaCoO_3$ are sensitive to the conditions of preparation and probably also to the degree of oxidation. However, there are some basic features which are independent of sample history.

2.1. $T<200$ K

Neutron diffraction studies (Menyuk *et al.* 1967) of $LaCoO_3$ have established a space group $R\bar{3}c$ at low temperatures, with two indistinguishable Co ions per unit cell. There is no evidence of long-range antiferromagnetic ordering. In this region, the magnetic susceptibility behaviour (Figure 3) is sensitive to sample history. At the lowest temperatures, a Curie-like behaviour is observed due to the presence of small amounts of paramagnetic species. Subtracting out the contribution from this species, the residual susceptibility, χ_{act}, shows linear $\log(\chi_{act}T)$ vs. $1/T$ behaviour implying an activated diamagnetic to paramagnetic spin-state transition with an activation energy of ~0·01 eV and a Curie constant of ~1·0 ($\chi_{act}T \sim 1{\cdot}0$ when $T \to \infty$) suggesting an $S=1$ for the paramagnetic species. This can happen if we have intermediate spin Co^{iii} ($t_{2g}^5e_g^1$). The latter would suggest $R\bar{3}m$ symmetry which is inconsistent with the neutron diffraction data. The activated nature of the behaviour at low temperatures is brought out rather clearly (Vasanthacharya & Ganguly 1983, Madhusudan *et al.* 1980b) when some of the Co ions are replaced by Al^{3+} or Ga^{3+} ions. ^{57}Co Mössbauer spectroscopic studies (Bhide *et al.* 1972, Rao & Bhide 1973) of $LaCoO_3$ (Figure 4) show that up to 200 K, there are only Co^{3+} ($t_{2g}^4e_g^2$) and Co^{III} species, the Co^{3+}/Co^{III} ratio increasing with temperature up to 200 K with an activation energy of ~0·01 eV. The persistence of a Curie-like behaviour at the lowest temperatures is puzzling as the effect becomes more prominent as the stoichiometry is improved (Vasanthacharya & Ganguly 1983). A theoretical model for spin-state transitions (Ramasesha *et al.* 1979) suggests that an off-centre lattice mode can mix the low- and high-spin states so that a non-zero population of high-spin states may be present even at 0 K.

The magnetic susceptibility behaviour of the other rare-earth cotbaltites,

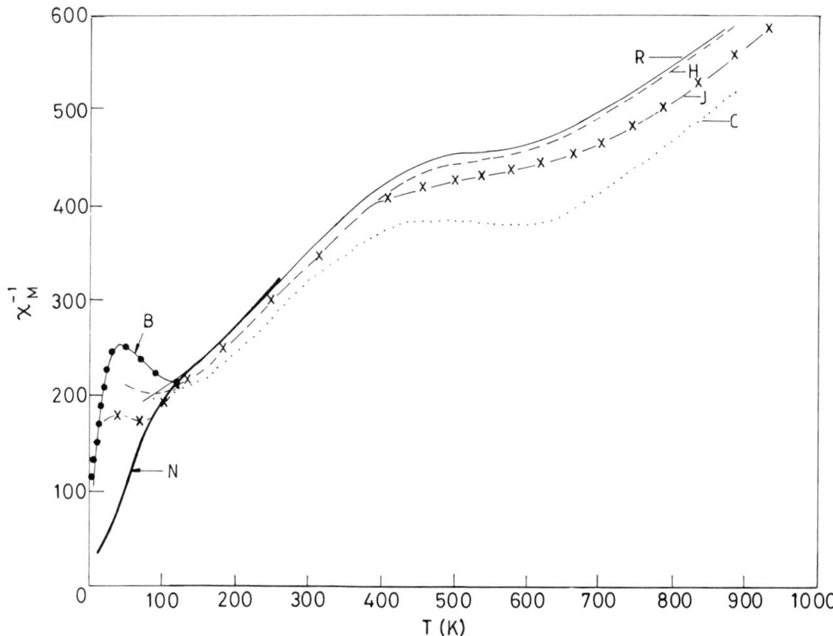

Figure 3. χ_M^{-1} vs. T plots for $LaCoO_3$ as reported by several workers: B, by Blasse (from Jonker 1966); C, Bhide *et al.* (1972); H, Heikes *et al.* (1964); J, Jonker (1966); N, Naiman *et al.* (1963); R, Raccah & Goodenough (1967).

$LnCoO_3$, is similar after the contribution due to the Ln ions is subtracted out (Madhusudan *et al.* 1980a). The activation energy for the spin-state transition, however, increases with the decreasing size of the rare-earth ion.

Electrical resistivity of $LaCoO_3$ at the lowest temperatures is greater than 10^5 Ω cm. The log ρ vs. $1/T$ plot is nearly linear between 80 K and 200 K with an apparent activation energy of 0·1 eV (Figure 5). Conductivity in this temperature range may be dominated by the spin-state transition mentioned earlier.

2.2. *200 K < T < 650 K*

Magnetic susceptibility in this region is nearly temperature independent, exhibiting a plateau-like behaviour in the χ^{-1} vs. T plot between 400 K and 650 K (Figure 3). From the magnitude of the susceptibility at 650 K ($\chi T_{650} \sim 1{\cdot}5$, $C_{Co^{3+}} = 3{\cdot}0$), it is assumed that half the Co ions are in the high-spin Co^{3+} state and the other half in the low-spin Co^{III} state, the two spin states ordering themselves in different sublattices. Mössbauer studies (Figure 4) show a relative decrease in the Co^{3+} population in this temperature region, which has been interpreted as due to the formation of charge-transfer (divalent and tetravalent cobalt) states which would have isomer shifts close to low-spin Co^{III}.

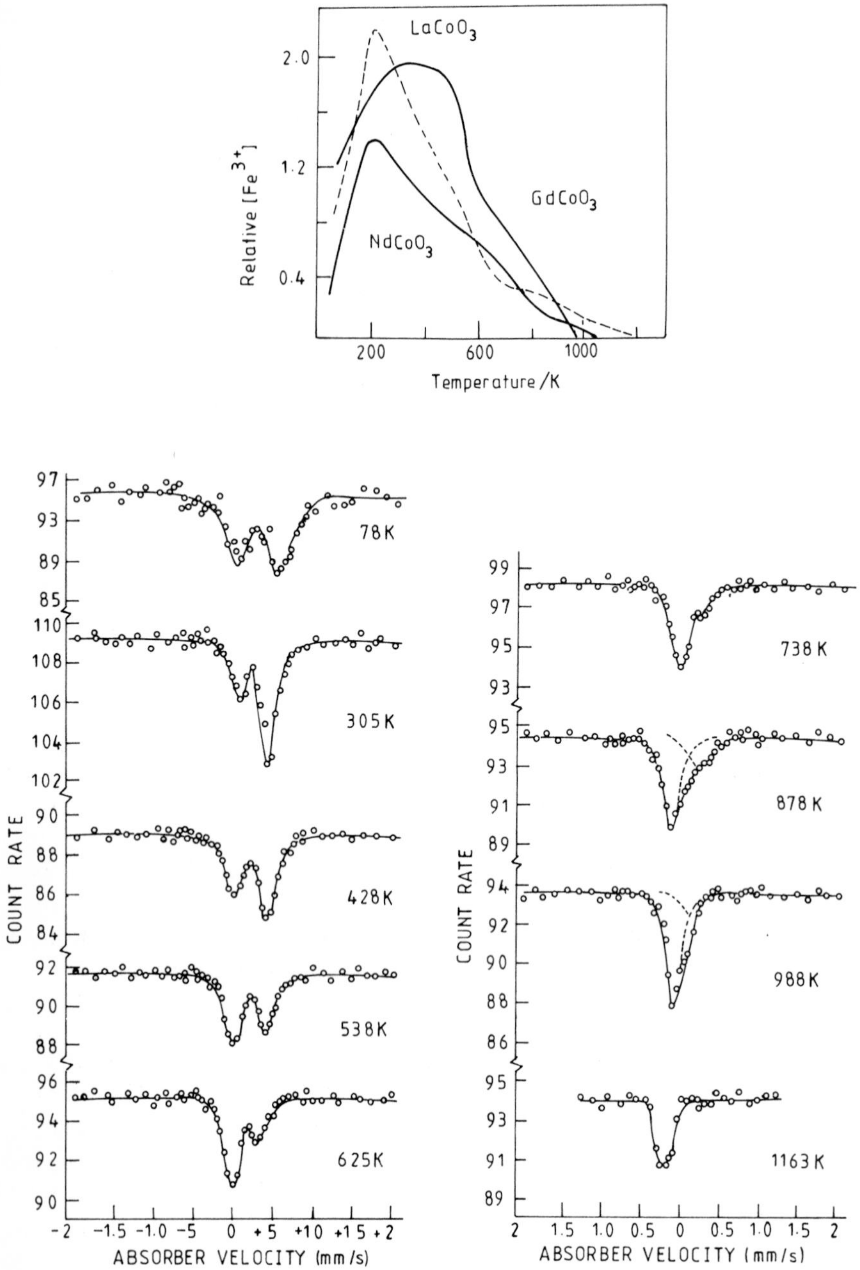

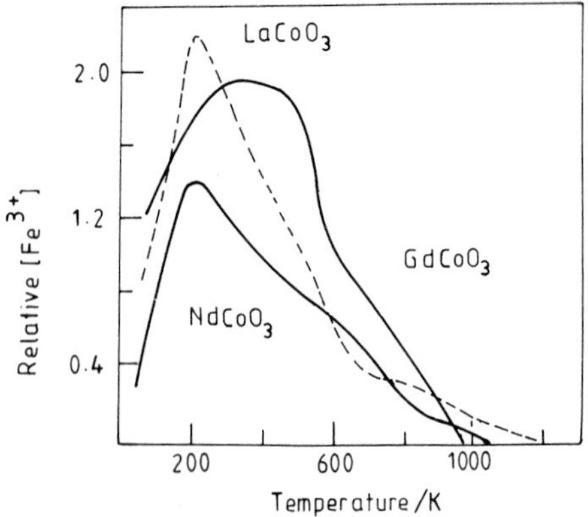

Figure 4. Series of Mössbauer spectra of $LaCoO_3$:^{57}Co matched against $K_4Fe(CN)_6 \cdot 3H_2O$ single-crystal absorber showing coexistence of Fe^{III} and Fe^{3+}. Inset: Fe^{3+}/Fe^{III} ratio in $LaCoO_3$, $NdCoO_3$ and $GdCoO_3$ (from Bhide *et al.* 1972 and Rajoria *et al.* 1974).

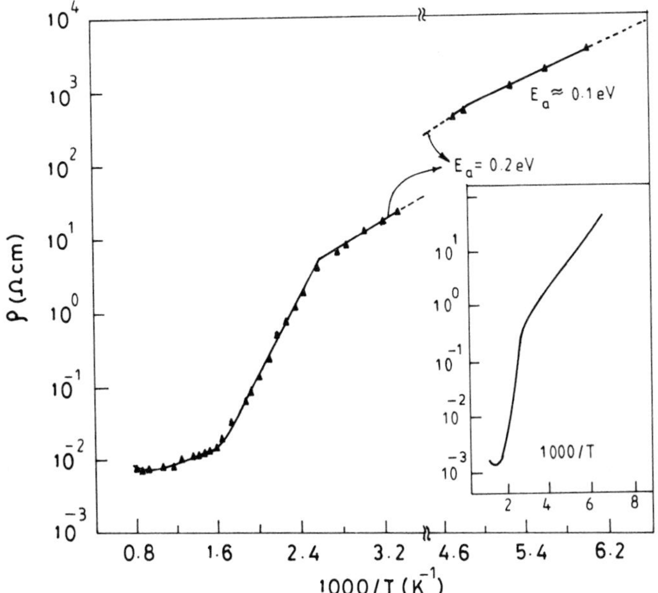

Figure 5. Log ρ vs. $1/T$ plot of polycrystalline $LaCoO_3$ (from Bhide *et al.* 1972). Inset from Heikes *et al.* 1964.

The electrical resistivity decreases sharply in this temperature range, the decrease being most prominent between 400 and 650 K. This is consistent with the formation of charge-transfer states. Since the e_g electrons forming σ bonds with the anion p_σ orbitals would be expected to be more itinerant, one would expect charge transfer involving the e_g electrons to be more favoured than that involving the t_{2g} electrons.

Crystal structure refinement of X-ray data by Raccah & Goodenough (1967) showed that in this temperature range, an $R\bar{3}$ component is introduced into the $R\bar{3}c$ character in as much as the reliability index of $R\bar{3}$ becomes progressively better. These authors therefore describe the region between 400 K and 650 K as one in which there is short-range order between the cobalt ions in the two spin states. Recent neutron diffraction studies (Thornton *et al.* 1982) and X-ray diffraction studies (Rao *et al.*, unpublished) support such a conclusion.

2.3. 650 K < T < 1200 K

In this temperature range, the magnetic susceptibility shows a Curie behaviour with a μ_{eff} corresponding closely to two unpaired electrons per cobalt. The resistivity in this region (10^{-2}–10^{-3} Ω cm) is surprisingly low for such localized moment behaviour to exist although a similar behaviour is found in

La_2NiO_4 (Ganguly & Rao 1973, Smolenskii *et al.* 1963). The temperature dependence of the resistivity in this region is dependent on sample history (Jonker 1966, Bhide *et al.* 1972, Thornton *et al.* 1982). Raccah & Goodenough (1967) observed that the crystal structure above 650 K is $R\bar{3}$ although recent neutron diffraction studies (Thornton *et al.* 1982) suggest the $R\bar{3}$ symmetry to be preferred over $R\bar{3}c$ only at 648 K (while at higher temperatures the $R\bar{3}c$ symmetry is to be preferred). The significance of the conclusion from neutron diffraction studies is not clear. Differential thermal analysis studies have usually shown an endothermic transition (Raccah & Goodenough 1967, Bhide *et al.* 1972) around 610 K, although this has recently been disputed (Thornton *et al.* 1982).

Mössbauer studies (Figure 4) show a continuous decrease in the high spin Co^{3+} population in this range. These results suggest thermal disordering above 650 K of the divalent–tetravalent ordered states of Co ions arising from charge transfer, resulting in the formation of Co^{iii} ($t_{2g}^5 e_g^1$) ions with the e_g electrons in a broad, nearly itinerant σ^* band. The evolution of the σ^* band in the 700–1000 K region (due to cation–anion orbital overlap) is also indicated by the decrease in the Mössbauer centre shift.

2.4. $T > 1210$ K

Electrical resistivity measurements by several workers have shown that $d\rho/dT$ becomes abruptly positive above 1000 K. The resistivity at the transition is $\sim 10^{-2}\ \Omega$ cm. An endothermic first-order transition with a large enthalpy change (~ 5 kJ mol^{-1}) is also observed around 1200 K (Raccah & Goodenough 1967, Bhide *et al.* 1972). A single ^{57}Co Mössbauer resonance (corresponding to the band state of d electrons) is seen above the transition. While the isomer shift varies continuously through this transition, the Lamb–Mössbauer factor (area under the resonance) increases at the transition. The large enthalpy (or entropy) change of the transition at 1200 K was suggested to be entirely electronic in origin (Raccah & Goodenough 1967) without any contribution from configurational factors. The transition was therefore considered to arise from the change of localized d electrons to itinerant electrons and this was taken as corroborating the idea that the two limiting descriptions of d electrons, namely, ligand field and band limits, correspond to distinct thermodynamic states. Recent studies (Rao, C.N.R., 1984, unpublished results) have shown that there is a change in crystal symmetry at the transition, the oxide becoming cubic. Such a symmetry change would contribute to some configurational ΔS. Several repeated measurements by Rao and co-workers and others have shown the occurrence of the first-order transition in $LaCoO_3$ and other rare-earth cobaltites at ~ 1200 K. In the light of these studies, the recent report of Thornton *et al.* (1982) on the absence of a transition in $LaCoO_3$ is rather intriguing.

The various features of the thermal evolution of cobalt states and the associated phase transition of $LaCoO_3$ are indeed most interesting and reveal the

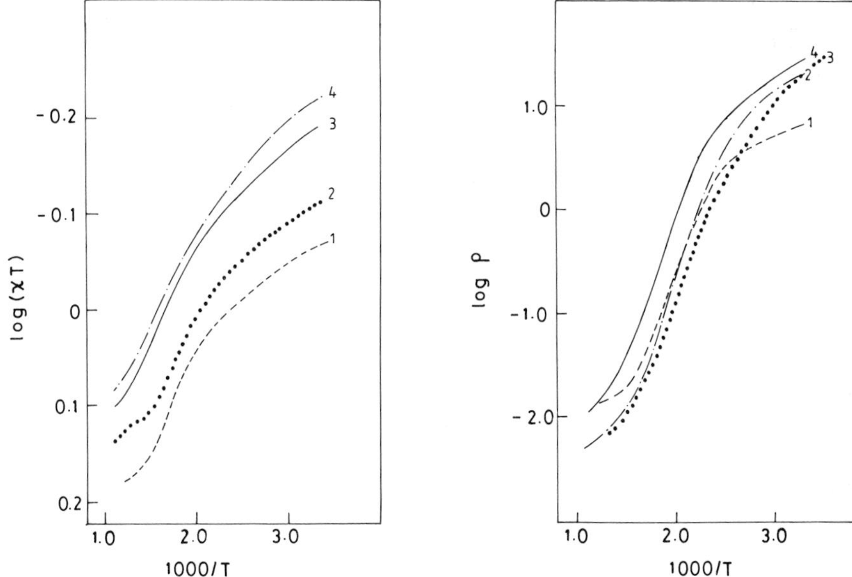

Figure 6. (*a*) Log (χT) vs. $1/T$ and (*b*) log ρ vs. $1/T$ plots of $LaCo_{0\cdot95}B_{0\cdot05}O_3$. 1. $LaCoO_3$; 2. B=Cr; 3. B=Al; 4. B=Ga (from Madhusudan *et al.* 1980b).

intimate relationship between the electronic configuration of the cobalt ions and the electrical properties of the oxide. A comparison of the log ρ vs. $1/T$ plots with the log χT vs. $1/T$ plots (Figure 6) of several $LaCo_{0\cdot95}B_{0\cdot05}O_3$ compounds (B=Al, Ca, Fe, Cr, Mn) shows striking similarities (Madhusudan *et al.* 1980b) and suggests a correlation between the number of paramagnetic species (χT is related to the number of paramagnetic species through the Curie law $\chi T = C = N\mu_{\text{eff}}^2/3k$) and the electrical conductivity. A second point to ponder is whether the imminent metallization of $LaCoO_3$ at high temperatures controls the electronic configuration of the Co ions at low temperatures or vice versa. The charge-transfer states in $LaCoO_3$ postulated around 650 K could be viewed as a disproportionation of the trivalent Co ions to divalent and tetravalent Co ions. Such a disproportionation is expected in narrow-band systems just before the onset of metallic behaviour.

3. The $La_{1-x}Sr_xCoO_3$ system

The $La_{1-x}Sr_xCoO_3$ system is a novel one where the black insulating $LaCoO_3$ (($x=0$) is converted to a metallic phase with increase in x; the oxide with $x=0\cdot5$ is a ferromagnetic metal with bronze lustre. Since the properties of this oxide were first reported in an early paper (Jonker & van Santen 1953), there have been many developments in the understanding of metal–insulator transitions brought about by localization due to disorder as well as of the concept of minimum metallic conductivity. These ideas were first used in the $La_{1-x}Sr_xCoO_3$ system by Rao *et al.* (1975) who used Mössbauer spectroscopy extensively for the investigation of the problem.

The low-temperature (4·2 K) resistivity of $La_{1-x}Sr_xCoO_3$ varies from $\rho > 10^5\ \Omega$ cm for $x = 0$ to $\rho \lesssim 10^{-3}\ \Omega$ cm for $x = 0{\cdot}5$, the change from a negative to a positive temperature coefficient of resistivity occurring around $x \sim 0{\cdot}3$. The lattice parameters of the rhombohedral unit cell ($R\bar{3}c$ space group) show a small but definite discontinuity in the range $0{\cdot}1 < x < 0{\cdot}15$ as does the paramagnetic Curie temperature (Bhide *et al.* 1975); this behaviour has been considered to be due to chemical inhomogeneities corresponding to a segregation into Sr^{2+}-rich and La^{3+}-rich regions within the perovskite phase (Raccah & Goodenough 1969, Goodenough 1972, Bhide *et al.* 1975). The tetravalent cobalt ions (holes) introduced by the Sr^{2+} ions are considered to be delocalized over the eight cobalt nearest neighbours of the Sr^{2+} ion. The critical concentration at which all the cobalt ions will be connected would correspond to $x = 0{\cdot}125$ as indeed found experimentally (Bhide *et al.* 1975). This would account for the discontinuities found in the range $0{\cdot}1 < x < 0{\cdot}15$. At low Sr^{2+} concentrations ($x < 0{\cdot}05$), molecular complexes of eight cobalt ions surrounding each Sr^{2+} ion are proposed to be formed. It would seem that the Sr^{2+} and La^{3+} ions are ordered in the lattice and evidence for such ordering has been obtained from electron diffraction studies of the $x = 0{\cdot}5$ sample (Gai & Rao 1975). The existence of La^{3+}-rich and Sr^{2+}-rich phases seems to argue favourably for the coexistence of localized and collective electron phases.

Mössbauer spectra of $La_{1-x}Sr_xCoO_3$ samples (Bhide *et al.* 1975) with $x \leqslant 0{\cdot}125$ show two resonances (Figure 7) almost at the same position as those in $LaCoO_3$ except for a slight broadening of the low-velocity resonance which may be attributed to a superposition of the tetravalent Co resonance on the Co^{III} resonance. With increasing x, the relative Co^{3+}/Co^{III} intensity ratio decreases up to $x \sim 0{\cdot}05$ and then increases again for $x = 0{\cdot}10$. For small x it may therefore be assumed that the high-spin Co^{3+} ions are converted to tetravalent Co while at higher x, the low-spin ions are converted to tetravalent Co; thus, with increase in x the crystal field on the Co ions increases, favouring the low-spin state. Another explanation would involve a time-averaged configuration associated with Sr^{2+}-rich clusters making a contribution so that the area between the two peaks is enhanced. The time-averaged configuration is clearly seen in the $x \geqslant 0{\cdot}125$ samples (Figure 7), the isomer shift of the observed resonance lying between that characteristic of high-spin Co^{3+} and low-spin $Co^{IV}(t_{2g}^5)$. The observation of a single peak at $x = 0{\cdot}125$ provides direct evidence for the formation of an impurity band at this concentration.

In the range $0{\cdot}125 \leqslant x \leqslant 0{\cdot}5$, a ferromagnetic component is seen at low temperatures. A single six-finger pattern is seen in the Mössbauer spectrum at 78 K for the $x = 0{\cdot}5$ sample with an internal field of 320 ± 20 kOe (Figure 8) and an isomer shift of $0{\cdot}25 \pm 0{\cdot}4$ mm s^{-1} (relative to 310 ESS). This value

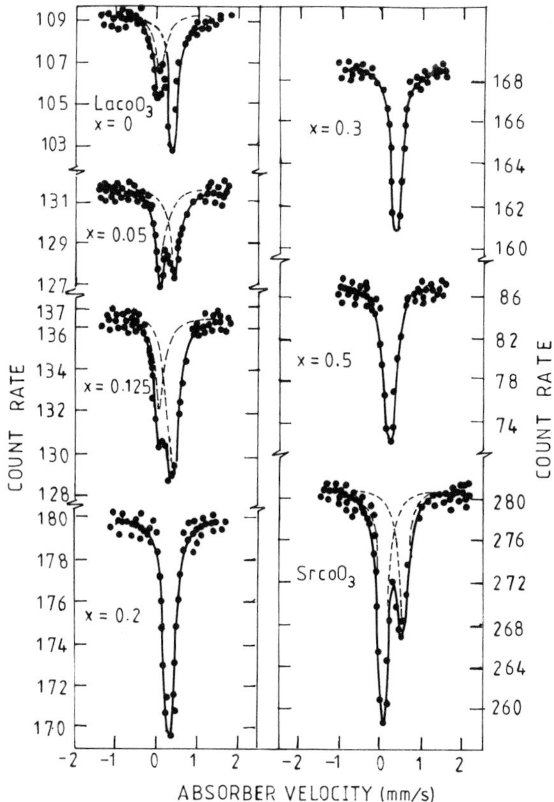

Figure 7. Typical Mössbauer spectra of $La_{1-x}Sr_xCoO_3$ ($0 \leqslant x \leqslant 0{\cdot}5$) at room temperature when matched against $K_4Fe(CN)_6 \cdot 3H_2O$ single-crystal absorbers (from Bhide *et al.* 1975).

suggests a chemical environment intermediate between high-spin Co^{3+} (H_{int} = 500–560 kOe, $\delta = 0{\cdot}40$–$0{\cdot}60$ mm s^{-1}) and low-spin Co^{IV} ($H_{int} = 250$–300 kOe, $\delta = -0{\cdot}33$–$0{\cdot}20$ mm s^{-1}). The Curie temperature of $La_{0{\cdot}5}Sr_{0{\cdot}5}CoO_3$ is 232 K.

The paramagnetic line in the Mössbauer spectra (Figure 8) at low temperatures of the $x > 0{\cdot}125$ samples is associated with superparamagnetic clusters. The presence of superparamagnetism even in the $x = 0{\cdot}5$ sample is a bit intriguing as electron diffraction studies have clearly established an ordering (Gai & Rao 1975) of La^{3+} and Sr^{2+} ions so that it is unlikely that there are localized and collective electron regions in the $x = 0{\cdot}5$ sample. It has been found recently that a well-annealed $x = 0{\cdot}5$ sample shows little or no paramagnetic component and the spectrum is characteristic of a well-ordered ferromagnet. These studies show that the $x > 0{\cdot}3$ samples may be described as itinerant-electron ferromagnets. The reflectivity vs. energy curve of the $x = 0{\cdot}5$ sample shows a minimum around 2·30 eV, marking the plasma frequency in this metal and accounting for the bronze lustre (Bhide *et al.* 1975).

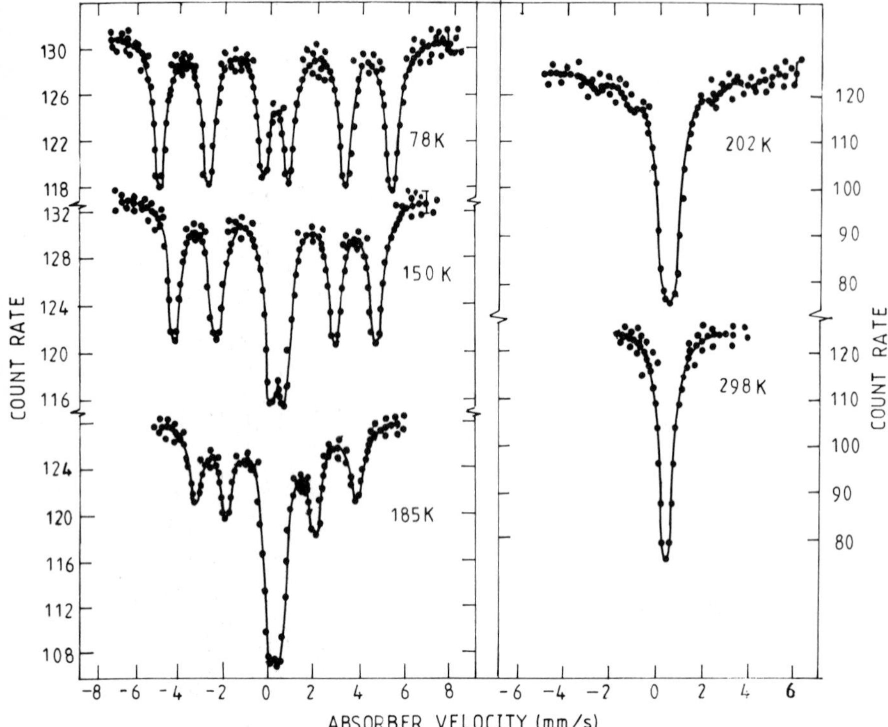

Figure 8. Variation of Mössbauer spectra as a function of temperature of $La_{0.5}Sr_{0.5}CoO_3$ (from Bhide *et al.* 1975).

Electronic and magnetic properties of several series of the $Ln_{1-x}Sr_xCoO_3$ family (Ln = Pr, Nd and other rare earths) have been investigated in detail (Rao *et al.* 1977). All these series of oxides show itinerant d electron behaviour above a critical value of x accompanied by ferromagnetism at low temperatures. The value of conductivity when $d\rho/dT$ changes sign at the critical value of x is about the same as in $La_{1-x}Sr_xCoO_3$. The Curie temperature increases with increase in x and decreases with the decrease in the size of the rare-earth ion. Replacement of La by Ba favours the itinerant d electron behaviour relative to Sr^{2+}; Ca^{2+} is less favourable than Sr^{2+}.

Two models have been considered to account for the properties of $La_{1-x}Sr_xCoO_3$; (i) a double exchange mechanism (Zener 1951, Zener & Heikes 1953) which gives the correct sign of the magnetic interaction and the correct order of magnitude of the conductivity, and (ii) itinerant electron ferromagnetism. The observation of one set of six-finger patterns with a hyperfine field and isomer shift between those characteristic of Co^{3+} and Co^{IV} tends to support the Zener mechanism. The criterion that the initial and final states in the double-exchange mechanism should be degenerate are met for e_g electron transfer by the states given below:

$$Co^{3+}(t^4_{2g}e^2_g)\text{–}O^{2-}\text{–}Co^{iv}(t^4_{2g}e^1_g) \leftrightarrow Co^{iv}(t^4_{2g}e^1_g)\text{–}O^{2-}\text{–}Co^{3+}(t^4_{2g}e^2_g)$$

$$Co^{iii}(t^5_{2g}e^1_g)\text{–}O^{2-}\text{–}Co^{IV}(t^5_{2g}e^0_g) \leftrightarrow Co^{IV}(t^5_{2g}e^0_g)\text{–}O^{2-}\text{–}Co^{iii}(5^5_{2g}e^1_g)$$

In the first, the saturated value of the magnetisation at 0 K should correspond to $\sim 3{\cdot}5\ \mu_B$ while in the latter the expected value is $1{\cdot}5\ \mu_B$. Magnetization studies (Goodenough 1971) at low temperatures give a value of $1{\cdot}5\ \mu_B$ supporting the latter model. The main objection to the application of the Zener model is that there are no discontinuities in the electrical transport properties at the magnetic ordering temperature (Bhide *et al.* 1975).

In the itinerant-electron model (Goodenough 1972, Bhide *et al.* 1975), σ^* and π^* bands are considered to be formed at high impurity concentrations (of Sr^{2+}) from acceptor levels (Figure 9). The formation of such bands generates a spontaneous ferromagnetism so long as the σ^* band remains less than quarter filled and the π^* band more than three-quarters filled. The saturation magnetization of $1{\cdot}5\ \mu_B$ led Goodenough to suggest that there are $0{\cdot}5\ \sigma^*$ electron per mole or one e_g electron per Co pair. This model is then similar to the Zener double-exchange model involving Co^{iii} and Co^{IV} ions.

Ferromagnetic resonance (f.m.r.) studies (Bahadur *et al.* 1979) on $Ln_{1-x}Sr_xCoO_3$ have shown that irrespective of the nature of the Ln ion or the value of x in the metallic state, the g_{eff} value is around 1·25; Co^{3+}, Co^{4+}, Co^{IV}, Co^{iv}, Co^{2+} and Co^{II} ions are expected to have g_{eff} values of 3·5, 2·0, 2·0, 1·6, 4·3 and 2·2, respectively. Bahadur *et al.* (1979) have given arguments to show that a g_{eff} value of 1·25 is satisfied by an intermediate Co^{iii} configuration.

The point that arises from the f.m.r. studies is that the tetravalent Co ions in the $x \geqslant 0{\cdot}3$ compositions are in the low-spin $(t^5_{2g}e^0_g)$ configuration forming a broad π^* band in which the single hole is itinerant. The intermediate spin Co^{iii} ion seems to be localized. In $LaCoO_3$ it was assumed, on the other hand, that at high temperatures the formation of Co^{iii} ions makes the system itinerant. This apparent contradiction is resolved if we remember that in $LaCoO_3$ the electrons of Co^{iii} ions are itinerant only at high temperatures. In the $x = 0{\cdot}5$ sample we may visualize a matrix consisting of Co^{IV} ions which because of the higher oxidation state broadens the π^* band to sustain the itinerant electron behaviour as well as ferromagnetism. The e_g electrons of the Co^{iii} ion may be considered to be hopping from site to site, being localized at each site by intra-atomic exchange of the $S = 1$ (Co^{iii}) configuration. In agreement with such a conclusion, it is found from recent studies (Taguchi *et al.* 1980) that $x > 0{\cdot}5$ compositions are also ferromagnetic as well as metallic so that the itinerant electrons need only be associated with the π^* band, the ferromagnetic interactions arising from the band being more than three-quarters filled.

Rao *et al.* (1975) have examined changes in the electrical conductivity of $Ln_{1-x}Sr_xCoO_3$ as a function of x in terms of localization and Mott's concept of

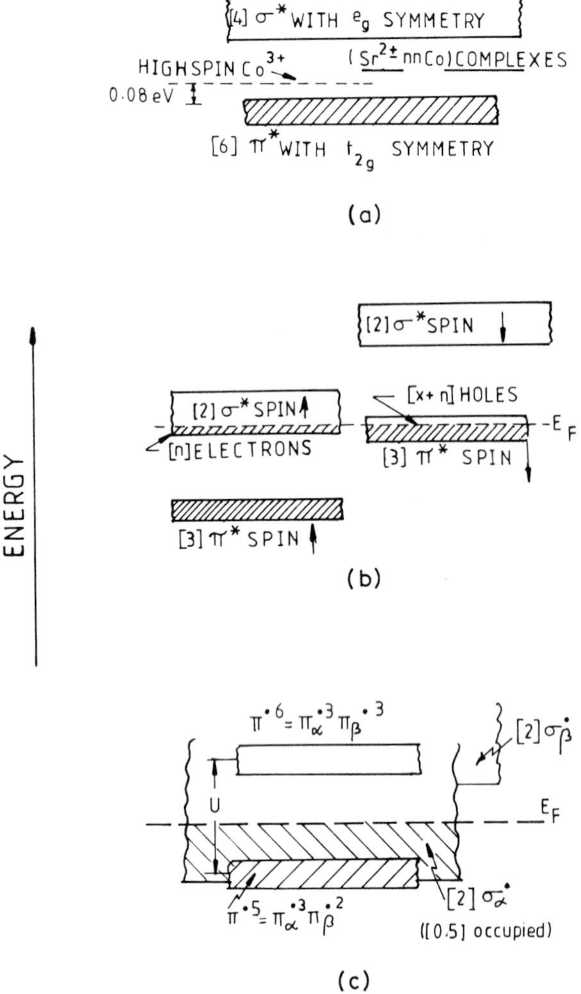

Figure 9. (*a*) Cobalt 3d bands for La-rich region showing Sr^{2+} impurity level; (*b*) cobalt 3d bands for ferromagnetic $La_{0\cdot5}Sr_{0\cdot5}CoO_3$ when there is no correlation splitting of π^* bands; (*c*) schematic cobalt 3d bands for ferromagnetic $La_{0\cdot5}Sr_{0\cdot5}CoO_3$ if π^* bands are split by electron correlations (after Goodenough 1971, Bhide *et al.* 1975).

minimum metallic conductivity, σ_{min} (Mott 1967, 1968, 1969, 1981, 1984a, 1984b; Mott & Davis 1971, Mott *et al.* 1975). They point out that the value of resistivity for the composition x_0 at which the temperature coefficient of resistivity changes sign is $\sim 2\times10^{-3}\,\Omega\,\mathrm{cm}$ (Figure 10) and is close to the value expected from the concept of σ_{min}.

Rao & Om Parkash (1977) have observed several other interesting correlations. In the composition range $0 < x \lesssim 0{\cdot}1$, disorder arising from Sr^{2+} ions

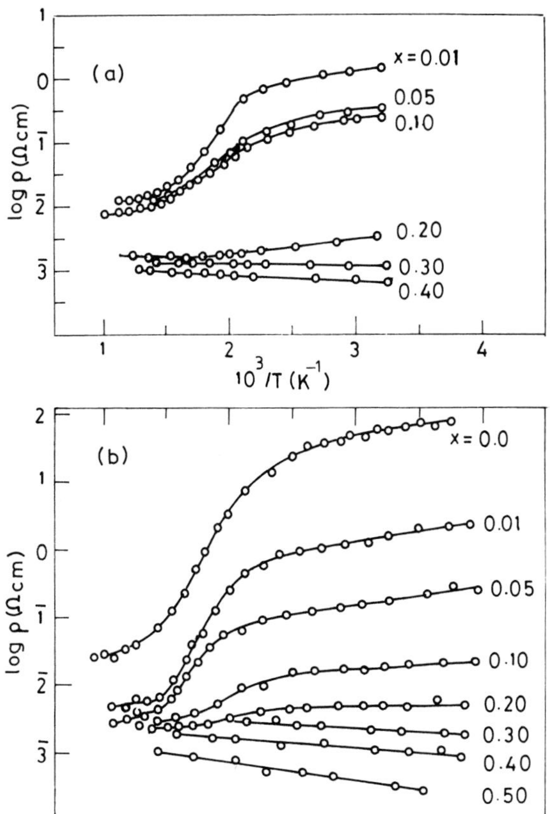

Figure 10. Plots of log ρ vs. $1/T$ for (*a*) $Pr_{1-x}Sr_xCoO_3$ and (*b*) $Nd_{1-x}Sr_xCoO_3$ (from Rao & Om Parkash 1977).

could induce Anderson localization of carriers. In the hopping regime, the activation energy varies as,

$$E_a \approx (x_0 - x)^{1\cdot 6 \pm 0\cdot 3}$$

The value of the exponent is consistent with that predicted (1·8) from the models of Anderson localization (Mott *et al.* 1975), but the uncertainty is too large to rule out percolation and other models (Cohen 1971). The d.c. conductivity in these systems show a linear log ρ vs. $T^{-1/4}$ behaviour at low temperatures up to a limiting temperature T_A; T_A decreases with increasing x. The frequency (ω) dependence of the conductivity was marked for temperatures well below T_A. In $Nd_{0\cdot 99}Sr_{0\cdot 01}CoO_3$, the value of the exponent s in the relation $\sigma \propto \omega^s$ is found to be 0·8 at 105 K and 0·6 at 85 K compared to the predicted value of 0·8 (Mott *et al.* 1975).

Thermopower (S) measurements (Rao & Om Parkash 1977) in these materials show a significant change in the sign of the slope of S vs. T plots near T_A (Figure 11). Since in the hopping regime, S is expected to increase with T while for hopping at the mobility edge, E_c, S is expected to decrease with T (Mott *et al.* 1975), the change in slope could indicate a change in conduction mechanism, the high-temperature behaviour reflecting transport at E_c. Interestingly, the temperature at which S shows a maximum corresponds closely to the temperature at which the relative Co^{3+}/Co^{III} ratio obtained from Mössbauer studies (Figure 6) shows a maximum.

An idea of the time-scale of hopping of electrons from site to site has been brought from Mössbauer studies. It has been observed in Mössbauer spectra of $Ln_{1-x}Sr_xCoO_3$ and $Ln_{1-x}Sr_xFeO_3$ systems (Bhide *et al.* 1975, Rao *et al.* 1975, Rao & Om Parkash 1977, Joshi *et al.* 1979), that distinct resonances due to different oxidation states (or electron configurations) are obtained in the regime where the electrical conductivity shows behaviour typical of Anderson localization. In the Anderson localized regime, the time-scale of hopping is therefore expected to be $\sim 10^{-7}$ s. The low value of resistivity of the $Ln_{1-x}Sr_xCoO_3$ samples (0·01–0·1 Ω cm) in the Anderson localized region, however, suggests hopping times of the order of 10^{-11} to 10^{-12} s. An explanation for the discrepancy has been proposed by Rao *et al.* (1975). An electron in a

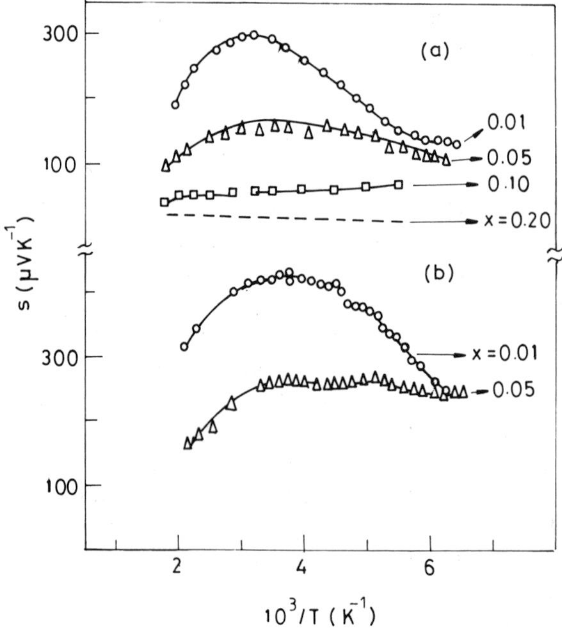

Figure 11. Plots of S vs. $1/T$ for (*a*) $Nd_{1-x}Sr_xCoO_3$ and (*b*) $Pr_{1-x}Sr_xCoO_3$ (from Rao & Om Parkash 1977).

weakly localized state with the wavefunction $\exp(-\alpha r)$ will only affect the state of cobalt ions by ($\sim a\alpha^3$) in a random sense, and to allow it to change completely $(1/a\alpha)^6$ jumps will be needed which could well account for the discrepancy. Such a tentative explanation seems to fail when the same model is applied to $Ln_{1-x}Sr_xFeO_3$ as in this system the resistivity is $\sim 10^2\ \Omega$ cm (Joshi *et al.* 1979), in agreement with the time-scales obtained from Mössbauer studies. The influence of clusters of collective electron states in the $Ln_{1-x}Sr_xCoO_3$ compounds which may coexist with localized states on the electrical conductivity may therefore have to be considered.

4. The $La_{1-x}Sr_xVO_3$ system

The $La_{1-x}Sr_xVO_3$ system is similar to the $La_{1-x}Sr_xCoO_3$ system, in as much as it undergoes a smooth insulator–metal transition (Dougier & Casalot 1970, Dougier & Hagenmuller 1975, Sayer *et al.* 1975) as a function of x (around $x = 0{\cdot}22$–$0{\cdot}25$). This behaviour was first interpreted in terms of Anderson localization by Mott (Mott 1969, 1973). $LaVO_3$ ($x = 0$) is made up of V^{3+} ($3d^2$) ions and each Sr^{2+} ion introduces a V^{4+} hole. For low concentrations, the hole would be bound to the Sr^{2+} but would necessarily be itinerant if it escaped from it. For higher concentrations, holes can form a degenerate gas and one has to ask whether states at the Fermi energy are localized in the Anderson sense. Mott (1973) suggested that in the intermediate range of concentrations disorder arising from a random distribution of Sr^{2+} and La^{3+} induces Anderson localization of carriers in the crystal and predictions of the Anderson model would therefore be valid.

It has been shown (Sayer *et al.* 1975) that the low-temperature conductivity in the nonmetallic regime conforms to a variable-range hopping behaviour with exhibiting linear $\log \rho$ vs. $T^{-1/4}$ plots. The fit is good for $x = 0{\cdot}05$ for $T < 120$ K and less appropriate for higher x. For highly doped samples, the frequency dependence (ω^s) has a value $s \sim 0{\cdot}5$ (instead of the expected $s = 0{\cdot}8$) with a pronounced temperature dependence of s just as in $La_{1-x}Sr_xCoO_3$. Sayers *et al.* (1975) suggested that the a.c. conductivity is determined by locally ordered chains of sites which are not of microscopic dimensions and that the d.c. conductivity could be due to uncorrelated regions surrounding such clusters.

A considerable uncertainty ($\pm 0{\cdot}05$) exists in the value of x_0 corresponding to the metal–nonmetal transition. The value of the exponent for the variation of the activation energy E_a in the expression $E_a = (x_0 - x)^n$ is found to be $1{\cdot}8 \pm 0{\cdot}5$ with $x_0 = 0{\cdot}25$ (Sayer *et al.* 1975). However, just as in $La_{1-x}Sr_xCoO_3$, the uncertainty in the value of n is too large to discount more general percolation models (Cohen 1971) predicting n to be around 1·6. The experimental resistivity of the $x = x_0$ sample is around $2 \times 10^{-3}\ \Omega$ cm just as in $Ln_{1-x}Sr_xCoO_3$. Sayers *et al.* (1975) have come to the conclusion, based on the consideration of the probable Bohr radius of the impurity state ($\sim 2{\cdot}2$ Å) and the large temperature dependence of the a.c. conductivity, that there is a shift in the Fermi level towards the valence band with a tail of localized states extending from the valence band into the band gap.

The possible density-of-states distribution in $La_{1-x}Sr_xVO_3$ has been examined (Edgell *et al.* 1984) by He(I) photoelectron spectroscopy. The spectra terminate in a sharp edge at the Fermi energy with an edge-width governed by the Fermi–Dirac distribution. The height of the Fermi edge discontinuity is $0{\cdot}20 \pm 0{\cdot}05$ relative to the total d-band height. This value is remarkably close to estimates from the resistivity data of Sayer *et al.* (1975) and from $1/f$ noise measurements (Prasad *et al.* 1979). However, the schematic density of states profile for $La_{1-x}Sr_xVO_3$ obtained by Egdell *et al.* (1984) is not consistent with that proposed by Sayer *et al.* (1975). According to Egdell *et al.* (1984) the metal–insulator transition does not occur in an impurity band separated from the 3d band.

5. $LaNi_{1-x}M_xO_3$ (M = Cr, Mn, Fe or Co)

It is well known that $LaNiO_3$ is metallic. The log ρ vs. T plots of some of the members of the $LaNi_{1-x}M_xO_3$ family are shown in Figure 12. In the $LaNi_{1-x}M_xO_3$ system (M = Cr, Mn, Fe or Co), there is a critical value, x_c, at which the temperature coefficient of resistivity changes sign so that formally x_c defines the concentration at which there is a metal–insulator transition (Ganguly *et al.* 1984). The value of x_c depends on the M ion and is in the range 0·03–0·05, 0·05–0·10, 0·25–0·35 and 0·35–0·5 for M = Mn, Cr, Fe and Co, respectively. The extrapolated value of the resistivity at x_c is in the range $\sim 2 \times 10^{-3}\ \Omega$ cm just as in $La_{1-x}Sr_xVO_3$ and $La_{1-x}Sr_xCoO_3$.

The variation in ρ and the magnetic ordering temperature with composition are shown in Figures 13 and 14 respectively (Ganguly *et al.* 1984, Vasanthacharya *et al.* 1984b).There seems to be a parallelism between the trends in the magnetic ordering temperature and the resistivity when M = Fe or Cr but not when M = Co. In the case of M = Mn or Cr, the metallic compositions do not show evidence for long-range order down to 12 K. Except when M = Fe, the other systems show a divergence in the susceptibility at low temperatures for intermediate values of x indicating ferro- or ferri-magnetic interactions. However, measurements of the magnetization of the M = Co, Mn and Cr members reveal that the magnetization does not saturate at 4·2 K even at 25 kOe and that the value of magnetization is non-integral (0·3 to 1·0 μ_B per mole). The M = Mn samples show evidence for spin-glass like behaviour when $x = 0{\cdot}10$ (Vasanthacharya *et al.* 1984a).

In order to account for the magnetic properties it would be necessary to have an estimate of the relative position of the $3d^n$ levels of the M^{3+} cation with respect to the Ni^{3+} $3d^n$ level, but such a quantitative estimate is not available to us. Mizushima *et al.* (1979) have employed photoconductivity and e.s.r. measurements to locate the energy levels of $3d^n$ ions in TiO_2 (Figure 15). One may anticipate a qualitatively similar picture for the positions of the energy levels of

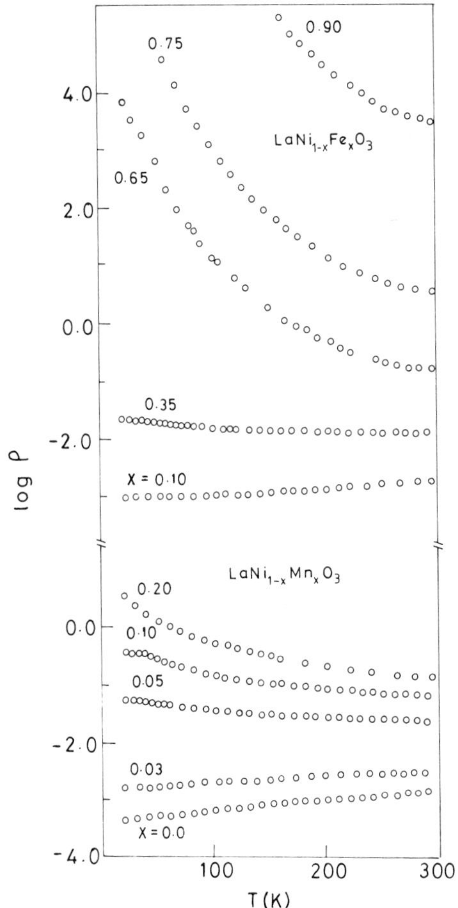

Figure 12. Log ρ vs. T for $LaNi_{1-x}Mn_xO_3$ and $LaNi_{1-x}Fe_xO_3$ (from Vasanthacharya *et al.* 1984a and Ganguly *et al.* 1984).

the $3d^n$ ions in perovskites. From these data we see that the Mn^{3+}, Cr^{3+} and Co^{3+} levels are quite well removed energetically from the Ni^{3+} level while the Fe^{3+} level is close. Thus, when M=Cr, Mn or Co, a seemingly reasonable explanation would be to assume that there is no significant overlap of the $3d^n$ levels of the M ion and the Ni^{3+} ions. X-ray photoelectron and ultraviolet photoelectron studies (Madhusudan *et al.* 1979) show that the 3d levels may be considered to be rigid in the solid solutions of perovskites. Thus, as x increases one could assume that the σ^* band of $LaNiO_3$ becomes narrow enough to sustain spontaneous ferromagnetism. The close proximity of the Fe^{3+} level could result in antiferromagnetic interactions. When M^{3+} = Mn, Cr or Co, the number of e_g electrons is less than one so that even if there is overlapping of levels, the σ^* band

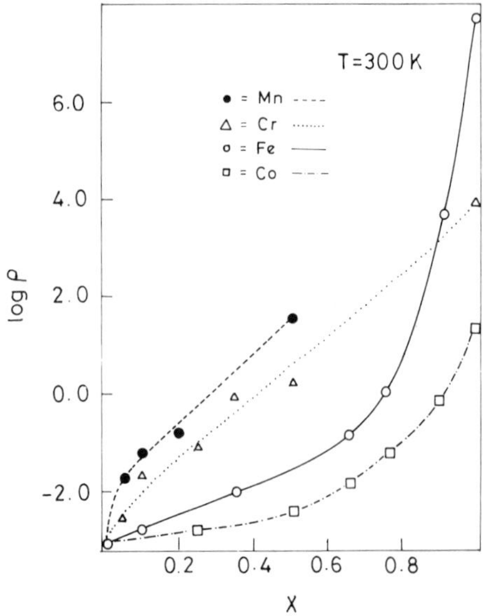

Figure 13. Log ρ vs. x for $LaNi_{1-x}M_xO_3$ samples (M = Cr, Mn, Fe, Co) (from Ganguly *et al.* 1984).

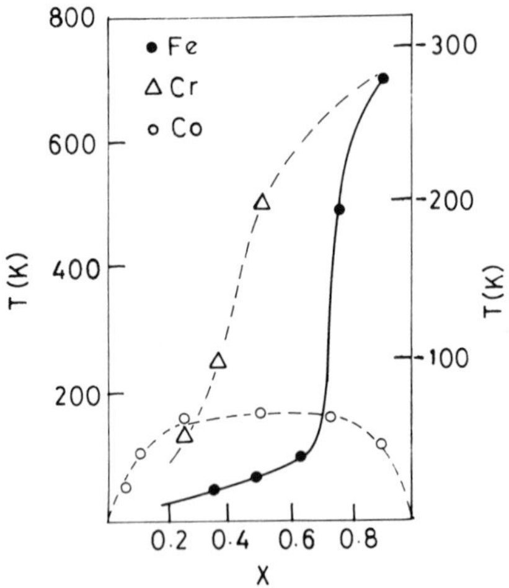

Figure 14. Variation of magnetic ordering temperature as a function of x for $LaNi_{1-x}M_xO_3$ samples (M = Fe, Cr, Co) (from Vasanthacharya *et al.* 1984b).

will be less than quarter filled and ferromagnetism would be anticipated; in the case of Fe^{3+} with two e_g electrons, the σ^* band would be more than quarter filled so that one may anticipate antiferromagnetic correlations. However, in such a picture, the magnetization values should reflect the full magnetic moments of the M ions and this is not observed.

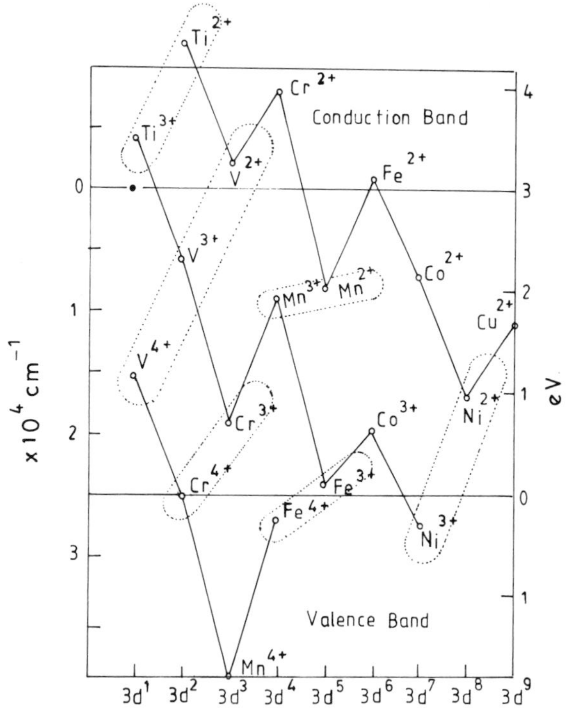

Figure 15. Location of $3d^n$ energies relative to the edges of the valence and conduction band in TiO_2 (taken from Mizushima *et al.* (1979) and used with permission).

6. Layered perovskite oxides

The layered perovskite La_2NiO_4 with the K_2NiF_4 structure has been reported to show a continuous insulator–metal transition around 550 K (Ganguly & Rao 1973) in as much as $d\rho/dT$ changes sign smoothly at this temperature (Figure 16). Theoretical interest in this system had initially been centred on the conjecture that the d_{z^2} electron of the Ni^{2+} ion may be localized while the $d_{x^2-y^2}$ electron may be itinerant (Goodenough 1974, Singh *et al.* 1984). Recently it has been recognized (Rao *et al.* 1984) tha the properties of La_2NiO_4 are particularly sensitive to the oxygen stoichiometry. The early studies were on air-fired ceramic samples which always contained small amounts (5–10%) of Ni^{3+}. By a controlled-atmosphere synthesis, single crystals as well as powders of stoichiometric La_2NiO_4 have been prepared. Electrical resistivity of the stoichiometric samples is an order of magnitude higher than that of the air-fired ceramic samples. There is also considerable anisotropy in the resistivity, $\rho_\parallel$ and $\rho_\perp$, parallel and perpendicular to the *c*-axis; $\rho_\perp$ is 10–100 times less than that of $\rho_\parallel$ (Figure 16) and drops sharply around 500 K indicating an insulator–metal

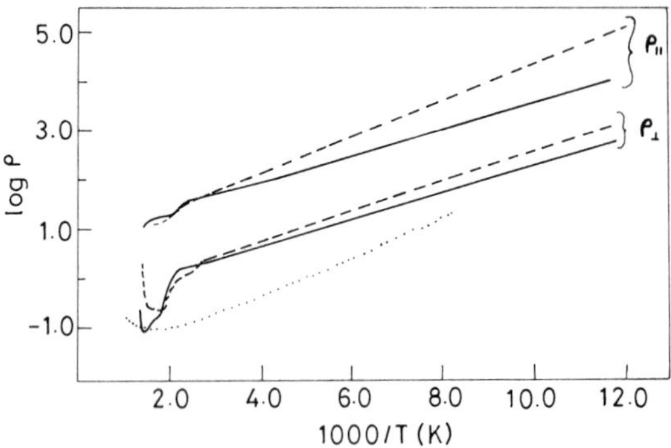

Figure 16. Log ρ vs. $1/T$ plots of La_2NiO_4; (- - - -) single crystals as grown by skull melting in air, (——) single crystals grown by skull melting and annealed at 1470 K for 125 hours in CO_2. $\rho_\perp$ is for current along the basal plane and $\rho_\parallel$ is along the *c*-axis (from Rao *et al.* 1984); (······) polycrystalline La_2NiO_4 fired in air (Ganguly & Rao 1973).

transition (Figure 16). The value of the resistivity after the transition is $\sim 2 \times 10^{-1}$ Ω cm which is two orders of magnitude higher than that at which the three-dimensional perovskites mentioned earlier exhibit change of sign of $d\rho/dT$ during the composition-controlled insulator–metal transitions.

It is interesting that none of the oxides with the layered K_2NiF_4 structure show a truly metallic behaviour at all temperatures. La_2CuO_4, which had originally been considered to be a metal, has a room-temperature resistivity of 10^{-1} Ω cm (Ganguly & Rao 1973). At low temperatures, La_2CuO_4 shows a semiconducting behaviour (Singh *et al.* 1984). Even the metallic three-dimensional perovskites such as $LaNiO_3$, $La_{0.5}Sr_{0.5}CoO_3$ and $La_{0.5}Sr_{0.5}VO_3$ when incorporated into the K_2NiF_4 structure as $SrO.LaNiO_3$, $SrO.La_{0.5}Sr_{0.5}CoO_3$, $SrO.La_{0.5}Sr_{0.5}VO_3$ and so on, show resistivities which are 10^2–10^4 times higher than the parent perovskite phases although magnetically the behaviour is similar (Singh 1982, Demazeau *et al.* 1976, Shinike *et al.* 1977). $BaPbO_3$ is a metal with $\rho \sim 10^{-4}$ Ω cm while Ba_2PbO_4 which has the K_2NiF_4 structure is a very good insulator ($\rho > 10^6$ Ω cm) (Ikushima & Hayakawa 1966, Wagner & Binder 1958). The observed difference in the resistivity behaviour between the three-dimensional and two-dimensional perovskite structures is interesting and worthy of further exploration.

7. General considerations

A common feature emerging from the above studies is that in all the oxide materials investigated by us in which there is strong coulomb interaction between the charge carriers, $d\rho/dT$ changes sign around a common value of conductivity ($\sim 5 \times 10^2$ (Ω cm)$^{-1}$). This value is independent of the details of the M–NM transition. In the theories developed by Mott (1967, 1981, 1984; Mott & Kaveh 1983), there could exist a minimum value of the conductivity ρ_{min} around which $d\rho/dT$ changes sign. This value of the conductivity is given by,

$$\sigma_{min} = Ce^2/\hbar a \simeq \tfrac{1}{3}g^2e^2/\hbar a \tag{1}$$

where $C = 0{\cdot}025$–$0{\cdot}05$ and $g = (N(E_F)/N(E_F)_{free}) \simeq 1/3$. Equation (1) gives the value found by us if we assume that $a = 4$ Å. The derivation of the above equation was based on the premise that the mean-free-path l can never be less than the interatomic spacing so that the condition for σ_{min} is that $k_F l = 1$. Equation (1) can be derived from considerations of the area of the Fermi sphere under these conditions. Prasad *et al.* (1979) measured the electrical excess of $1/f$ noise of $La_{1-x}Sr_xVO_3$ in order to obtain an estimate of the quantity g. In the metallic phase, the value of g^2 obtained by them was 0·1, in close agreement with theory.

Despite the close agreement of experimental data with the expected value of σ_{min}, recent scaling theories and experimental data (Abrahams *et al.* 1979, Rosenblaum *et al.* 1980, Mott 1984a,b) have questioned the concept of σ_{min} in as much as there is no discontinuous transition from the metallic to the insulating phase near 0 K (an insulator being defined as one which has zero conductivity at 0 K). We are not able to comment on this aspect as our measurements were carried down to only 12 K. Some indirect evidence that the transition may be continuous is obtained from the fact that in the $LaNi_{1-x}M_xO_3$ system, the conductivity data between 12 K and 300 K could be fitted satisfactorily to a single empirical equation of the type (Ganguly *et al.* 1984),

$$\sigma = A\exp(-E_a/k(T+\theta)) \tag{2}$$

where θ is positive and increases with decrease in E_a. The above equation employs a quasi-Boltzman distribution function (Sales & Viswanathan 1976) with a $T_{eff} > T$. A quasi-Boltzman distribution function of the form $n_i = n_0\exp(-E_a/k(T-T_0))$ has been used (Vogel 1921, Fulcher 1925) to describe the thermodynamic glass transition temperature T_0; T_0 is the temperature up to which the configurational entropy is zero so that the effective 0 K is shifted to T_0 and hence $T_{eff} < T$. When $T_{eff} > T$, there is a finite entropy at $T = 0$ K associated with collective electrons. It is tempting to make a connection between θ and the bandwidth of the collective electrons in the same manner as in the case of valence fluctuating systems (Sales & Viswanathan 1976). We note (Ganguly *et al.* 1984) that as $E_a \rightarrow 0$, $\theta \rightarrow 600$–700 K; it is possible that there is a minimum bandwidth W_{min} before the system becomes metallic. The empirical fit of equation (2) suggests that at $T = 0$ there is a finite conductivity and the material may not be a real insulator.

It might be instructive to consider the upper limit of the conductivity when the conduction is by diffusion processes with a zero activation energy. In this limit, the diffusion coefficient becomes $D \approx a^2\nu$ where ν is the attempt frequency.

One may estimate v from the bandwidth or the nearest-neighbour exchange integral as was done by Zener (Zener 1951) whose arguments, although applied initially to ferromagnetic systems characterized by double exchange, is nevertheless of general interest. Zener (1951, also see Zener & Heikes 1953) pointed out that the exchange energy E_{exch} determines the rate at which an electron jumps from one ion to its neighbour so that one obtains a frequency $E_{exch} = \hbar v$ for a fraction x of charge carriers. The following expression for the conductivity is obtained from Einstein's diffusion equation:

$$\sigma = \frac{xE_{exch}}{2\pi kT} \times \frac{e^2}{\hbar a} \tag{3}$$

for x charge carriers. Although the above expression gives a temperature dependence, the term $(xE_{exch}/2kT)$ may be equated to C and if E_{exch} is of the order of kT, the value of the conductivity is similar to that given by equation (1) for σ_{min}. Thus, the maximum conductivity obtained for diffusion conditions is close to that obtained from the band picture considerations and it will not be possible by conductivity measurements alone to distinguish between band-like conduction and diffusion.

The question that has to be answered is why the temperature coefficient of resistivity, or $d\rho/dT$, should change sign when the conductivity reaches the value given by σ_{min}. There is a formal similarity between this and the problem associated with the Mooij criterion (Mooij 1973) for amorphous metals. For amorphous metals, the Mooij criterion states that there is a critical value of the resistivity at which the temperature coefficient of resistivity changes. This value of resistivity in amorphous metals is of the order of 200 $\mu\Omega$ cm which is roughly one tenth the value found in our system ($\sim$2000 $\mu\Omega$ cm). It seems therefore that a criterion similar to the Mooij criterion may be used for composition-controlled M–NM transitions in the oxide systems. That is, a value of 2000 $\mu\Omega$ cm could be the critical value for these oxide systems. There are systems described in the literature which show a positive temperature coefficient of resistivity at high temperatures and are therefore considered to be metallic. In the perovskite related oxide system these include oxides such as $LaTiO_3$, La_2NiO_4 and La_2CuO_4 (Maclean & Greedan 1981, Ganguly & Rao 1973, Singh *et al.* 1984). All these systems have a $\rho > 2000$ $\mu\Omega$ cm and it is not surprising that at low temperatures they should show a negative temperature coefficient of resistivity and behave as semiconductors. Even $LaCoO_3$ which has a resistivity greater than 2000 $\mu\Omega$ cm at high temperatures ($\sim$10 000 $\mu\Omega$ cm) shows a nearly continuous semiconductor–metal transition.

We have observed that by defining a time-scale of the residence time of an electron at the nucleus, we obtain numbers which provide a criterion for the existence of a critical value for the resistivity. According to this criterion, in the metallic state the residence time of an electron should be much less than the vibration period of the atom. We therefore define a minimum residence time of 10^{-14} s since lattice vibration periods are of the order of 10^{-13} s in the oxides. This time can therefore be taken as the minimum relaxation time τ_{min}. The expression for the conductivity in the Boltzmann formula at this value of τ_{min} may be related to σ_{min} as

$$\sigma_{min} = ne^2 \frac{\tau_{min}}{m^*} \tag{4}$$

For the value of σ_{min} obtained in our studies, we obtain $m^* = 10m_0$ or a bandwidth of $\sim 0{\cdot}1$ eV. There is evidence in the literature that near the M–NM transition in oxides of 3d transition elements, the effective mass is close to $10m_0$. Magnetic susceptibility studies on $LaNiO_3$ and $La_{1-x}Sr_xVO_3$ (Vasanthacharya *et al.* 1984a, Dougier & Hagenmuller 1975) give a value of the temperature-independent susceptibility close to 600–700 emu mol^{-1} which when interpreted in terms of the enhanced susceptibility in the Brinkman–Rice model (Goodenough *et al.* 1973, Mott 1972) yield an effective mass of $10m_0$. Egdell *et al.* (1984) based on fits of the valence band ultraviolet photoelectron spectra suggest an $m^* = 10m_0$.

The critical concentration, n_c of charge carriers at which the metal–insulator transition occurs is given by (Edwards & Sienko 1978),

$$n_c^{1/3} a_H \sim 0{\cdot}26 \tag{5}$$

The volume occupied by the atoms with Bohr radii, a_H, is roughly 10% of the total volume. If we assume that the Bohr radius of the transition metal ion is less than the interatomic spacing, equation (5) implies that only 10% of the total charge carriers need to be involved for the system to appear metallic. In Mott's interpretation of the Brinkmann–Rice model (Mott 1972), the effective mass is the inverse of the fraction of charge carriers so that a $m^* = 10m_0$ corresponds to 10% of the atoms being charge carriers. This is what is observed. The implication would be that the change in sign of $d\rho/dT$ is a consequence of the residence time of the electron *vis-à-vis* the lattice vibration period as the bandwidth is increased. In the oxides when the effective mass is $10m_0$, we would therefore expect a polaron model to account for the metal–insulator transitions rather satisfactorily.

If we associate τ_{min} with a minimum bandwidth W_{min} by the uncertainty relation (Vasanthacharya *et al.* 1984a, Ganguly *et al.* 1984), $\tau_{min} \sim (\hbar/W_{min})$, we obtain for σ_{min},

$$\sigma_{min} = ne^2 \times \frac{\hbar}{m^* W_{min}} \tag{6}$$

With $W_{min} = \hbar^2 k_F^2/2m^* = \hbar^2(\pi/a)^2/2m^*$, we obtain,

$$\sigma_{min} = \frac{2ne^2}{\hbar} \times \frac{a^2}{\pi^2} \tag{7}$$

With $n=(1/a^3)$ we obtain,

$$\sigma_{min}=\frac{2}{\pi^2}\frac{e^2}{\hbar a} \tag{8}$$

However, if only a fraction f of the n charge carriers are conducting, then $n=f/a^3=1/m^*a^3$ where $m^*=1/f$ (Mott 1972) and σ_{min} is then given by

$$\sigma_{min}=\frac{2}{m^*\pi^2}\frac{e^2}{\hbar a} \tag{9}$$

With $m^*=10m_0$ we obtain a value of σ_{min} close to that obtained by Mott (equation (1)). The above equation, based on the consideration of $\tau_{min}\sim 10^{-14}$ s, suggests a large polaron and the implication is that a critical concentration of large polarons makes the system metallic.

In Figure 17, we show the σ vs. donor concentration behaviour for a wide range of materials exhibiting M–NM transitions (Ganguly *et al.* 1984, Edwards 1984). The materials range from the prototype low-electron-density materials (such as Ge:Sb and Si:P) through to metal–ammonia solutions and supercritical (expanded) mercury at high temperatures. Despite the apparently diverse nature of the M–NM transitions considered, in terms of the global or macroscopic

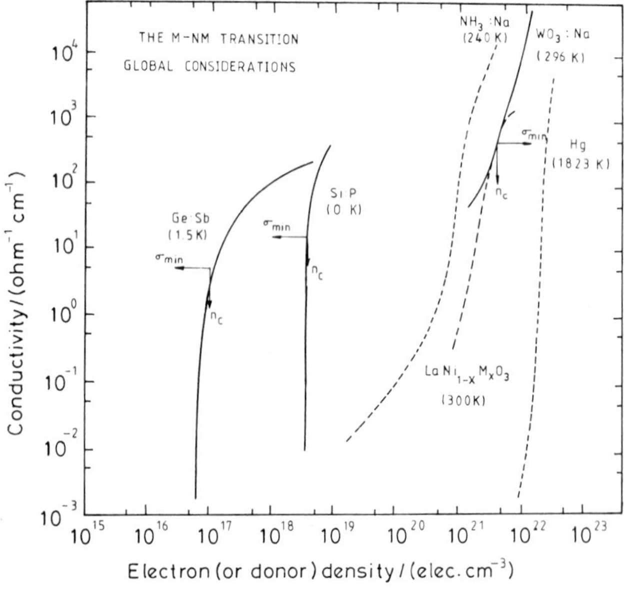

Figure 17. Log σ vs. log x plots of several systems as marked (modified version of figure in Ganguly *et al.* 1984).

considerations, the d-electron metal oxide systems correlate with other high-electron-density materials such as sodium tungsten bronzes and metal–ammonia solutions.

Acknowledgement

The authors thank the University Grants Commission for support of this research.

References

Abrahams, E., Anderson, P.W., Licciardello, D.C. and Ramakrishnan, T.V., 1979, *Phys. Rev. Lett.*, **42**, 673.

Anderson, P.W. and Hasegawa, H., 1955, *Phys. Rev.*, **100**, 675.

Bahadur, D., Kollali, S., Rao, C.N.R., Patni, M.J. and Srivatsava, C.M., 1979, *J. Phys. Chem. Solids*, **40**, 981.

Bhide, V.G., Rajoria, D.S., Rama Rao, G. and Rao, C.N.R., 1972, *Phys. Rev.* B, **6**, 1021.

Bhide, V.G., Rajoria, D.S., Rao, C.N.R., Rama Rao, G. and Jadhao, V.G., 1975, *Phys. Rev.* B, **12**, 2832.

Cohen, M.H., 1971, *Phys. Today*, **24**, 26.

Demazeau, G., Pouchard, M. and Hagenmuller, P., 1976, *J. Solid St. Chem.*, **18**, 159.

Dougier, P. and Casalot, A., 1970, *J. Solid St. Chem.*, **2**, 396.

Dougier, P. and Hagenmuller, P., 1975, *J. Solid St. Chem.*, **15**, 158.

Edwards, P.P. and Sienko, M.J., 1978, *Phys. Rev.* B, **17**, 2573.

Edwards, P.P., 1984, in *Physics and Chemistry of Electrons and Ions in Condensed Matter*, edited by J.V. Acrivos, N.F. Mott and A.D. Yoffe (Dordrecht: Reidel).

Egdell, R.G., Harrison, M.R., Hill, M.D., Porte, L. and Wall, G., 1984, *J. Phys. C: Solid St. Phys.*, **17**, 2889.

Fulcher, G.S., 1925, *J. Am. Ceram. Soc.*, **8**, 339.

Gai, P. and Rao, C.N.R., 1975, *Mater. Res. Bull.*, **10**, 787.

Ganguly, P. and Rao, C.N.R., 1973, *Mater. Res. Bull.*, **8**, 405.

Ganguly, P., Om Parkash and Rao, C.N.R., 1976, *Phys. Stat. Solidi*, **36**, 669.

Ganguly, P., Vasanthacharya, N.Y., Rao, C.N.R. and Edwards, P.P., 1984, *J. Solid St. Chem.*, **54**, 400.

Goodenough, J.B., 1971, *Mater. Res. Bull.*, **6**, 967.

Goodenough, J.B., 1972, *Prog. Solid St. Chem.*, **5**, 145.

Goodenough, J.B., 1974, in *Solid State Chemistry*, edited by C.N.R. Rao (New York: Marcel Dekker).

Goodenough, J.B., Mott, N.F., Demazeau, G., Pouchard, M. and Hagenmuller, P., 1973, *Mater. Res. Bull.*, **8**, 647.

Heikes, R.R., Miller, R.C. and Mazelsky, R., 1964, *Physica*, **30**, 1600.

Ikushima, H. and Hayakawa, S., 1966, *Solid-St. Electron.*, **9**, 921.

Jonker, G.H., 1966, *J. Appl. Phys.*, **36**, 1031.

Jonker, G.H. and van Santen, J.H., 1950, *Physica*, **16**, 337, 599.

Jonker, G.H. and van Santen, J.H., 1953, *Physica*, **19**, 120.

Joshi, V., Om Parkash, Rao, G.N. and Rao, C.N.R., 1979, *J. Chem. Soc. Faraday Trans.* II, **75**, 1199.

Maclean, D.A. and Greedan, J.E., 1981, *Inorg. Chem.*, **20**, 1025.

Madhusudan, W.H., Jagannathan, K., Ganguly, P. and Rao, C.N.R., 1980a, *J. Chem. Soc. Dalton Trans.*, 1397.
Madhusudan, W.H., Kollali, S. Sarode, P.R., Ganguly, P., Hegde, M.S. and Rao, C.N.R., 1979, *Pramana*, **12**, 317.
Madhusudan, W.H., Vasanthacharya, N.Y. and Ganguly, P., 1980b, *Ind. J. Chem.*, **19A**, 1037.
Menyuk, N., Dwight, K. and Raccah, P.M., 1967, *J. Phys. Chem. Solids*, **28**, 549.
Mizushima, K., Tanaka, M., Asai, A., Iida, S. and Goodenough, J.B., 1979, *J. Phys. Chem. Solids*, **40**, 1129.
Mooij, J.H., 1973, *Phys. Stat. Solidi* a, **17**, 521.
Mott, N.F., 1967, *Adv. Phys.*, **16**, 49.
Mott, N.F., 1968, *J. Non. Cryst. Solids*, **1**, 1.
Mott, N.F., 1969, *Phil. Mag.*, **19**, 835.
Mott, N.F., 1972, *Adv. Phys.*, **21**, 785.
Mott, N.F., 1981, *Phil. Mag.* B, **44**, 265.
Mott, N.F., 1984a, *Int. Rev. Phys. Chem.*, **4**, 1.
Mott, N.F., 1984b, *Rep. Prog. Phys.*, **47**, 909.
Mott, N.F. and Davis, E.A., 1971, *Electronic Processes in Non-Crystalline Materials* (2nd edn) p. 144 (Oxford: Clarendon Press).
Mott, N.F. and Kaveh, M., 1983, *Phil. Mag.* B, **47**, 577.
Mott, N.F., Pepper, M., Pollitt, S., Wallis, R.H. and Adkins, C.J., 1975, *Proc. R. Soc.* A, **345**, 169.
Naiman, C.S., Gilmore, R., Di Bartolo, B., Linz, A. and Santoro, R., 1965, *J. Appl. Phys.*, **36**, 1044.
Prasad, E., Sayer, M. and Noad, J.P., 1979, *Phys. Rev.* B, **19**, 5144.
Raccah, P.M. and Goodenough, J.B., 1965, *J. Appl. Phys.*, **36**, 1031.
Raccah, P.M. and Goodenough, J.B., 1967, *Phys. Rev.*, **155**, 932.
Raccah, P.M. and Goodenough, J.B., 1969, *J. Appl. Phys.*, **39**, 1209.
Ramasesha, S., Ramakrishnan, T.V. and Rao, C.N.R., 1979, *J. Phys. C: Solid St. Phys.*, **12**, 1307.
Rajoria, D.S., Bhide, V.G., Rao, G.R. and Rao, C.N.R., 1974, *J. Chem. Soc. Faraday Trans. II*, **70**, 512.
Rao, C.N.R. and Bhide, V.G., 1973, *Proc. 19th Conf. on Magnetism and Magnetic Materials, Boston* (Am. Inst. Phys.), p. 504.
Rao, C.N.R. and Om Parkash, 1977, *Phil. Mag.*, **35**, 1111.
Rao, C.N.R., Bhide, V.G. and Mott, N.F., 1975, *Phil. Mag.*, **32**, 1277.
Rao, C.N.R., Buttrey, D., Otsuka, N., Ganguly, P., Harrison, H.R., Sandberg, C.J. and Honig, J.M., 1984, *J. Solid St. Chem.*, **51**, 266.
Rao, C.N.R., Om Parkash, Bahadur, D., Ganguly, P. and Nagabushana, S., 1977, *J. Solid St. Chem.*, **22**, 353.
Rosenbaum, T.F., Andres, K., Thomas, G.A. and Bhatt, R.N., 1980, *Phys. Rev. Lett.*, **43**, 1723.
Sales, B.C. and Viswanathan, R., 1976, *J. Low. Temp. Phys.*, **23**, 449.
Sayer, M., Chen, R., Fletcher, R. and Mansingh, A., 1975, *J. Phys. C: Solid St. Phys.*, **8**, 2059.
Shinike, T., Sakai, T., Adachi, G. and Shiokawa, J., 1977, *Mater. Res. Bull.*, **12**, 831.
Singh, K.K., 1982, *PhD Thesis*, Indian Institute of Science, Bangalore.
Singh, K.K., Ganguly, P. and Goodenough, J.B., 1984, *J. Solid St. Chem.*, **52**, 254.
Smolenskii, G.A., Yudin, V.N. and Sher, E., 1963, *Sov. Phys.—Solid St.*, **4**, 2452.
Taguchi, H., Shimada, M. and Koizumi, M., 1980, *J. Solid St. Chem.*, **33**, 169.
Thornton, G., Tofield, B.C. and Williams, D.E., 1982, *Solid St. Commun.*, **44**, 1213.
Vasanthacharya, N.Y. and Ganguly, P., 1983, *Bull. Mater. Sci.*, **5**, 307.

Vasanthacharya, N.Y., Ganguly, P., Goodenough, J.B. and Rao, C.N.R., 1984a, *J. Phys. C: Solid St. Phys.*, **17**, 2745.
Vasanthacharya, N.Y., Ganguly, P. and Rao, C.N.R., 1984b, *J. Solid St. Chem.*, **53**, 140.
Vogel, H., 1921, *Z. Phys.*, **22**, 645.
Wagner, G. and Binder, H., 1958, *Z. Anorg. Allgem. Chem.*, **297**, 328.
Zener, C., 1951, *Phys. Rev.*, **82**, 403.
Zener, C. and Heikes, R., 1953, *Rev. Mod. Phys.*, **25**, 191.

CHAPTER 13

cerium under high pressure

A.K. Singh and T.G. Ramesh
Materials Science Division, National Aeronautical Laboratory, Bangalore 560017, India

1. Introduction

Among the elemental solids cerium is unique in many respects. Under moderately high pressure cerium undergoes a first-order transition involving no change in crystal structure. The phase boundary in the temperature–pressure plane terminates at a critical point. These are the first experimental observations of an isostructural transition and the phase boundary in a solid–solid transition terminating at a critical point. The melting of cerium under pressure is anomalous. Under different conditions of temperature and pressure, cerium undergoes a number of structural transitions. At least six allotropes of high purity cerium have been identified; these exhibit anomalous magnetic and electronic properties, Kondo-like behaviour and superconductivity.

The isostructural transition in cerium has attracted the attention of theoretical physicists as well. The simple electron-promotion model, the electron–hole interaction model, the virtual bound-state model, Mott delocalization of the 4f electron and the Kondo model are some of the theories put forward to explain the isostructural transition. It is generally recognized that the fascinating properties of cerium under pressure basically stem from the changes in the character of the 4f electrons. The precise nature of the transition however is still a matter of dispute.

The experimental and theoretical aspects of the cerium problem have been discussed in a number of review articles (Koskenmaki & Gschneidner 1978, Robinson 1979, Liu 1981). Attempts have been made in this paper to discuss the important experimental results with an emphasis on the isostructural transition, and to review the theories which have been put forward from time to time to explain the experimental results.

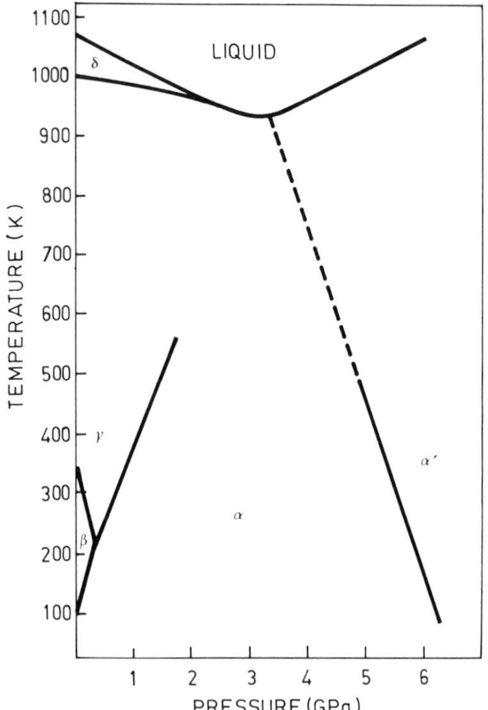

Figure 1. The phase diagram of cerium.

2. The p–T phase diagram

The phase diagram of cerium is shown in Figure 1. High-purity cerium under normal temperature and pressure conditions is face-centred cubic (γ-phase). Under pressure the γ-phase transforms to another face-centred cubic phase (α-phase) which has a much smaller unit cell volume than that of the γ-phase. The other crystalline modifications are the δ-phase (body-centred cubic) occurring at high temperature, the β-phase (double hexagonal close packed) observed below room temperature and the α'-phase above 5 GPa.

2.1. *Melting under pressure*

The high-temperature region of the phase diagram up to 7 GPa was studied by Jayaraman (1965) using a double-stage piston-cylinder apparatus to generate pressure, and a thermal-arrest technique to detect the phase changes. The melting point of cerium decreases with increase in pressure with an initial slope of $-47\ \mathrm{K\ GPa^{-1}}$, and reaches a local minimum at 3·3 GPa and 935 K. Above 3·3 GPa, the melting point increases with increase in pressure.

The slope of the melting line can be discussed using the Clausius–Clapeyron equation,

$$\frac{dT}{dp} = \frac{\Delta V}{\Delta S}$$

where ΔV and ΔS are respectively the changes in volume and entropy associated with melting. The slope of the melting line for a normal solid is positive, since ΔV is positive (ΔS is always positive). The empirical laws (Simon's melting law; Kraut & Kennedy 1966) and equation derived from the Lindemann melting equation (Vaidya & Raja Gopal 1966) suggest that the slope of the melting line should be positive. In this respect the melting line of cerium is abnormal up to 3·3 GPa. The negative dT/dp implies that the liquid phase is denser than the solid phase. The reason for such behaviour is not known. Vaidya (1979) showed that the slope of the melting line can be negative if melting of the solid is accompanied by an electronic transition. It is not unlikely that in cerium an electronic transition occurs during melting and leads to a negative slope.

2.2. $\gamma \rightarrow \alpha$ transition

The first indications of the $\gamma \rightarrow \alpha$ transition at room temperature came from the early work of Bridgman. The measurements as a function of pressure of the specimen volume (Bridgman 1927) and the electrical resistance (Bridgman 1951) suggested a discontinuous transition at about 0·7 GPa. The X-ray diffraction work of Lawson & Tang (1949) clearly established that the $\gamma \rightarrow \alpha$ transition does not involve a change in crystal structure; both the phases are face-centred cubic, the unit cell volume of the α-phase at 1·5 GPa being 16·5% smaller than that of the γ-phase at one atmosphere.

These results formed the basis of many more detailed studies. Poniatovskii (1958) examined the nature of the $\gamma \rightarrow \alpha$ transition up to nearly 2 GPa and 600 K by measuring the latent heat of transition by differential thermal analysis. The study revealed many interesting features of the transition. Most important of these is the observation that the transition pressure increases with temperature at a rate of 0·0043 GPa K^{-1}. Further, the latent heat of transition gradually decreases as the temperature is increased, and above 553 K the transition is not observed. The transition gradually changes its nature from discontinuous to continuous as the temperature is increased, and the phase boundary terminates at a critical point with coordinates 1·8 GPa and 553 K. This observation has been supported by many independent experiments. The measurement of the change in volume associated with the transition (Beecroft & Swenson 1960) at different temperatures (Figure 2) indicates that the magnitude of the volume change decreases with increase in temperature. An extrapolation of the volume change data (up to 575 K) indicates that the volume change associated with the transition

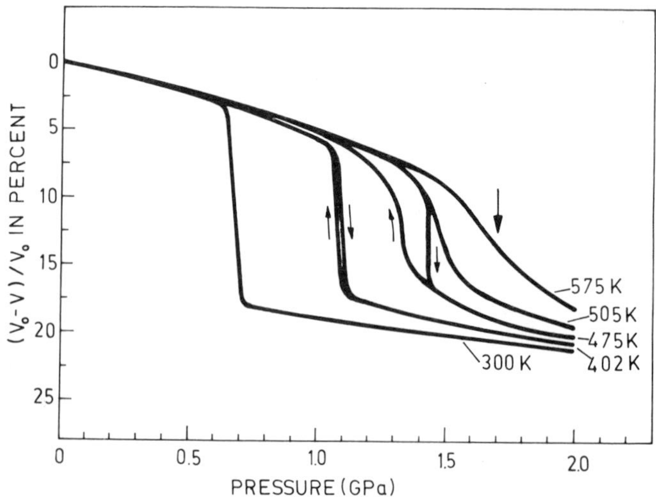

Figure 2. The pressure–volume isotherms of cerium. The volume change associated with the $\gamma\rightarrow\alpha$ transition is 14·1% at 300 K, and decreases with increase in temperature. Above 500 K, the transition resembles a continuous transition. The arrow on the isotherm at 575 K indicates the transition pressure. The difficulty in locating the transition pressure at higher temperatures is obvious (adapted from Beecroft & Swenson 1960).

vanishes at 630 K, the corresponding transition pressure being 2 GPa. Davis & Adams (1964) measured the separation of (111) diffraction lines from the γ- and the α-phase by recording the X-ray diffraction lines with a specially designed apparatus. These measurements suggest that the $\gamma\rightarrow\alpha$ transition pressure increases at a rate of 0·0039 GPa K^{-1}, and the critical point is in the vicinity of 2·0–2·2 GPa and 620–670 K. The pressure dependence of the electrical resistance at various temperatures was measured by Jayaraman (1965). The isotherms are shown in Figure 3. The magnitude of the resistance drop associated with the $\gamma\rightarrow\alpha$ transition decreases with increase in temperature. These data indicate that the slope of the $\gamma\rightarrow\alpha$ phase boundary is 265 K GPa^{-1}, and the critical point is located at 1·75 GPa and 550 K.

The results of the measurements by various investigators are summarized in Table 1. The values for the slope of the $\gamma\rightarrow\alpha$ phase boundary differ considerably. The intrinsic errors in the experiments, sample purity and contamination of the sample during handling and high-temperature experiments are some of the factors responsible for large scatter in the results. The discontinuity in the measured property at the $\gamma\rightarrow\alpha$ transition gradually loses sharpness as the temperature is raised, and becomes ill-defined in the vicinity of the critical point. The determination of the transition pressure in such cases is difficult, and to a great extent subjective. There are many other sources of error. For example, if the experiments are done with increasing pressure alone, the $\gamma\rightarrow\alpha$ transition pressure is obtained, which is nearly 0·1 GPa higher than the equilibrium $\gamma\rightleftharpoons\alpha$

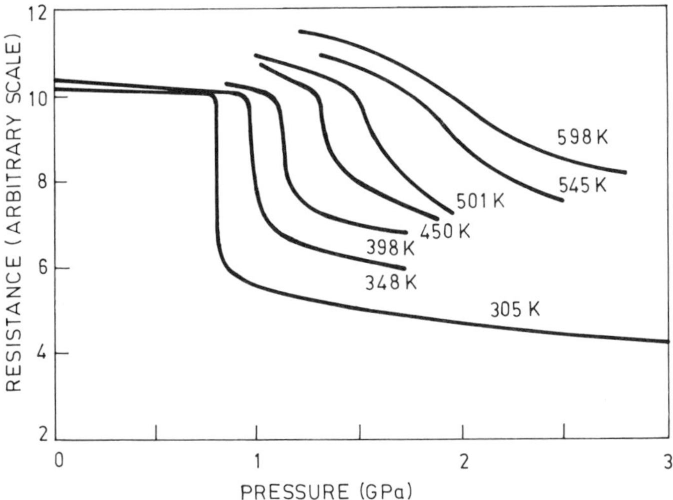

Figure 3. The pressure–electrical resistance isotherms of cerium. The data were obtained with increasing pressure (adapted from Jayaraman 1965).

transition pressure. Since the width of the $\gamma \rightleftharpoons \alpha$ hysteresis (to be discussed later in this section) decreases rapidly with increase in temperature, the $\gamma \rightarrow \alpha$ transition pressure at higher temperatures is closer to the equilibrium pressure than it is at lower temperatures. This can lead to an over-estimation of the slope of the phase boundary by about 10 per cent if the experiments are performed with increasing pressure only. The scatter in the coordinates of the critical point as reported by various investigators is still worse. The reported values for the critical temperature range from 480 K to 670 K and for the critical pressure from 1·45 to 2·2 GPa. It is suggested here that the best estimate of the slope of the $\gamma \rightleftharpoons \alpha$ phase boundary is 230 K GPa^{-1}, and the best estimates of the coordinates of the critical point are 2 GPa and 600 K. Obviously, more precise location of the critical point is needed.

The existence of a critical point implies that by suitably varying the pressure and temperature, a continuous transition from one phase to another can be effected without crossing the phase boundary. This is possible only if the two phases possess the same internal symmetry and differ only quantitatively, as in the case of liquid and gas. However, the phases such as liquid and crystalline solid (or two crystalline phases with different crystal structures) differ qualitatively. During a transition involving two such phases, the symmetry of one or the other phase is retained. A continuous transition from one phase to another is not possible, and therefore a critical point cannot exist for such phases (Landau & Lifshitz 1968). The solid–solid transitions normally involve a change in crystal structure, and the phase boundary cannot terminate at a critical point; it either goes to infinity or terminates at a triple point. The $\gamma \rightarrow \alpha$ transition, however, does not involve a change in crystal structure, and therefore the presence of a critical point is not forbidden by symmetry considerations.

The X-ray diffraction studies (Schuch & Sturdivant 1950) at 90 K and one atmosphere indicated the presence of a face-centred cubic structure with an anomalously small lattice parameter (4·82 Å), in addition to the normally contracted γ-phase. The unit cell volume of the new phase is 16·5% smaller than that of the γ-phase. The new phase is identified with the α-phase. This is further supported by the experimental observation that the $\gamma \rightleftharpoons \alpha$ phase boundary intersects the temperature axis at 150 K, indicating thereby that the α-phase can be obtained by cooling at one atmosphere (Herman & Swenson 1958, Likhter *et al.* 1958). The occurrence of the α-phase at one atmosphere and low temperatures has been confirmed by many subsequent studies (McHargue & Yakel 1960, Gschneidner *et al.* 1962, Rashid & Altstetter 1966).

2.3. Structural transitions

At one atmosphere and 1000 K, the γ-phase transforms to a body-centred cubic structure (δ-phase). The γ–δ phase boundary has a slope of -14 K GPa^{-1}, and intersects the melting line at 2·6 GPa and 947 K (Jayaraman 1965).

Transitions at low temperatures

The measurement of the physical properties of cerium (Trombe 1934, 1944, Trombe & Foex 1944) established that three phases are observed on cooling below room temperature. Systematic studies by McHargue & Yakel (1960) indicated that the γ-phase begins to transform at 260 K to a double hexagonal close-packed structure (β-phase). At 100 K the remaining γ-phase transforms to the α-phase. At still lower temperatures, the β-phase also tends to transform to the α-phase. These observations were supported by later studies (Gschneidner *et al.* 1962, Rashid & Altstetter 1966). The β-phase is metastably retained at room temperature and transforms back to the γ-phase only above 440 K. The lattice parameters of the β-phase at 296 K are: $a = 3{\cdot}673$ Å and $c = 11{\cdot}802$ Å. The β-phase is quite sensitive to mechanical treatment; plastic deformation inhibits the transformation to β-phase, and in an extreme case it may even cause the β-phase to transform back to the γ-phase. Plastic deformation below 100 K tends to form the α-phase (McHargue & Yakel 1960). The amount of β-phase in a specimen increases on thermal cycling. The details of these effects are given by Gschneidner *et al.* (1962), Rashid & Altstetter (1966) and Wilkinson *et al.* (1961).

Transition above 5 GPa

The α-phase at room temperature transforms to another phase, termed α'-phase, at high pressures. The $\alpha \rightarrow \alpha'$ transition pressures reported by various investigators differ considerably: 9 GPa (Bridgman 1952), 6–6·5 GPa (Stager & Drickamer 1964), 5 GPa (Wittig 1968) and 5·1 GPa (Schaufelberger & Merx 1975). The accepted value of the $\alpha \rightarrow \alpha'$ transition pressure is that of King *et al.* (1970) who measured the transition pressure as a function of temperature in the range 77–480 K. The transition at 297 K occurs at $5{\cdot}58 \pm 0{\cdot}05$ GPa. Further, the transition pressure decreases with temperature at a rate 0·0035 GPa K^{-1}. The $\alpha \rightarrow \alpha'$ phase boundary in the region 77–480 K, when extended to the high-temperature region, passes through the minimum of the melting curve.

The crystal structure of the α'-phase has been investigated by many workers. From X-ray diffraction experiments Franceschi & Olcese (1969) concluded that the α'-phase was face-centred cubic with about 4% smaller volume than that of the parent α-phase. The measurement of lattice parameters up to 10 GPa indicated that the α'-phase was incompressible! A further evaluation of this work is not possible as the detailed diffraction data are not available. McWhan (1970) found that the diffraction data at 6·5 GPa obtained by him could be explained on the basis of a hexagonal cell ($a = 3{\cdot}16 \pm 0{\cdot}01$ Å and $c = 5{\cdot}20 \pm 0{\cdot}02$ Å); it was suggested that the structure was distorted hexagonal close-packed. This analysis, however, has a serious drawback in that two reflections (103 and 112) are not observed. Schaufelberger & Merx (1975) recorded diffraction data at pressures up to 10 GPa and noticed that the α'-phase appears above 5·1 GPa. The diffraction lines were explained assuming hexagonal close-packed structure. Schaufelberger (1976) succeeded in recording a large number (nearly 30) of diffraction lines from the α'-phase. On the basis of this analysis a hexagonal close-packed structure has been proposed for the α'-phase. At 5·2 GPa the α- and α'-phases are found to coexist. The reported lattice parameters are $a = 3{\cdot}19 \pm 0{\cdot}01$ Å and $c = 5{\cdot}21 \pm 0{\cdot}02$ Å for the α'-phase, and $a = 4{\cdot}630 \pm 0{\cdot}007$ Å for the α-phase. These data suggest that the $\alpha \rightarrow \alpha'$ transition is associated with a volume decrease of 7·3%.

The most convincing crystal structure analysis of the α'-phase is by Ellinger and Zachariasen. In a publication dealing with the preliminary results, Ellinger & Zachariasen (1974) suggest that the stable structure above 5·8 GPa is orthorhombic, having the α-uranium structure. They also report a metastable phase (α'') which is monoclinic. Zachariasen & Ellinger (1977) realized that the diffraction patterns above 5·8 GPa in fact arise from the mixture of two phases. On cycling the pressure, one set of lines becomes spotty and gains intensity on further pressure-cycling; this set arises from the α'-phase. The second set of lines, arising from the α''-phase, remains smooth, indicating that the crystallite size does not increase on pressure-cycling. The monoclinic structure of the α''-phase is in fact a distortion of the face-centred cubic structure involving a small shift (about 0·1 Å) of atoms. The $\alpha \rightarrow \alpha'$ transition, on the other hand, requires a major rearrangement of atoms. It is reasonable to expect that the α''-phase will form from the α-phase with greater ease than the α'-phase. The single-phased sample of α'' can be obtained by rapid pressurization. The unit cell parameters at 5·8 GPa are $a = 3{\cdot}049$ Å, $b = 5{\cdot}998$ Å, $c = 5{\cdot}215$ Å for the α'-phase, and $a = 4{\cdot}762$ Å,

$b=3{\cdot}170$ Å, $c=3{\cdot}169$ Å, $\beta=91{\cdot}73°$ for the α''-phase. The volume change at the $\alpha\rightarrow\alpha'$ transition is 1·1%.

It is clear from the analysis given by Zachariasen (1977), and Zachariasen & Ellinger (1977) that their X-ray diffraction data are basically the same as those obtained by McWhan (1970) and Schaufelberger (1976). The differences lie in the interpretation of the diffraction data. The objection against the orthorhombic indexing raised by Schaufelberger & Merx (1975), and Schaufelberger (1976) is based on the fact that it gives a much poorer 'figure of merit' than the hexagonal indexing. The orthorhombic indexing, on the other hand, gives much better agreement between the observed and the calculated position of the diffraction lines. It may be noted that the two structures predict different values for the volume change associated with the $\alpha\rightarrow\alpha'$ transition; the value is 7·3% for the hexagonal close-packed structure (Schaufelberger 1976), and 1·1% for the orthorhombic structure (Zachariasen & Ellinger 1977). Macroscopic measurements, made with the sliding anvil apparatus, of the volume change associated with the $\alpha\rightarrow\alpha'$ transition (Bocquillon *et al.* 1978) give a value of 1·5%. This value supports the finding of Zachariasen and Ellinger that the structure of the α'-phase is orthorhombic. It is likely that Franceschi & Olcese (1969) got the α''-phase, a slight distortion of the face-centred cubic structure, and crudely indexed the diffraction pattern on the basis of a face-centred cubic structure; the zero compressibility of the high-pressure phase, however, remains unexplained. The precise measurement of the diffraction intensities, which has been not possible so far because of very high background and preferred-orientation effects in the high-pressure X-ray experiments, will help resolve much of the ambiguity.

An interesting point that emerges from these studies pertains to the effect of non-hydrostatic pressures on the crystal structure of the high-pressure phase. The studies by Zachariasen & Ellinger (1977) were made with a diamond anvil apparatus, in which the specimen is placed directly between the diamond tips and pressurized. The stress system experienced by the specimen has a large shear component, and is far from being hydrostatic. Schaufelberger (1976) used a hexahedral press with a solid pressure-transmitting medium to pressurize the specimen. The pressure is nearly hydrostatic in this apparatus. The phases obtained under two different pressure conditions are identical, judged from the diffraction patterns they produce. It appears that the lack of truly hydrostatic pressure does not affect the structure of the high-pressure phase; what is likely to be affected is the kinetics of the transformation (Singh 1983).

Endo *et al.* (1977) reported a body-centred tetragonal phase of cerium which exists above 12·2 GPa at room temperature. Making use of the drop in electrical resistance at the $\alpha'\rightarrow$tetragonal cerium transition (Fujioka *et al.* 1977), the temperature-dependence of the transition pressure between room temperature and 600 K has been determined (Endo & Fujioka 1979). The slope of the phase boundary is $-75{\cdot}6$ K GPa^{-1}. The straight line representing the phase boundary intersects the melting curve in the region of a broad minimum. Assuming that no new phase appears at higher temperatures, two triple points can be expected in a narrow range near the broad minimum; one triple point involves liquid, α-phase and α'-phase, and the other liquid, α'-phase and the tetragonal phase. The existence of two triple points has to be experimentally confirmed.

2.4. Some further comments

The discontinuous transition in general exhibits hysteresis. The equilibrium transition pressure has to be exceeded appreciably before the transition proceeds with detectable kinetics. The forward and the reverse transitions occur at two different pressures. The equilibrium transition pressure is taken as the average of the forward and reverse transition pressures. The various aspects of the hysteresis and the kinetics of pressure-induced transitions have been reviewed by Singh (1983). The $\gamma \rightleftharpoons \alpha$ transition exhibits a hysteresis of about 0·2 GPa. The hysteresis widths derived from the data of various investigators are listed in Table 1. The hysteresis decreases with increase in temperature and becomes small above 500 K (Livshits *et al.* 1960). The $\alpha \rightleftharpoons \alpha'$ transition exhibits a large hysteresis. The hysteresis widths reported by King *et al.* (1970) are 2·5 GPa at 77 K, 1·3 GPa at 297 K, and 0·7 GPa at 480 K. In this case also, the hysteresis width decreases at higher temperatures. The X-ray diffraction experiments at room temperature (Zachariasen & Ellinger 1977) indicate that the α'-phase coexists with the α-phase at pressures as low as 2·7 GPa. Obviously, under certain experimental conditions the $\alpha \rightleftharpoons \alpha'$ hysteresis can be much larger than those reported by King *et al.* (1970).

Table 1. The transition pressure, hysteresis width, critical pressure, critical temperature, slope of the phase boundary and volume change at the transition for the $\gamma \rightleftharpoons \alpha$ transition. The units of pressure and temperature are GPa and K respectively.

p_{tr}	Hysteresis	p_c	T_c	dT/dp	$(\Delta V/V_0)_t$	Method	Reference
0·74[a]	—	—	—	255	—	Volumetric	Bridgman (1927)
1·22	—	—	—	—	0·08	Volumetric	Bridgman (1948)
0·67	0·2	—	—	—	—	Resistivity	Bridgman (1951)
0·80[a]	—	1·8	553	230	—	DTA	Poniatovskii (1958)
0·82	—	—	—	237	0·08	Volumetric	Likhter *et al.* (1958)
0·69	0·25	—	—	217	0·12–0·16	Volumetric	Herman and Swenson (1958)
0·68	—	2·0	630	236	0·14	Volumetric	Beecroft and Swenson (1960)
0·69	0·16	—	—	224	—	Resistivity Volumetric	Livshits *et al.* (1960)
0·68	0·2	—	—	—	0·14	Volumetric	Gschneidner *et al.* (1962)
0·76	—	2·0–2·2	620–670	255	—	X-ray	Davis and Adams (1964)
0·80[a]	0·2	1·75	550	265	—	Resistivity	Jayaraman (1965)
0·71	—	1·45	480	245	—	Volumetric	Kutsar (1973)
0·67	0·2	—	—	—	0·13	Volumetric	Singh (1980)

[a] indicates measurements with increasing pressure only.

The $\alpha' \rightarrow$ tetragonal transition is also expected to exhibit hysteresis. The transition pressures reported by Endo & Fujioka (1979) appear to have been obtained with increasing pressure and therefore do not represent equilibrium transition pressures. Further, if the hysteresis of the $\alpha' \rightleftharpoons$ tetragonal transition also decreases with increasing temperature, the slope of the phase boundary reported by Endo & Fujioka (1979) is over-estimated for the reasons discussed in §2.2.

The $\gamma \rightarrow \alpha$ transition exhibits a time-dependent nature. On the application of a small increase in pressure near the transition pressure, the $\gamma \rightarrow \alpha$ transition starts but tails off and does not run to completion even after long waiting periods. A pressure increment is necessary to drive the transition further (Poniatovskii 1958). This behaviour is observed in experiments with solid as well as liquid pressure-transmitting media.

The properties of cerium are sensitive to impurities. Conflicting results were obtained in the early experiments (Bridgman 1927, 1948, 1951) because of the varying degree of sample purity. The experiments by Livshits *et al.* (1960) indicate that the $\gamma \rightarrow \alpha$ transition pressure increases with increasing impurity content, though the temperature dependence of the transition pressure remains practically unaffected. The impurities also tend to stabilize phases which are normally absent in high purity samples. Rashid & Altstetter (1966) detected 14 phases which are not allotropes of cerium but are stabilized by impurities.

3. Properties of cerium under pressure

3.1. Equation of state

Volumetric measurements (Bridgman 1927, 1948; Singh 1980) indicate that the equation of state of cerium is anomalous. The compressibility in the γ-phase is 0·052 GPa^{-1} at one atmosphere, which is very large as compared with that of the neighbouring elements. Further, the compressibility increases with the increase in pressure and reaches a value of 0·084 GPa^{-1} just before the $\gamma \rightarrow \alpha$ transition. The $\gamma \rightarrow \alpha$ transition is associated with a volume change of $\sim 13\%$; the values obtained by various investigators are listed in Table 1. The compressibility in the α-phase is 0·06 GPa^{-1} at 1 GPa and decreases rapidly with increase in pressure. The pressure derivative of the bulk modulus is abnormally large (~ 10) in the pressure range 1–2 GPa, and only above 2 GPa does it assume a value (~ 4) appropriate to normal solids. The abnormal compression behaviour is also supported by the measurement of elastic constants as function of pressure (Voronov *et al.* 1960). In the γ-phase, Young's modulus, bulk modulus and Poisson's ratio decrease, and the shear modulus slightly increases with increase in pressure. At the $\gamma \rightarrow \alpha$ transition all the moduli increase discontinuously, and continue increasing with further increase in pressure. The pressure derivative of the bulk modulus is ~ 8. The Poisson's ratio remains practically constant after

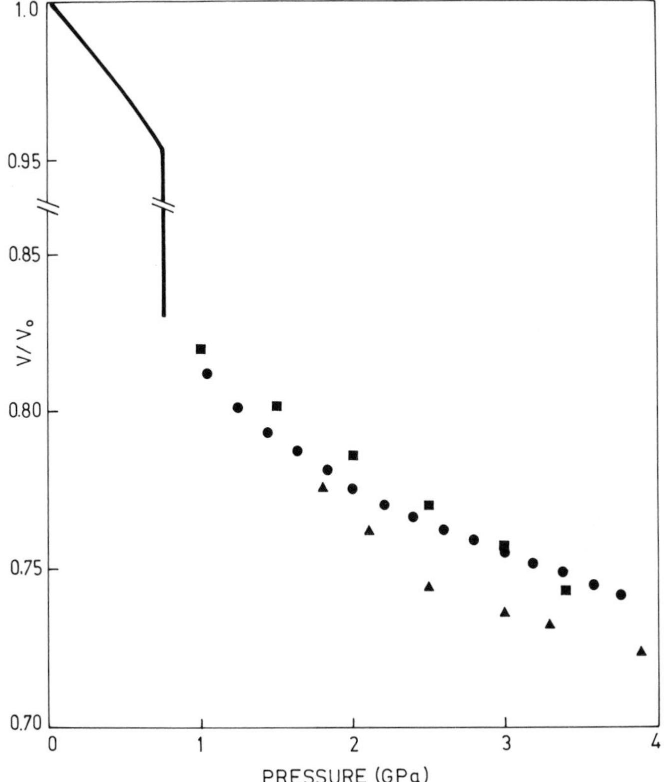

Figure 4. The pressure–volume relation of cerium up to 4 GPa. ■ Gschneidner (1964), ▲ Schaufelberger & Merx (1975), ● Singh (1980).

the $\gamma \rightarrow \alpha$ transition. The pressure–volume data up to 4 GPa are shown in Figure 4.

The low bulk modulus and its large pressure derivative in the pressure range 1–2 GPa may be an intrinsic property of the α-phase, or may arise because of some other factor such as an incomplete $\gamma \rightarrow \alpha$ transition. The γ-phase (coexisting with the α-phase) will transform to the α-phase with increasing pressure, giving rise to a volume decrease in addition to the normal compression of the α-phase. Unlike volumetric measurements on the bulk specimen, X-ray diffraction under pressure can be used to measure the compressibility of the individual phases. Thus, the X-ray measurements can be used to decide whether or not the anomalous compression behaviour is an intrinsic property of the α-phase. Detailed X-ray diffraction measurements of the compressibility, unfortunately, are not available. The limited data of Davis & Adams (1964), Schaufelberger & Merx (1975), and Zachariasen & Ellinger (1977) indicate that the abnormal behaviour of the pressure–volume curve in the pressure range 1–2 GPa is an intrinsic property of the α-phase.

3.2. Transport properties

Electrical resistivity

The electrical resistance of γ-cerium at room temperature increases with pressure, the pressure coefficient of resistance ($\Delta\rho/p\rho_0$) being 0·045 GPa^{-1}. At the $\gamma \rightarrow \alpha$ transition the drop in resistance is about 40%. The pressure coefficient of resistance in the α-phase is 0·13 GPa^{-1} (Bridgman 1927, 1951). The resistance–pressure curve shows a large curvature in the range 0·8–2·5 GPa, and is featureless above 2·5 GPa. At the $\alpha \rightarrow \alpha'$ transition the resistance shows a 6% increase (King *et al.* 1970). The resistance slightly decreases in the range 5–10 GPa. At the $\alpha' \rightarrow$ tetragonal phase transition (12 GPa) a change in slope of the resistance–pressure curve is observed (Fujioka *et al.* 1977). The temperature coefficient of resistance as a function of pressure has been studied by Itskevich (1962), Ramesh (1974a), Leger (1976) and Bastide *et al.* (1978a). In the γ-phase, the temperature coefficient of resistivity is $8{\cdot}7 \times 10^{-4}$ K^{-1} at one atmosphere, and decreases with pressure slightly. In the α-phase, it is about 50×10^{-4} K^{-1} immediately after the $\gamma \rightarrow \alpha$ transition, and decreases with increasing pressure.

The temperature dependence in the range 1·2 to 80 K at pressures ranging from 0·3 to 5 GPa have been measured by Katzman & Mydosh (1972). The spin-fluctuation resistivity derived from these data exhibits a T^2-dependence. The coefficient of the T^2 term is found to decrease rapidly with increasing pressure. Oomi & Mitsui (1976) reported similar results on pure cerium and cerium–lanthanum alloys in the α-phase. The coefficient of the T^2 term, however, is about one half of that obtained by Katzman & Mydosh (1972). Brodsky & Friddle (1973) observed that an annealed cerium sample, held at pressures in the range 1·0–1·8 GPa and quenched from room temperature to 77 K, contains small amounts of magnetic impurities arising from the clusters of β-phase. The electrical resistance of such a sample varies as T^2. On the other hand, if the cerium sample at a pressure between 1·0 and 1·8 GPa is slowly cooled to 77 K, then the α-phase free from any magnetic impurity is obtained. The resistivity of the pure α-phase obtained in this manner shows a T^5-dependence in the temperature range 10–30 K, and a linear temperature dependence above 160 K. The presence of the β-phase cannot be ruled out in the work of Katzman & Mydosh (1972) and Oomi & Mitsui (1976). However, because of the subtraction procedure adopted by Katzman & Mydosh (1972) to derive the spin-fluctuation resistivity, their results are not likely to be affected significantly by the presence of small amounts of β-phase.

Thermoelectric power

The pressure dependence of the thermoelectric power, S, of cerium at 300 K is shown in Figure 5. Ramesh *et al.* (1974b), and Ramani & Singh (1979) find that the thermoelectric power of cerium is about 8 μV K^{-1} at one atmosphere and

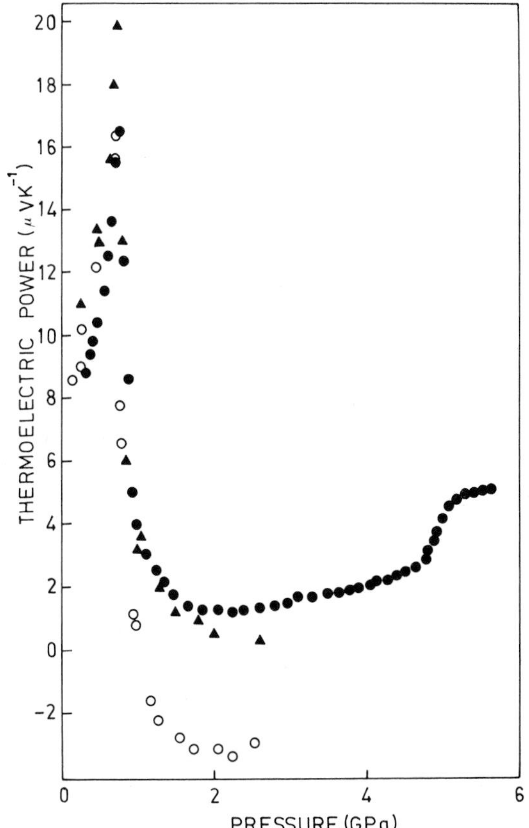

Figure 5. The pressure-dependence of the thermoelectric power of cerium at 300 K. ○ Khvostantsev *et al.* (1973). ▲ Ramesh *et al.* (1974b), ● Ramani & Singh (1979). The datum point of Khvostantsev *et al.* (1973) at 1 GPa has been omitted.

rises to about 20 μV K^{-1} just before the $\gamma \rightarrow \alpha$ transition. A large decrease in S occurs at the transition. In the pressure range 0·8 to 2 GPa, the TEP-pressure curve shows a large curvature, and above 2 GPa it increases smoothly. A jump of 2 μV K^{-1} occurring at 5 GPa is associated with the $\alpha \rightarrow \alpha'$ transition. The difference between the pressure dependence of S obtained by Khvostantsev *et al.* (1973) and that discussed above is only apparent, and arises because of a different sign convention used by Khvostantsev *et al.* (1973). Ramesh *et al.* (1974b), and Ramani & Singh (1979) use the convention that the thermoelectric power is positive if the junction at higher temperature is at a higher potential. For comparison, the data in Figure 1(*b*) of Khvostantsev *et al.* (1973) are plotted with changed sign. The three sets of data show basically the same trend, but quantitatively diverge above 1 GPa.

The temperature dependences of the thermoelectric power at different pressures have been measured by Ramesh *et al.* (1974c). The temperature coefficient of S is negative for the γ-phase, and positive for the α-phase. The pressure effect can be simulated by alloying cerium with thorium. The thermoelectric power–temperature data on some of these alloys (Zoric & Parks 1976) are reminiscent of the data on pure cerium.

Hall effect

The measurements by Kevane *et al.* (1953) indicate that the Hall coefficient of γ-cerium is $1{\cdot}8 \times 10^{-4}\,\mathrm{cm}^{-3}\,\mathrm{C}^{-1}$ at room temperature and increases with decreasing temperature, reaching a maximum value of $4{\cdot}5 \times 10^{-4}\,\mathrm{cm}^{3}\,\mathrm{C}^{-1}$. At 110 K, the Hall coefficient drops to a value of $1{\cdot}5 \times 10^{-4}\,\mathrm{cm}^{3}\,\mathrm{C}^{-1}$. This drop is associated with the temperature-induced $\gamma \rightarrow \alpha$ transition. Below 110 K, the Hall coefficient increases rapidly with decreasing temperature. The results of the measurement of the Hall coefficient as a function of pressure (Likhter & Venttsel 1962) are shown in Figure 6. The reported value of the Hall coefficient at one atmosphere is $2 \times 10^{-4}\,\mathrm{cm}^{3}\,\mathrm{C}^{-1}$, and increases with increasing pressure linearly up to the $\gamma \rightarrow \alpha$ transition pressure. The $\gamma \rightarrow \alpha$ transition is accompanied by a large decrease in the Hall coefficient. The Hall coefficient–pressure curve shows a

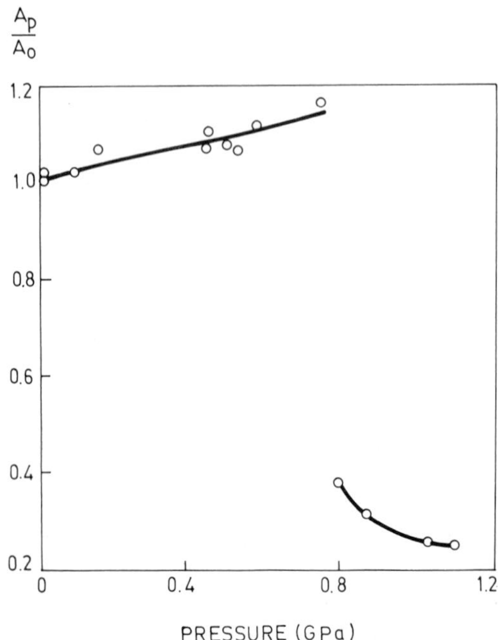

Figure 6. The Hall coefficient of cerium relative to the value at one atmosphere as a function of pressure (adapted from Likhter & Venttsel 1962).

large curvature immediately after the transition. The pressure dependence of the Hall coefficient resembles that of the electrical resistivity.

Kondo-like anomalies

It is well known that some concentrated rare-earth systems, such as $CeAl_2$ and $CeAl_3$, exhibit a Kondo-like logarithmic decrease in the resistivity in a certain temperature range. This behaviour is similar to that observed in dilute systems (Coqblin *et al.* 1976, Bredl *et al.* 1978). The current experimental and theoretical situation is well documented in several articles (Maple *et al.* 1977, Doniach 1977). The physical reasons for the similarity in the behaviour of concentrated rare-earth systems and alloys containing rare-earth impurities are two-fold. Firstly, the spatial extent of the 4f wavefunction is extremely small, being a minute fraction of the interatomic spacing so that the direct exchange coupling via the overlap of the 4f wavefunctions of the neighbouring atoms is negligible. Secondly, in a certain temperature range, the incoherent scattering of the conduction electrons by the 4f shells predominates. Thus, at high temperatures the Kondo-compounds mimic the resistivity behaviour of the dilute systems.

Gschneidner *et al.* (1976) observed that β-cerium exhibits an anomalous resistivity behaviour as a function of temperature. These authors separated the magnetic contribution to the resistivity by subtracting from the total resistivity the phonon contribution. This procedure is based on the assumption that the phonon contribution to the resistivity in β-cerium is the same as that of the α-lanthanum whose structure is identical to that of the β-cerium but devoid of the effects due to 4f electrons. The magnetic contribution to the resistivity thus obtained is found to decrease logarithmically with temperature in the range 80–300 K. The well known formula for Kondo scattering is found to fit the experimental data (Liu *et al.* 1976).

Ramesh & Shubha (1980) showed that γ-cerium too exhibits a Kondo-like anomaly in the high-temperature region. The phonon contribution to the resistivity was experimentally estimated by studying the behaviour of face-centred cubic lanthanum (β-phase), which was prepared by pressure-quenching the normally occurring hexagonal phase (Piermarini & Weir 1964). Figure 7 shows the plot of $\Delta\rho$, the difference between the resistivities of cerium and lanthanum, against ln T, at different pressures. The $\Delta\rho$–ln T plot is a straight line. This feature is characteristic of Kondo behaviour. Further, the slope of the plot changes from -13 $\mu\Omega$ cm at 0·183 GPa to -20 $\mu\Omega$ cm at 0·73 GPa. These results clearly show that the antiferromagnetic exchange integral increases with increasing pressure.

The temperature dependences of the thermoelectric powers of γ-cerium and β-lanthanum at different pressures have been measured by Shubha & Ramesh (1981). Figure 8 shows the data up to 470 K. The temperature behaviour of the TEP of γ-cerium is similar to that predicted theoretically taking into account

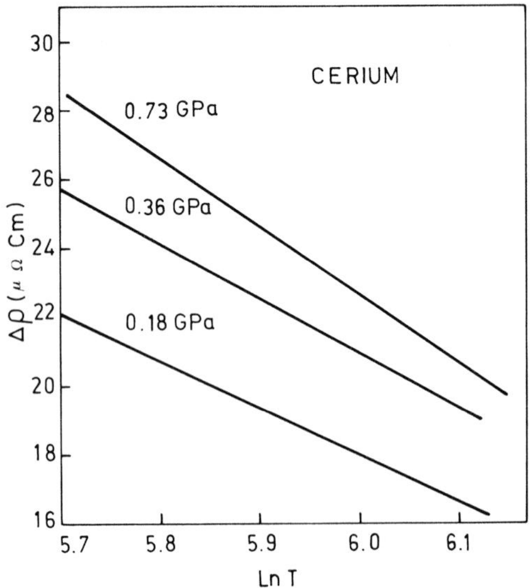

Figure 7. The plot of $\Delta\rho$ against ln T. $\Delta\rho$ is the difference between the electrical resistivities of cerium and lanthanum (cubic phase) under the same temperature and pressure (adapted from Ramesh & Shubha 1980).

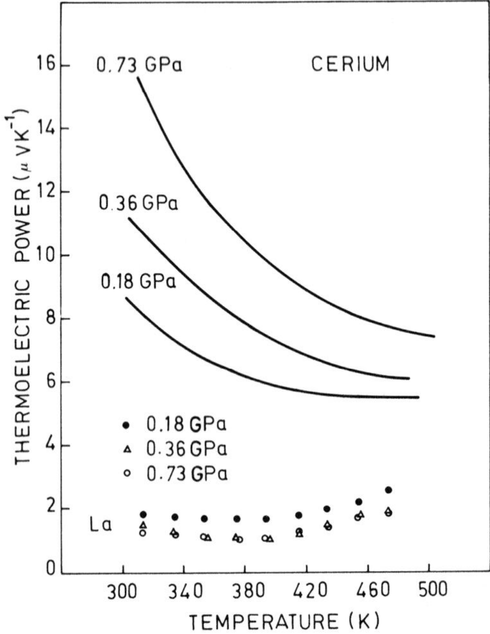

Figure 8. The temperature dependence of the thermoelectric power of γ-cerium and lanthanum (cubic phase) at different pressures (adapted from Shubha & Ramesh 1981).

Kondo scattering (Suhl & Wong 1967). It has been noted that the isobars in Figure 8 are part of a universal curve but with T_K, the characteristic Kondo temperature, shifting to higher temperature at higher pressures.

Superconductivity

Experiments on α-cerium, formed by pressurizing γ-cerium to 1·0 GPa, indicate the absence of superconductivity down to 0·3 K (Smith 1965, Phillips *et al.* 1968). It has been suggested that though no localized magnetic moment exists in the α-phase, some bound-state character of the 4f electrons still persists, which may be removed completely by further increasing the pressure. Wittig (1968) did observe that the α'-phase above 5 GPa becomes superconducting at 1·7 K. The superconducting transition temperature decreases with increasing pressure. Since the localized magnetic moments are not present in the α-phase, the occurrence of superconductivity is not ruled out. Brodsky & Friddle (1973) observed that under certain experimental conditions the α-phase shows traces of magnetic impurity arising from the clusters of β-phase, and suggested that this might have prevented the observation of superconductivity in the earlier studies (Smith 1965, Phillips *et al.* 1968). Probst & Wittig (1975) found that the α-phase at 2·0 GPa becomes superconducting below 20 mK.

3.3. Magnetic susceptibility

The magnetic susceptibility of cerium as a function of temperature and pressure has been measured by MacPherson *et al.* (1971), and the results are shown in Figure 9. The γ-phase, which is characterized by a localized moment, exhibits the well known Curie behaviour with temperature. The magnetic susceptibility of the γ-phase decreases with pressure in the pre-transition region. The $\gamma \rightarrow \alpha$ transition is accompanied by a large decrease in the susceptibility. The effect of pressure is again considerable in the α-phase; the susceptibility decreases by nearly 20% in the pressure range 0·8 to 1·8 GPa. The susceptibility of the α-phase is nearly independent of temperature in the range 100–300 K, suggesting that α-cerium is basically a Pauli paramagnet. Using the available electronic specific heat data, MacPherson *et al.* (1971) found for the un-renormalized exchange enhancement a value of ~ 8. This exchange enhancement, and the molar susceptibility itself, are very large and are comparable only with those of palladium.

3.4. Specific heat

The specific heat of cerium under pressure has been measured up to 2·0 GPa at 300 K (Bastide *et al.* 1978b). The experimental data are shown in Figure 10. The

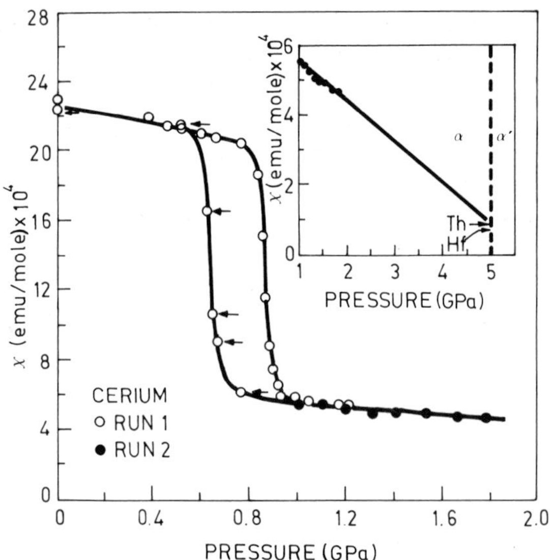

Figure 9. The pressure dependence of magnetic susceptibility of cerium at room temperature. The inset shows the extrapolation of the data to the $\alpha \rightarrow \alpha'$ transition pressure (adapted from MacPherson *et al.* 1971).

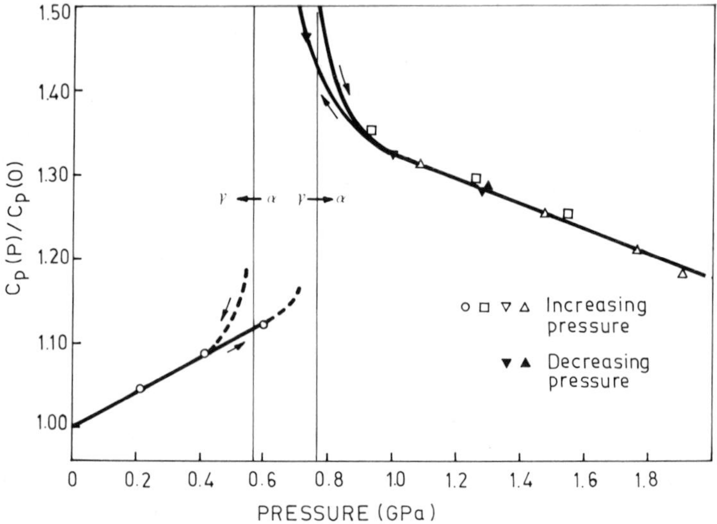

Figure 10. The pressure dependence of the specific heat relative to the value at one atmosphere (adapted from Bastide *et al.* 1978b).

specific heat increases linearly with increase in pressure in the γ-phase, shows a large jump across the $\gamma \rightarrow \alpha$ transition, and finally decreases with increasing pressure in the α-phase. The electronic specific heat, obtained by subtracting from the measured specific heat the lattice contribution and the dilatation term, also shows behaviour similar to that shown in Figure 10. The coefficient of electronic specific heat is 7·4 mJ $mol^{-1}K^{-2}$ at one atmosphere, and increases linearly with pressure to ~16 mJ $mol^{-1}K^{-2}$ just before the $\gamma \rightarrow \alpha$ transition. The transition is accompanied by a large increase in the coefficient of electronic specific heat, reaching a value of ~33 mJ $mol^{-1}K^{-2}$ at 1·0 GPa. The increase of pressure in the α-phase results in a decrease of electronic specific heat coefficient; typically the value at 2·0 GPa is ~23 mJ $mol^{-1}K^{-2}$.

3.5. Some further results

A number of other experiments using neutron diffraction, Compton scattering, photoemission, and so on, have yielded valuable information about the behaviour of 4f electrons in cerium. The results of such experiments are briefly described in the following sections.

Neutron diffraction

The measurement of inelastic neutron scattering from cerium at 300 K has been made by Rainford *et al.* (1977), in the γ- and α-phases. In the γ-phase, the energy width of the quasi-elastic peak increases with pressure, indicating an enhancement in the exchange parameter. They found no trace of paramagnetic scattering in the α-phase. This has now been interpreted to mean that probably the line width in the α-phase is an order of magnitude larger than in the γ-phase. Neutron scattering measurements on cerium–thorium alloys (Shapiro *et al.* 1977) have clearly shown that the energy width of the quasi-elastic peak increases gradually from ~18 meV to ~23 meV as the $\gamma \rightarrow \alpha$ phase boundary is reached. There is a precipitous increase in the spin-fluctuation energy (> 70 meV) in the α-phase. The energy scale for the spin-fluctuation energies obtained from these experiments has provided an important input to a recent theoretical model (Allen & Martin 1982).

Compton scattering

The changes in the localization of 4f electrons and in the density of conduction electrons are expected to result in characteristic changes in the electron momentum distribution. Kornstadt *et al.* (1980) measured the Compton profile in both γ- and α-cerium, and observed that any kind of promotion of the 4f electron to the conduction band is in striking contradiction to their experimental data. The interpretation of the differences in Compton profiles of the γ- and α-

phases requires a knowledge of the electronic wavefunctions in the two phases. In the absence of data on the wavefunctions from band-structure calculations, these authors used the renormalized free-atom approach for the s electron, and a tight-binding model for the d- and f-electrons. Good agreement with the experimental data is obtained if it is assumed that the number of conduction electrons does not change during the $\gamma \rightarrow \alpha$ phase transition.

Photoemission studies

Over a number of years a large body of theoretical work, based on the Friedel–Anderson virtual-bound-state model, has been applied to explain the anomalous properties of cerium. A common feature of these theoretical models is the hypothesis that the 4f level resides in close proximity to the Fermi energy. Typically, the 4f state is assumed to lie within 0·1 eV of the Fermi energy and the hybridization width, due to mixing with the 6s–5d states of the conduction band, is assumed to be about 0·02 eV. A variety of photoemission experiments have been carried out to probe the position of the 4f level in γ- and α-cerium (Hufner & Steiner (1982) and references therein).

Recent ultraviolet photoemission studies on cerium in both the γ- and the α-phase, and on a cerium–thorium alloy (Wieliczka *et al.* 1982, 1984, Martensson *et al.* 1982) have shown the existence of two 4f-related features in the valence band spectra, one at 2 eV and the other near the Fermi level. The main results of the high resolution studies by Wieliczka *et al.* (1984) are that in the γ-phase the two 4f-related features are located at $-0{\cdot}2$ and $-2{\cdot}0$ eV, while in the α-phase these features are located at the Fermi energy and at $-2{\cdot}1$ eV. The peak at $-2{\cdot}0(-2{\cdot}1)$ eV in $\gamma(\alpha)$-phase is a 'direct' peak and is a common feature of almost all the photoemission experiments. The peak close to the Fermi energy is due to a 'shake-down' process. Liu & Ho (1982) have shown that the two-peak structure of the 4f electron photoemission spectrum may arise from two kinds of final states: those with and those without screening by a cerium 5d electron in an impurity state. The important finding of these photoemission experiments is that at the $\gamma \rightarrow \alpha$ phase transition, the position of the 4f direct peak remains almost unchanged, while the intensity of the structure near the Fermi energy increases in α-cerium relative to that in γ-cerium. These results imply that the 4f electron hybridizes more in the α-phase than in the γ-phase. The theoretical analysis of the two-peak structure by Liu & Ho (1982) also shows that there is very little difference in 4f occupation between the two phases of cerium.

De Haas–van Alphen experiments

The de Haas–van Alphen experiments have not been performed on pure cerium in the α-phase because of the difficulty in preparing single crystals. However, $CeSn_3$, which has properties very similar to those of α-cerium, has been recently investigated (Johanson *et al.* 1981a,b). The results indicate that the Fermi surface geometry for $CeSn_3$ is vastly different from that of $LaSn_3$, a reference system with no anomalous behaviour; and the effective masses of the charge carriers in $CeSn_3$ are an order of magnitude larger than the corresponding values in $LaSn_3$. The implication of these results is that the 4f states hybridize strongly with the itinerant band states. The correlation between the large values of electronic specific heat and high effective masses in $CeSn_3$ are indicative of the coherent hybridization between the 4f states and conduction electron states to form new Block states with well defined energy and wavevector. The impurity type of virtual-bound-state model is not relevant, at least at low temperatures, because no de Haas–van Alphen oscillation will be observed due to the resonance scattering of the conduction electrons by the 4f virtual bound state.

Positron annihilation

The positron lifetime and the angular correlation of the positron-annihilation radiation have been measured by Gustafson *et al.* (1969) and Gempel *et al.* (1972). The lifetime in the α-phase, produced by application of pressure, was 8% shorter than that in the γ-phase. The angular correlation of positron-annihilation radiation in the α-phase (obtained by application of pressure) was slightly broader than that in the γ-phase. Exactly opposite effects were observed in the α-phase obtained by lowering the temperature; the lifetime was 5% longer and correlation sharper than the corresponding quantities in the γ-phase. The observed changes in the lifetime and the angular correlation in going from the γ- to the α-phase are considerably smaller than those predicted by the promotion of 4f electron.

4. Theoretical aspects

Zachariasen (quoted by Lawson & Tang (1949)), and independently Pauling (quoted by Schuch & Sturdivant (1950)) suggested that the $\gamma \rightarrow \alpha$ transition involves 4f–5d electron promotion. Based on this idea several theoretical models, often termed 'electron-promotion models', have been developed to explain the $\gamma \rightarrow \alpha$ transition, and the anomalous behaviour of the α-phase. In an alternative approach, the $\gamma \rightarrow \alpha$ transition has been viewed as a Mott transition. More recently, the 'Kondo volume-collapse model' has met with considerable success in explaining many features of the phase diagram. The various models have been discussed briefly in this section. A more detailed account can be found in review articles by Lawrence *et al.* (1981), and Liu (1980, 1981).

4.1. Electron-promotion models

Blandin–Coqblin–Friedel model

The basic ingredients of this theory (Blandin *et al.* 1965, Coqblin & Blandin 1968) are: (i) the conduction band in cerium is derived from the 6s and 5d electrons, (ii) the 4f level in the γ-phase lies close (within 0·1 eV) to the Fermi energy, and the localized 4f state is hybridized with the 6s–5d states forming a virtual-bound-state with a width $\sim$0·01 eV. This theory is essentially a virtual-bound-state description in the spirit of the Friedel–Anderson model, and takes into account the coulomb and the exchange interaction between the localized electrons. The dominant mechanism for the first-order $\gamma \rightarrow \alpha$ transition, however, is a compression shift interaction. The lattice elastic energy is explicitly included in the free-energy expression, and its pressure or volume dependence taken into account. Physically, this means that if the compressibility of the lattice is large enough, the T_{2g} states which form the bottom of the conduction band are lowered relative to the 4f virtual-bound-state, favouring excitation of the 4f electrons to the conduction band. This results in a volume collapse of the cerium ion. The large compressibility itself is a consequence of volume collapse so that this feedback mechanism leads to a catastrophic volume discontinuity. This model explains the termination at a critical point of the $\gamma \rightarrow \alpha$ phase boundary. It also predicts a change in valency of the cerium ion from 3·1 in the γ-phase to 3·7 in the α-phase. The pressure dependences of the specific heat (Bastide *et al.* 1978b), the temperature coefficient of electrical resistivity (Bastide *et al.* 1978a), and the thermoelectric power (Ramesh *et al.* 1974b) have been explained on the basis of this model.

Ramirez–Falicov–Kimball model

In this model the driving force for the transition is provided by the coulomb interaction between the 4f electrons and the conduction electrons (Ramirez & Falicov 1971, Falicov & Kimball 1969, Ramirez *et al.* 1970). The effective energy gap between the localized 4f level and the Fermi energy becomes carrier dependent,

$$E_{\text{eff}} = E_g - 2Gn$$

where E_g is the true gap, G is the short-range, screened, electron–hole interaction parameter, and n is the density of electrons excited to the conduction band. The decrease in the energy gap promotes more 4f electrons to the conduction band. This 'runaway' process can make the effective gap go to zero catastrophically, provided the value of G is large and positive. If G is below a critical value, then a continuous transition takes place. This model explains the existence of the γ–α critical point, the pressure dependence of the lattice parameter and the temperature dependence at different pressures of the paramagnetic susceptibility.

Hirst model

Hirst (1970, 1972, 1974) argued that a configuration-based description is more appropriate to the highly correlated 4f electrons in rare-earth systems. He argued that the two major contributions to the energy of the $4f^n$ configuration are the binding energy of a 4f electron to the nucleus and the closed shell core, and the coulomb repulsion between the 4f electrons. This leads to a parabolic dependence on n (the number of 4f electrons) of the energy of 4f state. One may ordinarily expect that the energy difference between the lowest two configurations, say $4f^n$ and $4f^{n-1}$, would be a sizeable fraction of the coulomb repulsion energy. In a solid, if $4f^n$ and $4f^{n-1}$ lie very close in energy, a situation called an interconfiguration fluctuation (ICF) state is realized. A configurational change from $4f^n \rightarrow 4f^{n-1}$ implies that one 4f electron is transferred to the 6s–5d conduction band at the Fermi energy. In this model the $\gamma \rightarrow \alpha$ transition occurs due to the adjustment of the Fermi energy so that the configurations $4f^n$ and $4f^{n-1}$ become degenerate.

Hirst (1974) included both the electronic and the elastic energy terms in the expression for the free energy. The electronic energy term specifically includes the Falicov–Kimball screening interaction, and the energy required to promote a given fraction of 4f electrons to the conduction band. The elastic energy contribution is essentially the same as that used by Coqblin & Blandin (1968) in discussing the compression shift mechanism. An analysis of the thermodynamic stability of the lattice suggests that the screening energy and the lattice compressibility are the 'accelerating' terms in the sense that large values of these favour a discontinuous transition. The fact that the Fermi energy increases with the increase in electron density gives rise to a 'decelerating' term, and favours a continuous transition. In an actual case, the nature of the transition is decided by the relative magnitudes of the 'accelerating' and 'decelerating' terms. The stiffening of the lattice subsequent to the volume collapse prevents the discontinuous $\gamma \rightarrow \alpha$ transition from proceeding all the way to a pure configurational state $4f^0$. Maple & Wohlleben (1971), and Wohlleben & Coles (1973) have used the ICF model to explain the magnetic behaviour of the α-phase.

The Hirst model does not make any prediction of the $\gamma \rightarrow \alpha$ phase boundary terminating at a critical point. This is because of the neglect of the entropy terms in the free-energy expression. Jefferson (1976) has generalized the Hirst model to finite temperatures by including the entropy term.

Coherent hybridization model

The Hirst model discussed above is essentially an impurity model for a concentrated system. Since the inherent assumption of this model is that the valence fluctuations on different rare-earth ions are uncorrelated, it cannot really describe the ground-state properties. In the coherent hybridization models, the f levels mix with the conduction-band states with phase correlation acquiring an itinerant character. The ground state is thus described in terms of a two-component Fermi liquid model (Varma 1976). The Fermi level intersects the narrow hybridized 4f band which has a low degeneracy temperature. Varma & Yafet (1976) and Robinson (1976) have shown that the magnetic susceptibility under these conditions remains finite as temperature tends to zero. Another interesting feature of this model is that a small energy gap appears in the k-space. However, it is not clear where the Fermi energy lies in relation to this energy gap.

Doniach (1974) has developed an interesting model for some actinide compounds, which should also be applicable to the rare-earth systems. In this model, as the temperature increases there is a rapid transition from the coherently hybridized state to the virtual-bound-state regime. It is now generally recognized that for temperatures less than the hybridization energy, the coherent-hybridization model is most appropriate, while the virtual-bound-state description is adequate at higher temperatures.

Some comments

The electron-promotion models have been successful in explaining the variation of many physical properties across the $\gamma \rightarrow \alpha$ transition, and the termination of the $\gamma \rightarrow \alpha$ phase boundary at a critical point. The basic assumption underlying the electron-promotion models, that the 4f level lies close to the Fermi energy, is supported neither by the band-energy calculations (Herbst *et al.* 1978) nor by the photoemission experiments (§3.5). One of the important predictions of the electron-promotion models, that under pressure the 4f electron occupation number decreases (and correspondingly the conduction electron number increases), contradicts the results of the positron annihilation and the Compton scattering (§3.4) experiments. The band-structure calculations (Glötzel 1978) indicate that the f-electron occupation number in the α-phase is close to unity. The band calculations and the melting point correlations (Hill & Kmetko 1975) give similar f-electron occupation numbers in the γ- and α'-phases. The Mott transition model and the Kondo volume-collapse model discussed below are not based on a small energy gap between the 4f level and the Fermi energy.

4.2. Mott transition model

Johansson (1974) proposed that the $\gamma \rightarrow \alpha$ transition is a Mott transition (Mott 1961). The localized 4f states broaden under pressure. This broadening reduces the intra-atomic coulomb interaction energy, which in turn aids further broadening of the 4f state. The γ-phase is on the low-density side of the Mott transition. In the α-phase the 4f electrons delocalize to form a narrow 4f band. Unlike in the electron-promotion models, the 4f level in the γ-phase or the 4f band in the α-phase can be situated far from the Fermi energy. Johansson (1974) used Hubbard's formalism of the Mott transition (Hubbard 1964) to discuss the $\gamma \rightarrow \alpha$ transition. The condition for the Mott transition for the non-degenerate case is that the bandwidth, W, nearly equals U, the energy of polar state formation. In the

γ-phase, W is much less than U, favouring a localized state. In the Mott transition model, the increased binding energy due to the itinerant 4f state is responsible for the large density difference between the γ- and the α-phases. Further, as the symmetry is not broken in the localized–itinerant transition, it is quite conceivable that at a higher temperature the transition ends up at a critical point. If the narrow 4f band in the α-phase intersects the Fermi energy, then the large electronic specific heat and the enhanced Pauli paramagnetism are naturally explained. This model is consistent with the results of the photoemission and the positron annihilation experiments. The band-structure calculations (Glötzel 1978) which indicate that the 4f bandwidth is of the order of 1 eV in the α-phase, with an occupation number close to unity, support Johansson's model.

4.3. Kondo volume-collapse model

Coqblin & Blandin (1968) were the first to put forward the hypothesis that a transition from the local moment to a spin-compensated state provides the driving force for the $\gamma \rightarrow \alpha$ transition. Edelstein (1968) from his measurements on the temperature dependence of the magnetic susceptibility also concluded that α-cerium is in a spin-compensated state, and estimated that T_K, the Kondo temperature, in the α-phase, is around 1600 K. Notwithstanding the large uncertainties involved in estimating T_K, it is interesting to note that T_K is of the order of T_c, the critical temperature for the γ–α phase boundary. If α-cerium is in a spin-compensated state with high T_K, one can easily visualize how it can have a reduced moment without a large change in the conduction electron density.

The well known Anderson model (Anderson 1961) has both charge and spin degrees of freedom. Krishnamurthy *et al.* (1980) have studied in detail the case of an impurity f state which is non-degenerate except for the spin degrees of freedom. These authors show that the magnetic susceptibility is equivalent to a spin $\frac{1}{2}$ Kondo problem with a characteristic scale of energy T_K of the form

$$T_K = 0.364 \varepsilon_f J^{1/2} \exp(-1/J)$$

for all $J \leqslant 0{\cdot}5$ where J is the effective Kondo coupling constant, ε_f is the position of the 4f state relative to the Fermi energy. J is related to the width, Δ, of the hybridized 4f state, to ε_f and to the coulomb energy U through the relation

$$J = 2\,\Delta/\pi\varepsilon_f + 2\,\Delta/\pi(\varepsilon_f + U)$$

The values of 100 K and 1000 K for the γ- and α-phases respectively lead to $J_\gamma \simeq 0{\cdot}25$ and $J_\alpha \simeq 0{\cdot}5$ (Allen & Martin 1982). The photoemission experiments on γ- and α-phases of cerium imply that $\varepsilon_f \simeq 2$ eV in both phases (see §3.5). The increase of J in going from the γ- to the α-phase is then explained by the increase in the hybridization width, Δ. Croft *et al.* (1981), while analysing their photoemission results on $CeAl_2$ and related alloys, noted that if one wishes to retain Anderson-model ideas to explain volume-dependent properties of cerium, the effects due to 4f broadening must be considered more important than the shifts in the 4f energy. In the light of these ideas, the earlier interpretation that the Kondo interaction gets enhanced under pressure in cerium compounds (Coqblin *et al.* 1976) and in γ-cerium (Ramesh & Shubha 1980) due to the shift of the low-lying 4f state towards the Fermi energy, appears to be erroneous. The Kondo model can also explain the resistivity behaviour across the $\gamma \rightarrow \alpha$ transition. It is clear from Figure 7 that magnetic contribution to the resistivity in γ-cerium is 28 $\mu\Omega$ cm. The loss of this contribution in the spin-compensated α-phase is of the right order of magnitude to account for the precipitous decrease of resistivity at the $\gamma \rightarrow \alpha$ transition.

Allen & Martin (1982) developed the Kondo volume-collapse model which not only provides a semiquantitative description of the $\gamma \rightarrow \alpha$ transition but makes new predictions as well on the phase diagram in the negative pressure region. They addressed themselves to the central issue of accounting for the large volume change at the phase transition without having recourse to the idea of 4f$\rightarrow$5d electron promotion. The equation of state is determined by the Gibbs free energy,

$$G = U - TS + pV$$

where the terms have their usual significance referred to a cerium atom. At the $\gamma \rightarrow \alpha$ transition, the change in each of these terms has been listed by Koskenmaki & Gschneidner (1978). Typically at $T = 300$ K, $p \simeq 7$ kbar, $\Delta V = V_\gamma - V_\alpha = 0 \cdot 15 V_\gamma$, $\Delta S \simeq 1{\cdot}54 K_B$, $p\,\Delta V \simeq 28$ meV, $T\,\Delta S \simeq 38$ meV and $\Delta U = 10$ meV. These authors argued that for the cerium problem, the Gibbs free energy could be modelled as consisting of a normal part, typical of rare-earths, plus an anomalous term. G was chosen to be of the form

$$G = \tfrac{1}{2} B_N V_N (\bar{V} - 1)^2 + F_K + pV$$

Here the first term represents the normal contribution, with the bulk modulus (B_N) and the volume (V_N), chosen as the average of neighbouring rare-earth elements; $B_N = 28$ GPa, $V_N = 36$ Å^3 and $\bar{V} = V/V_N$. The second term is the anomalous term and represents the extra contribution to the free energy due to the coupling of the 4f state with the conduction electrons. The normal term leads to an increase in the internal energy of the compressed solid so that at the phase transition $\Delta U_N = -51$ meV for $\Delta \bar{V} = 0{\cdot}15$. The anomalous term, they argued, must then account for $\Delta U_A = 60$ meV and $\Delta S \simeq 1{\cdot}54 K_B$, so that $\Delta U = \Delta U_N + \Delta U_A \simeq 10$ meV. The value of $\Delta U_A \simeq 60$ meV was proposed to be due to the changes in the binding energy of the singlet Kondo state. It is well known that the Kondo interaction makes an anomalous contribution to the ground

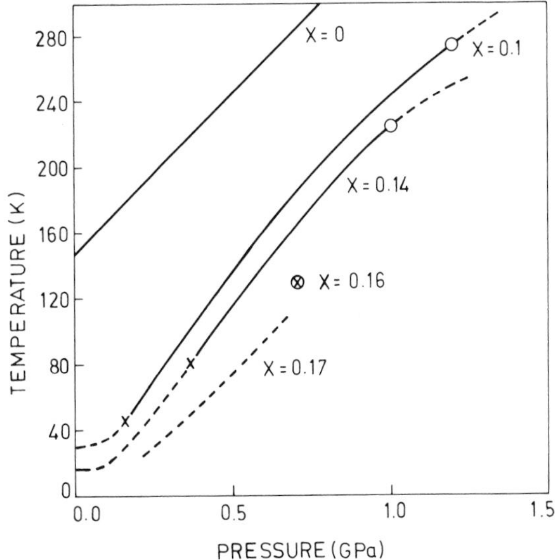

Figure 11. The p–T phase diagram of the $Ce_{0\cdot 9-x}La_xTh_{0\cdot 1}$ system. The upper and lower critical points are marked with open circles and crosses respectively. The two critical points come closer with increasing value of x, and at $x = 0{\cdot}16$, there is a single critical inflection point. Above $x = 0{\cdot}16$, the transition is continuous (adapted from Lawrence *et al.* 1984).

state energy, $U_K \simeq T_K$ (Yoshida & Yoshimori 1973). The inelastic neutron scattering experiments (see §3.5) have shown that the width of the quasi-elastic peak (roughly of the order of T_K) increases by nearly 60 meV in going from the γ- to the α-phase. The change in the binding energy (~ 60 meV) of the Kondo singlet state is of the right order of magnitude to account for the transition. The loss of spin degree of freedom in the α-phase compared to the configuration $J = 5/2$ appropriate to the γ-phase, easily accounts for the entropy change at the phase transition. The $\gamma \rightarrow \alpha$ transition is thus viewed as occurring between two states of vastly different Kondo temperatures, namely, $T_K \simeq 100$ K in γ-cerium to $T_K \simeq 1000$ K in α-cerium.

The volume dependence of the anomalous term, F_K, plays the crucial role in the phase transition. Allen & Martin (1982) used the spin $\frac{1}{2}$ Kondo model, and using the resonant level expression for F_K (T, J) (Schotte & Schotte 1975) obtained the pVT relation. The phase diagram in the p–T plane has a new feature: for pure cerium at negative pressures, there exists a second critical point. Thomson *et al.* (1983), and recently Lawrence *et al.* (1984), have experimentally confirmed this interesting possibility in the expanded system $Ce_{0\cdot 9-x}La_xTh_{0\cdot 1}$. The results of these experiments are summarized in Figure 11. With increasing lanthanum concentration, the two critical points approach each other and at a particular concentration coalesce into a single critical inflection point. This is the first system with the phase boundary terminating at two critical points in the p–T plane.

Acknowledgement

The authors wish to thank C. Divakar, M. Mohan and V. Shubha for their help during the preparation of this article.

References

Allen, J.W. and Martin, R.M., 1982, *Phys. Rev. Lett.*, **49**, 1106.
Anderson, P.W., 1961, *Phys. Rev.*, **124**, 41.
Bastide, J.P., Loriers-Susse, C., Massat, H. and Coqblin, B., 1978a, *High Temp.–High Press.*, **10**, 419.
Bastide, J.P., Loriers, C., Massat, H. and Coqblin, B., 1978b, in *Rare Earths and Actinides*, edited by W.D. Corner and B.K. Tanner (Bristol, London: The Institute of Physics), p. 66.
Beecroft, R.I. and Swenson, C.A., 1960, *J. Phys. Chem. Solids*, **15**, 234.
Blandin, A., Coqblin, B. and Friedel, J., 1965, in *Physics of Solids at High Pressures*, edited by C.T. Tomizuka and R.M. Emrick (New York: Academic), p. 233.
Bocquillon, G., Epain, R. and Loriers, C., 1978, *J. Appl. Phys.*, **49**, 4431.
Bredl, C.D., Steglich, F. and Schotte, K.D., 1978, *Z. Phys.* B, **29**, 327.
Bridgman, P.W., 1927, *Proc. Am. Acad. Arts Sci.*, **62**, 207.
Bridgman, P.W., 1948, *Proc. Am. Acad. Arts Sci.*, **76**, 71.
Bridgman, P.W., 1951, *Proc. Am. Acad. Arts Sci.*, **79**, 149.
Bridgman, P.W., 1952, *Proc. Am. Acad. Arts Sci.*, **81**, 167.
Brodsky, M.B. and Friddle, R.J., 1973, *Phys. Rev.* B, **7**, 3255.
Coqblin, B. and Blandin, A., 1968, *Adv. Phys.*, **17**, 281.
Coqblin, B., Bhattacharjee, A.K., Cornut, B., Gonzalez-Jimenez, F., Iblesias-Sicardi, J.R. and Julien, R., 1976, *J. Magn. Magn. Mater.*, **3**, 67.
Croft, M., Weaver, J.H., Peterman, D.J. and Franciosi, A., 1981, *Phys. Rev. Lett.*, **46**, 1104.
Davis, B.L. and Adams, L.H., 1964, *J. Phys. Chem. Solids*, **25**, 379.
Doniach, S., 1974, in *The Actinides: Electronic Structure and Related Properties*, edited by J.B. Darby and A.J. Freeman (New York: Academic), p. 51.
Doniach, S., 1977, *Valence Instabilities and Related Narrow Band Phenomena*, edited by R.D. Parks (New York: Plenum), p. 169.
Edelstein, A.S., 1968, *Phys. Rev. Lett.*, **20**, 1348.
Ellinger, F.H. and Zachariasen, W.H., 1974, *Phys. Rev. Lett.*, **32**, 773.
Endo, S. and Fujioka, N., 1979, *Phys. Lett.*, **70A**, 475.
Endo, S., Sasaki, H. and Mitsui, T., 1977, *J. Phys. Soc. Japan*, **42**, 882.
Falicov, L.M. and Kimball, J.C., 1969, *Phys. Rev. Lett.*, **22**, 997.
Franceschi, E. and Olcese, G.L., 1969, *Phys. Rev. Lett.*, **22**, 1299.
Fujioka, N., Endo, S. and Kawai, N., 1977, *Phys. Lett.*, **60A**, 340.
Gempel, R.F., Gustafson, D.R. and Willenberg, J.D., 1972, *Phys. Rev.* B, **5**, 2082.
Glötzel, D., 1978, *J. Phys. F: Metal Phys.*, **8**, L163.
Gschneidner, K.A., Jr., 1964, in *Solid State Physics*, edited by F. Seitz and D. Turnbull, Vol. 16 (New York: Academic), p. 307.
Gschneidner, K.A., Jr., Elliot, R.O. and McDonald, R.R., 1962, *J. Phys. Chem. Solids*, **23**, 555.
Gschneidner, K.A., Jr., Burgardt, P., Legvoid, S., Moorman, J.O., Vyrostek, T.A. and Stassis, C., 1976, *J. Phys. F: Metal Phys.*, **6**, L49.
Gustafson, D.R., McNutt, J.D. and Roelling, L.O., 1969, *Phys. Rev.*, **183**, 435.
Herbst, J.F., Watson, R.E. and Wilkins, J.W., 1978, *Phys. Rev.* B, **17**, 3089.
Herman, R. and Swenson, C.A., 1958, *J. Chem. Phys.*, **29**, 398.
Hill, H.H. and Kmetko, E.A., 1975, *J. Phys. F: Metal Phys.*, **5**, 1119.

Hirst, L.L., 1970, *Phys. Kondens. Mater.*, **11**, 255.
Hirst, L.L., 1972, *Adv. Phys.*, **21**, 759.
Hirst, L.L., 1974, *J. Phys. Chem. Solids*, **35**, 1285.
Hubbard, J., 1964, *Proc. R. Soc.* A, **281**, 401.
Hufner, S. and Steiner, P., 1982, *Z. Phys.* B, **46**, 37.
Itskevich, E.S., 1962, *Sov. Phys.—JETP*, **15**, 811.
Jayaraman, A., 1965, *Phys. Rev.*, **137**, A179.
Jefferson, J.H., 1976, *J. Phys. C: Solid St. Phys.*, **9**, 269.
Johanson, W.R., Crabtree, G.W., Koelling, D.D., Edelstein, A.S. and McMasters, O.D., 1981a, *J. Appl. Phys.*, **52**, 2134.
Johanson, W.R., Crabtree, G.W., Edelstein, A.S. and McMasters, O.D., 1981b, *Phys. Rev. Lett.*, **46**, 504.
Johansson, B., 1974, *Phil. Mag.*, **30**, 469.
Katzman, H. and Mydosh, J.A., 1972, *Phys. Rev. Lett.*, **29**, 998.
Kevane, C.J., Levgold, S. and Spedding, F.H., 1953, *Phys. Rev.*, **91**, 1372.
Khvostantsev, L.G., Vereshchagin, L.F. and Shulika, E.G., 1973, *High Temp.–High Press.*, **5**, 657.
King, E., Lee, J.A., Harris, I.R. and Smith, T.F., 1970, *Phys. Rev.* B, **1**, 1380.
Kornstadt, U., Lasser, R. and Lengeler, B., 1980, *Phys. Rev.* B, **21**, 1898.
Koskenmaki, D.C. and Gschneidner Jr., K.A., 1978, *Handbook of the Physics and Chemistry of Rare Earths*, edited by K.A. Gschneidner Jr. and L. Eyring (Amsterdam: North Holland), Vol. 1, Chap. 4.
Kraut, E.A. and Kennedy, G.C., 1966, *Phys. Rev. Lett.*, **16**, 608.
Krishnamurthy, H.R., Wilson, K.G. and Wilkins, J.W., 1980, *Phys. Rev.* B, **21**, 1044.
Kutsar, A.R., 1973, *Phys. Metals Metall.*, **33**, 197.
Landau, L.D. and Lifshitz, E.M., 1968, *Statistical Physics*, 2nd edition (London: Pergamon), p. 363.
Lawrence, J.M., Riseborough, P.S. and Parks, R.D. 1981, *Rep. Prog. Phys.*, **44**, 1.
Lawrence, J.M., Thomson, J.D., Fisk, Z., Smith, J.L. and Batlogg, B., 1984, *Phys. Rev.*, **29**, 4017.
Lawson, A.W. and Tang, T.Y., 1949, *Phys. Rev.*, **76**, 301.
Leger, J.M., 1976, *Phys. Lett.* A, **57**, 191.
Likhter, A.I. and Venttsel, V.A., 1962, *Sov. Phys.—Solid St.*, **4**, 352.
Likhter, A.I., Riabinin, Iu.N. and Vereshchagin, L.F., 1958, *Soviet Phys.—JETP*, **6**, 469.
Liu, S.H., 1980, in *Science and Technology of Rare Earth Materials*, edited by E.C. Subbarao and W.E. Wallace (New York: Academic), p. 121.
Liu, S.H., 1981, *Physics of Solids Under High Pressures*, edited by J.S. Schilling and R.N. Shelton (Amsterdam: North Holland), p. 327.
Liu, S.H. and Ho, K.M., 1982, *Phys. Rev.* B, **26**, 7052.
Liu, S.H., Burgardt, P., Gschneidner, K.A., Jr. and Legvoid, S., 1976, *J. Phys. F: Metal Phys.*, **6**, L55.
Livshits, L.D., Genshaft, Yu.S. and Ruabinin, Yu.D., 1960, *Phys. Metals Metall.*, **9**, 82.
MacPherson, M.R., Everrett, G.E., Wohlleben, D. and Maple, M.B., 1971, *Phys. Rev. Lett.*, **26**, 20.
Maple, M.B., Delong, L.I., Fertig, W.A., Johnston, D.C., McCallum, R.W. and Shelton, R.N., 1977, *Valence Instabilities and Related Narrow Band Phenomena*, edited by R.D. Parks (New York: Plenum), p. 17.
Maple, M.B. and Wohlleben, D.K., 1971, *Phys. Rev. Lett.*, **27**, 511.
Martensson, N., Reihl, B. and Parks, R.D., 1982, *Solid St. Commun.*, **41**, 573.
McHargue, C.J. and Yakel, H.L., Jr., 1960, *Acta Metall.*, **8**, 637.
McWhan, D.B., 1970, *Phys. Rev.* B, **1**, 2826.
Mott, N.F., 1961, *Phil. Mag.*, **6**, 287.
Oomi, G. and Mitsui, T., 1976, *J. Phys. Soc. Japan*, **41**, 705.

Piermarini, G.J. and Weir, C.E., 1964, *Science*, **144**, 69.
Phillips, N.E., Ho, J.C. and Smith, T.F., 1968, *Phys. Lett.*, **27A**, 49.
Poniatovskii, E.G., 1958, *Sov. Phys.—Dokl.*, **3**, 498.
Probst, C. and Wittig, J., 1975, in *Low Temperature Physics-LT 14*, edited by M. Krusius and M. Vuorio (New York: Elsevier), Vol. 5, p. 453.
Rainford, B.D., Buras, B. and Lebech, B., 1977, *Physica*, **86–88B**, 41.
Ramani, G. and Singh, A.K., 1979, *Solid St. Commun.*, **29**, 583.
Ramesh, T.G., 1974a, *Resonance Scattering in the Study of Solid and Liquid States*, thesis submitted to Mysore University, India, p. 167.
Ramesh, T.G. and Shubha, V., 1980, *J. Phys. F: Metal Phys.*, **10**, 1821.
Ramesh, T.G., Reshamwala, A.S. and Ramaseshan, S., 1974b, *Pramana*, **2**, 171.
Ramesh, T.G., Reshamwala, A.S. and Ramaseshan, S., 1974c, *Solid St. Commun.*, **15**, 1851.
Ramirez, R. and Falicov, L.M., 1971, *Phys. Rev.* B, **3**, 2425.
Ramirez, R., Falicov, L.M. and Kimball, J.C., 1970, *Phys. Rev.* B, **2**, 3383.
Rashid, M.S. and Altstetter, C.J., 1966, *Trans. Met. Soc. AIME*, **236**, 1649.
Robinson, J.M., 1976, *AIP Conf. Proc.*, **29**, 319.
Robinson, J.M., 1979, *Phys. Rep.*, **51**, 1.
Schaufelberger, Ph., 1976, *J. Appl. Phys.*, **47**, 2364.
Schaufelberger, Ph. and Merx, H., 1975, *High Temp.–High Press.*, **7**, 55.
Schotte, K.D. and Schotte, U., 1975, *Phys. Lett.*, **55A**, 38.
Schuch, A.F. and Sturdivant, J.H., 1950, *J. Chem. Phys.*, **18**, 145.
Shapiro, S.M., Axe, J.D., Birgeneau, R.J., Lawrence, J.M. and Parks, R.D., 1977, *Phys. Rev.* B, **16**, 2225.
Shubha, V. and Ramesh, T.G., 1981, *J. Phys. F: Metal Phys.*, **11**, 191.
Singh, A.K., 1980, *High Temp.–High Press.*, **12**, 47.
Singh, A.K., 1983, *Bull. Mater. Sci.*, **5**, 219.
Smith, T.F., 1965, *Phys. Rev.* A, **137**, 1435.
Stager, R.A. and Drickamer, H.G., 1964, *Phys. Rev.* A, **133**, 830.
Thomson, J.D., Fisk, Z., Lawrence, J.M., Smith, J.L. and Martin, R.M., 1983, *Phys. Rev. Lett.*, **50**, 1081.
Trombe, F., 1934, *C.R. Acad. Sci., Paris*, **198**, 1951.
Trombe, F., 1944, *C.R. Acad. Sci., Paris*, **219**, 90.
Trombe, F. and Foex, M., 1944, *Ann. Chim.*, **19**, 416.
Vaidya, S.N., 1979, *High Temp.–High. Press.*, **11**, 335.
Vaidya, S.N. and Raja Gopal, E.S., 1966, *Phys. Rev. Lett.*, **17**, 635.
Varma, C.M., 1976, *Rev. Mod. Phys.* **48**, 219.
Varma, C.M. and Yafet, Y., 1976, *Phys. Rev.* B, **13**, 2950.
Voronov, F.F., Vereshchagin, L.F. and Goncharova, V.A., 1960, *Sov. Phys.—Dokl.*, **135**, 1280.
Wieliczka, D., Weaver, J.H., Lynch, D.W. and Olson, C.G., 1982, *Phys. Rev.* B, **26**, 7056.
Wieliczka, D.M., Olson, C.G. and Lynch, D.W., 1984, *Phys. Rev.* B, **29**, 3028.
Wilkinson, M.K., Child, H.R., McHargue, C.J., Koehler, W.C. and Woolan, E.O., 1961, *Phys. Rev.* **122**, 1409.
Wittig, J., 1968, *Phys. Rev. Lett.*, **21**, 1250.
Wohlleben, D.K. and Coles, B.R., 1973, in *Magnetism*, edited by H. Suhl (New York: Academic), Vol. 5, p. 3.
Yoshida, K. and Yoshimori, A., 1973, in *Magnetism*, edited by H. Suhl (New York: Academic), Vol. 5, p. 263.
Zachariasen, W.H., 1977, *J. Appl. Phys.*, **48**, 1391.
Zachariasen, W.H. and Ellinger, F.H., 1977, *Acta Crystallogr.* A, **33**, 155.
Zoric, I. and Parks, R.D., 1976, in *Valence Instabilities and Related Narrow Band Phenomena*, edited by R.D. Parks (New York: Plenum), p. 459.

CHAPTER 14

electrons in small metallic particles

M.R. Harrison† and P.P. Edwards

University Chemical Laboratory, Lensfield Road, Cambridge, CB2 1EW, UK

1. Introduction

Transitions from metallic to nonmetallic behaviour occur in a wide variety of condensed-phase systems as a characteristic parameter of the system or an external variable is continuously changed (Mott 1974, Friedman & Tunstall 1978, Acrivos *et al.* 1984). For example, it was recognized some time ago that decreasing the concentration of shallow donors or acceptors in a doped semiconductor could bring about such a change, as could the expansion of a metallic saturated vapour of a monovalent element such as caesium. In a group of papers beginning in 1948, Mott injected a qualitatively new feature into these studies when he pointed out that the transition from metal to nonmetal in an infinite system (i.e. one in which any surface considerations were negligible) might not be a *continuous* one, as implicitly assumed by earlier workers (Mott 1949). Thus the 'Mott transition' was envisaged as a sharp, discontinuous transition between the two limiting electronic states of matter. However, Mott also noted in 1961, that "... the sharp transition described here is only expected in an *infinite* lattice. It goes without saying that for a *finite* number of atoms there will be a gradual decrease in the weight of the ionized states in the wave function as the interatomic distance is increased or in other words a gradual transition ..." (Mott 1961).

It seems clear that in the infinite system originally considered by Mott, one is dealing with continuum physics in which the effective intersite interactions approach the thermodynamic limit. Thus one generally assumes, for example, that crystals of doped semiconductors are of dimensions considerably larger than other characteristic length scales in the system such as the de Broglie wavelength of the itinerant conduction electrons. However, following Mott (1961) we can propose that a clear distinction may exist between the electronic transitions

† Present address: GEC Research Laboratories, Hirst Research Centre, East Lane, Wembley, Middlesex, HA9 7PP.

associated with an infinite system, and the corresponding properties of a single small particle in isolation. In this chapter we are primarily interested in the latter problem of the electronic properties of small, isolated, metallic particles and how these properties vary with particle size. This important problem is much discussed in both the chemistry and physics literature, and the description of these particles generally involves the neglect of any interparticle interaction (Fröhlich 1937, Kubo 1962, 1967, Friedel 1977, Borel & Buttet 1981, Jortner 1984, Kubo *et al.* 1984).

An experimental example of the problem at hand is illustrated in Figure 1 which shows electron microscope photographs of submicron-sized colloidal gold particles of extremely well defined dimensions. By a suitable choice of

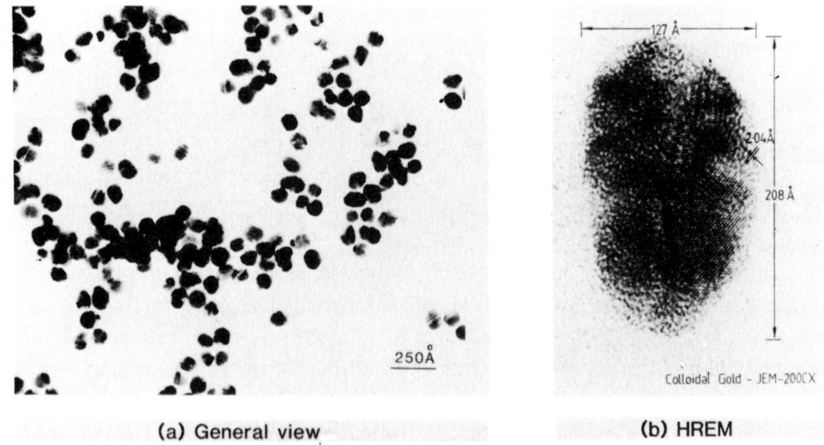

Figure 1. (*a*) Colloidal gold particles prepared from a chlorauric acid solution reduced with sodium citrate. (*b*) A high-resolution electron microscope picture of an individual gold particle (Curtis *et al.* 1985).

experimental conditions, it is possible to prepare colloidal particles with particle diameters of 20 Å upwards in which the interparticle interactions can (correctly) be neglected. Other representative systems of interest include: (i) clusters formed in supersonic beams of metal atoms (Jortner 1984, Knight *et al.* 1984), (ii) clusters formed by impregnating metals into porous oxide or zeolitic matrices (Jacobs 1982, Harrison *et al.* 1984), (iii) clusters formed via the controlled aggregation of metal atoms in low-temperature rare-gas matrices (Ar, Xe, etc.) (Ozin 1980, Thompson *et al.* 1983), (iv) inorganic cluster compounds, such as the transition metal carbonyl clusters of osmium, rhodium and platinum, representing clusters of metal atoms essentially isolated from adjacent clusters by an inert sheath of carbonyl ligands (Johnson 1980, Johnson *et al.* 1985).

The considerable technological interest in these and related systems stems from their use in a variety of applications, such as catalysis, photography, colouring, magnetic recording and ferrofluids, etc. Within the present context, these small metallic particles represent materials whose energy level structure can be varied by changing the effective particle or cluster size. Our purpose in this chapter, then, is to briefly summarize some important features of the electronic structure of small particles possessing a congested system of electron energy levels. We intend to focus attention primarily on two aspects. First, the changes in the energy level structure and electronic properties of individual metallic particles as their characteristic dimensions and relevant energies span typical length scales and energies associated with various electronic excitations. In particular, the fragmentation of a conducting, metallic particle must inevitably lead to the cessation of conducting behaviour. We are ultimately interested in any similarities or differences between this electronic transition within an individual particle and the more widely studied problem of the metal–nonmetal transition in (potentially) infinitely large macroscopic systems. Second, we focus on the intrinsic changes in the magnetic properties of small particles as one goes from a single atom to a bulk, macroscopic solid. We will attempt to illustrate how magnetic measurements represent an important and non-destructive probe of the electronic structure of microscopic particles.

2. The energy level structure in small particles

2.1. General considerations

First, we have to decide what constitutes metallic or nonmetallic behaviour in a small particle? In a recent elegant contribution, Perenboom *et al.* (1981) set out the broad features of the problem: "... When the number of atoms contained in a grain of solid matter is steadily reduced, it is plausible that in the course of this process a stage is realized when the particle does not behave like a smaller copy of the corresponding bulk solid anymore".

That a metal to nonmetal transition must occur as a result of the successive fragmentation of matter is unambiguous (Figure 2). Clearly, a stringent lower limit for this critical size must be a particle consisting of a single atom. However, one clearly wishes ultimately to probe this transition in more searching detail; in fact we will tentatively propose that two electronic transitions may be important in the present context.

It is useful to provide a qualitative description of particles, aggregates and clusters according to their (approximate) dimensions, and this is summarized in Figure 2. Here the appropriate compositional parameters are the particle nuclearity, n, and its diameter, d. In addition, some comments relating to the level structure and parentage of the various states of matter are indicated.

The general approach to the problem (Perenboom *et al.* 1981) usually revolves around the consequences of particle fragmentation within the framework of what one might term 'bulk' solid-state theory (Ziman 1960, Ashcroft &

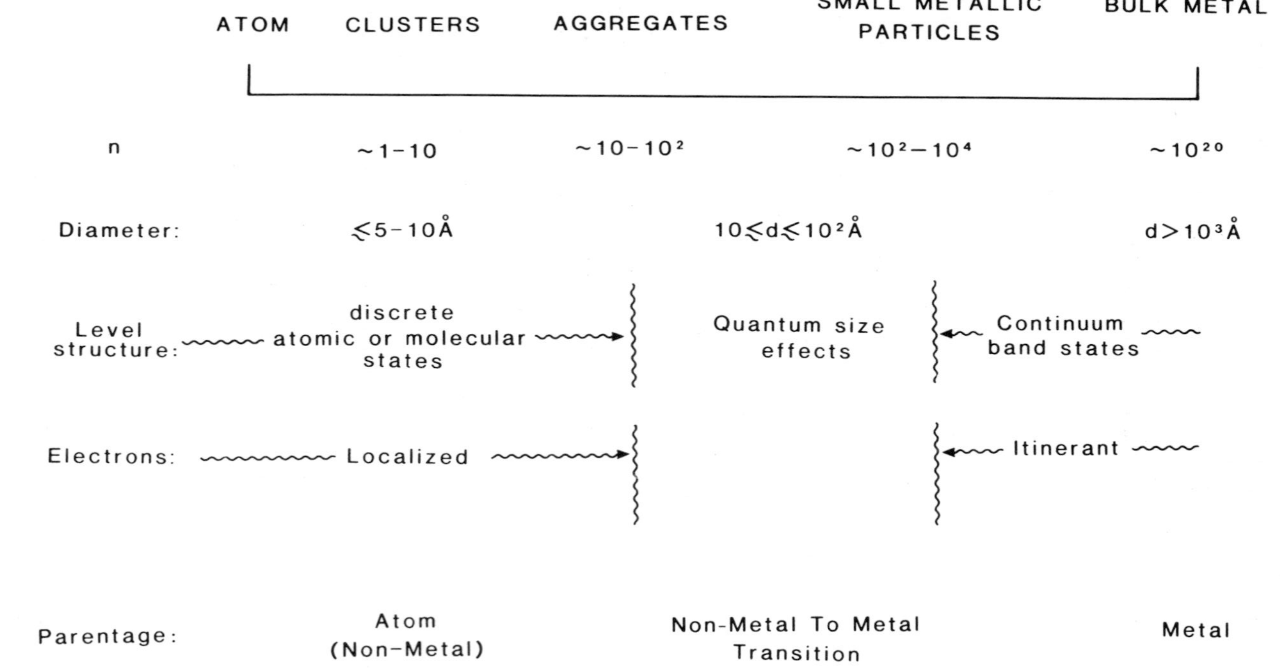

Figure 2. A possible classification of the various electronic regimes important in the evolution of clusters, aggregates and small metallic particles.

Mermin 1976). This is obviously inappropriate for particles consisting of 1 to 10 atoms, which are approaching the atomic or molecular limit. However, as pointed out elsewhere (Perenboom *et al.* 1981), it should give useful and intuitive information when the particle is produced by gradually reducing the size of a macroscopic solid. Such an approach from the bulk, metallic solid was first advanced by Fröhlich in 1937 and further elaborated by Kubo in 1962. We first return to our classification (Figure 2) and outline the various important electronic regimes.

2.2. *Clusters* ($n \sim 1$–10)

These involve atomic or molecular species which include van der Waals molecules for weak interactions, and ionic and covalent molecular structures for strong interactions. Here we anticipate a relatively smooth modification to discrete atomic states as the particle nuclearity increases. The molecular structures are generally reasonably well characterized (certainly for small n), although for larger clusters several isoenergetic isomers may exist. The essential electronic features in this regime are obvious and distinguishable atomic or molecular properties; typical examples might be the silver and alkali-metal clusters (Ag_3, Ag_5, Na_3, Na_5 etc.) in the rare-gas solids (Ozin 1980, Lindsay 1981), as well as the low-nuclearity transition-metal cluster carbonyls such as $Os_3(CO)_{12}$ shown in Figure 3 (Johnson *et al.* 1985). The cluster carbonyls of various transition elements (Rh, Os, Pt) have been shown by X-ray crystallography to exhibit many different geometries that are either fragments of extended metallic lattices (hcp, fcp, bcc) or resemble other types of close packing (polytetrahedral, icosahedral, etc.) that are theoretically favourable for particles containing a few dozen atoms (Johnson 1980, Johnson & Lewis 1981, Hoare & Pol 1972, 1975). These compounds represent cluster fragments effectively isolated from adjacent cluster units by inert sheaths of carbonyl ligands. Thus, there exists a marked difference between cluster carbonyl compounds and, for example, the ternary molybdenum chalcogenides (often called Chevrel phases, $M_xMo_6X_8$ where M represents a rare earth or other metallic element and X is S, Se or Te). For the latter systems, extensive intercluster interactions give rise to itinerant electron behaviour at ambient temperatures and superconductivity below 15 K (Yvon 1979).

2.3. *Aggregates* ($n \sim 10$–10^2)

The usefulness of molecular models and concepts begins to break down as the cluster nuclearity increases; here molecular structures may well be characterized by a very large number of isomeric forms. To illustrate, the number of atoms in a spherical sodium particle of diameter d (and bcc structure) is

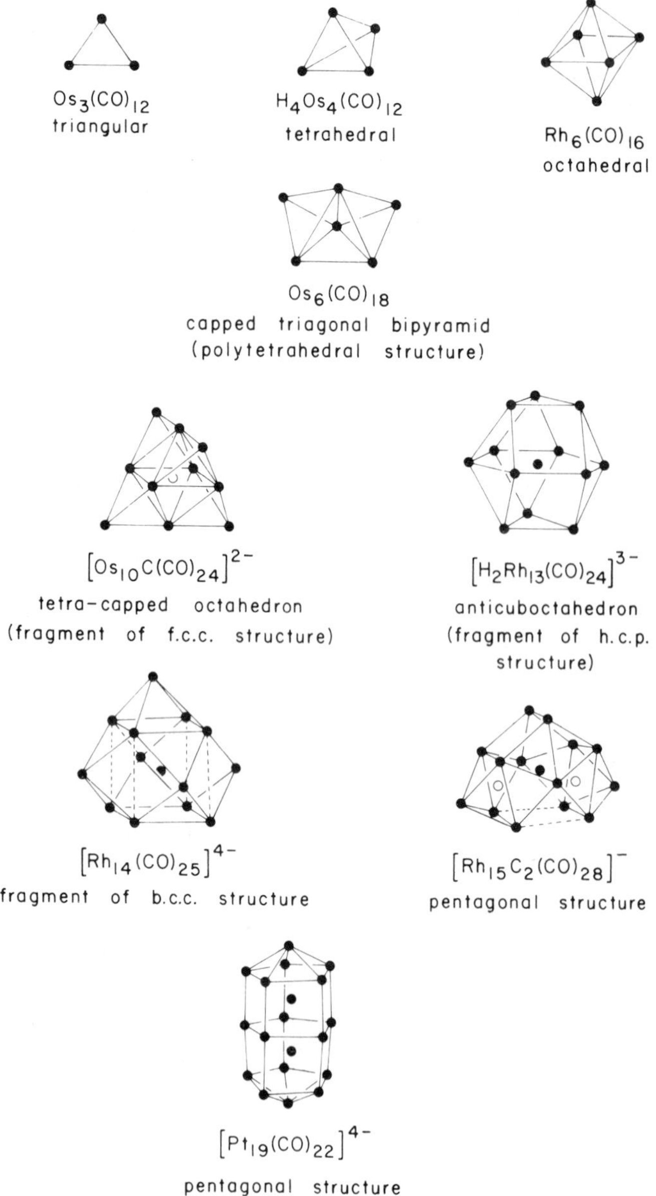

Figure 3. A representation of typical metal atom geometries in transition-metal carbonyl clusters. For clarity, the carbonyl ligands have been omitted. The open circles denote interstitial carbon atoms in the cluster compound. Taken from Johnson *et al.* (1985).

$\frac{4}{3}\pi(d/2)^3/(\frac{1}{2}a_0^3)$, where a_0 is the lattice constant (4·23 Å); for this metal the terminology aggregates ($n \sim 10$–10^2) encompasses particles of approximate diameter 10–20 Å. At this upper limit, the 'molecular' structure for the naked metal clusters is generally not known, but may be more akin to the short-range order found in liquid or disordered materials (Friedel 1977). Similarly, the precise details of the electronic properties of these aggregates are uncertain, although we may anticipate that the electronic structure can vary considerably with their shape and size. The large (molecular) transition metal carbonyl clusters (e.g. $H_2Os_{10}C(CO)_{24}$ and $[Pt_{19}(CO)_{22}]^{4-}$, Figure 3), may be judged to be at the lower limit of this category.

For large aggregates ($n > 10^2$), one also anticipates a gradual evolution of solid-state characteristics, such as a quasi-continuous density-of-states for electronic states, i.e. the potential appearance of a 'conduction' band and also the possibility of a lattice phonon spectrum (Jortner 1984). Clearly, however, the densities of electron carriers will still be insufficient to produce a genuine energy continuum. In addition, any view of these cluster states based on atomic properties is no longer applicable. It is perhaps now realistic to address any questions of parentage in terms of the electronic properties of the condensed, bulk metal—*but with several fundamental differences* (Fröhlich 1937, Kubo 1962, 1968, 1969, 1977).

For these small particles we might also anticipate an electronic transition (Figure 2) from 'metallic' behaviour to 'nonmetallic' behaviour. It has been customary to invoke the so-called Ioffe–Regel criterion in macroscopic systems for the condition for electron localization in the presence of scattering, namely,

$$k_F l \sim 1 \qquad (1)$$

where l is the electron mean free path and k_F is the electron wavevector (Ioffe & Regel 1960, Mott 1974). Wood & Ashcroft (1982) have noted that if we crudely interpret l as a characteristic small-particle dimension, then equation (1) can be used in the present context to consider the cessation of conducting behaviour in a small particle as a metal–nonmetal transition of the Ioffe–Regel sort. Clearly, even accepting the criticisms of equation (1) raised by Goëtze (1981), we ponder whether such a transition in a single particle may be different from the infinite lattice problem more generally associated with the metal–nonmetal transition (Mott 1974). It is interesting to note that Grannan *et al.* (1981) have reported the observation of a dielectric singularity, or catastrophe, at the percolation threshold of a granular composite of small silver particles randomly dispersed through a potassium chloride host. The dielectric catastrophe occurs quite close to the (classical) critical volume fraction of metal particles corresponding to the percolation threshold. However, it is clear that there are differences in the detailed critical behaviour between this classical percolation system, and the corresponding divergence in the dielectric constant of, for example, heavily doped Si:P (Capizzi *et al.* 1980). For the latter systems the dielectric enhancement of the transition is considerably larger than that expected from classical percolation theories, and is dominated by many-body quantum effects near the transition. These differences may indeed reflect the macroscopic *versus* microscopic physics of the transition in infinite systems (Mott 1961) and the corresponding transition in small metallic particles (Fröhlich 1937, Kubo, 1962, Wood & Ashcroft 1982).

2.4. *Small metallic particles* ($n \sim 10^2$–10^4)

In a bulk metal the one-electron energy levels are quasi-continuous and a sum over the electronic states is generally interpreted in terms of a density-of-electronic-states function which is independent of any boundary conditions imposed on the electron wavefunction. This standard picture must be modified for metallic particles of dimensions smaller than 10^4 Å (Figure 2). In this regime, the truncation of the wavefunction has two effects on the electronic properties. The first is a surface effect due to imposed boundary conditions. The fraction of surface atoms increases rapidly as the metal particle diameter decreases, and Figure 4 shows the effective fraction, f, of surface sodium atoms for particles of various diameters. Thus, for sodium particles less than ca. 100 Å, at least 10% of the constituent atoms are located on the particle surface. The second effect concerns the changeover in electron energy levels from quasi-continuous to discrete (Fröhlich 1937, Kubo 1962, 1967). For a particle containing N atoms, the spacing of adjacent energy levels denoted by Δ will be of order E_F/N, where E_F is the Fermi energy. If this energy gap for electronic excitation is comparable to some important characteristic energy, such as the thermal energy (kT), the Zeeman energy ($g_e\mu_B H$) or the energy of incident radiation ($\hbar\omega$) then interesting effects in the thermodynamic and electronic properties are to be expected. These are generally termed 'quantum size effects' and the location of the approximate onset of such effects is illustrated in Figure 2. Therefore, quantum-size effects in a nominally metallic sample will be severe when $\Delta > kT$, for example, so that for a sodium particle with $d \sim 100$ Å, the condition is satisfied below approximately 5 K; for a 50 Å sodium particle, these effects should be evident below 50 K, etc. Figure 4 also shows a plot of the average level spacing, Δ, as a function of the particle diameter, measured in terms of various energies (K, eV). The relevant values of the electron Zeeman energies are both X-, and Q-band microwave frequencies are also indicated.

Within this quantum size effect regime, magnetic susceptibility, magnetic resonance, optical and many other properties may be greatly different from bulk, macroscopic properties. This is especially the case for experiments carried out at low temperatures, such that $\Delta > kT$, or, in the case of magnetic measurements, at sufficiently weak external magnetic fields such that $\Delta > \hbar\omega$ (Figure 4). In addition, the dynamic response of electrons and nuclei in metallic particles at a frequency so low as to satisfy the inequality $\Delta > \hbar\omega$ may be completely different from those of the bulk materials (Kubo 1962, Kawabata 1970). As we shall see shortly, this is certainly the case for magnetic relaxation of

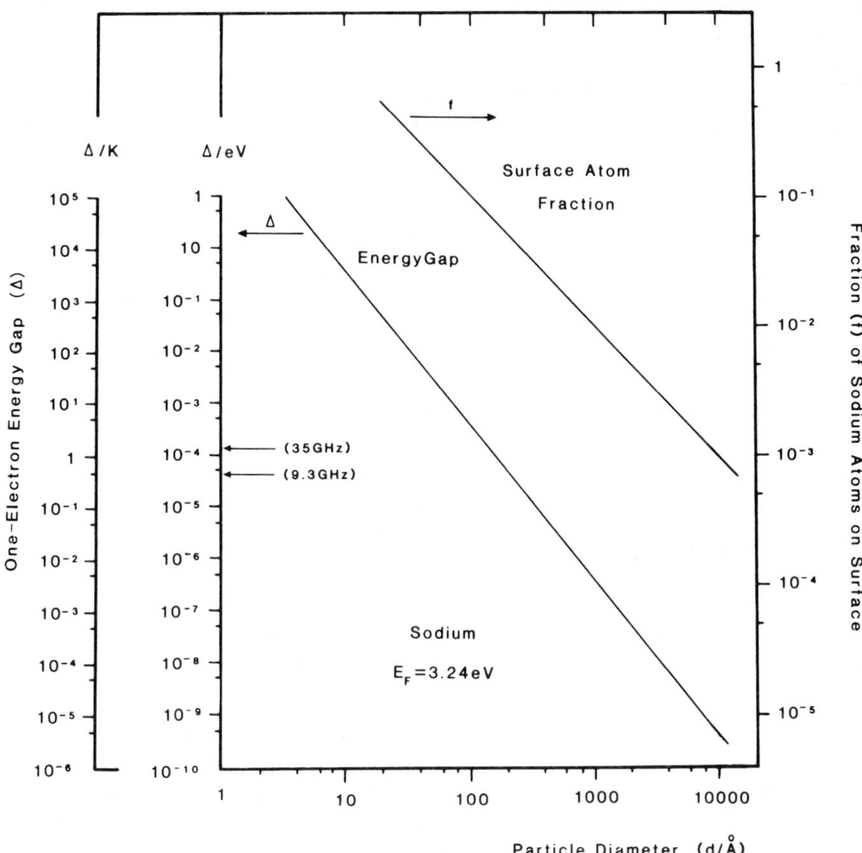

Figure 4. The fragmentation of bulk sodium; a plot of the average energy level spacing Δ as a function of the particle diameter, *d*. The figure also shows the calculated surface fraction, *f*, of atoms in a cluster as a function of the particle diameter.

electron spins in the quantum-size effect regime; here the normal (bulk) metallic spin relaxation properties are suppressed as the continuum band states break up into discrete, but congested, electronic levels. The electronic transition from the bulk to the quantum-size effect regime (Figure 2) also signals the appearance of large, spontaneous magnetism in normally feebly paramagnetic metals. We will have more to say about these matters in the following sections.

3. Magnetism and magnetic resonance in small particles

3.1. General considerations

In discussing the magnetic properties of small metallic particles at low temperatures, it is generally assumed that the electron number in each particle is constant. This innovation, due to Kubo (1962, 1967) has its 'atomic' counterpart in the Mott–Hubbard correlation energy, U (Mott 1974, Friedel 1984). The Kubo concept is based on pure electrostatics: the energy required to charge a sphere of diameter d with a charge e in a vacuum is $e^2/2\pi\varepsilon_0 d$, where ε_0 is the vacuum permittivity. The charge can only originate from the surrounding medium which is at a temperature T. Therefore when $kT(\sim 10^{-4}$ eV at 1 K, $\sim$0·03 eV at 300 K) is less than $e^2/2\pi\varepsilon_0 d$ ($\sim$0·2 eV for $d = 100$ Å), the probability for an electron to be captured by a particle is very small and charge fluctuations are therefore highly improbable. One might imagine, however, that for large metallic particles ($d > 10\,000$ Å) the changing energy is so small that this suppression of any fluctuation in the electron number might be overwhelmed; compare the analogous situation with the Mott–Hubbard U (Hubbard 1964a,b, Mott 1974, Edwards & Sienko 1983). However, for the present purposes we will suppose that, certainly at low temperatures, all metallic particles are electrically neutral. Thus, the problem of electrons in small particles relates to the number of particles containing an even number of electrons, and particles containing an odd number of electrons. This even/odd distinction, advanced by Kubo (1962), is generally adopted in almost all discussions relating to the properties of electrons in small metallic particles (Knight 1973, Marzke 1979, Perenboom *et al.* 1981, Friedel 1984).

3.2. *Electron spin susceptibility*

The charge conservation concept of Kubo complicates the exact computation of the Pauli spin paramagnetism, and indeed many other properties, since all quantities must be derived from a particle canonical ensemble, that is, the fraction of particles with an even and an odd number of electrons. A schematic energy level diagram with electron occupancies (at $T = 0$)

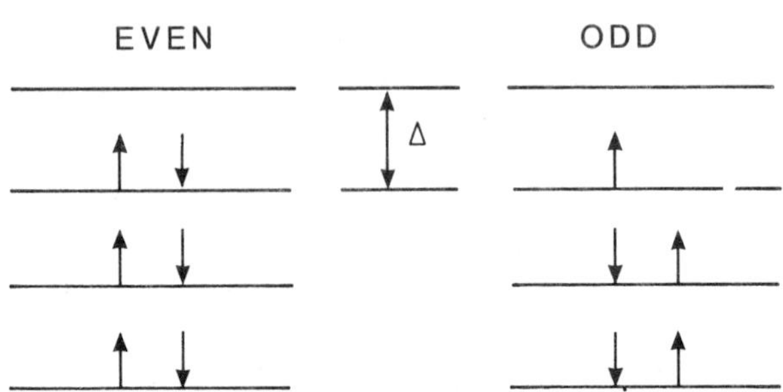

Figure 5. A schematic electron energy level diagram with electron occupancies at $T = 0$ K for small particles having an even and odd numbers of electrons (modified from Marzke 1979). The one-electron energy gap is identified as Δ, as previously illustrated in Figure 4.

for small particles having even and odd numbers of electrons is shown in Figure 5 (Marzke 1979). In this, only the levels near the highest occupied electronic state of energy are shown. The energy gap is the difference between the highest occupied state and the lowest unoccupied state; the most obvious parameterization is in terms of the one-electron energy gap, Δ, as given previously in Figure 4. In addition, for a collection of particles of various sizes (and shapes), the energy gap Δ will not be a fixed value but instead a random variable governed by some appropriate probability distribution. This leads to a statistical smearing-out of quantum size effects which clearly complicates both theoretical calculations and experimental interpretations. Much effort has been concentrated into the arduous task of calculating exactly the full temperature dependence of the spin susceptibility and other properties of all appropriate energy level distributions in metallic particles. The reader is referred to Perenboom *et al.* (1981) and to Kubo *et al.* (1984) for details of the various procedures. Here, we present only the major consequences of the probability distributions, including some discussion of experimental data.

A particularly striking even/odd effect occurs in the temperature dependence of the electron spin susceptibility. Kubo (1962) first showed that the high-temperature susceptibilities of *both* odd and even particles are equal, constant and Pauli-like, whereas at low temperatures the odd particles exhibit Curie-law paramagnetism of the unpaired spins and the even particles remain diamagnetic (Figure 6). Various probability distributions for Δ change the detailed temperature dependence, but the basic Kubo predictions remain qualitatively unchanged. The full temperature dependence is given in Figure 6, in terms of the change from the Pauli value (χ_p) in terms of the reduced energy (kT/Δ). The limiting behaviour may be briefly interpreted as follows. Each odd particle carries a magnetic moment of one Bohr magneton and has a Curie susceptibility. For the even particles, kT/Δ represents a fraction of 'dissociated' pairs. The susceptibility for even particles in Figure 6 drops gradually from the Pauli value at $kT/\Delta = 1$.

Monod & Millet (1976) measured conduction electron spin resonance (c.e.s.r.) of small silver particles prepared by simultaneous condensation of silver and benzene followed by heat treatment. The susceptibility of the conduction electrons can be deduced from the intensity of the resonance signal, and this showed a T^{-1} dependence at low temperatures. Similar observations of enhanced low-temperature susceptibilities have been reported for small particles of lithium (Borel & Millet 1977), magnesium (Millet & Borel 1981, Sako & Kimura 1984), platinum (Marzke *et al.* 1976, Gordon *et al.* 1977) and calcium (Sako & Kimura 1984). Clearly, magnetic impurities can play havoc with this type of measurement but all reports stress the great precautions workers take to eliminate such spurious results (see, for example, Perenboom *et al.* 1981). This illustrates one of the fundamental experimental problems in the study of small particles, namely, the production of pure, well-characterized samples. However, considerable progress has been made in this direction during the past few years; for example, in the area of supersonic beams (for a recent

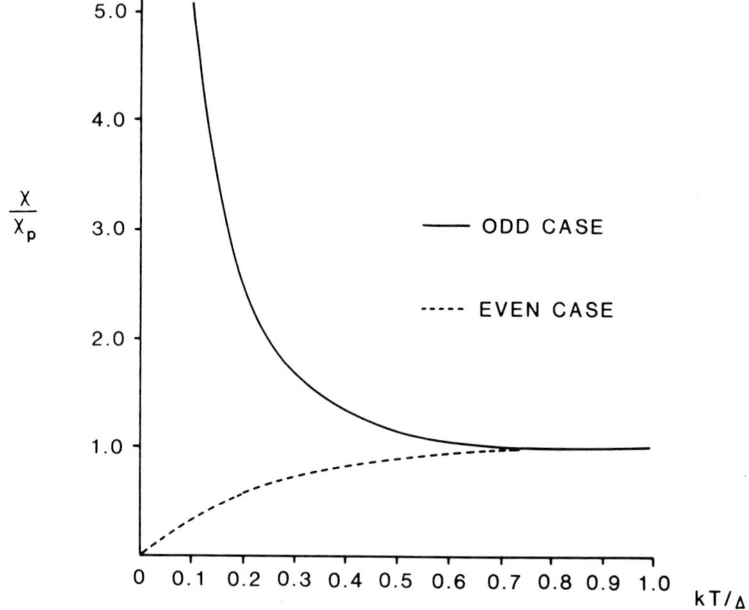

Figure 6. The magnetic susceptibilities of even and odd particles. The curves are normalized to the Pauli value, χ_P, at the high-temperature limit. The magnetism of the even particles vanishes at $T=0$ K. The odd particles have a Curie-like magnetism at low temperatures (modified from Denton *et al.* 1971).

review of selected aspects see Jortner 1984) and also in condensed-state systems composed of metallic particles impregnated in various host materials (Perenboom *et al.* 1981, Jacobs 1982). In addition, purely 'chemical' methods can be used to produce well defined, high-purity molecular clusters of the sort shown in Figure 3.

Johnson *et al.* (1985) have recently reported measurements of the magnetic susceptibilities of a series of cluster compounds of osmium (Figure 7) over the temperature range 1·5–300 K. These clusters are well characterized molecular species which effectively span the transition region between metal clusters and aggregates (Figure 2). With the exception of the hydrido cluster ($H_2Os_{10}C(CO)_{24}$, the magnetic susceptibilities of the entire series of clusters were rigorously temperature-independent down to 1·5 K. In these systems at least, the problems of possible paramagnetic contaminants can be ruled out. The observed paramagnetic excess molecular susceptibilities ($\chi_{\rm ems}$), plotted as a function of cluster nuclearity in Figure 7, have been determined from the total measured susceptibilities by subtraction of the ion-core and ligand contributions (Johnson *et al.* 1985). The observed susceptibilities are most naturally interpreted as arising from a Van Vleck paramagnetic contribution to the susceptibility. This contribution arises from the modification of the cluster wavefunction caused by

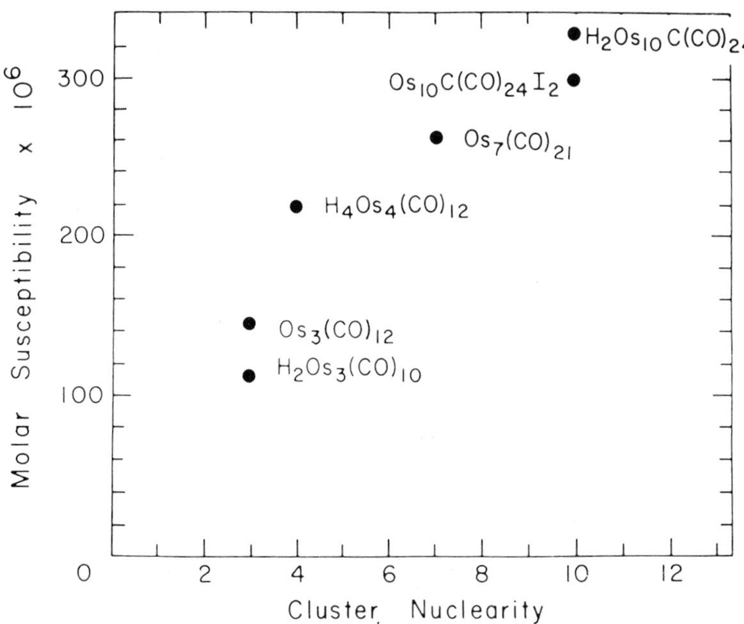

Figure 7. Paramagnetism in osmium cluster carbonyls: the variation of the high temperature (298 K) excess molecular susceptibility (χ_{ems}) for osmium cluster compounds of increasing nuclearity (Johnson *et al.* 1985).

the external magnetic field. The modified wavefunction is described by the admixture of excited states in which electrons are promoted from occupied bonding orbitals to the first unoccupied (antibonding) levels. The increase in χ_{ems} with cluster size (Figure 7) suggests that the energy gap separating filled from vacant orbitals decreases with increasing cluster size. Shiba (1976) and Sone (1976, 1977) have given a full quantum-mechanical treatment of spin–orbit coupling in their calculations of small particle susceptibility. In this, the susceptibility of even particles of heavy atoms can be understood most directly as the temperature-independent Van Vleck paramagnetism (Van Vleck 1932). It is interesting to note that this (second-order) Van Vleck contribution transforms smoothly from temperature-independent paramagnetism to a Curie-like paramagnetism as the characteristic electron energy gap becomes comparable to kT, the thermal energy (Johnson *et al.* 1985). Thus, predictions of enhanced paramagnetism in large clusters and small metallic particles arise both from molecular- and metallic-based theories of magnetism in this intermediate electron regime.

The electron spin susceptibility in small particles can also be measured by nuclear magnetic resonance (n.m.r.) spectroscopy. The polarization of unpaired electron spins in a magnetic field produces, by virtue of the hyperfine coupling of the electrons in s-states and the nuclear spins, an excess field at the nucleus in question (see, for example, Andrew 1969, or Slichter 1965). The n.m.r. in a paramagnetic metallic sample is then shifted relative to the value for the same nucleus in the absence of any electron spins. This shift, the so-called Knight shift ($\mathscr{K}$), is directly proportional to the paramagnetic spin susceptibility, χ_p.

If we designate the n.m.r. shift in a bulk metal by $\mathscr{K}_p$, where the Pauli spin susceptibility is appropriate, then for even particles at low temperatures, the Knight shift is given by (Knight 1973)

$$\mathscr{K} = \mathscr{K}_p \left\{ \frac{2{\cdot}86NkT}{E_F} \right\} \tag{2}$$

whereas for odd particles,

$$\mathscr{K} = \mathscr{K}_p \left\{ \frac{2E_F}{3NkT} \right\} \tag{3}$$

We point out that for a particle containing approximately 2000 atoms with a level spacing $(E_F/N) \sim 10^{-7}$ eV at 100 K, the factor in brackets in equation (3) is of order 30. Therefore, depending upon the even/odd parity of the particles (Figure 5), we may expect, respectively, either an n.m.r. with essentially zero Knight shift, or an n.m.r. line shifted by a very large factor. In both cases, the n.m.r. shift will thus be dependent upon temperature and particle size.

The first report of n.m.r. shifts in small metallic particles was for lithium particles in neutron-irradiated ^{7}LiH at room temperature. The n.m.r. spectrum from these specimens showed three lines, as in Figure 8, one of which had the 'normal' (bulk) Knight shift and was hence attributed to micron-sized globules of metallic lithium. A second line originated from the lithium of the parent compound LiH and a narrower absorption superimposed on the second at $\mathscr{K} = 0$ was attributed to small 'even' platelets of metallic lithium of some one or two atoms thick, and of the order of 10–20 atoms wide and long. More details on the n.m.r. experiments can be found elsewhere (Taupin 1966, 1967; see also Perenboom *et al.* 1981). Very careful measurements on copper particles with diameters as small as 25 Å have been reported by Kobayashi *et al.* (1972) and by Yee & Knight (1975). The latter authors prepared small copper particles in the size range 25–110 Å by flash evaporation of successive layers of copper on an SiO_2 substrate. In Figure 8 the derivative of the absorption signal is shown for copper particles with an average diameter of 100 Å. The observed n.m.r. signal from the small particles was broader than the bulk metal line, having a longer tail on the low-magnetic-field side and a peak shifted from the bulk position in the direction of the expected (bulk) salt position. The combination of these two spectral characteristics led to an assignment of an even-electron small copper particle. More detailed analyses and assessment of these experiments are given in the reviews by Knight (1973), Perenboom *et al.* (1981) and Kubo *et al.* (1984). A

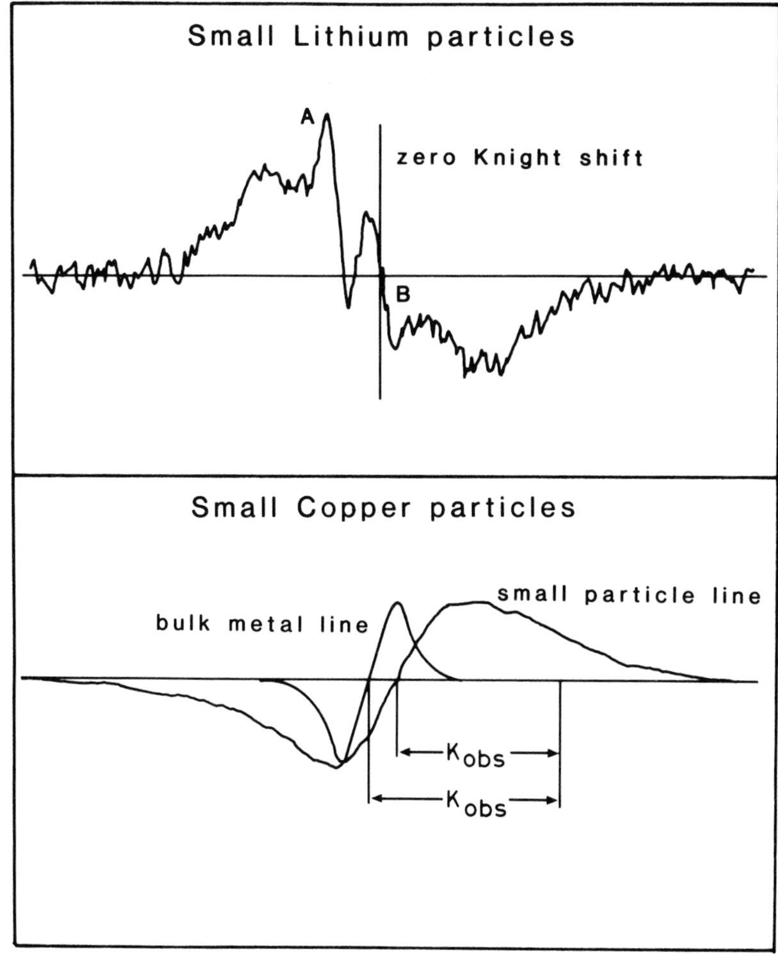

Figure 8. *Top:* the derivative of the nuclear magnetic resonance spectra for small particles of lithium. For lithium particles, the Knight shift of the resonance, labelled 'A', is equal to that for the bulk metals. The resonance labelled 'B', with zero Knight shift, is attributed to very small lithium particles containing an even number of electrons. (From Charvolin *et al.* 1966.) *Bottom:* the n.m.r. resonance of small copper particles with an average diameter of 100 Å again showing the reduction of the Knight shift $\mathscr{K}_{obs}$ observed in even particles. (From Yee & Knight, 1975.) The compilation of spectra is from Perenboom *et al.* (1981).

recent important development has been the observations of the ^{195}Pt Knight shift by van der Klink *et al.* (1984) and Makowka *et al.* (1982; see also Stokes *et al.* 1983) on small platinum particles on various substrates. Stokes *et al.* (1983) also compare their n.m.r. data on alumina-supported platinum samples with data taken on platinum carbonyl clusters (Figure 3) and conclude that the platinum atoms in the platinum carbonyl clusters behave very much like those on the surface of platinum catalyst particles.

3.3. Electron spin relaxation

The electron spin Zeeman energy at 9·3 GHz (X-band frequencies) is 4×10^{-5} eV. Therefore in the light of our earlier comments, one expects that the dynamic relaxation processes of the electron spins will be drastically modified by quantum size effects if the energy gap, Δ, exceeds these values (Kubo 1962, Kawabata 1970). Thus, the Elliott mechanism (Elliott 1954) for the relaxation of unpaired electron spins will be suppressed since the inelastic spin-flip scattering processes will be drastically reduced if the energy level spacings are much larger than the appropriate Zeeman energies. In addition, both the e.s.r. and n.m.r. experiments in bulk metallic systems and small particles reveal that the resonance lineshape is particularly sensitive to the mobility and spin relaxation properties of the unpaired electrons, as well as to the actual dimensions of the particle itself. We consider as representative the form of the conduction electron spin resonance (c.e.s.r.) signal in conducting small metallic particles.

The mobility of conduction electrons determines both the microwave skin depth (δ) and the electron mean-free-path (l). In addition, the c.e.s.r. lineshape is dependent upon T_D, the time taken for the conduction electron to diffuse through δ, and T_2, the electron spin–spin relaxation time. In both bulk and large-particle samples, these considerations are of cardinal importance in determining the form—indeed the very existence—of experimentally observable c.e.s.r. lines (Dyson 1955, Feher & Kip 1955). The classical skin depths and electron mean-free-path for potassium metal at 9·25 GHz are shown in Figure 9. Consider a metallic particle with a diameter of approximately 7000 Å. The c.e.s.r. spectra at different temperatures for a particle of this dimension, as well as for one with a diameter of approximately half this size are shown in Figure 10. The most noticeable features at low temperature in all cases are (i) a reduction in the c.e.s.r. linewidth, and (ii) the onset of an asymmetry (generally defined in terms of the intensity ratio, A/B, in the signal. Briefly, (i) is a manifestation of the fact that the classical skin depth decreases at low temperature, while the onset of asymmetry in the c.e.s.r. signals the condition $\delta < d$. Clearly, different sized metallic particles will show the asymmetry onset at different temperatures (Figure 9). At sufficiently low temperatures, the electron mean free path will exceed δ and the penetration of the microwave field is controlled by l in this 'anomalous skin-effect regime'. Clearly l is also limited by the particle dimensions, such that $l \leqslant d$.

The electron spin–spin and spin–lattice relaxation rates (T_2^{-1}, T_1^{-1} respectively) in bulk metals are extremely rapid and show a strong dependence upon both the measurement temperature and the spin–orbit coupling constant of the atomic core. For example, $T_1^{-1} = T_2^{-1} \simeq 10^6\ \mathrm{s}^{-1}$ in sodium metal at 20 K, and $T_1^{-1} = T_2^{-1} \simeq 10^9\ \mathrm{s}^{-1}$ for rubidium metal at the same temperature. The mechanism of electron spin relaxation in metals was first considered by Overhauser (1953) and subsequently by Elliott (1954), who showed that spin relaxation occurs through the combined action of the lattice vibration and the spin–orbit coupling. The spin–orbit coupling produces admixture of spin-up and

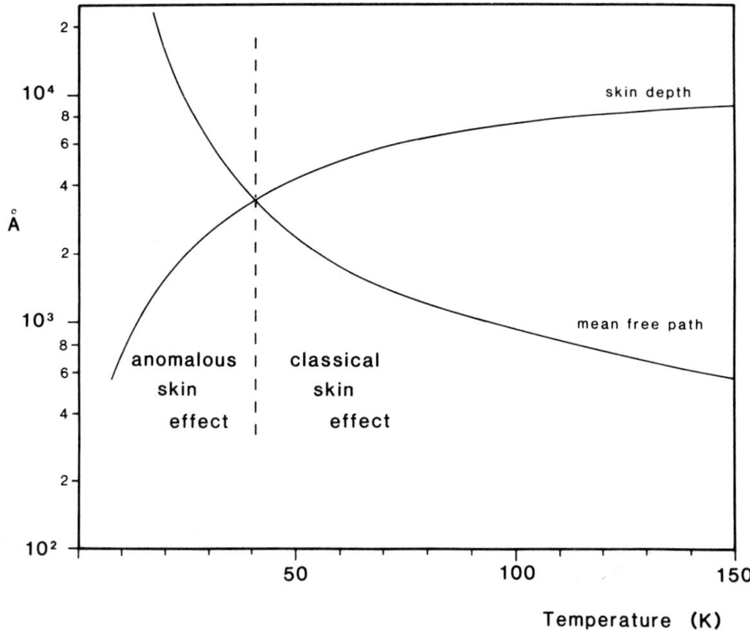

Figure 9. The classical skin depth δ and mean free path l *vs.* temperature in potassium metal. (Taken from Edmonds *et al.* 1984.)

spin-down wavefunctions of the conduction electrons. Thus the electron–phonon scattering process, which causes transitions between k states and hence gives rise to electrical resistivity, may simultaneously cause an electron spin-flip transition with probability approximately $(\Delta g)^2$, where Δg is the deviation of the electronic g-factor from the free-spin value (2·00231). Elliott's relation between the spin relaxation rate and the resistivity relation rate (τ_R^{-1}, which is proportional to the resistivity) is (Elliott 1954, Beuneu & Monod 1978, Monod & Beuneu 1979)

$$(T_1^{-1})_{ph} = (T_2^{-1})_{ph} = \alpha(\Delta g)^2 \tau_R^{-1} \tag{4}$$

where α is a numerical factor. This duality of electron spin relaxation and resistivity relaxation in metallic systems highlights an important feature of the use of c.e.s.r. as a simple, non-destructive technique for monitoring electronic structure. In addition, particle diameters in the range 10^2–10^4 Å are ideally suitable for study; this size range is particularly appropriate since it spans the characteristic sampling distance of the incident microwave field, given by the microwave skin depth (Figure 9). Thus the precise form of the c.e.s.r. spectrum is extremely sensitive to both the electrical resistivity within a particle and, indeed, the particle dimensions (Figure 10).

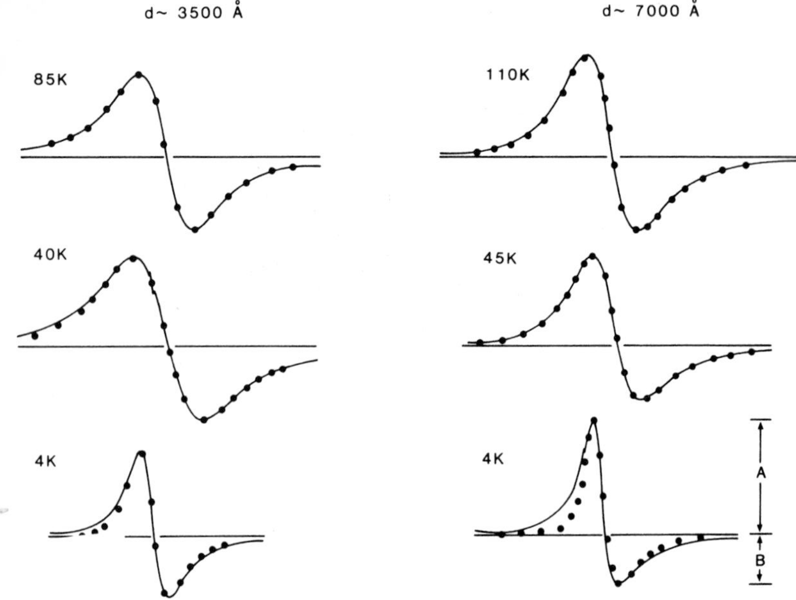

Figure 10. Derivative conduction electron spin resonance line shapes for potassium colloids at various temperatures. The dots represent the calculated lineshape function which is the sum of the dispersion and absorption parts of a Lorentzian resonance with various weighting coefficients. The discrepancy is most pronounced at the lowest temperatures where the anomalous skin effect ensures that the penetration of the microwave field is controlled by l (Figure 9). (Taken from Edmonds *et al.* 1984 and Guy *et al.* 1985.)

For metallic particles too large to exhibit any quantum-size effects at accessible temperatures ($d > 100$ Å), the observed electron spin relaxation rate is the sum of the temperature-dependent *intrinsic* rate via phonon scattering, $(T_1^{-1})_{ph}$, given by equation (4) and a temperature-independent *residual* rate which contains contributions from both impurities, $(T_1^{-1})_{imp}$, and from surface scattering of electrons, $(T_1^{-1})_s$:

$$T_1^{-1} = \alpha(\Delta g)^2 \tau_R^{-1} + (T_1^{-1})_{imp} + (T_1^{-1})_s \tag{5}$$

In general, the residual relaxation rate $[(T_1^{-1})_{imp} + (T_1^{-1})_s]$ is identified from measurements at very low temperatures, in a similar way to the well known Matthiess rule in resistivity scattering (Meaden 1966).

Kawabata (1970, 1977) has considered modifications.to equation (5) to treat surface relaxation in metallic particles. In this case it is the particle *surface* rather than the lattice phonon which breaks the translational invariance of the lattice and produces spin relaxation. The relaxation rate analogous to τ_R^{-1} in equation

(4) is now the frequency of collisions with the surface, given approximately by v_F/d where v_F is the Fermi velocity. Kawabata's result for particles sufficiently large that quantum-size effects do not occur is $(T_1^{-1}) \simeq \gamma(\Delta g)^2(v_F/d)$ where γ is a numerical constant of order unity. Thus, Kawabata predicts an increase in (T_1^{-1}), or equivalently an increase in the c.e.s.r. linewidth, ΔH, as the particle diameter is reduced from the bulk value. There are certainly reports of enhanced surface relaxation in 'colloidal' particles in the 100–10 000 Å range (Smithard 1974, Gordon 1976, see also Perenboom *et al.* 1981). However low-temperature broadening of c.e.s.r. lines can also arise from the presence of unpaired spins in the surrounding matrix, provided they are in close enough proximity to the particles to interact via exchange interactions with the conduction electrons (Knight 1973). In fact Guy *et al.* (1985) have recently reported c.e.s.r. data for sodium particles varying from 10^3 to 10^4 Å which indicate that, over this size range, surface scattering is not the primary mechanism contributing to the residual spin relaxation at low temperatures. Rather, it seems probable that impurity scattering, $(T_1^{-1})_{imp}$, is the dominant interaction in addition to the normal spin–phonon collision term $(T_1^{-1})_{ph}$.

The phonon scattering contribution, $(T_1^{-1})_{ph} = \alpha(\Delta g)^2\tau_R^{-1}$, is expected to be strongly temperature-dependent (via τ_R^{-1}) and equation (4) leads us to suppose that it will closely parallel the electrical resistivity of the bulk metal, or small particle, provided

$$\delta > l \tag{6}$$

and

$$l < d \tag{7}$$

That is, the samples are neither in the anomalous skin effect regime ($l > \delta$) nor in the low-temperature limit in which the form of the spectrum is dictated by the condition $l \sim d$.

To illustrate this duality of both scattering mechanism, we plot in Figure 11 the intrinsic spin–phonon contribution, $(T_1^{-1})_{ph}$, as a function of temperature for large particles ($d \sim 7 \times 10^3$ Å) of sodium, potassium and rubidium metals.

Several important points can be made regarding spin and resistivity scattering processes in the 'large metallic particle' range with $d \sim 10^2$–10^4 Å. As expected from equation (4), the enhanced spin–orbit interaction in the heavier alkali metals leads to considerably faster spin relaxation (via an increase in $(\Delta g)^2$ at *all* temperatures. In Figure 11 we also plot the temperature dependence of the ideal resistivities of the bulk alkali metals (Meaden 1966). There are clear parallels between the temperature dependence of spin- and resistivity-scattering processes for metallic particles sufficiently large to that quantum size effects do not come into prominence. In this regard, recent c.e.s.r. experiments with potassium colloids of dimensions down to 10^3 Å suggest that only minor changes, via small reductions in the characteristic Debye temperature, occur in the details of the

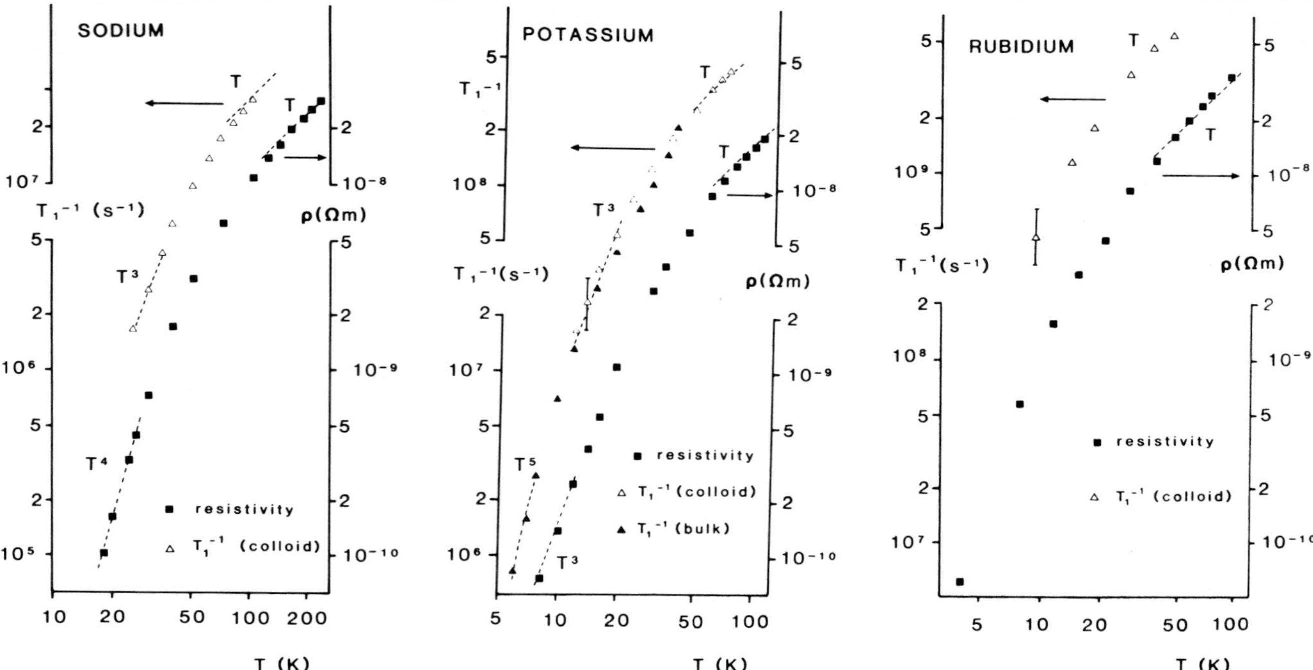

Figure 11. The temperature dependence of the electron spin–spin relaxation rate as a function of temperature for sodium, potassium and rubidium colloids; also included is the corresponding temperature dependence of the ideal electrical resistivity for the three metals. In all cases, temperature exponent is indicated. (Taken from Edmonds 1985.)

lattice phonon spectrum for particles down to these dimensions (Guy *et al.* 1985). Here again, the close similarity between spin and resistivity scattering processes appears to be maintained for particles not subject to quantum-size effects.

During the scattering of an electron spin by the lattice vibrations, the change in the Zeeman energy is usually taken from the kinetic energy of the high-velocity itinerant electrons. We have seen that this electron spin relaxation in a large metallic particle is very rapid, especially when the spin–orbit coupling constant of the metal core is large or at temperatures above the characteristic Debye temperature. For a particle in which the energy level structure is discrete, rather than continuous, this energy flow and concomitant electron spin relaxation is suppressed relative to that in large metallic particles (Kubo 1962, Holland 1967, Kawabata 1970).

In a theoretical treatment of c.e.s.r. in small particles, Kawabata (1970) has given two criteria for line narrowing due to level discreteness,

$$\hbar\omega/\Delta \ll 1 \tag{8}$$

$$\hbar/\tau\Delta \ll 1 \tag{9}$$

where τ is given by $d/v_{\mathrm{F}}(\Delta g)^2$. Thus, for the usual microwave frequencies, particles of dimensions below about 100 Å are required for the observation of quantum size effects. When both equations (8) and (9) are satisfied, the c.e.s.r. signal comes from particles with an odd number of conduction electrons. Thus, the c.e.s.r. signal is now characterized by a long T_1 (even for metals with a heavy atomic core, cf. equation (4)) and a concomitant narrow resonance. Kawabata's results for the c.e.s.r. linewidth as a function of particle diameter for potassium are shown in Figure 12 (Edmonds *et al.* 1984). For potassium particles examined at 9·3 GHz, the two criteria (equations (8) and (9)) are fulfilled when the diameter does not exceed about 500 Å. In this size range a Curie-like susceptibility would be anticipated (Figure 6) and surface relaxation is quenched by level separation such that the linewidth should vary as the square of the diameter, vanishing for the smallest particles. Thus a small alkali particle, well within the domain of quantum size effects, should produce a resonance with the experimental characteristics of a narrow linewidth (long T_1) and temperature-dependent susceptibility.

We expect, therefore, that in this size regime the duality of electron spin and resistivity scattering processes is almost certainly lost (compare Figure 11). Observations of c.e.s.r. signals from particles in this size range have been reported for Li, Na, K, Al, Ag, Au and Pt, Mg and Ca and claims of narrowing of c.e.s.r. lines due to level separation have been made for most of these metals (for reviews see Hughes & Jain 1979 and Perenboom *et al.* 1981).

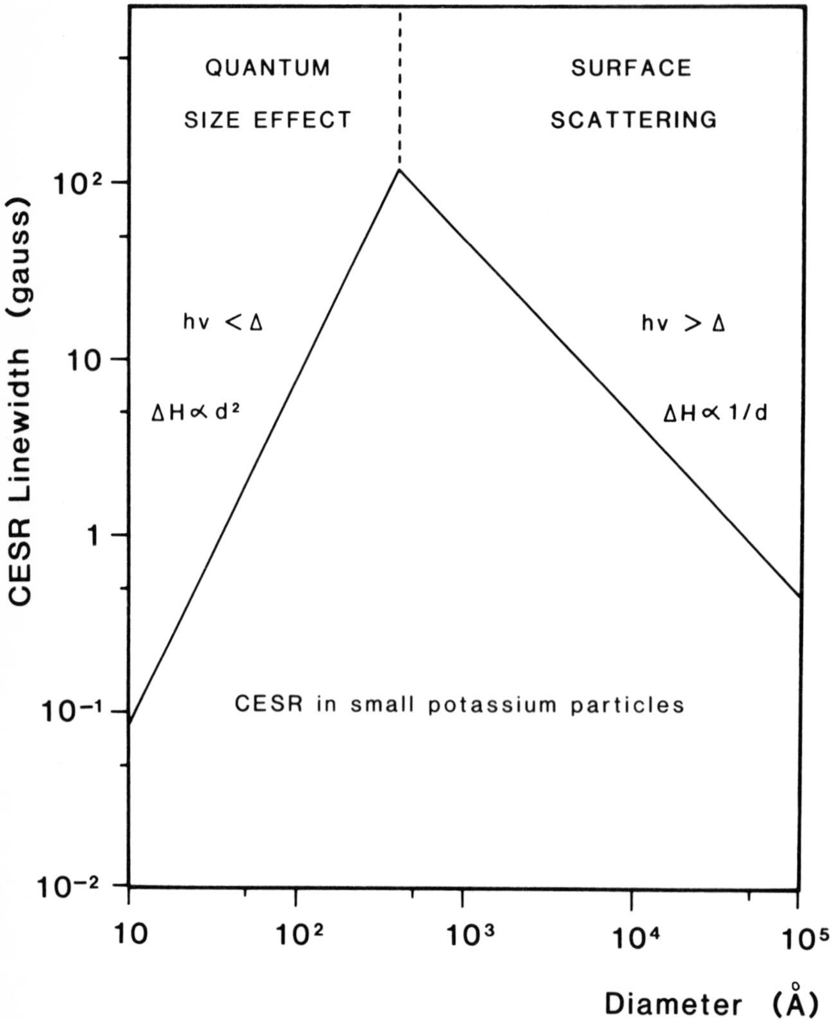

Figure 12. The predicted size dependence of the potassium particle conduction electron spin resonance linewidth calculated from the theory of Kawabata. The d^2 dependence in the smallest particles is a quantum-size effect. (Taken from Edmonds *et al.* 1984.)

3.4. Nuclear spin relaxation

The hyperfine interaction between the nuclear and electron spins causes the Knight shift in the n.m.r. signal (Knight 1956) and the off-diagonal elements provide the mechanism for nuclear spin relaxation in a paramagnetic system (Korringa 1950, Andrew 1969). Thus when an unpaired electron is in close proximity to a nucleus, it exerts a large, time-varying local magnetic field which may induce transitions between the nuclear spin levels. The energy emitted or absorbed by the electron (since the electron–nuclear interaction essentially short-circuits the direct nucleus–lattice interaction) occurs by adjustment of this kinetic energy. Thus the nuclear spin relaxation characteristics are determined by the distribution of energy levels. In bulk metals and large metallic particles, only those electrons near the Fermi level with an energy range kT contribute to this process, so that the nuclear spin relaxation rate is proportional to the temperature. Under these circumstances, the nuclear spin relaxation rate is given by the Korringa relation

$$(T_{1n}^{-1})_{\text{Korr}} = \left\{ \frac{9\pi A_{\text{iso}}^2}{\hbar E_{\text{F}}^2} \right\} kT \tag{10}$$

where A_{iso} is the isotropic hyperfine coupling constant (Korringa 1950). Kubo (1969) originally predicted that this Korringa relaxation process would be strongly suppressed in small particles since it is a direct spin-flip process in which the nuclear Zeeman energy is transferred to the conduction electrons. The appearance of a large energy gap (Δ) in the electronic spectrum would seem to make it difficult to conserve energy in such a process. There have been many extensive studies of n.m.r. in small particles in both the superconducting and normal states, and the reader is referred to the reviews by Knight (1973) and Perenboom *et al.* (1981) for extensive discussion of these experiments and their theoretical interpretation. However, in our view the experimental realization of substantially reduced nuclear relaxation rates in small particles is still somewhat uncertain. As Perenboom *et al.* (1981) point out "... For some samples, even a reduction of the relaxation time T_{1n} below the bulk value was noted; indicating the presence of an extra-nuclear relaxation mechanism probably connected with the presence of paramagnetic impurities at the surface of the particles". This statement once again highlights the considerable experimental problems in producing samples with maximum particle density and uniformity embedded in non-paramagnetic matrix materials (Knight 1973).

However, the emergence of enhanced nuclear relaxation in the small particles may also be linked to the appearance of spontaneous paramagnetism within the particles, as previously discussed (section 3.2). Moreover, if we compare the situation of nuclear and electron spin relaxation in microscopic small particles with the corresponding behaviour in (nominally) macroscopic systems crossing the metal–nonmetal transition, then certain important points emerge.

4. Electron and nuclear spin relaxation across the metal–nonmetal transition

There is perhaps a direct analogy to be investigated between the n.m.r. and e.s.r. properties of microscopic small particles and the extensively studied area of electron and nuclear spin relaxation across the metal–nonmetal transition in nominally macroscopic systems, such as doped semiconductors, metal–ammonia solutions, liquid semiconductors, etc. (Warren 1971, Edwards 1984a,b). This once again amplifies our comments at the beginning of this chapter concerning the metal–nonmetal transition in 'microscopic' and 'macroscopic' systems. Consider the situation for systems in which changes in composition density or bring about the transition between the metallic and the non-metallic regimes. In these, the magnetic relaxation properties of both electron and nuclear spins change quite dramatically across the metal–nonmetal transition. In essence, the transition in these macroscopic systems from itinerant to localized electron behaviour generally results in an increase in the nuclear spin relaxation rate, but a decrease in the electron spin relaxation rate (Edwards 1980). There is, as yet, no detailed unifying theory for describing the changes in relaxation behaviour across the transition region. However, following Warren's pioneering work in this area (Warren 1971), some progress has been made in at least identifying some of the physics responsible for the changes in behaviour. Basically, one starts from the description of the magnetic relaxation properties of electron and nuclear spins in the metallic state (as in the case of our earlier discussion on small particles). The magnetic effects arising from incipient electron localization at specific sites are then introduced in the guise of phenomenological enhancement or reduction parameters for both nuclear and electron spin relaxation rates, respectively.

Consider the case of n.m.r., where the onset of electron localization as one moves towards the metal–nonmetal transition leads to an enhancement of the relaxation rate relative to $(T_{1n}^{-1})_{\mathrm{Korr}}$. Warren (1971) defines a Korringa enhancement parameter, H, where

$$H=(T_{1n}^{-1})/(T_{1n}^{-1})_{\mathrm{Korr}} \tag{11}$$

and from general arguments

$$H \sim [\mathrm{T}_{\mathrm{nmr}}/\hbar N(E_{\mathrm{F}})] \tag{12}$$

where $\mathrm{T}_{\mathrm{nmr}}$ is identified as the average lifetime, or reside time, of an electron at a particular site. Therefore, electron localization, as reflected in an increase in $\mathrm{T}_{\mathrm{nmr}}$ naturally leads to en enhancement in nuclear relaxation rates above $(T_{1n}^{-1})_{\mathrm{Korr}}$. In the case of e.s.r., the onset of electron localization generally leads to a reduction in the spin relaxation rate relative to the normal spin–phonon value, $(T_1^{-1})_{\mathrm{ph}}$. Here we can define a reduction factor, K_0^2, where

$$K_0^2=(T_1^{-1})/(T_1^{-1})_{\mathrm{ph}} \tag{11}$$

and, by analogy with Warren's treatment of nuclear relaxation, we argue (Edwards 1984a,b) that $K_0^2<1$ is an indicator of the degree of electron localization as gauged from e.s.r. measurements.

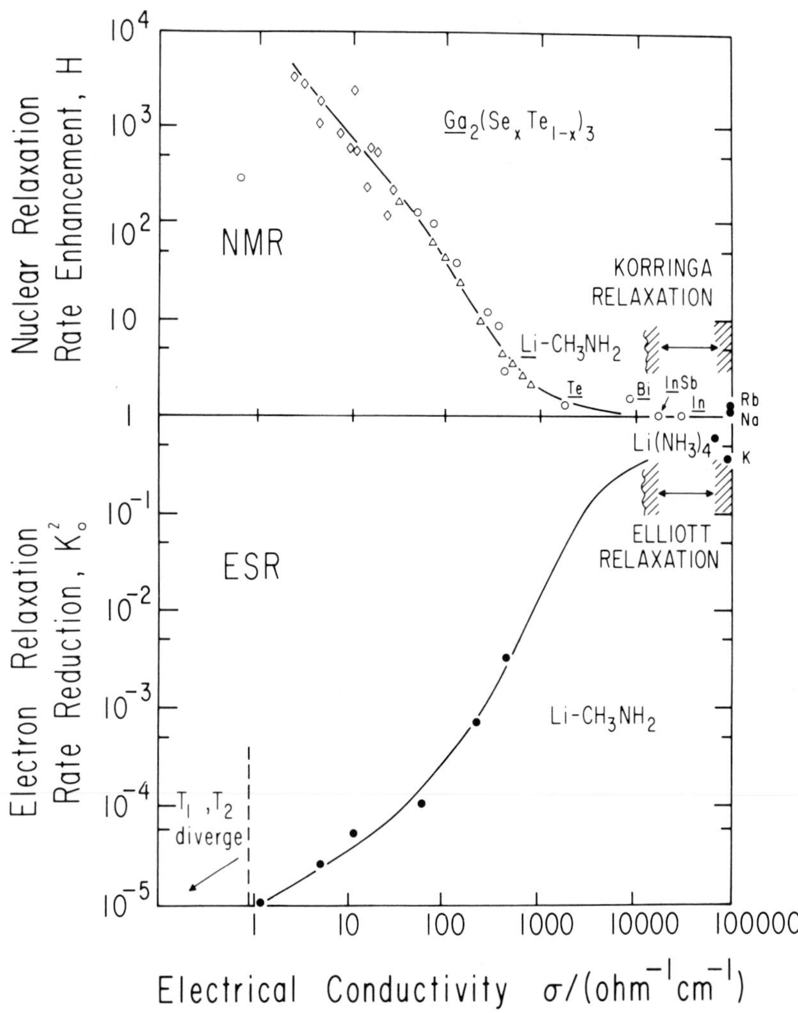

Figure 13. Electron and nuclear spin relaxation across the metal–nonmetal transition in lithium–methylamine solutions. The top half of the diagram shows a plot of the Korringa enhancement parameter (H) *vs.* electrical conductivity for various elements, the liquid alloy systems Ga_2Te_3 and $Ga_2(Se_xTe_{1-x})$ and lithium–methylamine solutions. The lower half of the diagram shows a plot of the Elliott reduction factor, K_0^2, *vs.* electrical conductivity for bulk metals, the nearly-free-electron compound $Li(NH_3)_4$ and lithium–methylamine solutions across the metal–nonmetal transition. Note that electron localization gives rise to an enhancement in the nuclear spin relaxation rate but a reduction in the electron spin relaxation rate; below $\sigma \sim 1\ (\Omega\ \mathrm{cm})^{-1}$, $T_1^{-1} \neq T_2^{-1}$ in lithium–methylamine solution.

In Figure 13, we present the conductivity dependence of both H and K_0^2 for a variety of systems traversing the metal–nonmetal transition. The data shown in the top half of Figure 13 for lithium–methylamine solutions clearly show how the Korringa enhancement parameter is correlated with electrical conductivity. On moving from the metallic to the nonmetallic regimes, one finds that the electron–nuclear reside time increases with a concomitant increase in the efficiency of the nuclear spin relaxation processes (Warren 1971). However, as noted elsewhere (Edwards 1984a,b), the precise form of electron localization may well lead to differing magnetic behaviour across the transition region. Therefore, we note that a major difference now emerges in the nuclear relaxation properties of the systems reviewed in Figure 13, and the anticipated behaviour in small metallic particles for dimensions down to $\sim 1/k_F$ (§3.4). The bottom section of Figure 13 shows the corresponding conductivity dependence of the electron relaxation reduction factor K_0^2 across the transition region in lithium–methylamine solutions. Electron spin relaxation rates in bulk metals and in concentrated metal solutions agree reasonably well with the Elliott relation ($K_0^2 \sim 1$). However, in the metal–nonmetal transition region there is evidence of a rather large departure from nearly free electron characteristics as the systems move into the 'diffusive' or 'strong-scattering' electronic regime for which the electron mean free path is comparable to the interdonor (lithium) distances. This reduction in electron spin relaxation rates seems to indicate an inhibition of the Elliott-type relaxation mechanism across the metal–nonmetal transition. A similar restriction is observed for small metal particles in the quantum size regime (§3.3).

5. Concluding remarks

In this chapter, we have summarized various aspects of the problem of electrons in small particles with a congested set of energy levels.

An intriguing problem centres on the possible analogy between the successive fragmentation of a microscopic metallic particle and the metal to nonmetal transition in nominally macroscopic systems. In this regard Wood & Ashcroft (1982) have discussed the evolution of the optical properties of metallic particles down to the metal–nonmetal transition which occurs for particle dimensions below k_F^{-1}. For sodium metal the cessation of conducting behaviour should occur for dimensions below 20 Å. We have discussed the interrelationship particles. The appearance of an electronic energy gap larger than some relevant energy (for example, the thermal energy, the Zeeman energy) in the small particle is expected to trigger substantial changes in the magnetic properties of the itinerant electrons. The further fragmentation of the particle down to diameters approaching k_F^{-1} should lead to even more drastic modifications of the electronic properties. A possible connection has been outlined between the magnetic properties of small particles and macroscopic systems across the metal–nonmetal transition.

References

Acrivos, J.V., Mott, N.F. and Yoffe, A.D. (eds), 1984, *Physics and Chemistry of Ions and Electronics in Condensed Matter* (Dordrecht: Reidel).

Andrew, E.R., 1969, *Nuclear Magnetic Resonance* (Cambridge: Cambridge University Press).

Ashcroft, N.W. and Mermin, N.D., 1976, *Solid State Physics* (New York: Holt, Rinehart & Winston).

Beuneu, F. and Monod, P., 1978, *Phys. Rev.* B, **18**, 2422.

Borel, J.-P. and Buttet, J. (guest editors), 1981, Small Particles and Inorganic Clusters, *Surface Sci.*, **106**, 1.

Borel, J.-P. and Millet, J.-L., 1977, *J. Physique Coll.*, **38**, C2–115.

Capizzi, M., Thomas, G.A., De Rosa, F., Bhatt, R.N. and Rice, T.M., 1980, *Phys. Rev. Lett.*, **44**, 1019.

Charvolin, J., Froidevaux, C., Taupin, C. and Winter, J.M., 1966, *Solid St. Commun.*, **4**, 357.

Curtis, A.C., Duff, D.G., Edwards, P.P., Jefferson, D.A. and Johnson, B.F.G., unpublished work.

Denton, R., Mühlschlegel, B. and Scalapino, D.J., 1971, *Phys. Rev. Lett.*, **26**, 707.

Dyson, F.J., 1955, *Phys. Rev.*, **98**, 349.

Edmonds, R.N., Edwards, P.P., Guy, S.C. and Johnson, D.C., 1984, *J. Phys. Chem.*, **88**, 3764.

Edmonds, R.N., 1985, Ph.D. Thesis, Cambridge.

Edwards, P.P., 1980, *J. Phys. Chem.*, **84**, 1215.

Edwards, P.P., 1984a, *J. Phys. Chem.*, **88**, 3772.

Edwards, P.P., 1984b, in Acrivos *et al.* (1984), p. 297.

Edwards, P.P. and Sienko, M.J., 1983, *Int. Rev. Phys. Chem.*, **3**, 83.

Elliott, R.J., 1954, *Phys. Rev.*, **96**, 266.

Feher, G. and Kip, A.F., 1955, *Phys. Rev.*, **98**, 337.

Friedel, J., 1977, *J. Physique* (*Coll.*), **38**, C2.

Friedel, J., 1984, in Acrivos *et al.* (1984), p. 45.

Friedman, L.R. and Tunstall, D.P. (editors), 1978, *The Metal–Nonmetal Transition in Disordered Systems* (Edinburgh: 19th Scottish Universities Summer School in Physics), p. 1.

Fröhlich, H., 1937, *Physica* (*Utr.*), **4**, 406.

Göetze, W., 1981, in *Recent Developments in Condensed Matter Physics*, Vol. 1, edited by J.T. Devreese (New York: Plenum), p. 133.

Gordon, D.A., 1976, *Phys. Rev.* B, **13**, 3738.

Gordon, D.A., Marzke, R.F. and Glaunsinger, W.S., 1977, *J. Physique* (*Coll.*), **38**, C2–87.

Grannan, D.M., Garland, J.C. and Tanner, D.B., 1981, *Phys. Rev. Lett.*, **46**, 375.

Guy, S.C., Edmonds, R.N. and Edwards, P.P., 1985, *J. Chem. Soc.*, *Faraday II* (in press).

Harrison, M.R., Edwards, P.P., Klinowski, J., Thomas, J.M., Johnson, D.C. and Page, C.J., 1984, *J. Solid St. Chem.*, **54**, 330.

Hoare, M.R. and Pal, P., 1972, *Nature* (*Phys. Sci.*), **236**, 35.

Hoare, M.R. and Pal, P., 1975, *Adv. Phys.*, **24**, 645.

Holland, B.W., 1967, in *Magnetic Resonance and Relaxation* (Colloque Ampère XIV) (Amsterdam: North-Holland), p. 468.

Hubbard, J., 1964a, *Proc. R. Soc.* A, **277**, 237; 1964b, *Proc. R. Soc.* A, **281**, 241.

Hughes, A.E. and Jain, S.C., 1979, *Adv. Phys.*, **28**, 717.

Ioffe, A.F. and Regel, A.R., 1960, *Prog. Semiconductors*, **4**, 237.

Jacobs, P.A. (editor), 1982, *Metal Microstructures in Zeolites—Preparation, Properties, Applications* (Amsterdam: Elsevier).

Johnson, B.F.G. (editor), 1980, *Transition Metal Clusters* (Chichester: Wiley).

Johnson, B.F.G. and Lewis, J., 1981, *Adv. Inorg. Chem. Radiochem.*, **24**, 225.

Johnson, D.C., Benfield, R.E., Edwards, P.P., Nelson, W.J. and Vargas, M.D., 1985, *Nature* (in press).

Jortner, J., 1984, *Ber. Bunsenges. Phys. Chem.*, **88**, 188.

Kawabata, A., 1970, *J. Phys. Soc., Japan*, **29**, 902.

Kawabata, A., 1977, *J. Physique (Coll.)*, **38**, C2.

Knight, W.D., 1956, in *Solid State Physics*, Vol. 2, edited by F. Seitz and D. Turnbull (New York: Academic Press).

Knight, W.D., 1973, *J. Vac. Sci. Technol.*, **10**, 705.

Knight, W.D., Clemenger, K., de Heer, W.A., Saunders, W.A., Chou, M.Y. and Cohen, M.L., 1984, *Phys. Rev. Lett.*, **52**, 2141.

Kobayashi, S., Takahashi, T. and Sasaki, W., 1972, *J. Phys. Soc., Japan*, **32**, 1234.

Korringa, A.J., 1950, *Physica (Utr.)*, **16**, 601.

Kubo, R., 1962, *J. Phys. Soc., Japan*, **17**, 975.

Kubo, R., 1969, in *Polarisation, Matière et Rayonnement*, Livre Jubilée en l'honneur de Professor A. Kastler (Presses Universitaires de France), p. 325.

Kubo, R., 1968/1969, Comments on Solid State Physics, **1**, 61.

Kubo, R., 1977, *J. Physique (Coll.)*, **38**, C2–69.

Kubo, R., Kawabata, A. and Kobayashi, S., 1984, *Ann. Rev. Mater. Sci.*, **14**, 49.

Lindsay, D.M., 1981, *Surface Sci.*, **106**.

Makowka, C.D., Slichter, C.P. and Sinfelt, J.H., 1982, *Phys. Rev. Lett.*, **49**, 379.

Marzke, R.F., 1979, *Catal. Rev.-Sci. Eng.*, **19**, 43.

Marzke, R.F., Glaunsinger, W.S. and Bayard, M., 1976, *Solid St. Commun.*, **18**, 1025.

Meaden, G.T., 1966, *Electrical Resistance of Metals* (London: Heywood).

Millet, J.-L. and Borel, J.-P., 1981, *Surface Science*, **106**, 403.

Monod, P. and Beuneu, F., 1979, *Phys. Rev.* B, **19**, 911.

Monot, R. and Millet, J.-L., 1976, *J. Physique (Paris)*, **37**, L–45.

Mott, N.F., 1949, *Proc. Phys. Soc.* A, **62**, 416.

Mott, N.F., 1961, *Phil. Mag.*, **6**, 287.

Mott, N.F., 1974, *Metal–Insulator Transitions* (London: Taylor & Francis).

Overhauser, A.W., 1953, *Phys. Rev.*, **89**, 689.

Ozin, G.A., 1980, in *Diatomic Metals and Metallic Clusters*, Faraday Symposia of the Royal Society of Chemistry, No. 14, p. 7.

Perenboom, J.A.A.J., Wyder, P. and Meier, F., 1981, *Physics Reports*, **78**, No. 2, 173.

Shiba, H., 1976, *J. Low Temp. Phys.*, **22**, 105.

Slichter, C.P., 1965, *Principles of Magnetic Resonance* (New York: Harper & Row).

Smithard, M.A., 1974, *Solid St. Commun.*, **14**, 407, 411.

Sone, J., 1976, *J. Low Temp. Physics*, **23**, 699.

Sone, J., 1977, *J. Phys. Soc., Japan*, **42**, 1457.

Stokes, H.T., Makowka, C.D., Wang, P.K., Serge, L., Rudaz, S.L., Slichter, C.P. and Sinfelt, J.H., 1983, *J. Molec. Catalysis*, **20**, 321.

Taupin, C., 1966, *C. R. Acad. Sci.* B, **262**, 1617.

Taupin, C., 1967, *J. Phys. Chem. Solids*, **28**, 41.

Thompson, G.A., Tischler, F. and Lindsay, D.M., 1983, *J. Chem. Phys.*, **78**, 5946.

van der Klink, Buttet, J. and Graetzel, M., 1984, *Phys. Rev.* B, **29**, 6352.

Van Vleck, J.H., 1932, *The Theory of Electric and Magnetic Susceptibilities* (Oxford: Oxford University Press).

Warren, W.W., Jnr., 1971, *Phys. Rev.* B, **3**, 3708.

Wood, D.M. and Ashcroft, N.W., 1982, *Phys. Rev.* B, **25**, 6255.

Yee, P. and Knight, W.D., 1975, *Phys. Rev.* B, **11**, 3261.

Yvon, K., 1979, in *Current Topics in Materials Science*, **3**, 53; and references therein.

Ziman, J.M., 1960, *Electrons and Phonons* (Oxford: Clarendon Press).

INDEX